Construction Scheduling with Primavera Project Planner®

SECOND EDITION

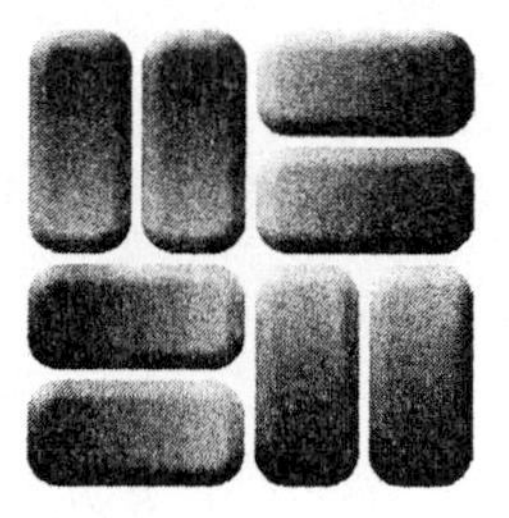

Leslie Feigenbaum

Texas A & M University

Upper Saddle River, New Jersey
Columbus, Ohio

Library of Congress Cataloging-in-Publication Data

Feigenbaum, Leslie.
Construction scheduling with Primavera Project planner / Leslie Feigenbaum.—2nd ed.
p. cm.
Includes index.
ISBN 0-13-092201-3
1. Building—Superintendence. 2. Production scheduling. 3. Primavera Project planner (P3) I. Title.

TH438.4 .F45 2002
690′.068—dc21

2001036230

Editor in Chief: Stephen Helba
Executive Editor: Ed Francis
Production Editor: Christine M. Buckendahl
Design Coordinator: Robin G. Chukes
Cover Designer: Linda Fares
Production Manager: Matt Ottenweller
Marketing Manager: Mark Marsden

This book was set in Times Roman by TechBooks, and was printed and bound by Courier/Kendallville, Inc. The cover was printed by The Lehigh Press, Inc.

Pearson Education Ltd., *London*
Pearson Education Australia Pty. Limited, *Sydney*
Pearson Education Singapore Pte. Ltd.
Pearson Education North Asia Ltd., *Hong Kong*
Pearson Education Canada, Ltd., *Toronto*
Pearson Educación de Mexico, S.A. de C.V.
Pearson Education—Japan, *Tokyo*
Pearson Education Malaysia Pte. Ltd.
Pearson Education, *Upper Saddle River, New Jersey*

10 9 8 7 6 5 4 3 2
ISBN: 0-13-092201-3

In memory of Jerry Trost, a friend, colleague, and mentor.

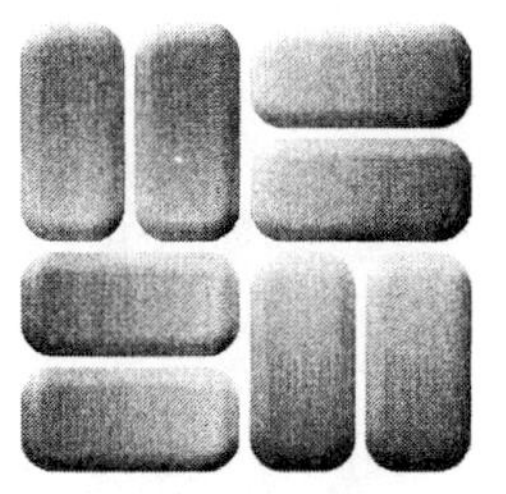

Preface

The intent of *Construction Scheduling with Primavera Project Planner* is to assist the user in developing a proficiency in construction planning and a working knowledge of Primavera Project Planner. In order for the reader to get the most out of this book, he or she should have an understanding of how construction projects are estimated and how they are assembled, and have a basic understanding of the Windows operating system. The chapter on estimating is included to develop terminology and to show the relationship between the estimate and the comprehensive construction plan. It is essential that the information developed for the estimate be presented in a fashion that can be easily used during the actual construction process.

This book is divided into three distinct sections. The first section addresses the development and manual calculations of the construction schedule. The middle section is an introduction to Primavera Project Planner. The final section addresses project controls and how Primavera Project Planner can be used to enhance decision making. This book is intended to impart a practical hands-on perspective to the software reference manuals. After completing this book, the reader should have developed a proficiency in Primavera Project Planner and in construction project controls. Furthermore, the user should be able to customize Primavera Project Planner as necessary so as to effectively gauge project performance.

The suggested exercises at the end of each chapter provide practical applications. These exercises are cumulative, and upon completion of all these exercises, the user will have gone through a complete project simulation. This simulation will include both planning the project and monitoring the project through actual construction. The chapter discussion together with the simulation will enable users of this text to develop the skills needed to effectively manage construction projects.

Brief Contents

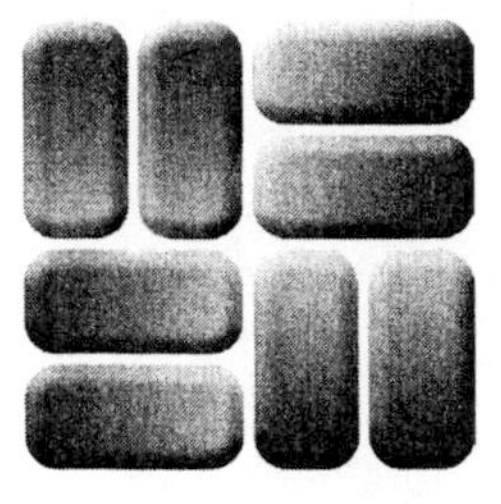

Contents

Overview

KEY TERMS

Project Management The comprehensive process of planning, directing, and controlling the construction project from its inception through its completion.

Project Controls The process of planning, directing, and controlling the project from the start of construction through its completion.

The Project Management Process

Effective project management is a comprehensive process that involves planning, directing, and controlling all of the construction operations and resources. This process begins with the receipt of the contract documents and extends through the completion and subsequent delivery of the project. The use of a comprehensive model maximizes the potential for completing projects within budget and on schedule.

In the simplest terms, project management involves developing a plan, initiating the execution of the plan, and comparing actual progress to planned progress on a continual and ongoing basis. By performing these comparisons, deviations can be quickly identified and a new plan can be developed to recover from any deviations. This new plan then becomes the benchmark for comparison with the actual progress. Figure 1-1 is a graphic representation of this process. The project execution plan is a dynamic document that must be updated regularly in order to reflect the best information available at any point in time. By continually updating and revising the plan it will reflect the most current and realistic roadmap for completing the project within the allotted time and budget. The project execution plan not only details the sequence of construction but also addresses financial and logistical aspects of the project. There is a clear relationship between the project sequence, project duration, and the ultimate project cost.

The objective of the comprehensive project management process is to complete the construction project on schedule and within budget. These objectives are attainable if the estimate is accurate and the right resources in the right quantity are at the right place at the right time. The ability to accomplish this balancing act is the most distinguishing

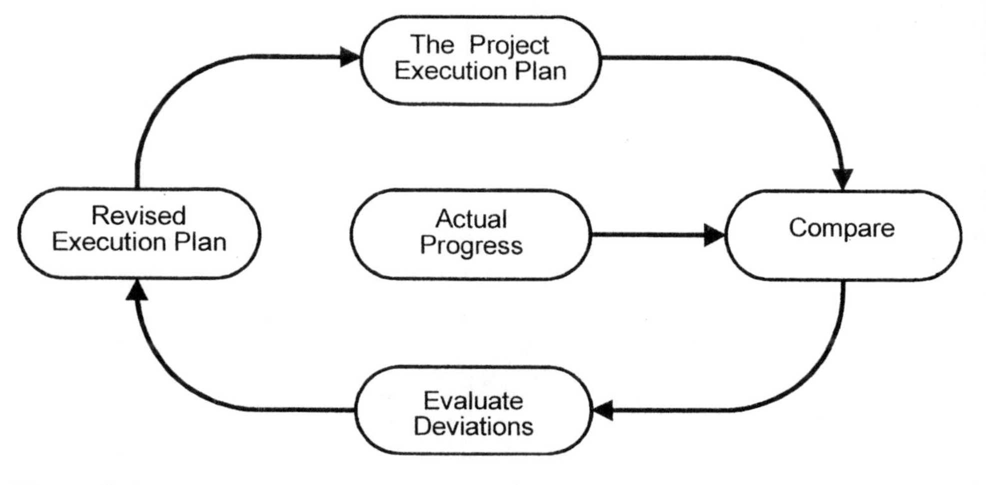

Figure 1-1
The Project Controls Process

factor between contractors. The contractor who is successful at accomplishing these objectives has a competitive advantage through a lower cost structure and ultimately has the opportunity to acquire more work at higher profit margins. For example, when erecting structural steel, the most efficient method would be to have the steel erected directly from the delivery truck. While this seems simple and logical, a substantial amount of planning would be required. The initial step would be to develop an erection plan with the erection contractor. This plan would detail when erection was to commence and in what sequence it was to proceed. Once this plan is developed, it must be communicated to the fabricator in the form of a delivery plan or schedule. Any discrepancies between the erection subcontractor sequence and the supplier's ability to deliver will have to be resolved. The contractor's role as facilitator would ensure that the structural steel is fabricated and shipped in the order it is to be erected. Through effective planning, coordination, and communication this task can be accomplished. If this planning does not take place, the fabricator most likely will fabricate the steel from the largest to the smallest members. While this would be a logical method of fabricating the steel, some of the lighter members used for bracing would not arrive when needed and the progress of the project would be impeded. If the steel is delivered and is not immediately needed it would then have to be unloaded, stored, and moved from the storage area when needed for erection. This would add unnecessary costs to the project and introduce the potential for damage and theft. While these added steps would not influence the ultimate quality of the completed structure, they clearly add cost to the project without adding value. If these costs were unanticipated in the estimate, the project profitability would be reduced. On the other hand, if these costs were included in the estimate, these higher costs might have been the factor that prevented the contractor from being awarded the project.

The cost of resources is relatively the same for all contractors. While suppliers may trim the cost from one contractor to another, the difference is typically nominal. The same holds true for craft labor. These persons will migrate from contractor to contractor and project to project based upon the prevailing wage. Unless there is an extreme shortage of craft labor, the wage differential between contractors is quite small. Since most resource costs are so similar for all contractors, it is the management of these resources that influences a particular contractor's cost structure and competitiveness.

This book will look at construction management as a process that begins with the contractor's receipt of the bid package and extends throughout the entire project. Particular emphasis will be placed on information management. Through effective information management the contractor can minimize the burdens on the field staff, maintain morale, reduce costs, and improve relationships with subcontractors and suppliers. When any information is generated it must be formatted and presented not only for its immediate purpose but also for its future needs. If this is done, rework will be minimized. For example, the project estimate is initially used to get work. However, the successful contractor will then use that document to order materials, determine craft needs, and control costs. If these future needs are taken into consideration when the estimate is produced, subsequent users will be able to quickly find and use the needed information. If this is not done, field personnel will be required to perform the takeoffs a second time in order to procure the needed materials. This exercise adds to the contractor's cost structure by requiring additional field staff as well as creating friction between the field and office staff.

Conclusion

The ability of contractors to distinguish themselves and obtain a competitive advantage is directly related to their ability to plan, direct, and control the construction process. An effective planning process is one that is comprehensive and continual. Through an iterative process deviations in the construction plan can be quickly identified and evaluated to determine their impact on the remainder of the project. Then the original plan can be revised to overcome these deviations. This new plan then becomes the basis for future evaluation of field progress. By employing this continual process of planning, analyzing, and planning again, the impact of deviations will be minimized through prompt corrective action. The pace of construction in the field is hectic, and the window of opportunity to impact the ultimate cost and schedule for the project is very limited. Therefore a system must be in place to quickly identify and take advantage of such opportunities.

Suggested Exercises

1. Identify a project to use for a project simulation. Since this project will be used for all of the suggested exercises throughout the book, it is essential to have a complete set of drawings and specifications. It would also be helpful to have access to an estimate for the project. Without this document, you will need to generate an estimate for the project. Strive to select a project that matches your knowledge of materials and methods and the needs of the project.
2. After identifying the sample project and securing a complete set of contract documents, take some time to familiarize yourself with the project. This can often be accomplished by reviewing the drawings and making a list of the types of systems that are found within the project. For example, identify the type of foundation system, structural system, and mechanical system.

The Estimate Process

KEY TERMS

Quantity Takeoff Estimated in-place quantities.

Workhour One person working for one hour.

Bare Labor Cost The labor cost for performing work that does not include any labor benefits, labor burdens, or mark-ups.

Crew An aggregation of specific crafts brought together to perform a specific task.

Crew Hour A crew of any size working for one hour.

Introduction

The estimate process is the initial phase in the comprehensive construction management process. While this phase is typically associated with acquiring work, the data gathered can and should be used throughout the project execution. Quantities developed from the quantity takeoff (QTO) can be used to purchase materials, develop the progress plan, and determine the craft labor needs. The quantity takeoff and the resulting estimate must be well organized so that this document can be effectively used by the multitude of persons who will need it during the life of the project. If this document is not well organized, subsequent users will have to regenerate the information, adding unneeded cost to the project. Figure 2-1 should serve as a guide to the estimating process.

Work Breakdown

Effective construction project management requires that the project be divided into a group of smaller subprojects. This breakdown is necessary because of the size and complexity of construction projects. The resulting subprojects can then be estimated, purchased, installed, and monitored effectively. This subdividing of the construction project is referred to as the work breakdown. Figure 2-2 is an excerpt from a work breakdown structure. The work breakdown structure is a hierarchical diagram that shows how each of the subprojects is tied to a larger subproject and ultimately how they comprise the entire project. While it would appear that the work breakdown structure is unique for each project, that is not the case. While all construction projects are unique, they have many common elements. The best approach is to enlarge an existing work breakdown structure as projects are estimated and ultimately executed. By using the common elements from project to project, valuable historical data can be accumulated and used to verify current project performance and develop labor productivity factors for future projects.

The lowest level in the work breakdown structure is the work package or activity. Several guidelines can be helpful when determining whether the project has been broken down sufficiently. Some of these guidelines include a work package being an item of work that is under the direction of a single supervisor; it is continuous, with a defined start and stop; or it is performed by a single unique crew. These should serve as guides with the ultimate criteria being a combination of these guidelines coupled with talent of the

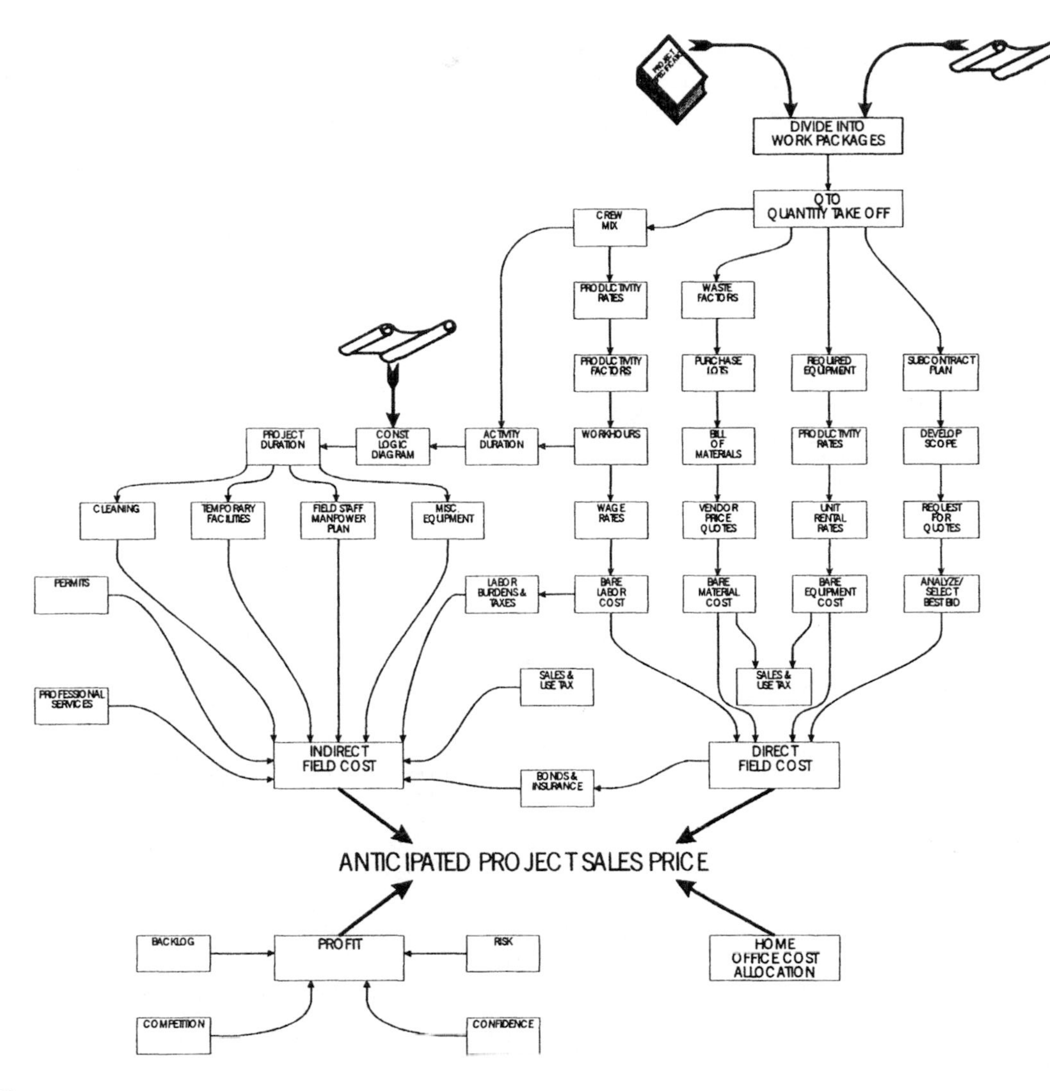

Figure 2-1
Construction Estimate Process

management team, the typical project size, the project complexity, and craft requirements. For example, in industrial construction it is a typical contractual requirement that pipe fitters have a specific certification prior to being allowed to weld on the project. In order to effectively complete the project it is necessary to have the right number of certified pipe fitters at the right time. The number of pipe fitters needed must be known in advance, so that they can be recruited before they are actually needed for the project. This demonstrates why the lowest level in the work breakdown structure—the work package—has to be an activity performed by a specific crew. The same level of detail is needed if the contractor uses union labor. With a union labor force, the job descriptions of each of the crafts are defined in the union contract. The contractor needs to know in advance what crafts in what quantities are needed, so arrangements can be made with the union to ensure that these persons will be available. The only way this information can be gathered is if the lowest level in the work breakdown structure is the work package; i.e., work being performed by a specific crew.

In contrast, open-shop contractors have a much broader definition of what each craft does. A person may be a rod buster in the morning and a carpenter in the afternoon. In this case, breaking the project down into work being done by a specific crew would be an unnecessary exercise. In these situations, the lowest level in the work breakdown structure can be whatever the estimator and project manager feel comfortable with.

Work Package Number	Level 1	Level 2	Level 3	Level 4
03	Concrete			
03-01		Foundations		
03-01-01			Spread Footings	
03-01-01-05				Excavation
03-01-01-10				Form
03-01-01-15				Reinforce
03-01-01-20				Pour and Finish
03-01-01-25				Strip Forms
03-01-01-30				Backfill
03-01-03			Continuous Footings	
03-01-03-05				Excavation
03-01-03-10				Form
03-01-03-15				Reinforce
03-01-03-20				Pour and Finish
03-01-03-25				Strip
03-01-03-30				Backfill
03-01-05			Drilled Piers	
03-01-05-05				Drill
03-01-05-10				Remove Soil
03-01-05-15				Reinforce
03-01-05-20				Pour
03-03		Walls		
03-03-01			Straight	
03-03-01-05				Reinforce
03-03-01-10				Form
03-03-01-15				Pour
03-03-01-20				Strip

Figure 2-2
Work Breakdown Structure

One of the limitations of estimating by work package is that the same types of materials may be used in multiple work packages. This can make it difficult to get price quotations that take advantage of quantity discounts. For example, concrete would be found in many work packages such as curbs, paving, foundations, and structural frames. With concrete in so many work packages, the estimator would have to accumulate all the concrete before requesting a price quotation. The same activity would have to be done for all materials. When estimates were performed manually this was a severe limitation; however, with the advent of computerized estimating systems such as *Timberline®* this limitation is removed. These estimating systems allow the user to perform quantity takeoffs by work packages and view the estimate in work package format or by material type.

Bare Labor Costs

To determine bare labor costs, the estimator needs to have completed the quantity takeoff, needs to know which crews will be performing which activities, and what crafts will be included on which crews, and needs to be aware of the prevailing wage rates for all of the crafts, and be familiar with the productivity rate for that specific crew. Information about the crews should come from historical data. If historical data is used for the prevailing craft wage rates, the user must be aware that inflation and competition make their accuracy and long-term validity questionable.

There are many factors that influence craft wage rates, such as market conditions, required skill, contractual requirements, and governmental regulations. If the contractor is a union contractor, the prevailing wage rate for a specific craft is set by the negotiated labor agreements. An open-shop contractor will have to survey local market conditions to determine what wage rates must be paid to attract and retain the needed craft persons.

Table 2-1
Craft Composition and Labor Cost

Quantity	Craft	Hourly Wage Rate ($ /Hr.)	$ per Crew Hour
5	Carpenters	$20.00	$100.00
3	Laborers	$10.00	$30.00
Total			$130.00

Historical data and local market information are by far the best sources of information concerning productivity rates and wage rates. However, a source such as *Mean's Cost Data®* also provides this information. When using third party databases, one must remember that the work methodology most likely will differ from the methods assumed by the contractor. This difference in how the work will be performed and consequently the price of the labor can have a profound impact on the cost of the project. It is essential to recognize and address the limitations of third party databases objectively.

Once the quantity takeoff for a work package has been completed and an appropriate crew determined it is necessary to determine the crew average wage rate. This average wage rate is determined by multiplying the specific craft wage rate by the number of persons of that craft on the crew. The sum of these individual craft wage rates is the cost for one crew hour. If the cost for one crew hour is divided by the number of persons on the crew, the crew average wage rate can be determined.

$$\text{Crew Average Wage Rate} = \frac{\text{\$ per Crew Hour}}{\text{People on the Crew}}$$

Formula 2-1

The best way to complete this calculation is to create a table similar to Table 2-1 for each of the anticipated crews. In this example, a crew of five carpenters and three laborers costs $130.00 for one crew hour.

Using the crew information from Figure 2-3 in conjunction with Formula 2-1, the crew average wage rate can be determined as

$$\text{Crew Average Wage Rate} = \frac{\text{\$130.00 per Crew Hour}}{\text{8 People on the Crew}}$$
$$= \$16.25$$

Productivity Rates

A labor productivity rate is the number of workhours it takes for a specific crew to place or install one unit of a particular item of construction. By multiplying the productivity rate for a specific item by the quantity takeoff, the workhours to perform that specific task can be estimated. Formula 2-2 shows how to calculate workhours given the quantity takeoff and the appropriate productivity rate.

$$\text{Workhours} = \text{Productivity Rate} \times \text{Quantity Takeoff}$$

Formula 2-2

Using a crew of five carpenters and three laborers with a productivity rate of .142 workhours per square foot of four-use concrete forms, the workhours for a typical job can be estimated. The assumed quantity takeoff of 2,000 square feet of contact area (SFCA) is multiplied by the productivity rate to come up to approximately 284 workhours to complete the task.

$$\text{Workhours} = .142\ \text{Workhours/SFCA} \times 2{,}000\ \text{SFCA}$$
$$= 284\ \text{Workhours}$$

The determination of crew productivity rates is an ongoing and complex operation. The productivity rates used in the estimate need to be based on sound factual information. The best source of productivity rates is recently completed projects. This information will reflect how a particular contractor actually executes work on projects. However, to gather this information the contractor must allocate the needed resources and have an information management system in place. If data from previous projects is to be used as a guide for getting new work, it is essential that this information be accurate. In practical terms, the time cards of the craft personnel in the field must accurately reflect what element of the project or work package they were building. In addition, the materials cost-coding scheme must be consistent. Through the use of cost codes and actual costs, quantities for specific items can be gathered and compared to the amount in the estimate. If the cost coding is not accurate, faulty information will be used to get future work. Consistency, accuracy, and validity must be of concern when gathering actual information for productivity rates. Furthermore, objectivity needs to be applied when using these rates. If an element of the estimate does not look correct, it must not be accepted at face value. Rather, the estimator should investigate the accuracy of that portion of the estimate.

Productivity Factors

Workhours from previous projects are an acceptable approximation if the working conditions are the same between the source of the productivity rates and the project currently being estimated. But it is often necessary to adjust the estimated workhours to reflect how the proposed work will differ from the previous projects.

Productivity factors are used to compensate for the difference between the current project and the "standard project." Some of the items that productivity factors compensate for are climate, experience level of the craft personnel, and the quality of the project management team. Quantifying productivity factors is a difficult task. How much will work slow down in an extremely cold climate? How many extra workhours will it take because the superintendent is relatively new at that position? The answers to these questions can only come from the experience of the estimator and project team. When the workhours are multiplied by the productivity factor, the result is adjusted workhours.

$$\text{Adjusted Workhours} = \text{Workhours} \times \text{Productivity Factor}$$

Formula 2-3

The standard condition has a productivity factor of one. If the productivity factor is greater than one, the number of required workhours will be greater than the standard condition. Conversely, if the productivity factor is less than one, the required workhours will be less than the standard condition.

In the previous example, the formwork is being constructed in an environment that is substantially worse than the standard condition. The estimator and project manager determine, from their experience, that it should take 8 percent more workhours to construct the required forms. Therefore, the assumed productivity factor is 1.08. Calculations show that these inclement conditions should increase the number of workhours required for the sample task from 284 workhours to slightly less than 307 workhours:

$$\text{Adjusted Workhours} = 284 \text{ Workhours} \times 1.08 \text{ Productivity Factor}$$
$$= 306.72 \text{ Workhours (use 307 Workhours)}$$

Once the anticipated workhours for a specific work item have been estimated, the bare labor costs for that work package can be determined. The bare labor cost for a work item is simply the adjusted workhours multiplied by the average wage rate for the specific crew that will be performing the work.

$$\text{Bare Labor Cost} = \text{Adjusted Workhours} \times \text{Crew Average Wage Rate}$$

Formula 2-4

The adjusted workhours found in the previous example and the crew specified in Table 2-1 can be used to estimate the bare labor cost of the formwork.

$$\text{Bare Labor Cost} = 307 \text{ Adjusted Workhours} \times \$16.25 \text{ Average Wage Rate}$$
$$= \$4{,}988.75$$

Activity Duration

Once the workhours for a specific item of work have been estimated and adjusted, the duration for that activity can be approximated. However, several considerations must be evaluated prior to performing the calculation. First, the length of the workday must be established. The typical construction workday is either 8 or 10 hours. There are valid arguments for using either of these workday lengths. With four 10-hour days comes the advantage of having only four morning and post-lunch startups, while the fifth day of the week is a convenient make-up day in the event one of the days is rained out. In addition, the fifth day can be used for overtime if necessary. However, the 10-hour day is long, and fatigue will lower productivity as the workday progresses. With five 8-hour days, there is an additional start-up, and if a make-up day is required it must be on the weekend. However, the potential for fatigue is lessened. The workday that best fits the needs of the employees, owner, subcontractors, and suppliers is the one that is appropriate.

If the definition of a work package is a specific portion of the project being completed by a specific crew, then calculating the activity duration is simplified. With that scenario Formula 2-5 should be used to estimate that activity duration.

$$\text{Activity Duration} = \frac{\text{Adjusted Workhours}}{\text{No. of Persons on the Crew} \times \text{Hours in the Workday}}$$

Formula 2-5

From the adjusted workhours in the previous example the activity duration can be calculated. If an 8-hour workday is assumed, the duration is estimated to be 4.80 days.

$$\text{Activity Duration} = \frac{307 \text{ Adjusted Workhours}}{8 \text{ People on the Crew} \times 8\text{-Hour Workday}}$$
$$= 4.80 \text{ days (use 5 days)}$$

For scheduling purposes, that duration would be rounded up to 5 days.

If a 10-hour workday is assumed, the duration would be roughly 4 days:

$$\text{Activity Duration} = \frac{306.72 \text{ Adjusted Workhours}}{8 \text{ People on the Crew} \times 10\text{-Hour Workday}}$$
$$= 3.84 \text{ days (use 4 days)}$$

When estimating the duration of an activity, it is recommended that the calculated duration be rounded up to the nearest whole day. Rounding up is encouraged since the overall calculation is not accurate enough to justify carrying the activity duration to parts of a day. In addition, scheduling in units of less than one day is rarely used. The only time using hours as a scheduling unit would be appropriate is if the project is of very short duration and its completion is highly critical. An example of this is in turn-around work. This type of work typically occurs in process plants, manufacturing plants, and power generation stations. These facilities are in use 24 hours a day and are rarely shut down. Typically all major maintenance and modification will take place annually during a one to two week window. Due to production schedules and contractual requirements, the shutdown time must be minimized. In these situations, the construction work is typically done in shifts working around the clock. Since this work is complicated, short term, and critical, scheduling by the hour might be appropriate.

Adjusting the Activity Duration

The mathematically correct activity duration often fails to account for the construction methodology and construction constraints. If the calculated duration is too long it can be decreased by increasing the number of crews assigned to that specific task. Increasing the number of crews working on a particular activity should not impact the cost of that activity, provided adequate space and resources are available to support this greater level of staffing. The previous example used a crew of eight persons working 8 hours a day. If that is changed to sixteen persons working 8 hours a day, the new duration would be 3 days.

$$\text{Activity Duration} = \frac{\text{306.72 Adjusted Workhours}}{\text{16 People on the Crew} \times \text{8-Hour Workday}}$$
$$= \text{2.36 days (use 3 days)}$$

However, one must be cautious not to overload a particular area of the project. If too many people are working in an area, labor productivity will decline, increasing the number of required workhours and ultimately increasing the cost for that element of the project.

An exception to using the calculated duration would be when the proposed construction methodology restrains the activity duration. For example, assume that some particular formwork is estimated to take 2 days based on the crew and required workhours. However, the estimator assumed that due to the quantity of the formwork and its repetitive nature it would be cost effective to perform these pours in four distinct operations and use the same formwork for all four of these operations. To assume a 2-day duration would be contrary to the planned methodology. Therefore, the assumed duration would be changed to 4 days to accommodate one day to install the forms for each of their four uses. In this example, the planned methodology would clearly take precedent over the calculated duration. If the work is to be performed by subcontractors it is important that they be consulted for needed duration information. This will provide the most accurate information as well as enhancing the subcontractors' commitment to completing the work as promised. This will make it easier for the contractor to hold the subcontractors accountable for the flow of the work.

All of the above examples and discussions have assumed that the work package is being performed by a specific crew. While that assumption helps determine the duration, it has many limitations. This is particularly true when the work package has a variety of related items performed by a variety of different crafts at different times. In this situation the scheduler, the estimator, and the rest of the project team must either make some assumptions about an average crew size or rely on their experience and intuition to determine the activity duration.

Bare Material Costs

To determine the bare material costs, the quantity takeoff, waste factors, and purchase lots must be known. Then the formula is

$$\text{Minimum Purchase Quantity} = \text{Quantity Takeoff} \times (1 + \text{Waste Factor})$$

Formula 2-6

The waste factors and purchase lots vary from material to material. Guidelines for waste factors can come from historical data or third party databases. Using Formula 2-6 with a quantity takeoff of 350 cubic yards of concrete with a 5 percent waste factor, the minimum purchase quantity for that item can be determined.

$$\text{Minimum Purchase Quantity} = 350 \text{ CY of Concrete} \times (1 + .05 \text{ Waste Factor})$$
$$= 367.5 \text{ CY (Use 368 CY)}$$

While 368 cubic yards of concrete may be the needed quantity, that certainly would not be the quantity that would be purchased. The concrete batch plant could not batch the concrete to that level of accuracy; moreover, concrete is typically sold by the load. The typical load size is somewhere between 6 to 8 cubic yards. Assuming a 6-cubic-yard load would require the purchase of 62 loads, or 372 cubic yards. Therefore, the quantity that would have to be purchased to pour this concrete would be 372 cubic yards. When the minimum purchase quantity is adjusted to reflect the quantity that will have to be purchased, the resulting quantity is called the bill of materials quantity. This quantity can than be submitted to potential vendors for unit price quotes. The price quotation can then be multiplied by the bill of materials quantity, and an estimate of the direct material cost can be determined:

$$\text{Bare Material Cost} = \text{Bill of Materials Quantity} \times \text{Unit Price}$$

Formula 2-7

Using the information from the previous example and Formula 2-7, the estimated concrete cost can be computed:

$$\text{Bare Material Cost} = 372 \text{ CY of Concrete} \times \$48.50 \text{ per CY}$$
$$= \$18{,}042$$

Bare Equipment Costs

Equipment costs can be determined in a number of different ways. One of the methods closely parallels the way labor costs were estimated. Through the use of productivity rates for specific pieces of equipment, their required hours can be calculated.

$$\text{Hours} = \text{Quantity Takeoff} \times \text{Specific Piece of Equipment Productivity Rate}$$

Formula 2-8

The required hours can then be multiplied by the hourly rental rate to determine their bare cost.

$$\text{Bare Equipment Cost}$$
$$= \text{Hours} \times \text{Hourly Rental Rate for the Specific Piece of Equipment}$$

Formula 2-9

The productivity rates can come from the equipment manufacturers, rental companies, or historical data. However, these productivity rates can be impacted by the state of repair of the equipment. If the equipment is well maintained it should operate at the manufacturer's quoted rates. However, if the equipment is not well maintained it will be subject to frequent breakdowns, increasing the number of hours that it will be required on the project and ultimately increasing the project cost.

The above approach works well if a specific piece of equipment is brought onto the project, used for a particular task, and then returned. However, this is not what typically happens in the construction process. Typically, equipment is brought on the project, used for a number of tasks, and then returned. The contractor must pay rent on the equipment not just while it is being used, but all the time it is on the construction job site. For example, a crane must be on the project site from when it is needed for the first lift until the last lift. The contractor will be charged rent for the entire intervening time. Therefore, another method for determining equipment cost is to develop an equipment plan. To do this, one must identify during the estimate phase what equipment will be required for what activities and then determine the number of days that it will be on the project. This will result in the days for each piece of equipment, which can then be multiplied by daily rental rates to determine the bare equipment costs.

Subcontract Costs

The first thing that must be done to determine subcontract costs is to determine what items will be subcontracted. This subcontracting plan should be developed early in the estimate process. There are many advantages as well as limitations to using subcontractors on a particular project. Some of the advantages are that subcontractors have a specific expertise, risk can be spread, and the contractor's overhead can be reduced. On the other hand, the craft persons provided by the subcontractor are under the direction and control of the subcontractor and not the general contractor. There may be disagreements as to what is in the subcontractor's scope of work and when that work will be performed. To minimize the difficulties of dealing with subcontractors, the general contractor should determine what items will be subcontracted and develop a scope of work for each of these. These work items, with their detailed scope of work, can then be submitted to potential subcontractors for price quotes. The contractor should remain in regular communication with the potential subcontractors during the bidding process to ensure that they bid the desired scope of work, that they will submit the bid on time, and to clarify any concerns. It would be nice if these quotations came in a few days prior to bid opening so that the general contractor could perform an evaluation prior to including them in the overall estimate. However, these bids are rarely received more that a few minutes prior to bid opening, making a thorough evaluation impossible. If communications are taking place throughout the bid process the likelihood of the subcontractor bidding the desired scope of work is improved and the subcontractor selection is simplified.

Indirect Field Costs

The labor, material, equipment, and subcontracts that are ultimately integrated into the physical construction project represent the direct field cost of the project. The costs that represent the items and activities that are required to support the field effort are referred to as indirect field costs (IFC). The majority of the indirect field costs are a function of time rather than any material quantity. The following items often contribute to the indirect field cost portion of the overall project costs.

Field Supervision

Superintendents
Job Engineers
Time Keepers
Project Controls
Material Controls

Temporary Facilities

Construction Fences
Temporary Enclosures
Dust Control
Erosion Control
Temporary Doors and Windows
Job Site Office

Temporary Utilities

Toilets
Electricity
Gas
Water
Lighting

Protection

Protecting the Completed Work
Protecting Adjacent Structures

Permits/Inspections

All Required Building Inspections
Materials Testing

Professional Services

Engineering
Surveying
Accounting
Legal

Labor Burdens and Taxes

Unemployment Insurance
FICA
Medicare
Vacation Benefits
Retirement Benefits

Bonds and Insurance

Builders Risk Insurance
Workers Compensation Insurance
Bid Bond
Payment Bond
Performance Bonds

A quick review of the list should confirm that the costs for most of these items are a function of the overall project duration. For example, the job site office is required from the beginning through the completion of the project. Depending upon the size of the project, the number of office trailers may vary over time. The indirect costs are a substantial percentage of the overall costs of the project, and the estimate of their cost requires the same level of diligence as do the direct costs. Since most of these costs are a function of time, a workable construction schedule is required to accurately estimate the cost of these items.

The cost of the home office is often included in the overall estimated cost of a project. These costs are typically an allocation of the cost of maintaining the home office, which is applied to all of the projects. By allocating this cost to all projects, a better picture of the true cost of executing a project can be developed.

Conclusion

The construction estimate process is the starting point for effective construction management. The estimate becomes the benchmark by which the actual project performance will be measured. In addition, this document is used to determine which materials will be required for actual construction and in what quantity. Furthermore, the direct field cost portion of the estimate is needed to estimate the project duration and the cost for all associated indirect costs. To meet these many objectives, the estimate must be organized in a fashion that will allow it to be used for a variety of needs. Advance planning is needed prior to beginning the estimate to determine what information will be needed in the future and how it should be formatted to facilitate the effective use of the estimate document. If this advance planning is performed prior to commencing the quantity takeoff, the usefulness of the estimate will be enhanced and the amount of time required in the field to compare the actuals to the plan and purchase materials will be reduced. Ultimately this will reduce construction costs for the project and lead to a competitive advantage for the contractor.

Suggested Exercises

1. Develop the work breakdown structure for your sample project. From a practical perspective, the number of individual work packages should be limited to between 90 and 120. These work packages should cover direct field cost items only.
2. Of the work packages identified in Step 1, determine which will be performed by subcontractors and which will be completed by the contractor's craft personnel. The items that are identified as subcontracted will then form the basis of the subcontracting plan. Select three of the subcontract work packages and write a request for quote to a vendor. Be sure to include a scope of work that will communicate to the subcontractor what portion of the project you intend to have them install.
3. Develop an activity list of the 90 to 120 activities that were identified in the first step. The table below should be used as a guide:

Activity #	Activity Description	Work-hours	Dur.	Labor $	Material $	Equipment $	Subcontract $	Total $

A spreadsheet program will work well for developing this activity list. This type of program allows for the needed calculations as well as providing a means to summarize the entire estimate. The total cost of all of the activities should be the total direct field costs of the project. For the initial development of this table the activity numbers should not be included. However, once the layout of the schedule has been completed they can be added and the list can be sorted in activity number order.

Scheduling Logic

KEY TERMS

Precedence Diagramming Method (PDM) A graphical scheduling methodology composed of activity nodes and relationship arrows that show the interrelations and constraints between activities.

Arrow Diagramming Method (ADM) A graphical scheduling methodology that is composed of activity arrows that are joined by nodes.

Critical Path The longest path through the schedule.

Critical Activity An activity that must be started and finished on a specific date in order not to extend the project duration.

Lag A period of no activity that must elapse between two events.

Scheduling

Construction scheduling, in general, is a graphic representation of a specific plan that will be used to assemble the project. Both the Arrow Diagramming Method (ADM) and Precedence Diagramming Method (PDM) are acceptable construction scheduling techniques. Of these two methods, PDM offers the user a more powerful methodology coupled with better and more understandable graphics. With the advent of personal computers and better trained construction managers, PDM has come forward as the most popular technique. While there are still contractors that use ADM, they are few and their numbers are declining. Therefore, all future discussions will deal exclusively with the techniques necessary to effectively utilize the Precedence Diagramming Method.

The PDM schedule consists of activity nodes and relationship arrows. Figure 3-1 is an example of how activities are related. In that figure if Activity B is the current activity, the preceding activity, Activity A, is referred to as its predecessor. Activity B's successor activity would be Activity C. In PDM scheduling it is common for an activity to have multiple predecessor and successor activities.

When analyzing a PDM schedule, the activities are always grouped together into activity pairs. Figure 3-2 demonstrates how the three activities from Figure 3-1 would be broken down into two distinct pairs of activities.

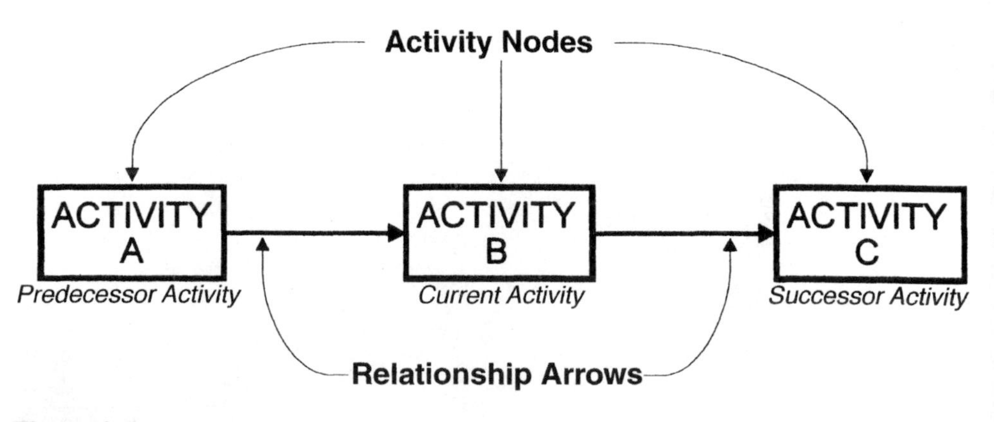

Figure 3-1
PDM Activity Terminology

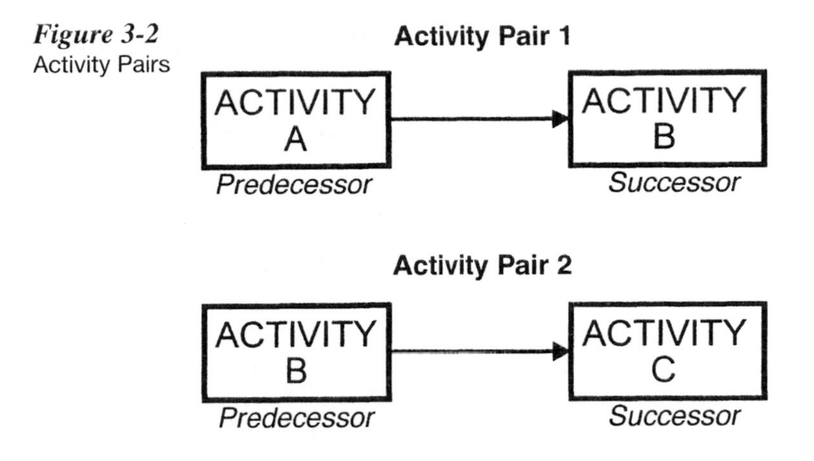

Figure 3-2
Activity Pairs

Every activity in the schedule must be given a unique activity identifier, and the node identifier at the tail of the relationship arrow should be less than the node identifier at the head of the relationship arrow. To meet these requirements it is recommended that activity numbers be used for all activities and that they be in increments of five or ten. The use of alphabetic activity identifiers become awkward past twenty-six, and if an activity needs to be added later all of the succeeding activities would have to be given new identifiers. If numbers are used, these two limitations are easily overcome. For example, Activity A could be Activity 10, Activity B could be Activity 20 and Activity C could be Activity 30. If it becomes necessary, at a later date, to add an activity between Activity 20 and Activity 30, it could become Activity 25.

If an activity does not have a predecessor, that activity could qualify to be the first activity on the project. Conversely, if an activity does not have a successor that activity could qualify to be the last activity in the schedule. When developing a construction schedule, caution needs to be exercised to ensure that activities without predecessors or successors could logically be either the first or last activity on the project.

Activity Relationships

There are four possible ways to relate any two activities in precedence diagramming. However, the *finish-to-start*, *start-to-start*, and *finish-to-finish* are the three that are of consequence. The *start-to-finish* relationship serves virtually no purpose in construction scheduling; this relationship has the activity pair headed in the wrong direction. Through the use of the finish-to-start, start-to-start and finish-to-finish relationships, one can realistically diagram the planned construction.

The relationship arrows can have lags associated with them. A lag signifies that there must be some predetermined waiting period between the activities. Lags are positive in almost all circumstances; however, there are occasions where a negative lag would be appropriate. When a negative lag is used, it is called a lead.

Finish-To-Start Relationship

The finish-to-start relationship, or the conventional relationship, is the most common relationship in construction scheduling. Figure 3-3 is a graphic representation of the finish-to-start relationship. With this relationship, the earliest that Activity B can begin is when Activity A is completed; however, Activity B does not have to start immediately upon the completion of Activity A.

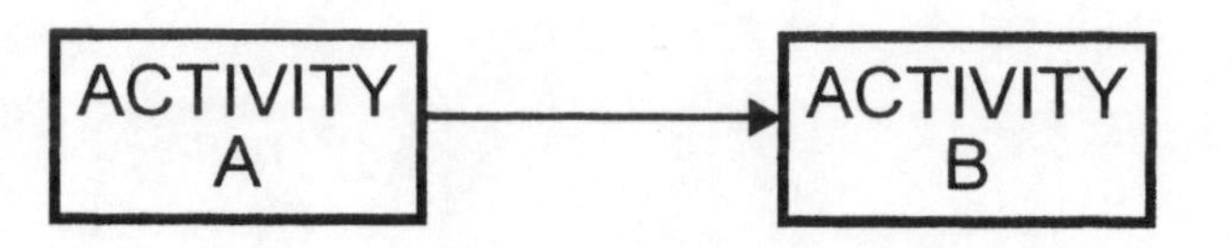

Figure 3-3
Finish-to-Start Relationship: Activity B can Begin Only After Activity A is Completed

If a lag is employed, it is typically designated with the symbol $L=$ directly above the relationship arrow. If no lag is placed on the relationship arrow, then the implied lag is zero. However, it is recommended that a lag always be placed above the relationship arrow, even if it is zero. This labeling technique will ensure that the schedule is internally consistent and more usable. Figure 3-4 is an example of finish-to-start relationship with a 2-day lag.

Figure 3-5 diagrams three steps required in pouring a slab on grade. First, all of the forms must be set, then all of the reinforcement must be placed, and finally the slab can be poured. In this example there is no overlap or concurrence between these three activities.

While there is no need for a lag between the activities in Figure 3-5, if that series of activities is expanded to include removing the formwork, a lag would be appropriate. Since the formwork cannot be removed until the concrete has had time to set and cure, a period of inactivity would be necessary. This period of inactivity can be designated by using a lag to denote the number of days required for the concrete to set and cure. In Figure 3-6, the slab is poured and then at least 2 days must elapse prior to starting the removal of the forms.

Start-to-Start Relationship

The start-to-start (SS) relationship is used to show how the start of one activity triggers the start of a successor activity. In Figure 3-7, Activity B can begin as soon as Activity A begins. Activity B would be referred to as the successor to Activity A. Figure 3-8 shows how a lag would be noted. In Figure 3-9 a start-to-start relationship is used to diagram the construction of multiple spread footings. When constructing multiple spread footings, it is unlikely that all of the excavations would be completed prior to commencing with the setting of the forms. A contractor constructing footings in that fashion would be taking unnecessary risks. If all of the excavations for the footings were completed and it was to rain, the contractor would have to either pump out these excavations, or wait for the water to evaporate or percolate into the soil. While the day it rained may not be charged against the number of days allowed to complete the project, the days used for pumping and drying would be charged as working days. In addition, the contractor would have to spend additional money on equipment and people to remove the water.

If these items were not included in the project estimate, they would reduce the anticipated profit on the project. The second unnecessary risk would be the safety of the workers. As long as there are open excavations on a site the potential exists for someone to fall into these pits and become injured. These injuries add nonproductive time and cost to the project, negatively impact the morale of the work force, and increase long-term costs through higher insurance premiums. To overcome these risks, the most appropriate approach would be to initiate formwork as soon as excavation is far enough along to support the formwork effort.

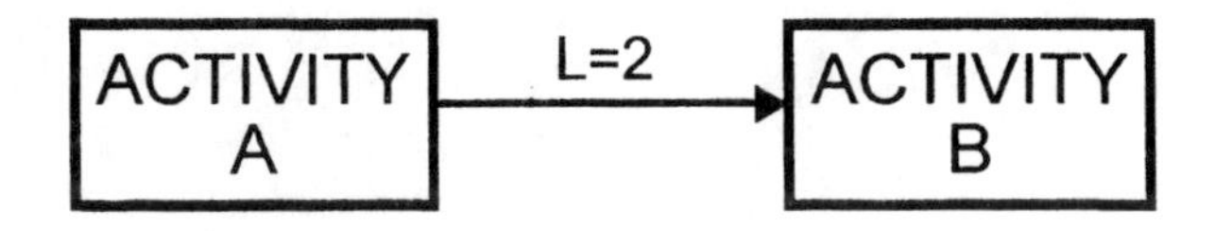

Figure 3-4
Finish-to-Start Relationship with a Lag: The Earliest that Activity B can Start is Two Days After Activity A is Completed

Figure 3-5
Finish-to-Start Activity Chain

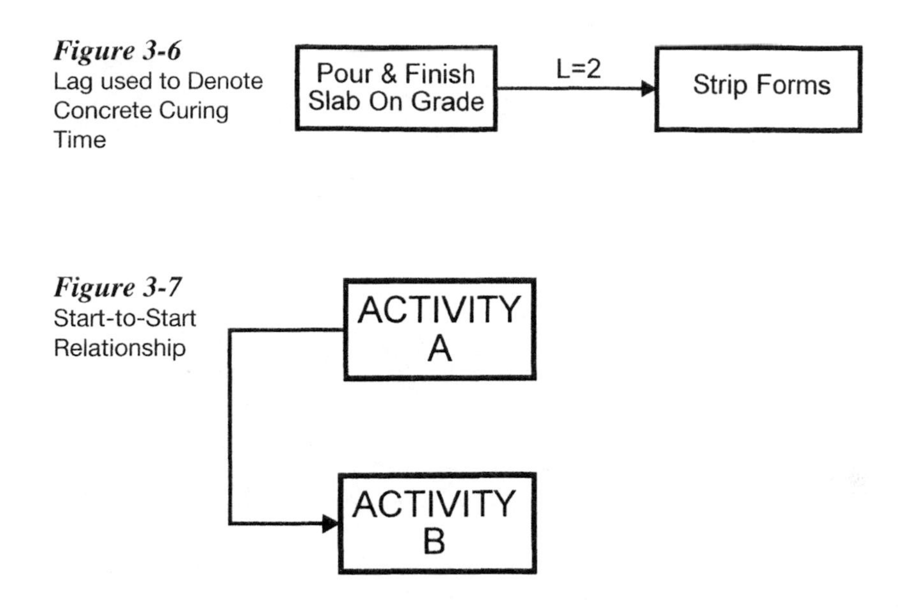

Figure 3-6
Lag used to Denote Concrete Curing Time

Figure 3-7
Start-to-Start Relationship

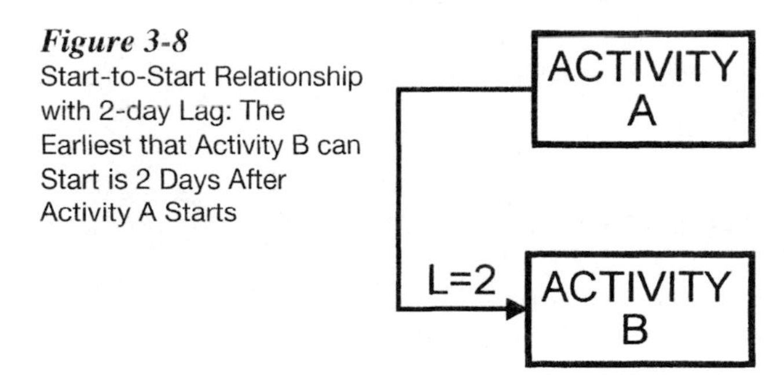

Figure 3-8
Start-to-Start Relationship with 2-day Lag: The Earliest that Activity B can Start is 2 Days After Activity A Starts

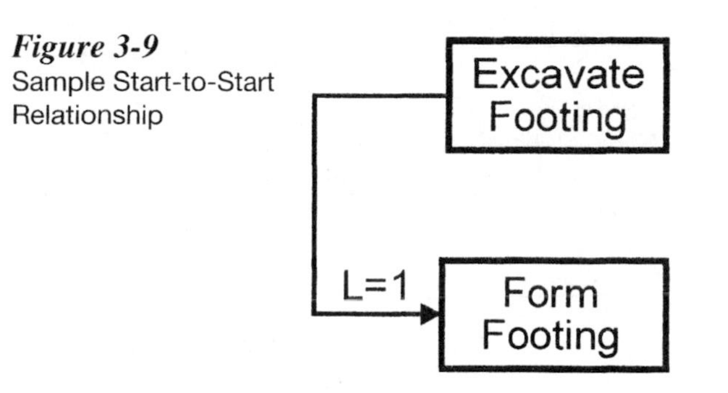

Figure 3-9
Sample Start-to-Start Relationship

Figure 3-10
Finish-to-Finish Relationship: Activity B can be Completed at the Same Time as Activity A

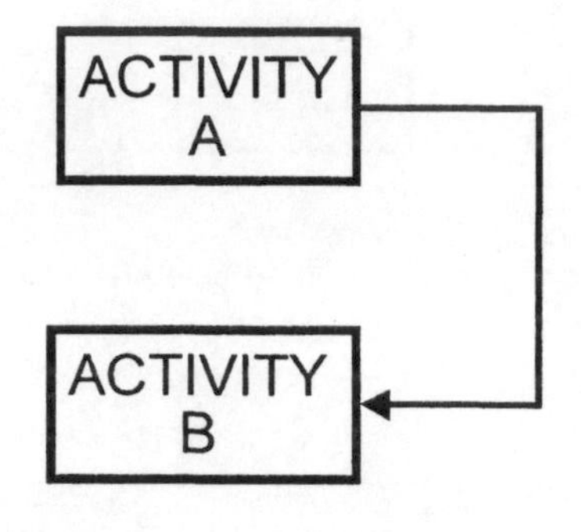

With a lag of one, formwork could begin one day after the excavation was begun and the two activities could run concurrently. On Day 1 excavation is started; on Day 2 the forms are placed in the excavations completed on Day 1 and the excavation work continues. Through this approach, these activities are concurrent, the work is more efficient, and the potential for rain damage is minimized. In all types of excavation it should always be the goal to begin excavation, perform what underground work is required, and backfill as rapidly as possible.

Finish-to-Finish Relationship

The finish-to-finish relationship, diagrammed in Figure 3-10, is typically used in conjunction with start-to-start relationships. With a finish-to-finish relationship, activities are concurrent but it is necessary for the initial activity to remain ahead of its successor activity. The completion of a successor activity is contingent on the completion of its predecessor.

Figure 3-11 shows how a lag would be delineated in this type of relationship. Activity A and Activity B are concurrent, but Activity B cannot be completed until at least 2 days have elapsed since the completion of Activity A. In Figure 3-9 it was shown how the excavation needed to start prior to the start of the setting of the footing formwork. However, it is necessary to show that the excavation must *always* be progressing ahead of the setting of the formwork. Figure 3-12 shows how that requirement is integrated into the schedule with the use of a finish-to-finish relationship with a lag of one.

Graphic Conventions

The construction logic diagram, with activities and multiple relationships, could quickly become an unmanageable document unless some graphic conventions are adopted to make this document usable. The first convention is that *relationship arrows do not cross each other*. If arrows are allowed to cross each other, the schedule will be too complicated and most likely only understood by the person that constructed it. That would limit its availability and usefulness. The crossing of arrows can be avoided through planning and by using connectors. Connectors are circles with some identifier placed in them. This designation is used to

Figure 3-11
Finish-to-Finish Relationship with a Lag: The Earliest that Activity B can Finish is 2 Days After Activity A is Completed

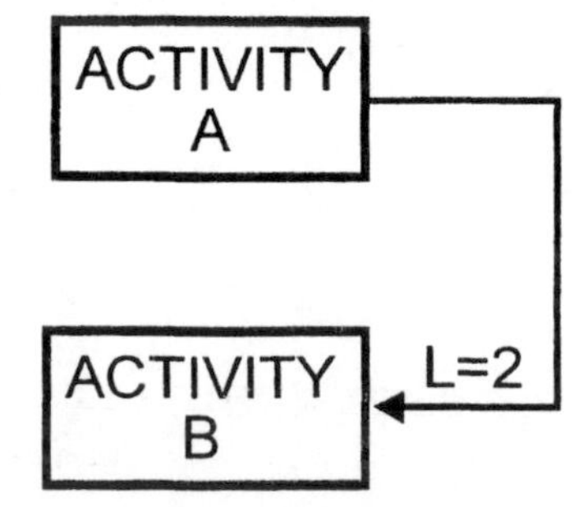

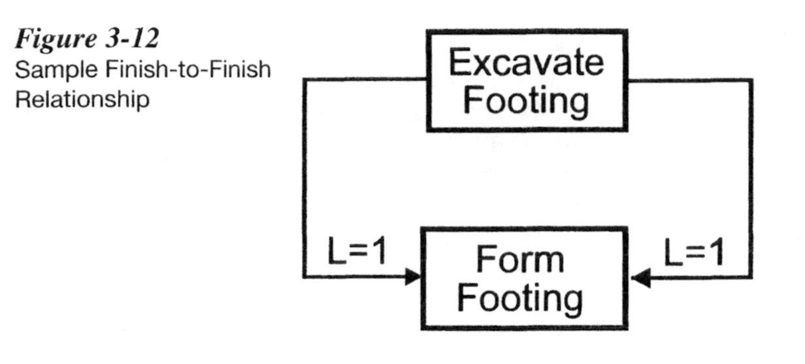

Figure 3-12
Sample Finish-to-Finish Relationship

show where a relationship breaks and restarts. Figure 3-13 delineates how Activity 20 has a finish-to-start relationship with Activity 40. In addition, Activity 30 has a finish-to-start relationship with Activity 60. Another use of connectors is to keep the schedule from becoming too linear. While a very long construction schedule may look impressive on the wall of the job trailer, it must help facilitate the actual construction process. Often when the construction schedule's physical size becomes intimidating, it tends to become a wall decoration, rather than a working project management tool. The sample project schedule found at the end of this chapter uses multiple connectors to keep the schedule as compact as possible.

Construction Logic

The development of the construction logic diagram should be undertaken by members of the project team, not by an individual. Bringing together a group of representative managers will provide varied insights into executing the project. If some of the team's members later become involved in the actual execution of the project, the added benefit of commitment and ownership of the plan accrues. By bringing all of the team members together prior to bidding and construction, a consensus can be reached on how best to execute this specific

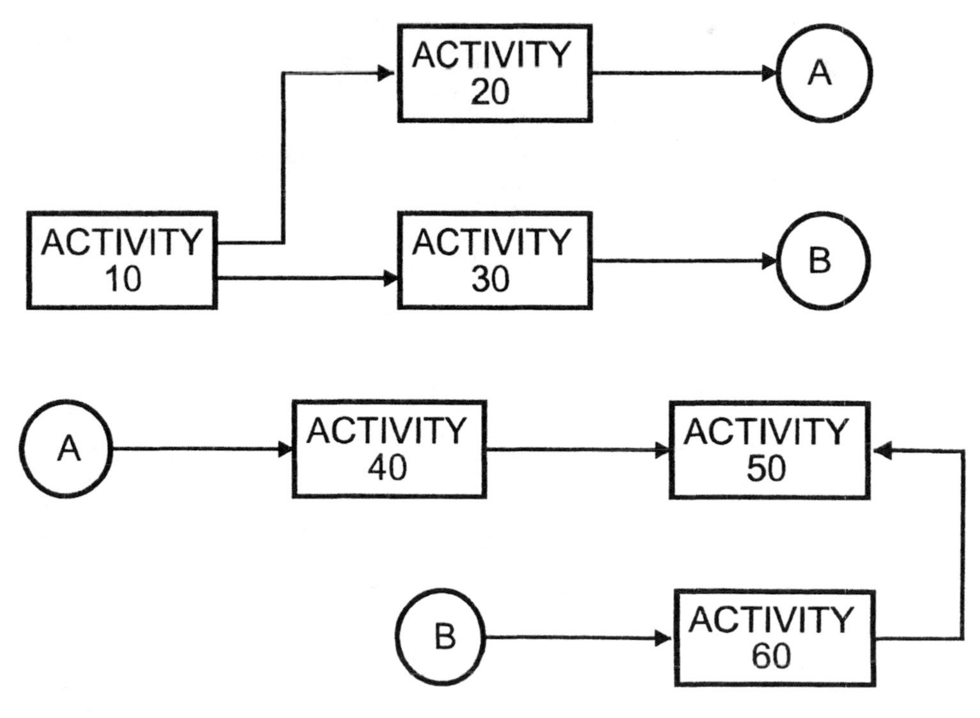

Figure 3-13
Adding Connectors to the Schedule

project. In addition, the construction execution plan will facilitate an accurate estimate of the indirect field costs.

To effectively develop a construction logic diagram the project team must have access to a complete set of contract documents and the estimate for the direct field cost portion of the project. There are several approaches that can be used in developing the initial construction logic diagram. One strategy can be particularly effective in organizing the project plan and building consensus at the same time. The first step in this process is to write each of the project activities on a small, separate piece of paper. After thorough discussion, tape all the slips to a wall in the order that the project team believes is the most effective method for constructing the project. By taping these small pieces of paper to a wall they can be moved and manipulated easily. Once the project team reaches consensus as to the best order, the construction logic diagram can be drawn.

There is rarely one best method for constructing a project. Instead multiple acceptable solutions could allow the project to be efficiently and effectively constructed. The plan that is adopted should accentuate the talents of the organization and maximize the use of the resources that are owned by the general contractor.

Lags

The duration of a lag should be the minimum number of days that must elapse between activities. This duration should be based on the physical constraints of the construction process. The simplest way to visualize the sequence of activities is with a bar chart schedule.

Figure 3-14 is the bar chart schedule for placing one of four runs of continuous footings for a building foundation. In that chart the activities are linear to prevent people and the excavation equipment from working on the same side simultaneously.

The two days of inactivity (days 11 and 12) are to allow time for the concrete to cure prior to removing the formwork. This example can be further enhanced by expanding it to show the construction of all sides of the foundation system. For that example the following activity list and durations would be used.

Excavation	16 Days
Form Strip Footings	12 Days
Reinforce Strip Footings	8 Days
Place and Finish Footings	4 Days
Strip Forms	4 Days

The bar chart schedule shown in Figure 3-15 is for constructing all four sides of the foundation. The amount of time that elapses from the beginning of the activities to their end is longer than the stated duration for all activities other than excavation. For example, the formwork begins on day 4 and finishes at the end of day 19 for a start to finish duration of 15 days, while the stated duration is only 12 days. The number of work days is still 12 days but there are interruptions that will occur during the completion of the activity.

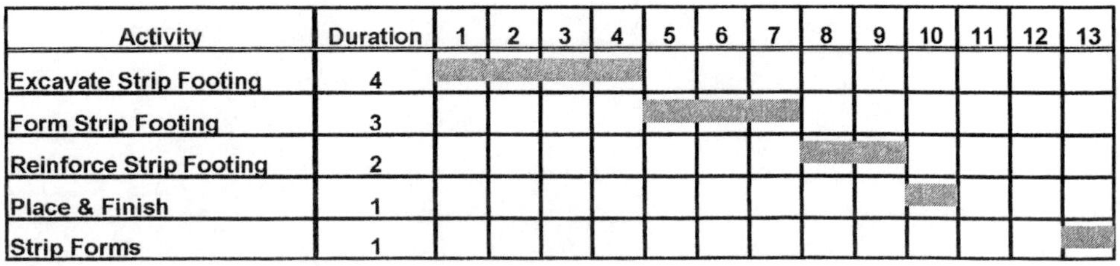

Activity	Duration	1	2	3	4	5	6	7	8	9	10	11	12	13
Excavate Strip Footing	4													
Form Strip Footing	3													
Reinforce Strip Footing	2													
Place & Finish	1													
Strip Forms	1													

Figure 3-14
Sample Schedule for Continuous Footings

Activity	Duration	1	2	3	4	5	6	7	8	9	10	11	12	13	14	15	16	17	18	19	20	21	22	23	24	25
Excavate Strip Footing	16																									
Form Strip Footing	12																									
Reinforce Strip Footing	8																									
Place & Finish	4																									
Strip Forms	4																									

Side 1
Side 2
Side 3
Side 4

Figure 3-15
Sample Schedule for Continuous Footings

These interruptions occur because of the need for a predecessor event to occur. This schedule also assumes that the required crafts, materials, and equipment are available in the required quantity at the required time. Figure 3-16 is the logic diagram that corresponds with the bar chart schedule. In that figure the 4-day lag on the start-to-start relationship between Activity 10 and Activity 20 allows enough time for the first side of the four-sided excavation to be completed. That would mean that at the beginning of the fifth day the excavation could begin on the second side and the formwork could begin on the first side.

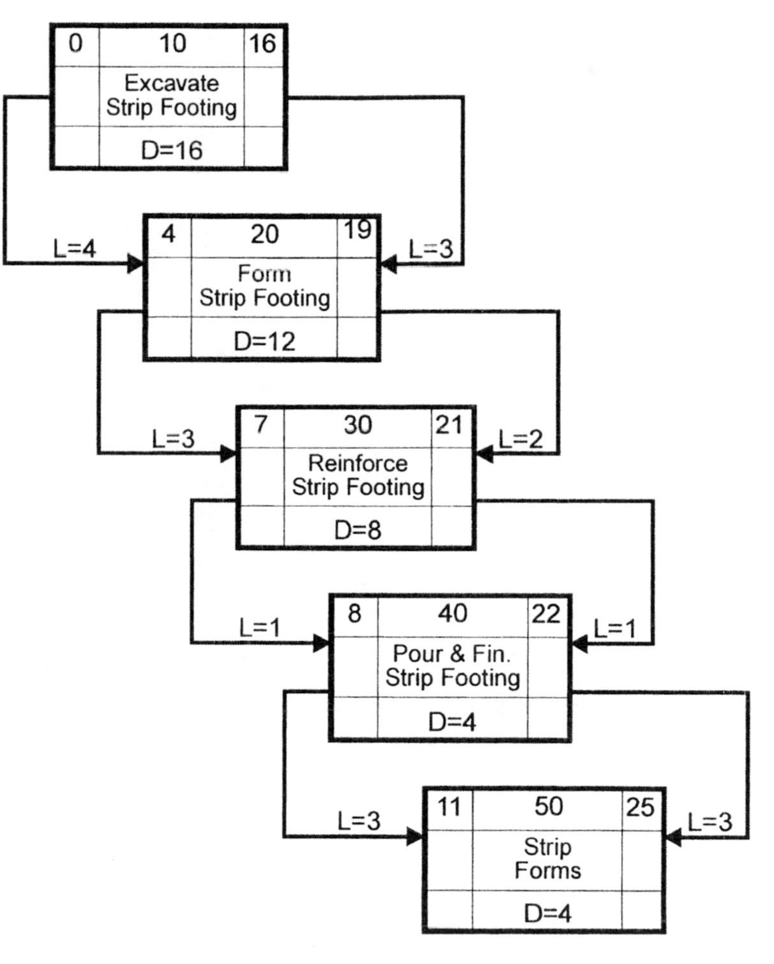

Figure 3-16
Sample Schedule for Continuous Footings

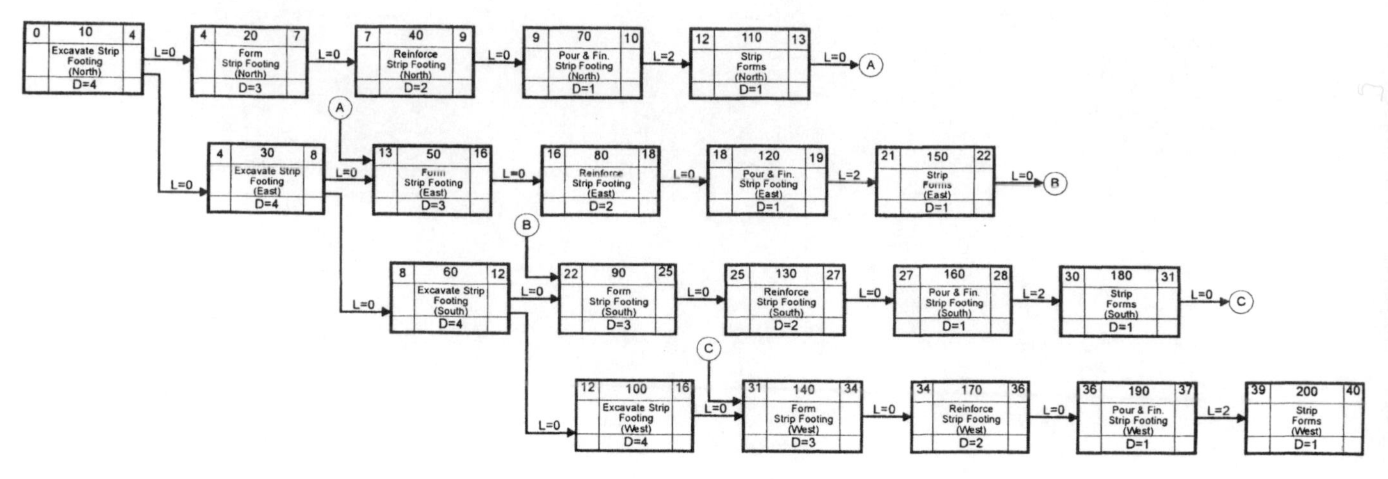

Figure 3-17
Schedule for Continuous Footings with Reused Forms

The finish-to-finish relationship on that pair of activities has a 3-day lag to ensure that the excavation is always one side ahead of the formwork. This corresponds with the bar chart in Figure 3-15. The same logic holds true throughout the network except for activity pair 40 and 50. The logic for the 3-day lag on the start-to-start relationship is one day to place the concrete and 2 days for the concrete to cure. This same logic helps explain the 3-day lag on the finish-to-finish relationship. One day is required to strip the forms and 2 days of curing time are needed. Once again, the bar chart in Figure 3-15 will help explain the required lags. While schedule calculations are not presented until the next chapter, Figure 3-16 does include a forward pass calculation that corresponds to the associated bar chart schedule.

Using the same example with the added criteria that the forms must be reused for each side changes the complexion and complexity of the schedule. In that scenario, the setting of all of the forms past the first side is contingent upon the forms being removed from the previous side and the excavation completed. Figure 3-17 is the logic diagram for that example. In addition, notice how connectors are used to reduce the size of the schedule. Figure 3-18 is the corresponding bar chart schedule.

By limiting the availability of forms the project duration was extended from 25 to 40 days. This should demonstrate how changing the availability or constraining the quantity of resources impacts the project duration. If a needed item is in short supply, there will be an impact on the project duration and ultimately on the cost of the project. It also points out the need for coordination between the people preparing the estimate and those producing the project schedule. The assumptions about how the project is to be executed from the estimate must be integrated into the construction schedule.

Activity	Duration
Excavate Strip Footing	16
Form Strip Footing	12
Reinforce Strip Footing	8
Place & Finish	4
Strip Forms	4

Days: 1 2 3 4 5 6 7 8 9 10 11 12 13 14 15 16 17 18 19 20 21 22 23 24 25 26 27 28 29 30 31 32 33 34 35 36 37 38 39 40 41

Side 1
Side 2
Side 3
Side 4

Figure 3-18
Schedule for Continuous Footings with Reused Forms

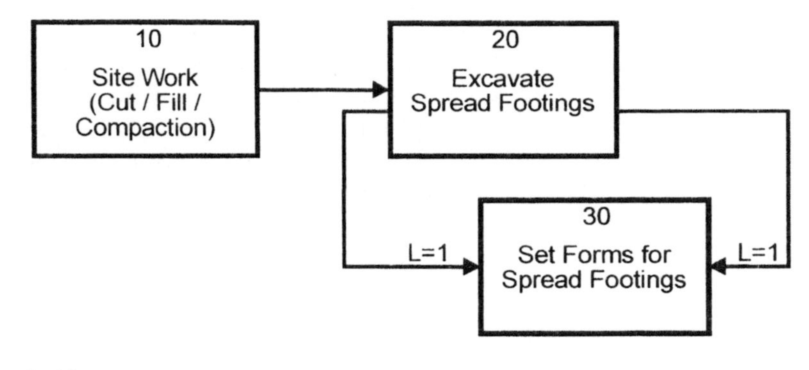

Figure 3-19
Sample Concurrent Activities

Related and Nonrelated Concurrent Activities

The start-to-start and finish-to-start relationships are both used to denote concurrent activities. If one of the activities is dependent upon the other, they are referred to as related concurrent activities. These activities should be joined with start-to-start relationships. In Figure 3-19, the site work is completed, the spread footing excavation begins, and the setting of the forms lags behind that activity by one day.

While the forms are being set in one location, excavation for other footings is proceeding. However, the setting of the forms for a particular footing is dependent upon the completion of the excavation for that specific footing. Due to the dependence between these activities, the start-to-start relationship is required.

If construction activities can occur at the same time but are independent, then finish-to-start relationships would be appropriate. Figure 3-20 is an example of how nonrelated concurrent activities would be denoted. In that figure the excavation for the site utilities could occur at the same time as the excavation for spread footings. However, the start of

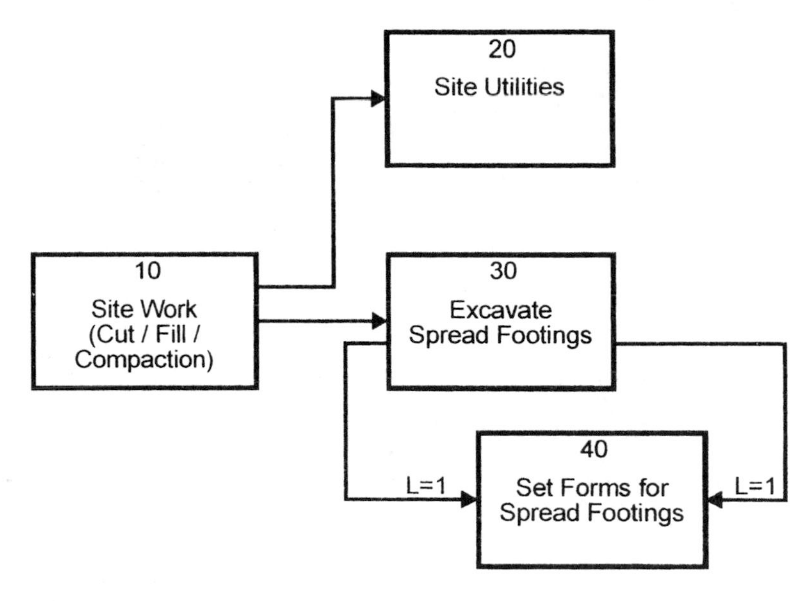

Figure 3-20
Sample Nonrelated Concurrent Activities

these activities and the rate of progress are independent. If independent activities are denoted on the schedule with start-to-start activities, a disproportionate number of activities will end up on the critical path. Activities which by their nature should not be on the critical path will end up there. While these activities may end up being critical they should not be on the critical path in the initial construction plan. For example, one would not expect to see activities such as striping the parking lot or hanging drapes on the critical path. If these type of activities end up on the critical path, the confidence in the schedule is diminished and its usefulness impaired.

Sample Schedule

The sample schedule at the end of this chapter is for a small office/warehouse building. The building's foundation system is spread footings to support all of the structural steel interior columns. The exterior of the building is load-bearing split-faced concrete masonry units (CMU) which rest upon a foundation wall on a continuous footing that circles the entire perimeter of the building. The office portion of the building is constructed of metal studs covered with drywall that is textured and painted.

This sample schedule (Figure 3-21) will be manipulated throughout the remainder of this book. It should serve as a guideline for effective project management and controls.

Conclusion

There are three distinct relationships that should be used to develop the graphic plan for the construction project. They are finish-to-start, start-to-start, and finish-to-finish relationships. These relationships in conjunction with lags develop the methodology for assembling the project. The logic that is used to associate the activities must recognize any constraints in the construction process. In addition, the assigned lags must be a function of the constraints imposed by the nature of the activities rather than based upon having these activities begin and end at a desired point in time.

Suggested Exercise

1. Using the activity list you developed in Chapter 2, write each of the activities and its associated duration on a post-it note. Then, on a large sheet of paper (24″ × 36″), begin to lay out the activities in the order in which you anticipate they will be executed. Connect all of the activities with the appropriate relationships and lags. The completed document should look like Figure 3-21.

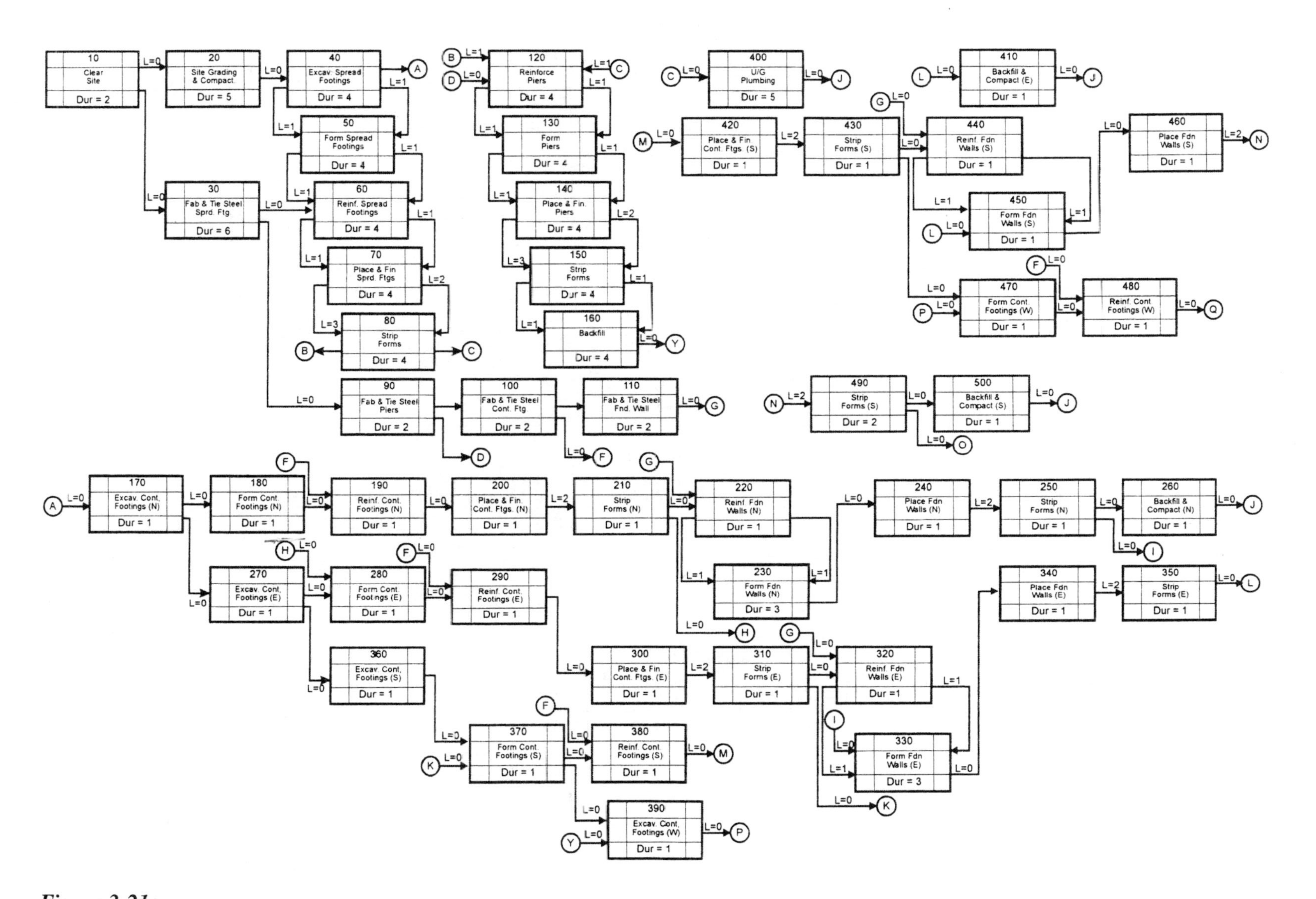

Figure 3-21a
Sample Schedule for Small Office/Warehouse Building

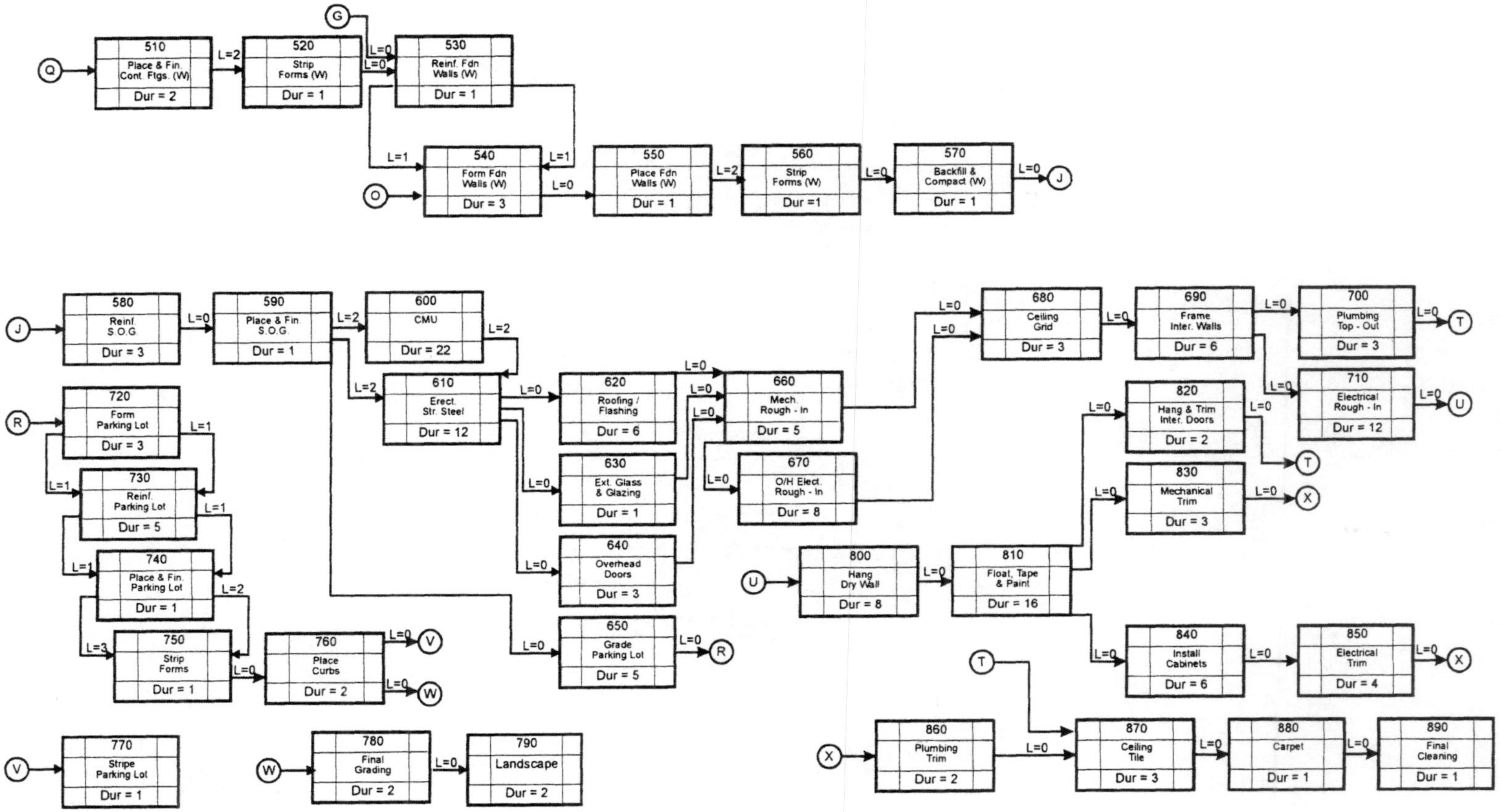

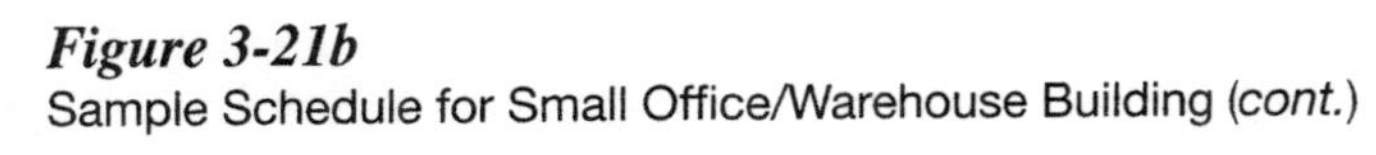

Figure 3-21b
Sample Schedule for Small Office/Warehouse Building (*cont.*)

Calculating the Project Schedule

KEY TERMS

Early Dates The earliest that a scheduled event can begin or end.

Late Dates The latest that a scheduled event can begin or end.

Forward Pass A calculation starting with the first activity and culminating with the last activity performed to find the early dates and the duration of a specific project.

Backward Pass A scheduling calculation done to determine the activity late dates. This calculation begins with the last activity and project duration and culminates with the first activity.

Float The number of days that an event or lag can be delayed or extended without impacting the completion of the project.

Project Days The number of days that a project, activity, or lag has or will be under way.

Early Start (ES) The earliest that an activity can start.

Early Finish (EF) The earliest that an activity can possibly finish.

Late Start (LS) The latest that an activity can start and not impact project completion.

Late Finish (LF) The latest that an activity can be completed without impacting the project completion.

Forward Pass

The forward pass is a series of calculations performed to determine the activities' early dates and the project duration. This process starts with the beginning activities and terminates with the ending activities. The project schedule can have multiple beginning and ending activities. The beginning activities are those that could be the first activities, and are identified by not having a predecessor. Conversely, the last activities on the project are those without a successor. In a graphical format, the beginning activities will have no relationship arrows coming into them and the ending activities will have no relationship arrows coming out of them.

In Figure 4-1, a two-activity schedule, Activity 10 would be the beginning activity and Activity 20 would be the ending activity. To begin the forward pass calculation, there must be an initial assumption that the project will start on Project Day 0. This is shown by entering a zero in the top left-hand corner of the node for Activity 10. If there were multiple starting activities, a zero would be entered on all of them. In Figure 4-1, if Activity 10 starts at Time 0 and lasts 3 days, then the earliest that this activity could be completed is Project Day 3. Therefore, a 3 is recorded in the top right-hand corner of the Activity 10 node. Since the relationship between these two activities has no lag, the early start of Activity 20 would be the early finish of Activity 10. If there were a lag, then the lag amount would have been added to the early finish of Activity 10 and the new amount would have been entered as the early start of Activity 20.

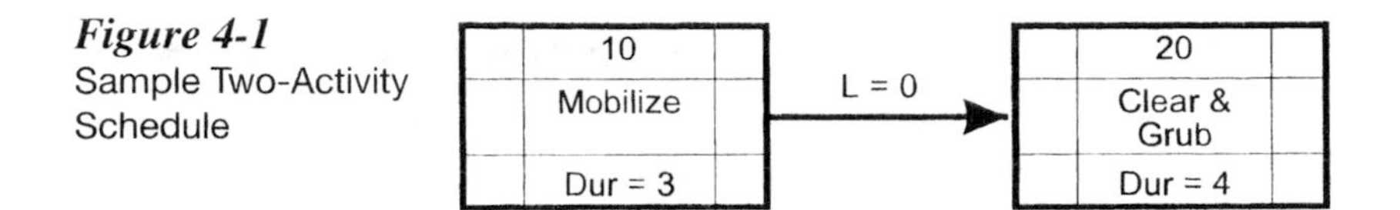

Figure 4-1
Sample Two-Activity Schedule

Finally, the earliest that Activity 20 can be completed is 4 days after it starts, or on Project Day 7. When the forward pass is completed, the early dates and project duration will be known. The project duration is the greatest early finish date of all of the ending activities. Figure 4-2 details all of these calculations and shows how they would be entered on the project schedule.

Once the schedule's early dates have been calculated, they can be entered in a table similar to Table 4-1. This tabular format presents the needed information in a fashion that can be understood by persons who may not be well versed in PDM scheduling but will be integrally involved in the project.

In Figure 4-3, two activities—a finish-to-start relationship and a start-to-start relationship—have been added. The early start of Activity 50 is found using the same process as was used for Activity 20. Since the earliest that Activity 20 can be completed is Project Day 7, the earliest that Activity 50 can begin is also Project Day 7. Furthermore, the earliest that Activity 50 can be completed is Project Day 10, which is 3 days after it begins. Even though the relationship between activities 50 and 60 is different from those that were previously discussed, the process for determining the early start dates is the same. The lag between these two activities states that the earliest that Activity 60 can begin is one day after Activity 50 begins. In terms of the forward pass calculation, that means that for the diagrammed start-to-start relationship, the early start of the successor node is the early start of the predecessor node plus the lag. In this example, the early start of Activity 60 is Project Day 8. This was found by taking the early start of Activity 50 (Project Day 7) and adding the one-day lag. The earliest that Activity 60 could end, if it starts on Project Day 8, is on Project Day 12. If these four activities comprised the entire schedule, the project duration would be 12 days, the largest early finish date.

In Figure 4-4, a finish-to-finish relationship with a lag of one has been added between Activity 50 and Activity 60. The process for finding the early finish date for Activity 60 is basically the same as in the previous example. Activity 60 can be completed as early as one day after the completion of Activity 50 (Project Day 11) or 4 days after it starts (Project Day 12). In this scenario, as with all forward passes, both of these criteria must be met, which would mean that the project duration is 12 days. The only way to ensure that both of these criteria are met is to always select the larger of all of the available criteria.

Figure 4-5 is an expanded schedule that serves as a good example of how to perform a forward pass. One of the unique features of this diagram are the two predecessors to Activity 60. If Activity 30 could be completed on Day 6 and there is no lag between Activity 30 and Activity 60, then it would appear that Activity 60 could begin on Day 6. However, the start-to-start relationship between Activity 50 and Activity 60 must be evaluated prior to making that statement. If Activity 50 begins on Day 7 and Activity 60 could start as early as one day later, then Activity 60 could begin on Day 8. Therefore, when there are multiple predecessors to the start of an activity, there will be an equal number of possible start dates.

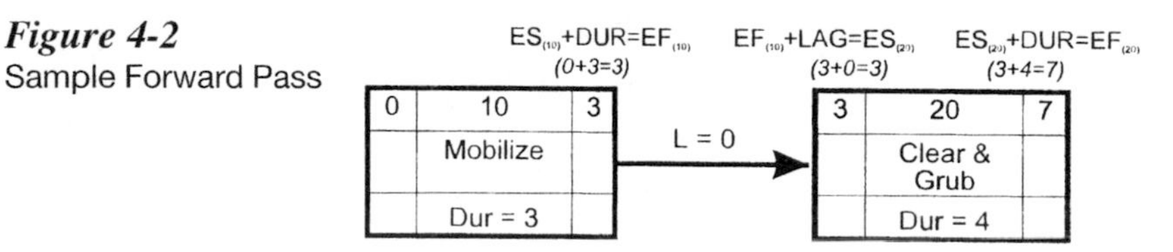

Figure 4-2
Sample Forward Pass

$ES_{(X)}$ = Early start of activity X
$EF_{(X)}$ = Early finish of activity X
DUR = Activity duration

Table 4-1
Early Date Table

Activity	Description	Duration	Early Start (ES)	Early Finish (EF)
10	Mobilize	3	0	3
20	Clear & Grub	4	3	7

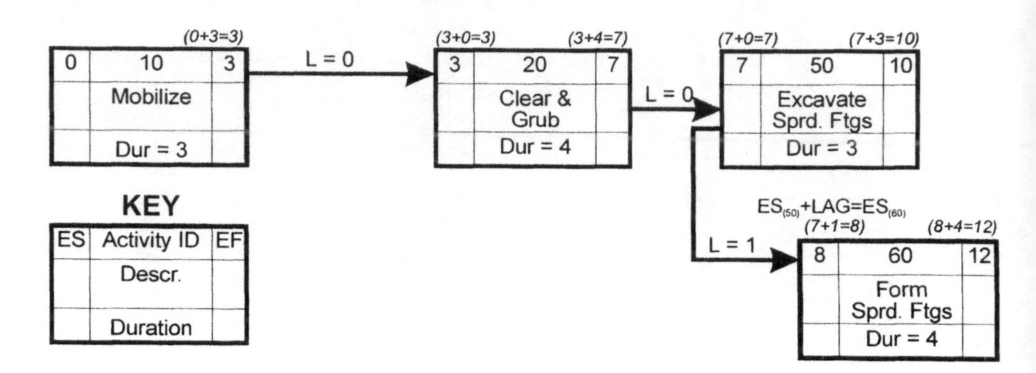

Figure 4-3
Sample Forward Pass with a Start-to-Start Relationship

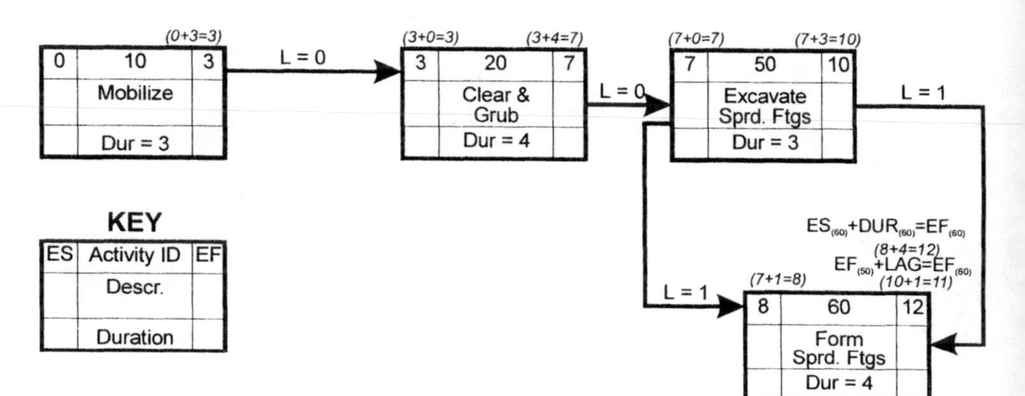

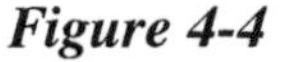
Figure 4-4
Sample Forward Pass with a Finish-to-Finish Relationship

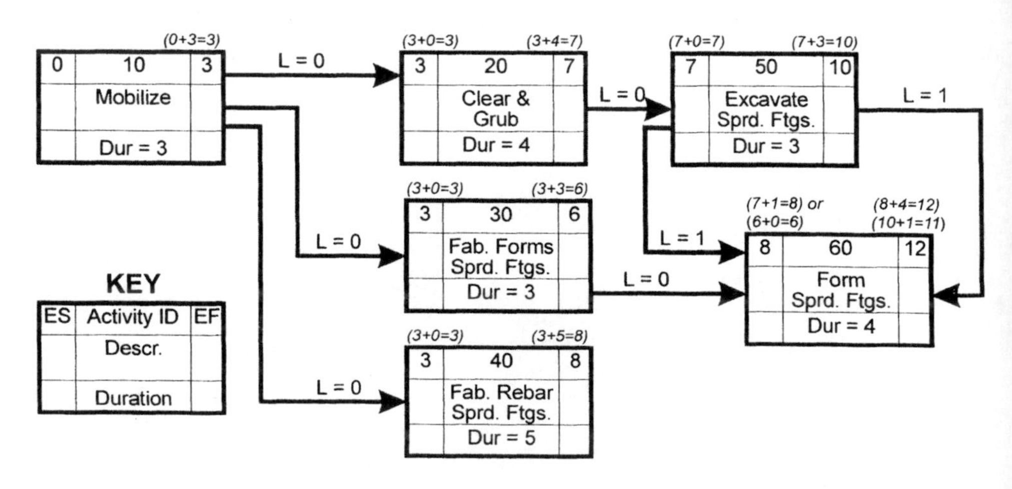

Figure 4-5

Sample Schedule with Completed Forward Pass

Which of these start dates is the early start of that activity? Since the start of Activity 60 is contingent upon the completion of Activity 30 and the start of Activity 50 and its one-day lag, the largest calculated early start date would prevail. From a practical perspective, the forms cannot be erected until they are fabricated and the required portion of the excavation has been completed. This reasoning holds true in all situations.

Therefore, Rule 1 for calculating a forward pass can be developed:

Rule 1
When performing a forward pass with multiple relationship arrows pointing to the start of an activity, the calculation that results in the largest number of project days is always selected.

Activity 60 has another distinction; it has two possible completion dates. If Activity 60 is started on Day 8 and lasts for 4 days, the earliest that the activity can be completed is Day 12. But, there is also the requirement that Activity 60 be completed at least one day after the completion of Activity 50. This requirement results in a possible early finish date of Project Day 11. Since both of these criteria must be met, selecting the largest number of project days is the only way to ensure that all of the criteria will be met, which facilitates the development of Rule 2:

Rule 2
When performing a forward pass with multiple alternatives for the early finish date (from multiple logic arrows and the early start plus duration calculation), always select the largest number of project days.

In order to ensure consistency in calculations, it is essential that the early starts for an activity always be calculated prior to its early finish. Therefore, Rule 3 is:

Rule 3
When performing a forward pass, always determine the early start date for an activity prior to determining its early finish date.

When the forward pass is completed, the early dates for all of the activities will be known as will the calculated project duration. In Figure 4-5 there were two activities without successors, thus qualifying them as ending activities. Activity 40 had an early finish of Day 8 and Activity 60 had an early finish of Day 12, making the estimated project duration 12 days. From Figure 4-5, Table 4-2 can be developed.

Backward Pass

The backward pass is basically the opposite calculation of the forward pass. This calculation begins with the ending activities and works backward through the schedule until it reaches the beginning activities. This series of calculations determines the late start and late finish dates for the individual activities.

Table 4-2
Early Date Table

Activity	Description	Duration	Early Start (ES)	Early Finish (EF)
10	Mobilize	3	0	3
20	Clear & Grub	4	3	7
30	Fab. Forms for Sprd. Ftgs.	3	3	6
40	Fab. Rebar for Sprd. Ftgs.	5	3	8
50	Excavate Sprd. Ftgs.	3	7	10
60	Form Sprd. Ftgs.	4	8	12

Just as with the forward pass calculations, there are a series of rules that must be followed in order to ensure a correct calculation.

Rule 1
When performing a backward pass, always set the late finish of all of the ending activities to the project duration.

Rule 2
When performing a backward pass with multiple relationship arrows, the calculation that results in the smallest number of project days is always selected.

Rule 3
When performing a backward pass with multiple alternatives (either from multiple logic arrows or late finish dates minus duration), always select the smallest number of project days.

Rule 4
When performing a backward pass, always determine the late finish date for an activity prior to determining the late start dates.

Using these four rules, a backward pass calculation can be performed which will result in the late dates for the project. Using the simple two-activity schedule in Figure 4-6 and following the four rules for a backward pass, the following logic would be used. First, the project duration of 7 days would be entered into the lower right-hand corner of Activity 20, the last activity. Given that the late finish of that activity is Project Day 7 and the duration is 4 days, then the latest Activity 20 can begin is on Project Day 3. Therefore, a 3 is entered

Figure 4-6
Backward Pass

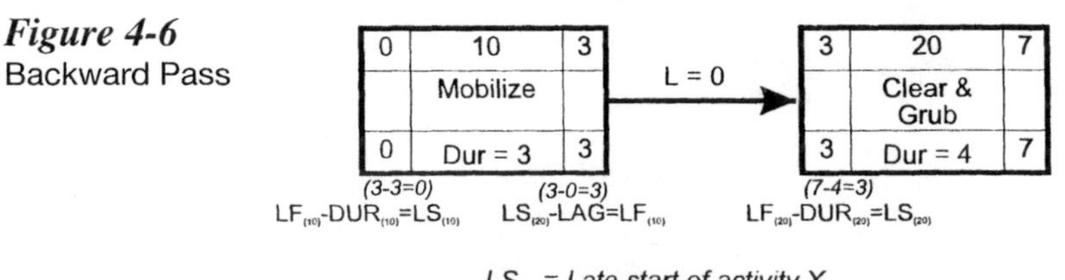

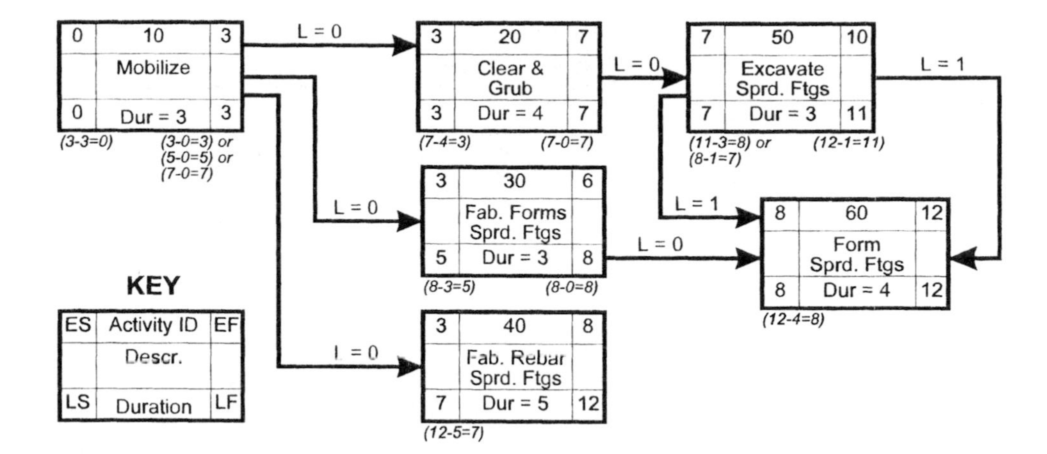

Figure 4-7
Completed Forward and Backward Pass

in the bottom left-hand corner of Activity 20. Since the lag between Activities 10 and 20 is zero, the subtraction of that lag from the late start of Activity 20 will result in a late finish date of Project Day 3 for Activity 10. Finally, the 3-day duration of Activity 10 is subtracted from the late finish date to obtain the late start date of Day 0.

Figure 4-7 can serve as an example of how a more complicated backward pass would be accomplished. In that example the project duration is 12 days and the late finish date is set to 12 on both of the ending activities, Activity 40 and Activity 60. Once the backward pass has been completed, the results can be entered in a table similar to Table 4-3.

In the previous examples, the late start and the early start dates for the first activity ended up being zero. Had there been multiple beginning activities, one of these activities would come back to zero. The "coming back to zero" should be a check for calculation errors. The only exception to this is when the project completion date that is used to start the backward pass is not the project duration that was developed from the forward pass. A logical reason for using a different late finish date for the last activity would be if the construction contract specifies a completion date. In that situation it might be appropriate to insert that date as the completion date rather than the one calculated from the forward pass. However, if the number of project days allowed from the assigned completion date is less than the number calculated from the forward pass, then the schedule needs to be revised. In that situation, the project will be behind schedule before it starts. In more practical terms, several of the late start dates will be before their corresponding early start dates.

Table 4-3
Early and Late Date Table

Activity	Description	Duration	Early Start (ES)	Early Finish (EF)	Late Start (LS)	Late Finish (LF)
10	Mobilize	3	0	3	0	3
20	Clear & Grub	4	3	7	3	7
30	Fab. Forms for Sprd. Ftgs.	3	3	6	5	8
40	Fab. Rebar for Sprd. Ftgs.	5	3	8	7	12
50	Excavate Sprd. Ftgs.	3	7	10	7	11
60	Form Sprd. Ftgs.	4	8	12	8	12

That would mean that these activities were behind schedule even though the project had not begun. That condition would invalidate the entire schedule. If that condition occurs, one or a combination of revised strategies needs to be adopted: modifying the logic, increasing the number and size of the crews, or extending the assigned completion date. Once the revision is completed, the forward pass should be recalculated to see if the new project duration falls on or before the contractually specified completion date. If not, the schedule must be revised again. This revision process must continue until the calculated number of days is less than or equal to the number of days allowed.

Float

The float in the schedule refers to the number of days that an activity, event, or lag can be delayed or extended without impacting the overall completion of that project. In precedence diagramming there are four different types of float. The first type is *starting float* (SF), which is the number of days that the start of an activity can be delayed without impacting the completion of the project or the late finish of the specific activity. Since the late start of an activity is the latest that an activity can start without impacting project completion and the early start is the earliest that an activity can commence, the difference between these two dates is the starting float.

$$\text{Starting Float} = \text{Late Start} - \text{Early Start}$$

Formula 4-1

Just as the start of an activity can be delayed and potentially not impact the completion of the project, the same is true of the completion of an activity. The *finish float* (FF) is used to describe the number of days that the completion of an activity can be postponed without impacting the completion of the project. Since the late finish of an activity is the latest that an activity can be completed and the early finish is the earliest that a specific activity can be completed, the difference between the two is the finish float.

$$\text{Finish Float} = \text{Late Finish} - \text{Early Finish}$$

Formula 4-2

The third type of float, the *task float* (TF), describes the number of days that the duration of a specific activity can be increased without impacting the project completion. If the early start for an activity is subtracted from the late finish of that activity the result is the longest possible duration for that activity. By subtracting this longest possible duration from the stated duration, the task float is found.

$$\text{Task Float} = \text{Late Finish} - \text{Early Start} - \text{Duration}$$

Formula 4-3

These three types of float can be shown either on the schedule itself (as in Figure 4-11) or in a chart similar to Table 4-4.

Relationship Float

Just as there was the potential for slack or free time in the start, finish, and duration of an activity, there is also the possibility that there could be slack in the lag assigned to a particular relationship. Since there are three distinct types of relationships, there are three

Table 4-4
Schedule Dates and Float

Activity	Duration	Early Start (ES)	Early Finish (EF)	Late Start (LS)	Late Finish (LF)	Starting Float LS-ES	Finish Float LF-EF	Task Float LF-ES-D
10	3	0	3	0	3	0	0	0
20	4	3	7	3	7	0	0	0
30	3	3	6	5	8	2	2	2
40	5	3	8	7	12	4	4	4
50	3	7	10	7	11	0	1	1
60	4	8	12	8	12	0	0	0

distinct formulas that must be used when calculating relationship float. While each of these formulas is unique, the logic is the same for all three. The first step is to determine the longest possible lag and then subtract that from the stated lag to determine how many days the lag can be extended.

For a start-to-start relationship, Formula 4-4 would be used to find relationship float.

$$\text{Relationship Float}_{(\text{Start-to-Start})} = LS_{(\text{Successor Node})} - ES_{(\text{Predecessor Node})} - \text{Lag}$$

Formula 4-4

Figure 4-8 (an excerpt of the schedule presented in Figure 4-7) is an example of how this formula would be applied. In this example, subtracting the late start of the activity pair successor node from the early start of the pair predecessor node results in the longest possible lag of one day. By subtracting this longest possible lag from the stated lag the relationship float is zero.

In a finish-to-finish relationship, Formula 4-5 would be used. Figure 4-9 is an example of this formula being applied.

$$\text{Relationship Float}_{(FF)} = LF_{(\text{Successor Node})} - EF_{(\text{Predecessor Node})} - \text{Lag}$$

Formula 4-5

Figure 4-8
Sample Relationship Float—Start-to-Start Relationship

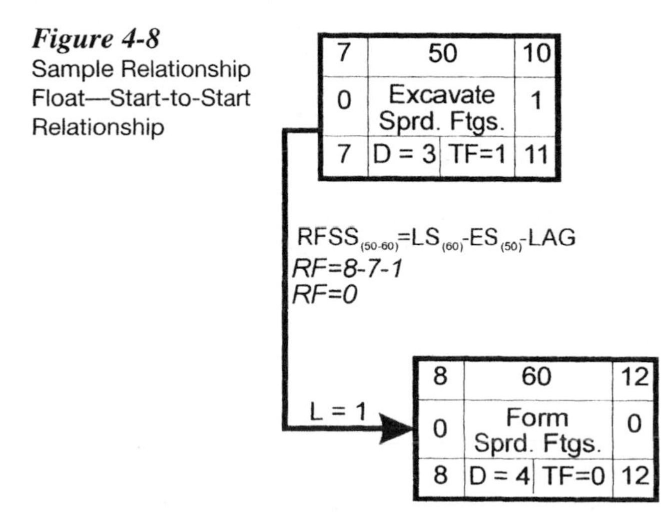

Figure 4-9
Sample Relationship Float—Finish-to-Finish Relationship

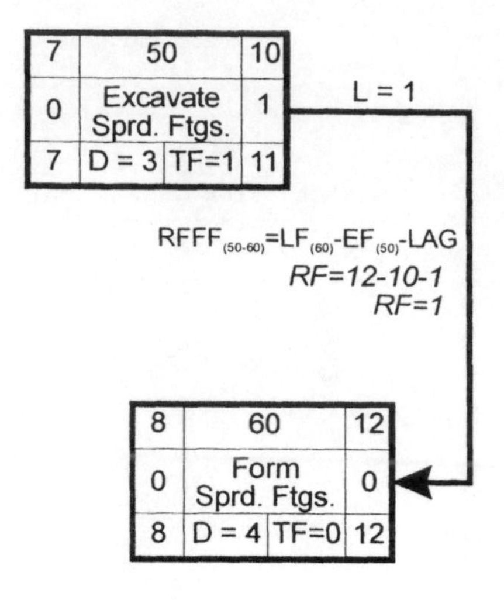

The finish-to-start relationship float would use Formula 4-6. Figure 4-10 is an example of how that formula would be used.

$$\text{Relationship Float}_{(FS)} = LS_{(\text{Successor Node})} - EF_{(\text{Predecessor Node})} - \text{Lag}$$

Formula 4-6

Critical Path

The critical path is the longest continuous path through the schedule. Unless an arbitrary end date is assigned, it also is the path of zero float. If an arbitrary end date is established, the critical path is the path with the least float. There may be multiple critical paths through a schedule, provided that the duration through each of these paths is the same. Figure 4-11 shows the critical path as well as all of the calculated information. In that example, Activities 10, 20, and 60 are critical since their starting floats, finish floats, and task floats are zero. The start of Activity 50 is critical since its starting float is also equal to zero. The relationships that are critical are the finish-to-start relationship between Activity 10 and Activity 20, the finish-to-start between Activity 20 and Activity 50 and the start-to-start between Activity 50 and Activity 60. These relationships are critical because the relationship floats for all three of these are zero.

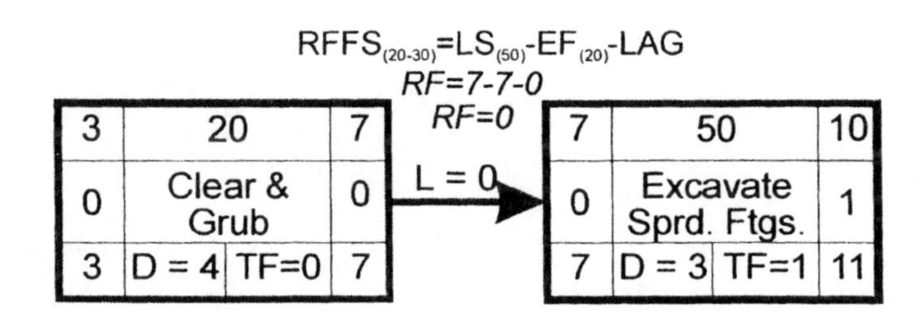

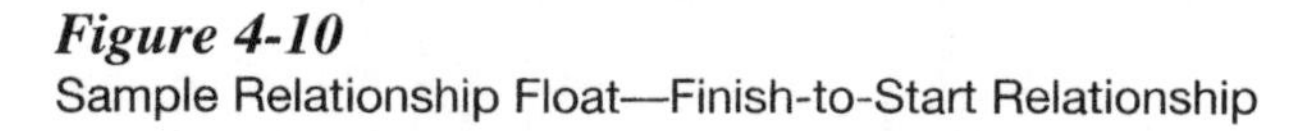

Figure 4-10
Sample Relationship Float—Finish-to-Start Relationship

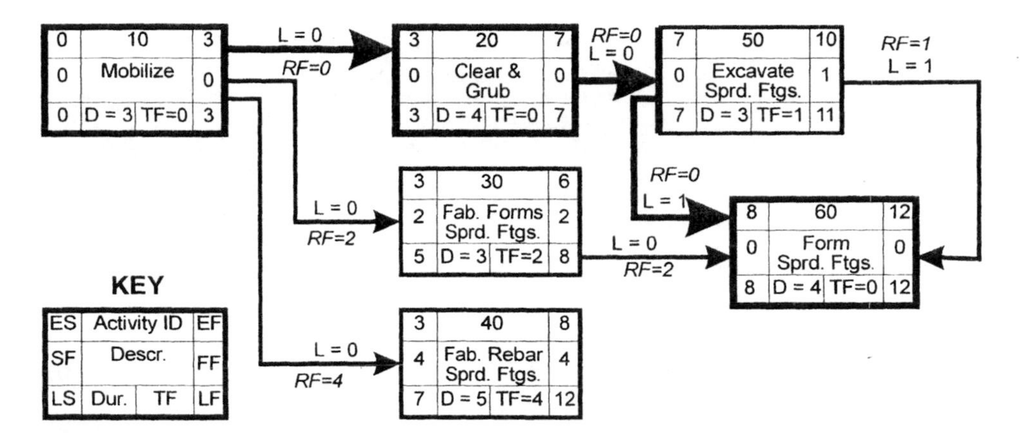

Figure 4-11
Critical Path

DECEMBER

Sun	Mon	Tues	Wed	Thurs	Fri	Sat
	1	2	3	4	5	6
7	8	9	10	11	12	13
14	Day 1 15	Day 2 16	Day 3 17	Day 4 18	19	20
21	Day 5 22	Day 6 23	Holiday 24	Holiday 25	26	27
28	Day 7 29	Day 8 30	Day 9 31			

JANUARY

Sun	Mon	Tues	Wed	Thurs	Fri	Sat
				Holiday 1	2	3
4	Day 10 5	Day 11 6	Day 12 7	8	9	10
11	12	13	14	15	16	17
18	19	20	21	22	23	24
25	26	27	28	29	30	31

Figure 4-12
Sample Project Calendar

Table 4-5
Project Tabular Schedule

Activity	Dur	ES	ES	LS	LS	EF	EF	LF	LF
10	3	0	15 Dec	0	15 Dec	3	17 Dec	3	17 Dec
20	4	3	18 Dec	3	18 Dec	7	29 Dec	7	29 Dec
30	3	3	18 Dec	5	23 Dec	6	23 Dec	8	30 Dec
40	5	3	18 Dec	7	30 Dec	8	30 Dec	12	7 Jan
50	3	7	30 Dec	7	30 Dec	10	5 Jan	11	6 Jan
60	4	8	31 Dec	8	31 Dec	12	7 Jan	12	7 Jan

Project Calendar

When calculating a construction schedule, the use of project days is necessary. But when executing a project, this term is of little value. "Starting an activity at the end of Project Day 10" has little meaning to most people. To add meaning, these days need to be converted into calendar days. The calendar-day format is universally understood. Telling a person that the activity needs to start on June 10 would be clearly understood and not subject to any different interpretation.

In order to convert project days into calendar days, there needs to be an understanding of the impact of starting the forward pass on Project Day 0. This assumption means that the time of day for all of the calculated workdays comes at the end of the workday. This allows for simplified calculations, but causes some difficulty. Telling someone that an activity begins at the end of the workday on June 10 does not make much sense. However, telling that same person that the activity begins at the beginning of the workday on June 11 is sensible. The end of the workday on June 10 and the beginning of the workday on June 11 are actually the same unless multiple shifts are being worked. Converting activity starts into calendar dates is achieved by adding one to the calculated activity start, so that all start dates represent the beginning of the day.

The first step in converting project days to calendar days is to establish a project calendar. This calendar is a sequential numbering of days that will be worked. Figure 4-12 is an example of what a project calendar should look like. In this example, the project commences at the beginning of the day on December 15 and has three holidays in addition to the weekends. As previously stated, the start dates need to be increased by 1 to make the conversion to calendar dates. Activity 10 began at the end of Project Day 0. By adding 1 to it, it now can be said that the scheduled start date is on December 15. That activity is completed at the end of Project Day 3, which corresponds to the end of the day on December 17.

Once the conversion is completed, it is recommended that these dates be recorded in a table. Table 4-5 shows how project days can be converted into a form that can be understood by everyone on the project.

Sample Schedule

Figure 4-13 is the sample schedule that was introduced in Chapter 3, now with a completed forward and backward pass.

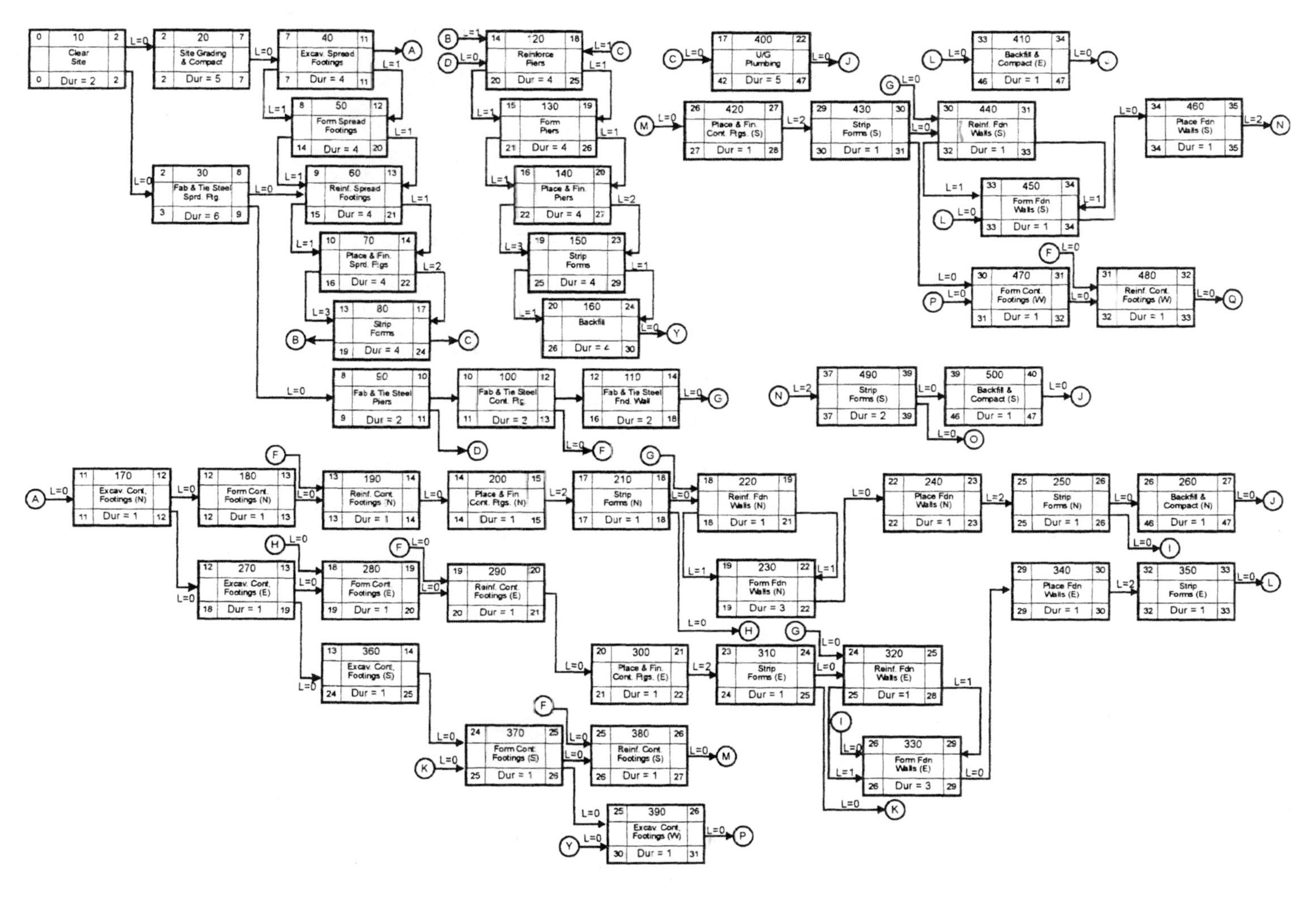

Figure 4-13a
Sample Project

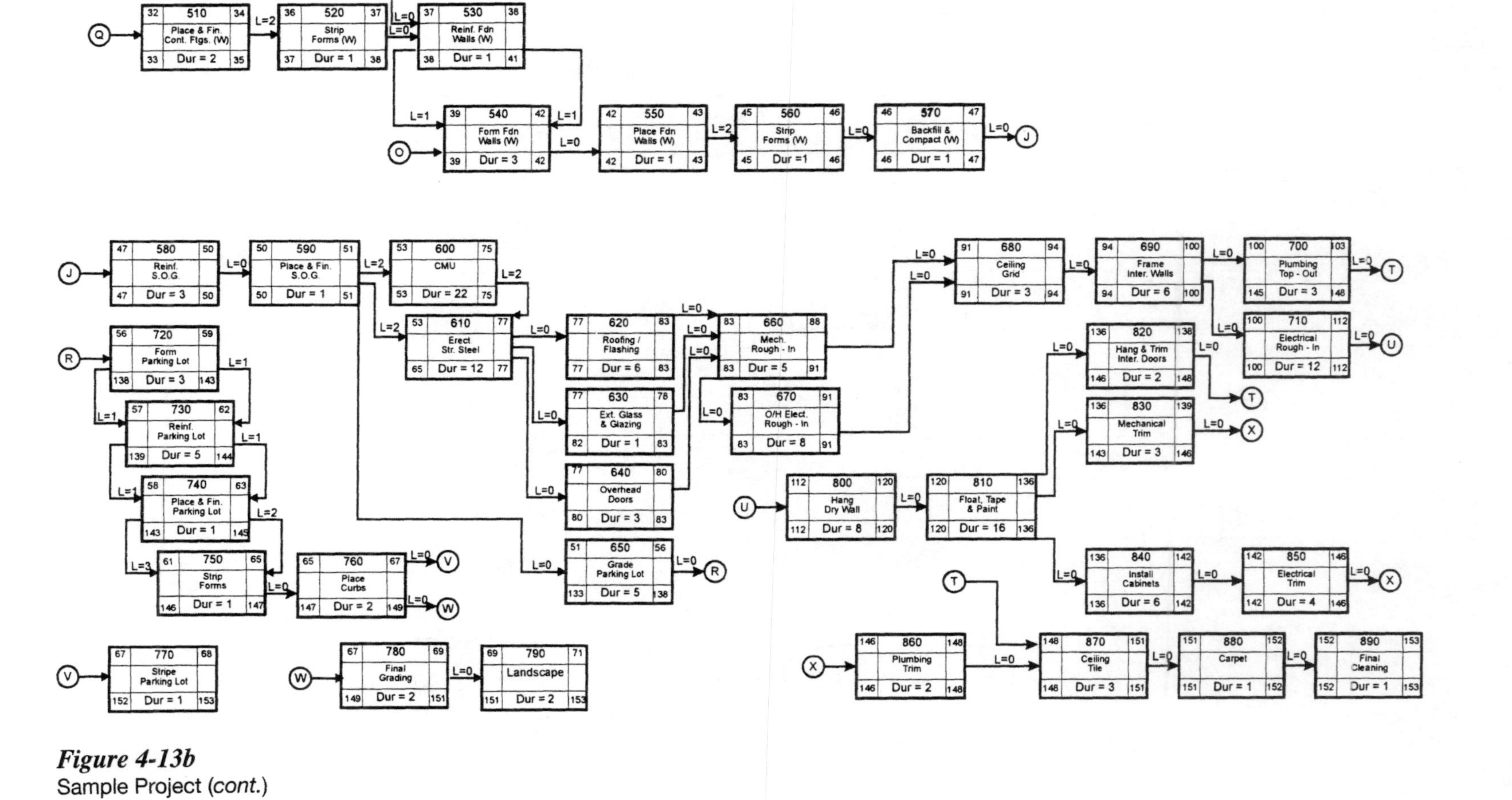

Figure 4-13b
Sample Project (*cont.*)

Conclusion

The combination of the forward and backward pass yields parameters for when activities can begin and end. If the activities' start dates are equal, then the start of that activity is critical. Similarly if the finish dates are equal, then the finish of the activity is critical. Critical activities or events are those happenings that must take place at a specific point in time or the completion of the project will be impacted. Those activities where this does not occur are referred to as having float. The float time is the measurement of flexibility that is available to the contractor to shuffle resources in order to maximize the efficiency of the construction process.

Suggested Exercises

1. Using the schedule you developed in Chapter 3, perform a forward and backward pass, doing all the required float calculations. Then construct and complete a table similar to the one shown below.

Activity ID	Activity Description	Activity Duration	ES	LS	SF	EF	LF	FF	TF

2. Get a calendar, assume a project start date and mark all of the nonwork periods for the next year. This calendar should look like Figure 4-12.
3. Convert the five earliest early start dates into calendar dates. Then convert the five latest late finish dates into calendar dates.

Creating and Saving Projects

Key Terms
Creating a New Project
Saving a Project
Opening an Existing Project
Backing up Projects
Restoring a Project
Copying a Project
Conclusion

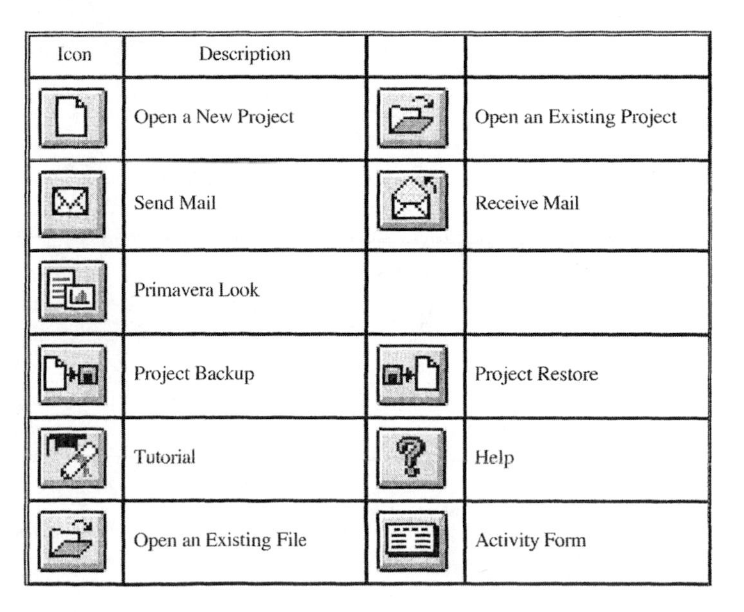

Icon	Description		
	Open a New Project		Open an Existing Project
	Send Mail		Receive Mail
	Primavera Look		
	Project Backup		Project Restore
	Tutorial		Help
	Open an Existing File		Activity Form

Icon Descriptions

KEY TERMS

Backup The process of creating a copy of all of the project files. This copy is typically stored on a diskette for transportability and security.

Restore The process by which a backup copy of the project files is loaded onto the computer's hard drive.

Creating a New Project

Primavera Project Planner is a widely used interactive construction scheduling/planning software. This program, with its graphical screens and output, simplifies the interpretation of the scheduling documents and ultimately leads to more effective project management.

Due to the ever-changing dynamics of a construction project, new information is continually becoming available. This information needs to be integrated into the construction schedule so opportunities and deviations can be identified and new plans developed. These dynamics require the schedule to be updated and recalculated on a regular basis. Through the use of a computer program such as Primavera Project Planner (P3) these calculations can be performed with thousands of activities in a matter of seconds. This speed allows construction managers to make decisions and plans based on the most current information. The better and more timely the information, the better the planning and decision making. This in turn increases the likelihood that the project will be completed within the allocated time and budget.

Screen Image 5-1 is the initial screen that is activated when Primavera Project Planner is started. This screen consists of three commands and a number of icons. The number of icons will vary depending on which other Primavera products are loaded on a particular workstation. The *File*, *Tools* and *Help* commands will unveil a series of options. Appendix A shows a menu tree for these commands. To create a new project, click on the new project icon (first from the left) or select the *File* command followed by the *New* option as shown in Screen Image 5-2. Either of these operations will open the *Add a New Project* window, as shown in Screen Image 5-3. The *Add a New Project* window has a number of fields that require input in order to correctly create a new project.

The *Current folder* is displayed at the top of the window and shows the drive and directory where the new project will be stored. In Screen Image 5-3 the *Current folder* is *C:\p3win\projects*. When Primavera Project Planner is installed on a stand-alone workstation the default project directory is *p3win\projects*. If the program is being run from a network or if the default project file location was changed during installation, then the *Current folder* will be some other location. To change the folder, click on the *Browse* button to open the *Browse* window as shown in Screen Image 5-4. From this window the user can browse the drives and their associated directories until the desired file location is found. Clicking on the desired folder followed by clicking the *OK* button will set that location as the *Current folder* and return to the program to the *Add a New Project* window. If the program is being run on a network, the user must ensure that he or she has full control access rights to the directory where the project is to be stored. This is a network issue that needs to be discussed with the network administrator. The *Project name* field, on the *Add a New Project* window, will accept any combination of four alphanumeric characters. This name will become the first four characters of all of the files that will be created for this project. The *Number/Version, Project title*, and *Company name* fields are cosmetic items and should be completed with the appropriate project specific information.

Screen Image 5-1
Initial Input Screen

Screen Image 5-2
File/New Command

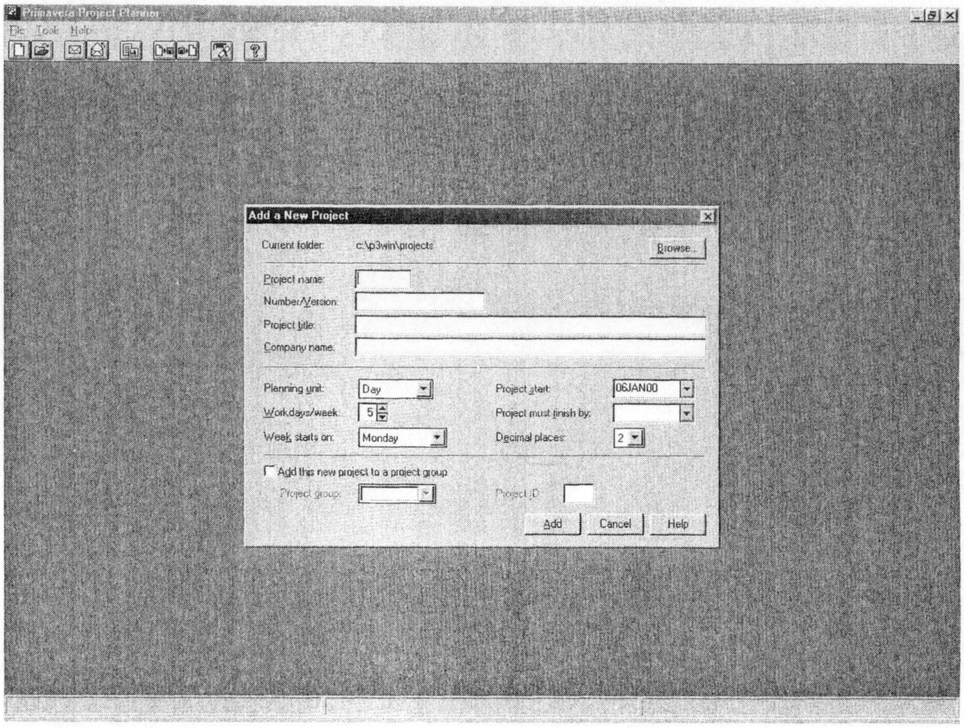

Primavera Project Planner

File Tools Help

Add a New Project

Current folder: c:\p3win\projects Browse...

Project name:

Number/Version:

Project title:

Company name:

Planning unit: Day

Workdays/week: 5

Week starts on: Monday

Project start: 06JAN00

Project must finish by:

Decimal places: 2

Add this new project to a project group

Project group:

Project ID:

Add Cancel Help

Screen Image 5-3
Add a New Project Window

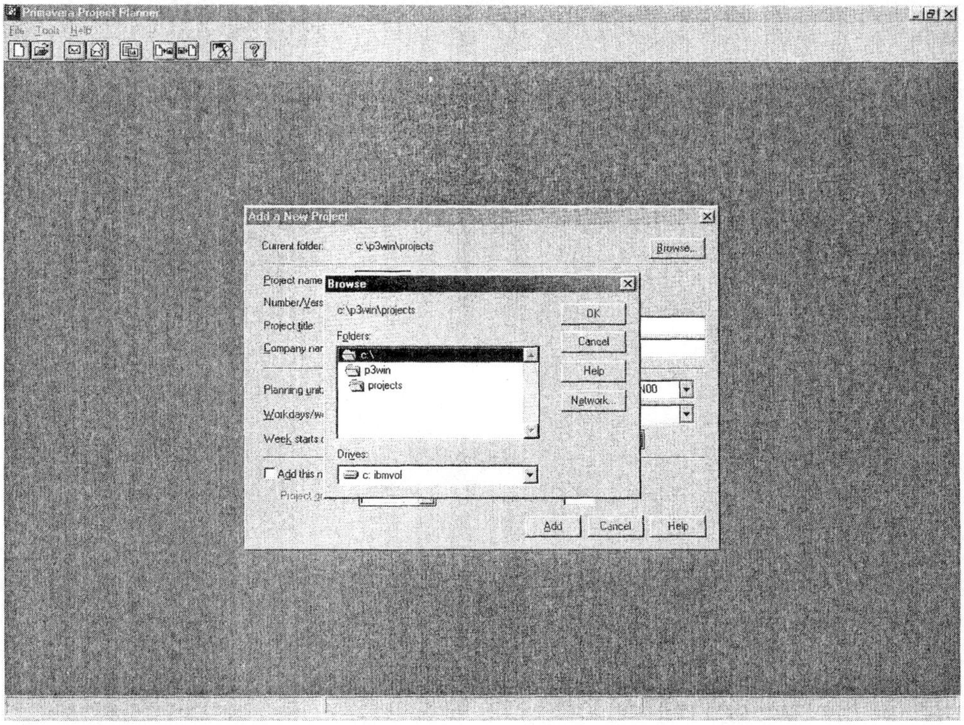

Primavera Project Planner

File Tools Help

Add a New Project

Current folder: c:\p3win\projects Browse...

Project name

Number/Vers

Project title

Company nar

Planning unit

Workdays/w

Week starts

Add this n

Browse

c:\p3win\projects

Folders:

c:\

p3win

projects

Drives:

c: ibmvol

OK

Cancel

Help

Network...

Add Cancel Help

Screen Image 5-4
Browsing For a New File Location

The *Planning unit* field is where the user specifies the units of duration for the activities. Primavera Project Planner supports activity duration units of hours, days, weeks, and months. The most common unit and default setting is days. However, if one of the other planning units is required, click on the down arrow adjacent to the *Planning unit* field, and select the desired unit.

When creating a new project, the default project start date is the current date. If the project start date is known, it should be entered into the *Project start* field. If not, some future date should be input. In either case, the project start date can easily be changed later when the exact start date is known. A graphical calendar can be activated to assist with determining the start date by clicking on the pull-down menu button adjacent to the *Project start* field as shown in Screen Image 5-5.

The *Workdays/week* and the *Week starts on* fields set up the basic parameters for the default project calendar. In most situations the workweek begins on Monday and is either 4 or 5 days long. It is important for the user to be consistent between this entry and how the durations for the individual activities were determined. If the individual activity durations were calculated using a 10-hour workday, then the 4-day workweek needs to be specified. If a portion of the project were to be performed on a 4-day workweek and another portion on a 5-day workweek, then multiple calendars would be required. The issue of creating multiple calendars will be discussed later. The *Project must finish by* field is optional and should be used if the project has a specified completion date. However, that field should be left blank for the initial input, regardless of whether the required completion date is known. If a required finish date is entered that is earlier than the early finish date of the last activity, as determined from the forward pass calculation, then some of the activities will have negative float. This would mean that these activities are behind schedule even though actual construction has not begun. By leaving this field blank, the forward pass calculation will determine the completion date. Then through an iterative process modifications can be made to the order of the activities or their durations until the planned completion date

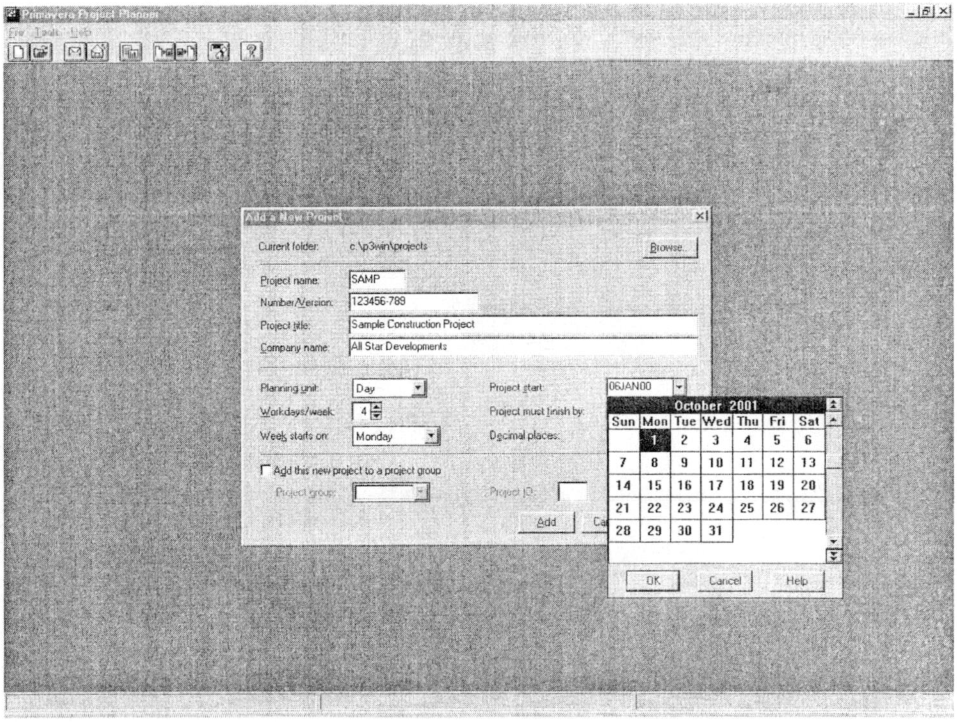

Screen Image 5-5
Pull-Down Calendar

is earlier than the required completion date. The *Project must finish by* date can easily be changed at any time. The *Decimal places* field is where the user specifies the number of decimal places to which the costs and resources will be computed. Rarely are the accuracy requirements so stringent that any decimal places are required, therefore 0 is the most common entry in this field. The bottom portion of the *Add a New Project* window contains options for linking projects. Primavera Project Planner allows for projects to be nested within a project group. This allows for information from multiple projects to be integrated into a single project. This feature is useful if managing multiple projects for a single client. By grouping the projects, each of the individual project schedules is managed at the project level and capable of being "rolled up" up into a single master project. This allows for project management on a number of different levels. When the master project is viewed, the first two characters of the activity ID are reserved for delineating which project the activity belongs to. The *Project ID* field is used to input that two-character identifier. Screen Image 5-6 is an example of a completed *Add a New Project* screen. If there is a mistake on this window, any of the information—except for the project name—can be changed in the future. When the input is acceptable, click on the *Add* button to activate the main input screen (Screen Image 5-7).

The main input screen is where the majority of input for the specific project will be entered. If all of the commands at the top of the screen are not highlighted, as shown in Screen Image 5-7, press the *Esc* key on the keyboard. This will activate all of the commands. Screen Image 5-8 shows the main input screen with all of the commands activated. Appendix B contains a menu tree for this screen. This screen can be made more user friendly by turning on the activity form area, which provides activity specific information as shown in Screen Image 5-9. The F7 key can be used to turn the activity form on and off, or press the activity form icon, or select the *View* command followed by the *Activity Form* option.

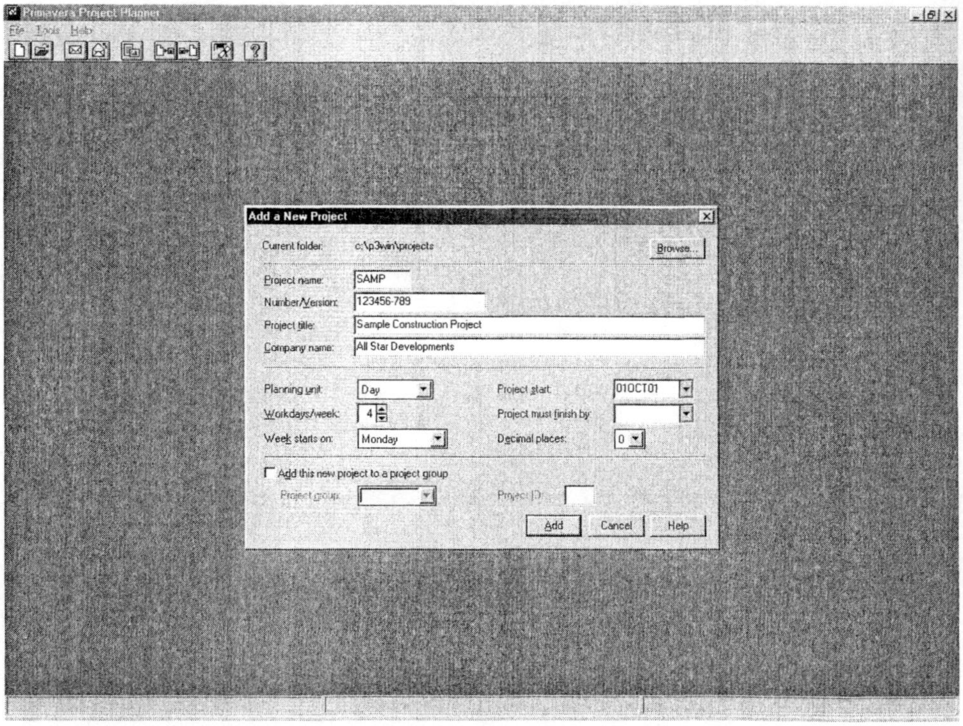

Screen Image 5-6
Completed Add a New Project Window

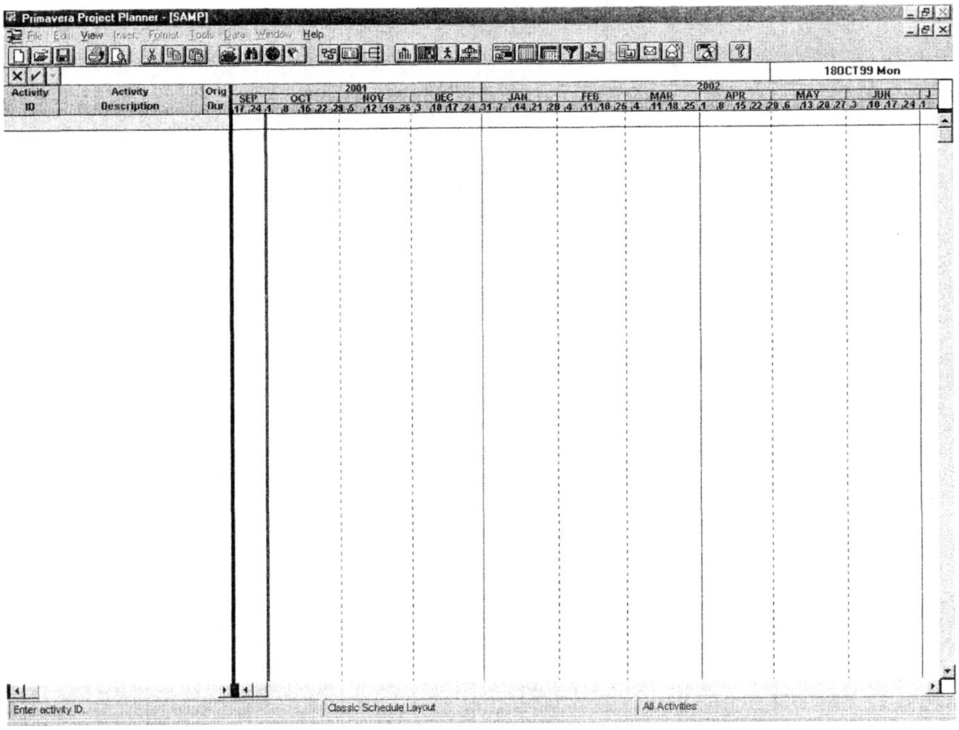

Screen Image 5-7
Main Input Screen—Commands Inactive

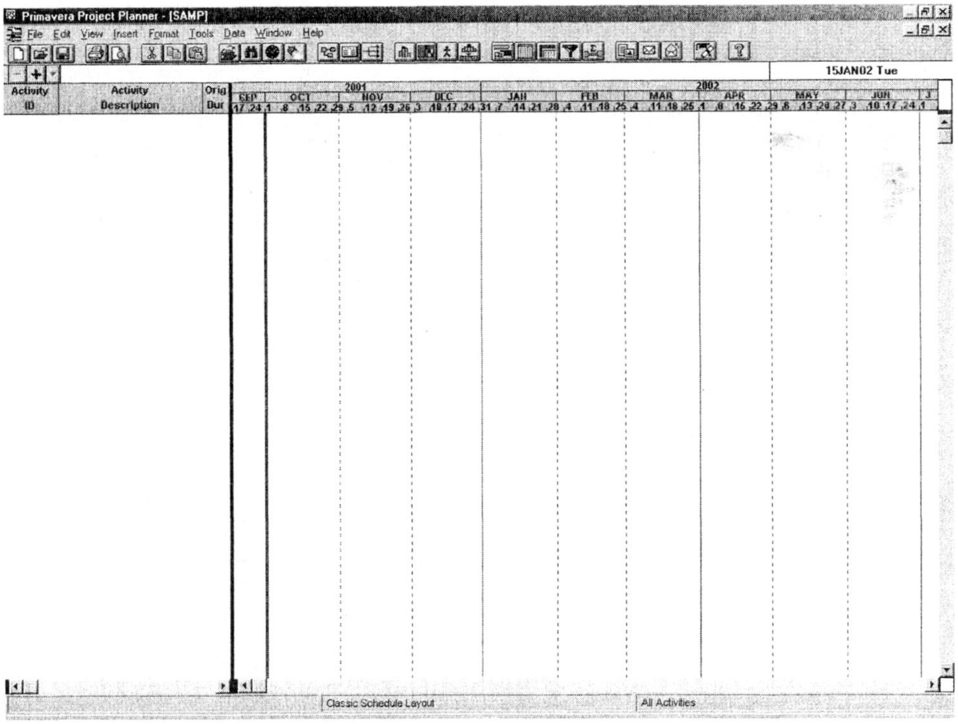

Screen Image 5-8
Main Input Screen—Commands Active

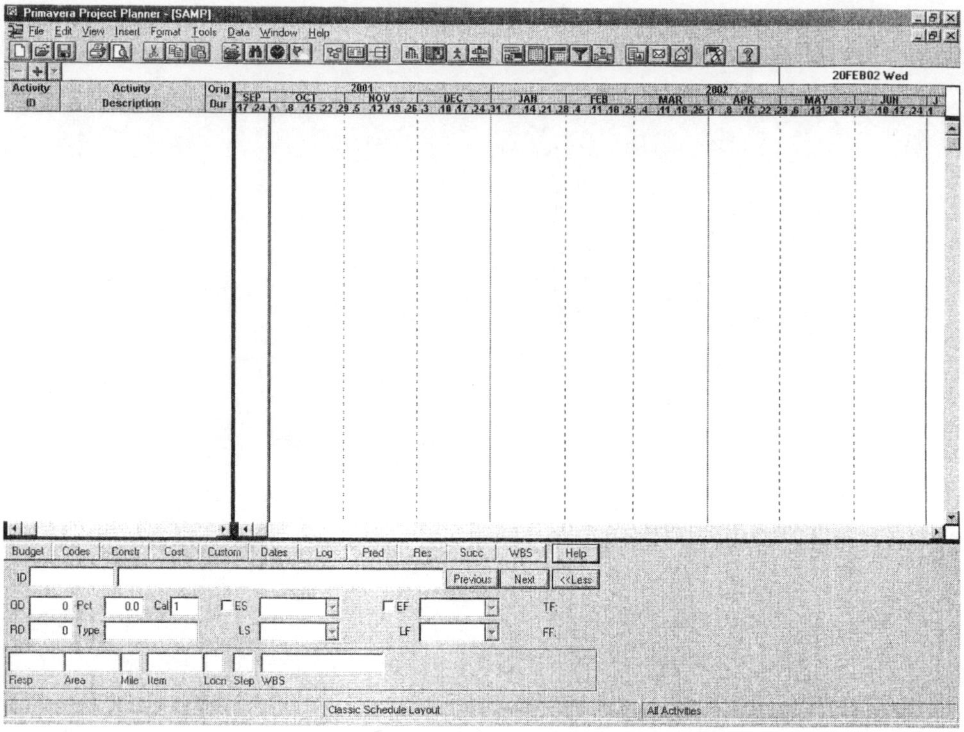

Screen Image 5-9
Active Activity Form

Saving a Project

Primavera Project Planner saves the work being performed on the project schedule on a continuous basis. As long as the user exits the program correctly, all the work performed during that session will be saved. To exit Primavera Project Planner, select *File* from the command line and select the *Close* option as shown in Screen Image 5-10. This will close the project, but you will not exit Primavera Project Planner. To close and exit the program, select the *Exit* option. Either of these options will cause the *Save Layout* window to appear as shown in Screen Image 5-11. This window has nothing to do with saving data but is about saving the layout in which the data is organized on the screen. To save the layout click *Yes*, otherwise click *No* to discard the layout. Unless that layout is particularly difficult, it is hard to justify the effort to keep track of all of the layouts.

Opening an Existing Project

When Primavera Project Planner is started, a screen similar to the one shown in Screen Image 5-12 should appear. To open an existing project use either the *File* command followed by the *Open* option (Screen Image 5-13) or click the open file folder icon. Either of these operations will open the *Open a Project* window, similar to the one shown in Screen Image 5-14. The top portion of that window addresses file location management. The bottom half of the screen is the list of projects that are stored within the specified directory. If the project schedules are stored in multiple directories, the project list will be for only those that are in the currently active directory. The entire list of projects can be viewed by using the slide bar to the right of the project titles.

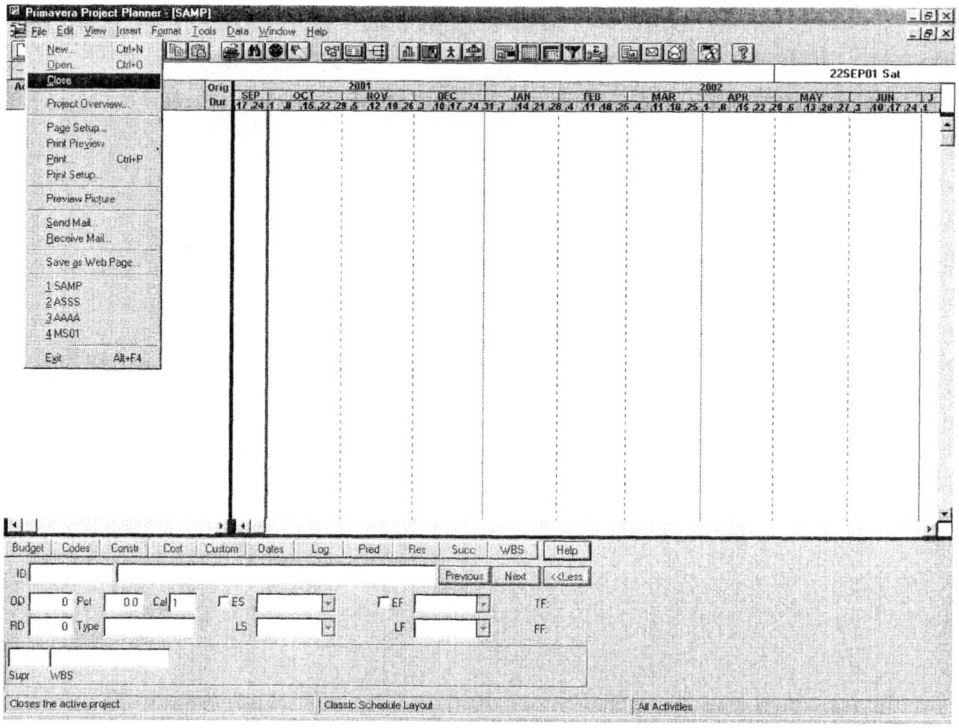

Primavera Project Planner - [SAMP]
File Edit View Insert Format Tools Data Window Help
New... Ctrl+N
Open... Ctrl+O
Close
Project Overview...
Page Setup...
Print Preview
Print... Ctrl+P
Print Setup...
Preview Picture
Send Mail...
Receive Mail...
Save as Web Page...
1 SAMP
2 ASSS
3 AAAA
4 MS01
Exit Alt+F4
22SEP01 Sat
Orig Dur
2001 SEP OCT NOV DEC 2002 JAN FEB MAR APR MAY JUN J
Budget Codes Constr Cost Custom Dates Log Pred Res Succ WBS Help
ID Previous Next <<Less
OD 0 Pct 0.0 Cal 1 ES EF TF:
RD 0 Type LS LF FF:
Supr WBS
Closes the active project | Classic Schedule Layout | All Activities

Screen Image 5-10
Closing a Project

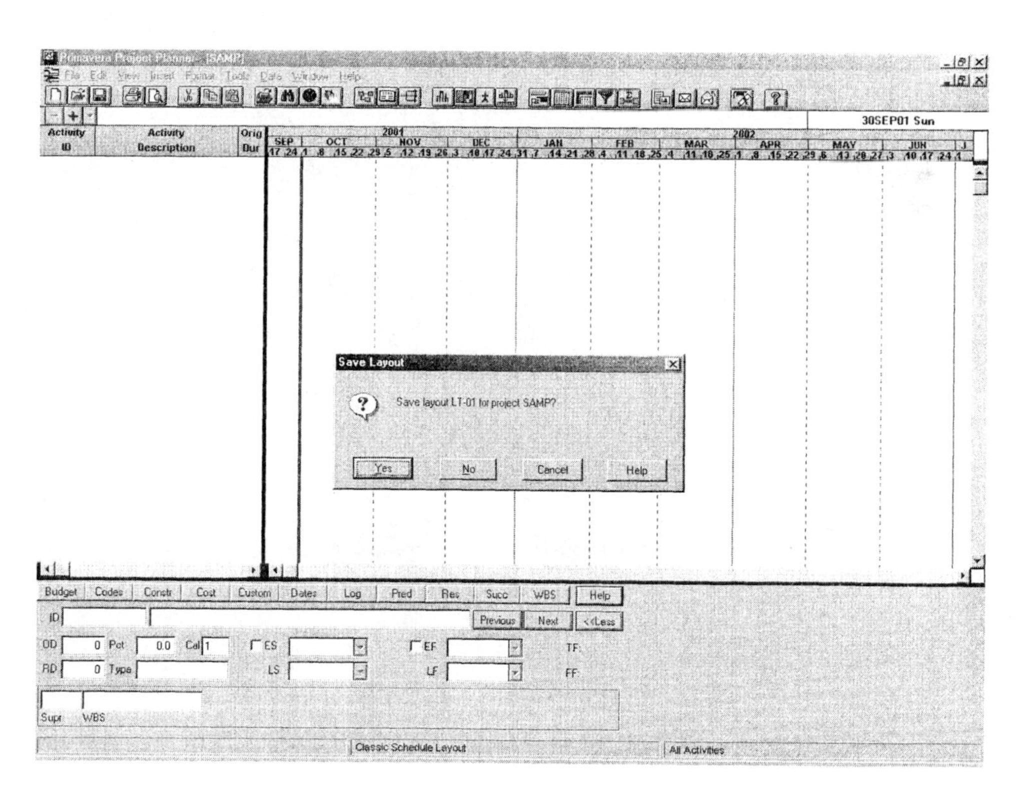

Screen Image 5-11
Saving Screen Layout

Screen Image 5-12
Initial Screen

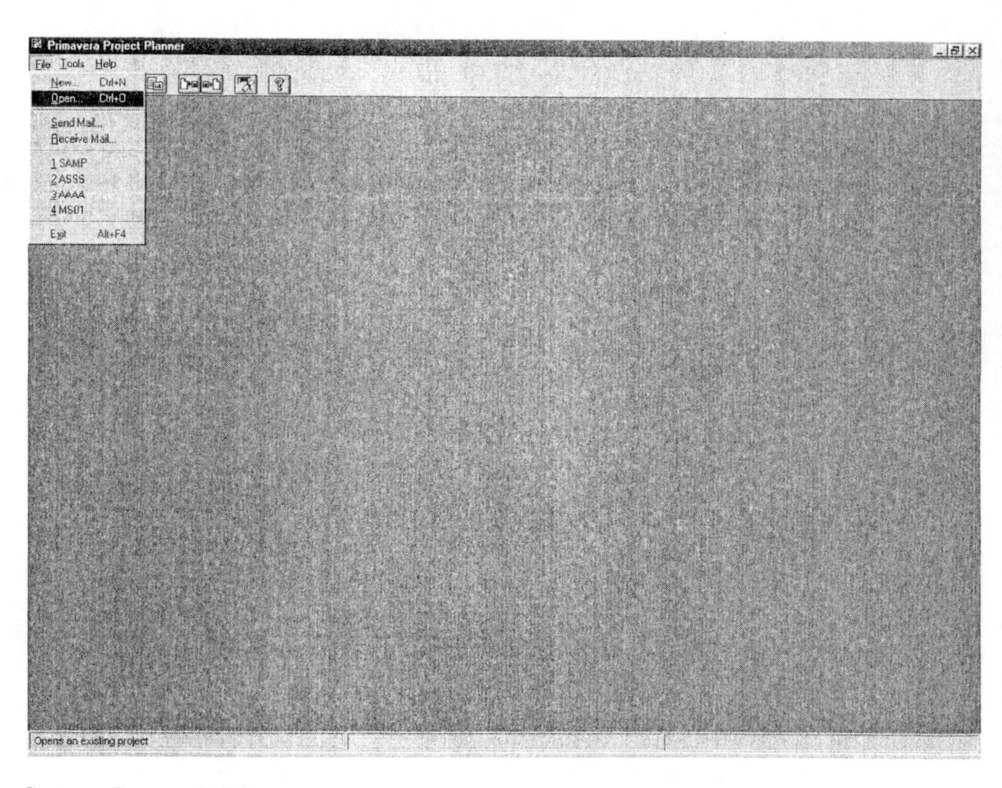

Screen Image 5-13
File Open Command

Screen Image 5-14
Open a Project Window

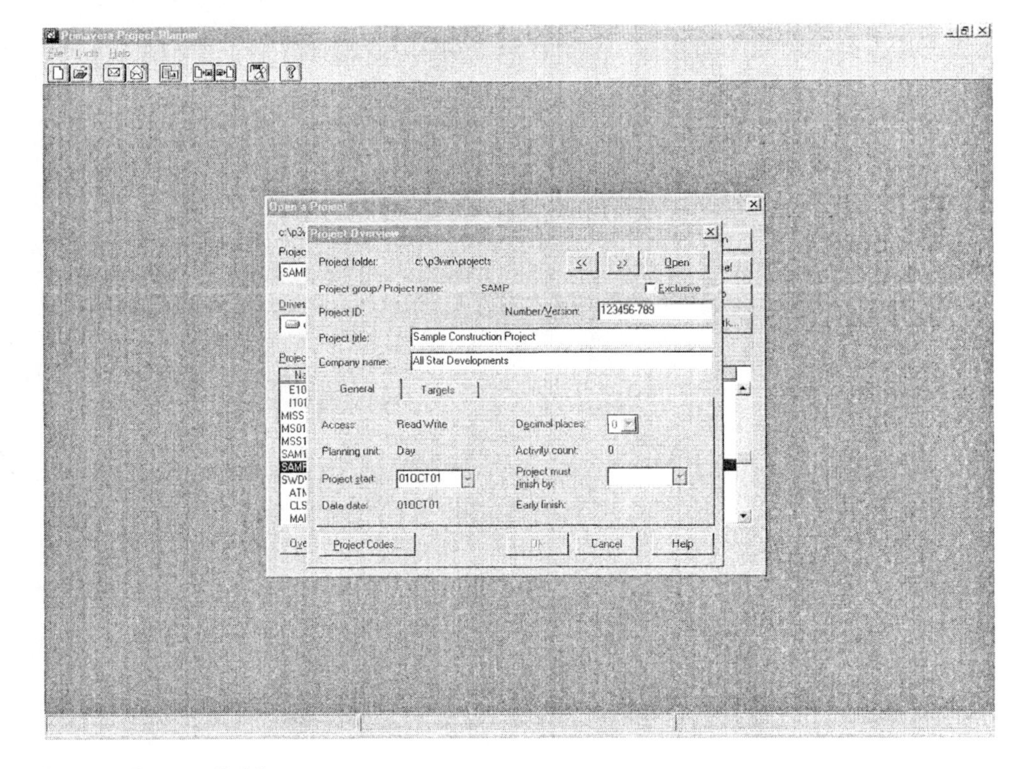

Screen Image 5-15
Project Overview Window

Clicking on any of the projects will highlight that one. Then the *Overview* button can be used to open the *Project Overview* window. This window gives specific information about the selected project. Screen Image 5-15 is the overview for project SAMP. From this window, the user can review the description of the project as well as modify the project start and required project finish date. To modify these dates, enter the desired dates in the appropriate field or move the pointer to that field and use the adjacent down arrow to activate the pull-down window.

Once the desired project has been located or the needed dates changed, clicking the *Open* button will activate that project. Within the *Project Overview* window the arrows adjacent to the *Open* button are used to move up and down the project list. The *Exclusive* box is typically left unchecked. In that position, the project can be shared in a network environment. In practical terms, multiple people can be updating the schedule at the same time. If the *Exclusive* box is checked, then only the project's owner/originator has access to the project. This feature is used for security or to change the *Activity ID* for a specific activity. If this box is checked, the user can change an *Activity ID*, otherwise a new activity must be added and the old one deleted. All other information about an individual activity can be changed while in the share mode. The *Project Codes* button, along with the items on the *Targets* tab, will be discussed later.

Backing Up Projects

To protect against potential hardware failure, it is recommended that the project schedule be backed up at the completion of any session during which modifications to the schedule were made. While Primavera Project Planner will allow for multiple projects to be backed up on the same disk, this is not a good practice. It is also not a good practice to save files from other applications on the same disk that is being used to back up Primavera Project Planner files. Unlike spreadsheet and word processing applications, where items are kept in single files, Primavera Project Planner creates a number of different files for a single project.

To create a backup copy of a project, select the *Tools* command followed by the *Project Utilities* and *Back Up* option, as shown in Screen Image 5-16. Within that window select the project to be backed up and the location for the backup. By clicking on the *Browse* button (see Screen Image 5-17) the location of where to store the backed up files can be changed. In addition, there is the *Compress files* option and the *Remove access list during backup* option. Unless the project is exceptionally large and will not fit on one disk, it is recommended that the compression option not be selected. The access list deals with who has rights to read and write to the project. The removal of this list will allow anybody to access the backed up files. The ownership of the project files is rarely an issue on a stand-alone computer system. However, this is an area of concern when multiple persons are working on the project from the same computer.

Typically, when multiple persons are working on the same project they are not sharing the same computer. Rather, they are networked and the project files are stored on a central server rather than on an individual's computer. In this environment, control of access to the project files is a more critical issue. However, network operating systems as well as Primavera Project Planner have built-in security systems. Furthermore, in these operating environments the files stored on the central server are typically backed up to a tape on a daily basis and the need for backing up to disks is minimal. In that environment the primary reason for backups are for archival purposes.

Clicking the *Back Up* button will initiate the backup process. However, the user will be prompted as to whether to erase the disk prior to backing up the files, as shown in Screen Image 5-18. If there are files from other applications or from other projects,

Screen Image 5-16
Initiating Backup Procedure

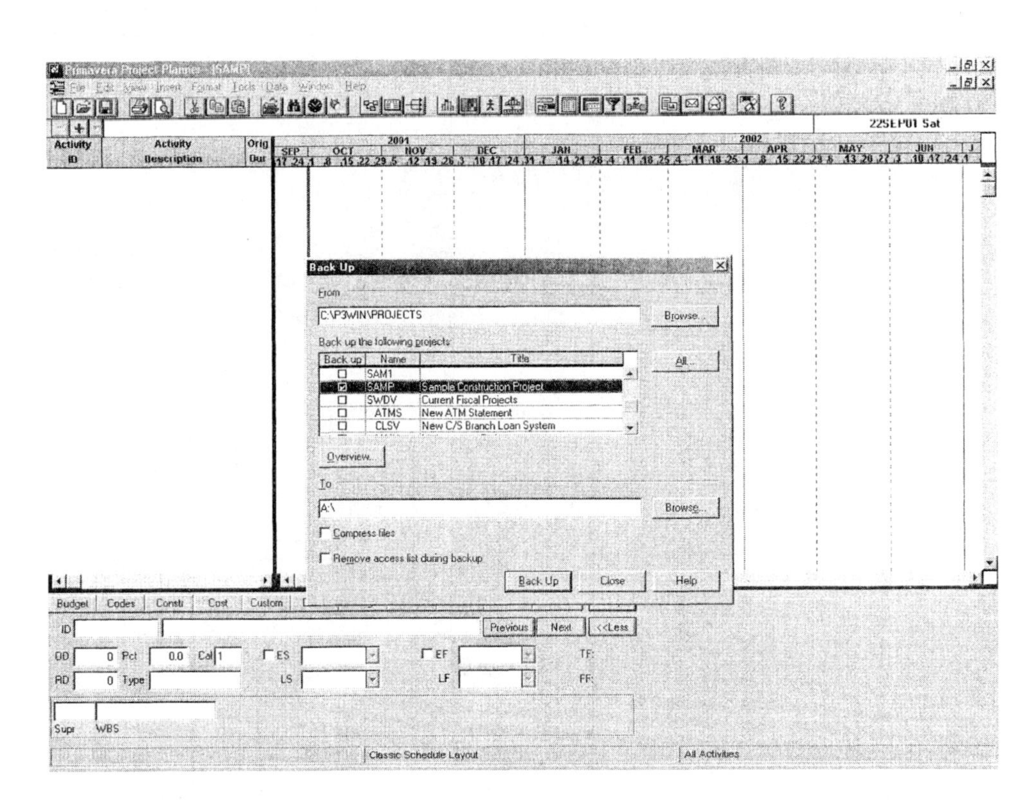

Screen Image 5-17
Back Up Window

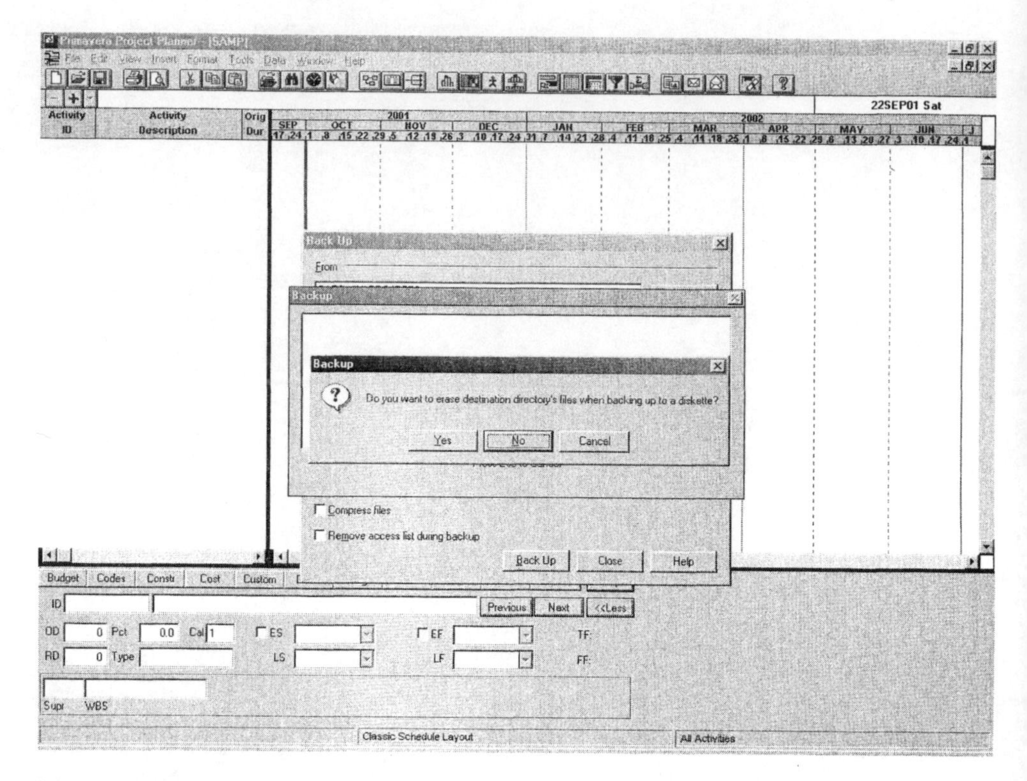

Screen Image 5-18
Backup Window

the *No* option should be selected. If the user uses one disk per project and uses that disk for Primavera Project Planner files only, then the *Yes* option is appropriate.

Restoring a Project

To restore a project, select the *Tools* command followed by the *Project Utilities/Restore* options as shown in Screen Image 5-19. This will activate the *Restore* window as shown on Screen Image 5-20. The *From* box designates the location of the files that are available to be restored. To change the file location, press the *Browse* button to the right of this field. The box below the *From* field contains the names of all of the files that are available to be restored from within the specified directory. The *To* box is used to specify onto which disk drive and directory the project being restored should be loaded. To change that location, press the adjacent *Browse* button. In Screen Image 5-20 the project that is to be restored is called SAMP and it is to be restored to *C:\P3WIN\PROJECTS.*

The *Restore report specification* option requires some additional explanation. When report specifications are developed, they are not project specific. Rather, they can be used on any of the project schedules. For example if a report was written to list the critical activities while working on project SAMP, those report specifications could be accessed when using any subsequent project. This feature makes Primavera Project Planner very flexible, but adds some confusion when restoring projects. Assume that project SAMP has been backed up to a disk and removed from the computer. At a later date it becomes necessary to restore this project. If the *Restore layout, tabular report and graphic report specification* option is selected, the current report specifications will be removed and replaced with those stored when project SAMP was backed up. This would cause all of the report specifications developed since SAMP was backed up to be permanently lost.

Screen Image 5-19
Initiating the Restore Procedure

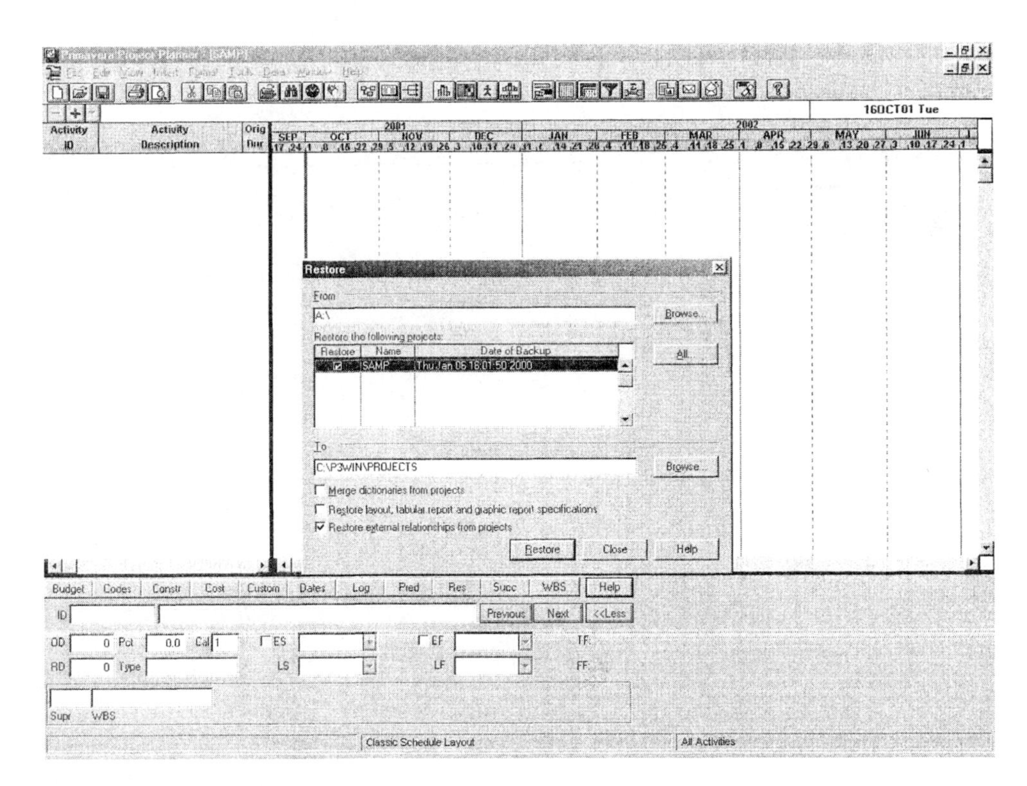

Screen Image 5-20
Restore Window

Primavera Project Planner - [SAMP]
File Edit View Insert Format Tools Data Window Help

Activity ID | Activity Description

02OCT01 Tue

2001 | 2002
NOV | DEC | JAN | FEB | MAR | APR | MAY | JUN

Schedule... F9
Level... Shift+F9
Global Change...
Update Progress...
Store Period Performance...
Tabular Reports
Graphic Reports
Look
InfoMaker
Update Data Dictionary...
Web Publishing Wizard
MPX Conversion Utility...
Project Utilities
Fragnet
Options
1 Project Check-in/Check-out
2 Monte Carlo

Copy...
Delete...
Merge...
Summarize...
Back Up...
Restore...
Targets...
Import...
Export...

Budget | Codes | Constr | Cost | Custom | Dates | Log | Pred | Res | Succ | WBS | Help
ID | Previous | Next | <<Less
OD 0 Pct 0.0 Cal 1 ES EF TF:
RD 0 Type LS LF FF:
Supr WBS
Copies project with a different name | Classic Schedule Layout | All Activities

Screen Image 5-21
Opening Copy Window

Primavera Project Planner - [SAMP]
File Edit View Insert Format Tools Data Window Help

Activity ID | Activity Description | Orig Dur

14OCT01 Sun

2001 | 2002
SEP | OCT | NOV | DEC | JAN | FEB | MAR | APR | MAY | JUN

Copy
From
c:\p3win\projects
Project group/ Project name: SAMP
Project ID: Planning unit: Day
Number/version: 123456-789
Title: Sample Construction Project
Company: All Star Developments
Browse...
Projects:
I101
MISS
MS01
MSS1
SAM1
SAMP
Filter: All - All Activities
To
c:\p3win\projects
Browse...
Project group: SAM2 Project name:
Planning unit: Day Project ID:
Number/version: 123456-789
Title: Sample Construction Project
Company: All Star Developments
Add the following characters: to the left / right of the current ID's
Copy | Cancel | Help

Budget | Codes | Constr | Cost | Custom
OD 0 Pct 0.0 Cal 1 ES EF TF:
RD 0 Type LS LF FF:
Supr WBS
Classic Schedule Layout | All Activities

Screen Image 5-22
Restore Window

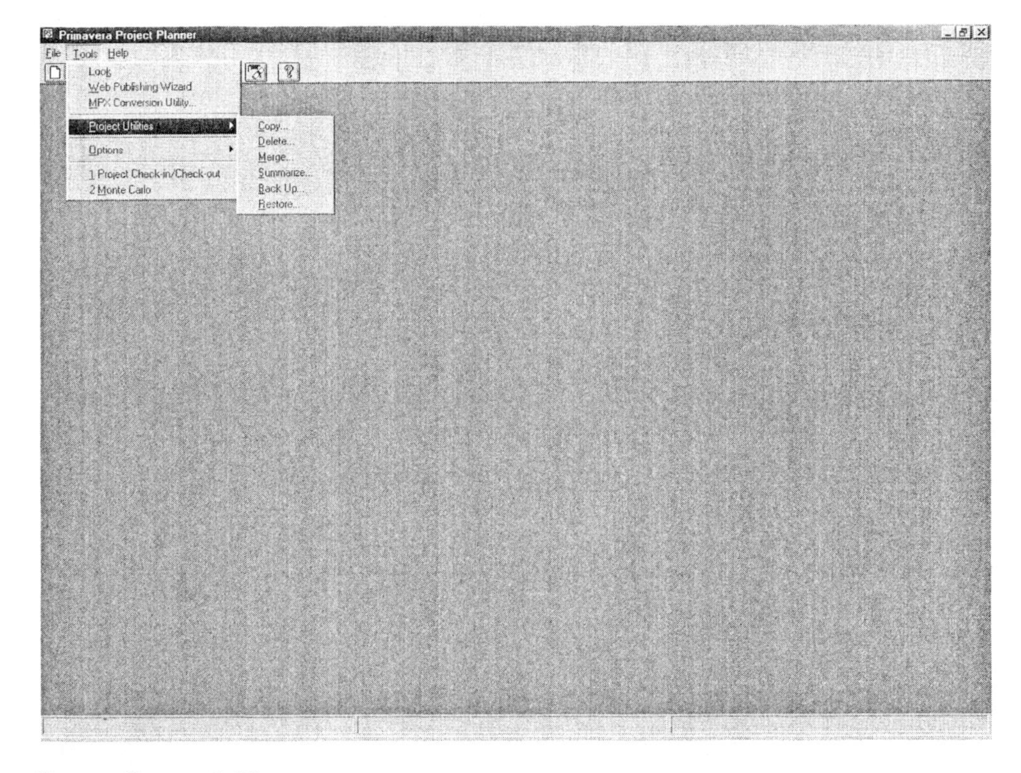

Screen Image 5-23
Restore Window

Copying a Project

In addition to creating backup copies of a project, the entire project can be copied. However, the copy must have a different project name. To make a copy of a project, select the *Project Utilities* and *Copy* options from within the *Tools* command as shown in Screen Image 5-21. These keystrokes will open the *Copy* window as shown in Screen Image 5-22. In that window, project *SAMP* is being copied to a new project named *SAM2*. Various cosmetic and date information can be changed during the copying process. These items can also be changed later via the *Project Overview* window. Projects can be backed up, restored, and copied only when they are closed. In Screen Image 5-23, there are no projects open; however the *Tools/Project Utilities* commands are available.

Conclusion

The opening, closing, and backing up of project specific information can be handled from a number of different places within the program. With the exception of the project name, any information about the project can be changed at any time. This allows for an iterative process of manipulating the dates, crews, and activity durations to have the schedule work within the allocated time.

Suggested Exercise

1. Using the procedures outlined in the chapter, create a new project, copy that project to a different name, and then back up and restore both of the projects.

Primavera Project Setup

- Key Terms
- Calendar Setup
- Activity Codes
- Project Codes
- Conclusion

KEY TERMS

Activity Codes Fields that may be customized in Primavera Project Planner, allowing a user to organize the project schedule in many different ways; for example, a schedule could be printed showing all activities a particular superintendent is responsible for.

Project Calendar A calendar that spans the project duration and shows holidays and other nonwork periods specific to a project. This calendar links the project days to calendar days.

Calendar Setup

Prior to entering any project specific information, the project calendar needs to be customized. The calendar data that was entered when the project was initiated created only the basic outline of the project calendar. In addition to that calendar there can be any number of different calendars to accommodate differing work schedules for specific activities. Initially, the basic project calendar needs to be modified to show holidays and other specific nonwork periods. All modifications to the calendars are made through the *Calendars* window. This window is opened by clicking the *Data* command and then select the *Calendars* option as shown on Screen Image 6-1. Screen Image 6-2 shows the initial settings for the *Calendars* window. The initial calendar is *Calendar ID 1* with no title as shown in Screen Image 6-2. Anytime that the title field is active, a descriptive title can be assigned. A

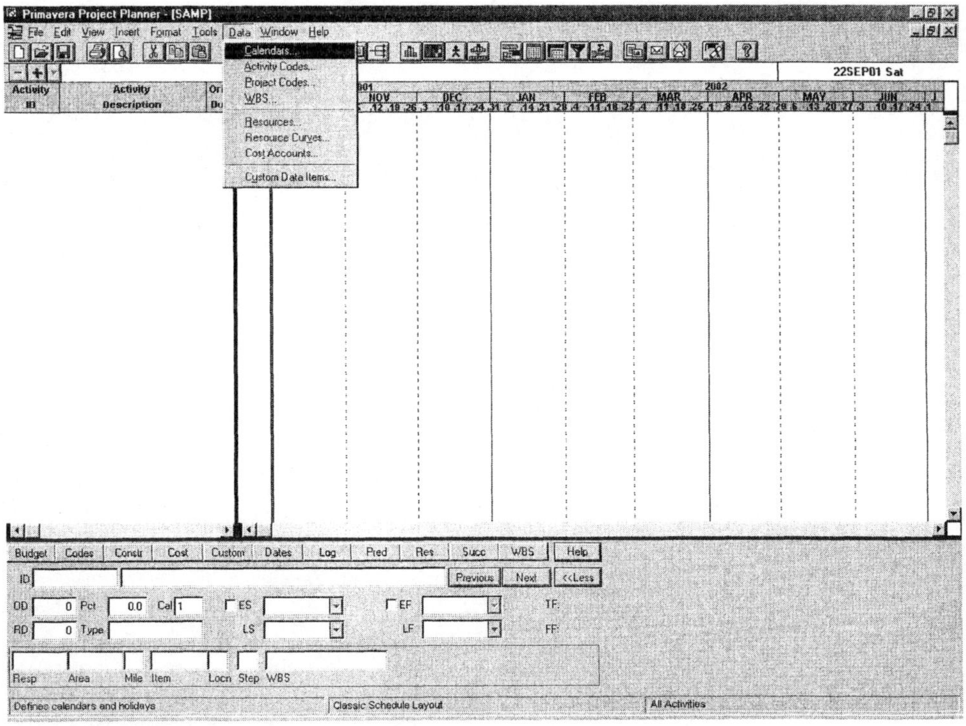

Screen Image 6-1
Activating the Project Calendar

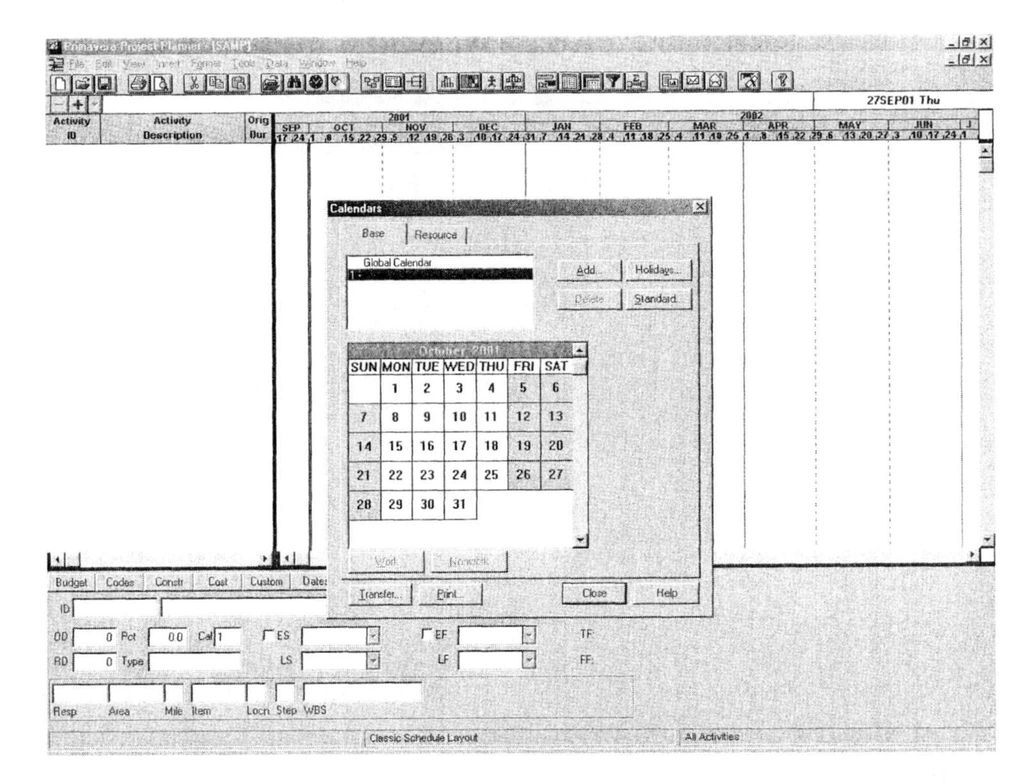

Screen Image 6-2
Calendars Window

descriptive title is important for the pull-down menus to be of any value. The *Standard* button opens the *Standard Daily Information* window as shown in Screen Image 6-3. This window allows the user to designate which days will be workdays for a specific calendar.

Clicking on the boxes adjacent to the days of the week will specify those days as workdays. In Screen Image 6-3, Monday through Thursday have been specified as workdays for Calendar 1. Once the workdays have been specified for each of the calendars, the *OK* button should be clicked to close the *Standard Daily Information* window.

In order for Primavera Project Planner to properly assign dates to activities, all of the nonwork periods need to be specified. The standard window established the workdays for a standard workweek for a calendar. However, holidays and any other exceptions to the standard workweek need to be added. The holidays can be input through selecting the date on the calendar or by entering the date from the keyboard or by a combination of both.

It is best to enter holidays that fall on the same day each year through the *Holidays List* window. To enter these dates, click on the *Holidays* button, which will open the *Holidays List* window as shown in Screen Image 6-4. From that window all of the repeating holidays can be entered and designated as repeating by placing a check in the repeating box that is adjacent to the holiday. The check is added by clicking in the repeating box and then clicking the check mark. Once the repeating holidays have been entered the *OK* button can be clicked and Primavera Project Planner will return to the *Calendars* window.

Screen Image 6-5 shows Christmas of 2001 as a nonwork period. The *R* in that box designates that day as a repeating holiday. From the *Calendars* window the remaining nonwork periods can be specified. Pointing and clicking on the desired day and then clicking the *Nonwork* button specifies that selected date as a nonwork period. Screen Image 6-6 shows how Thanksgiving 2001 would be marked. Screen Image 6-7 shows that holiday shaded, designating that date as a nonwork period. After all of the nonrepeating holidays have been designated, the user can return to the *Holidays List* window and see all of the desired holidays as shown in Screen Image 6-8.

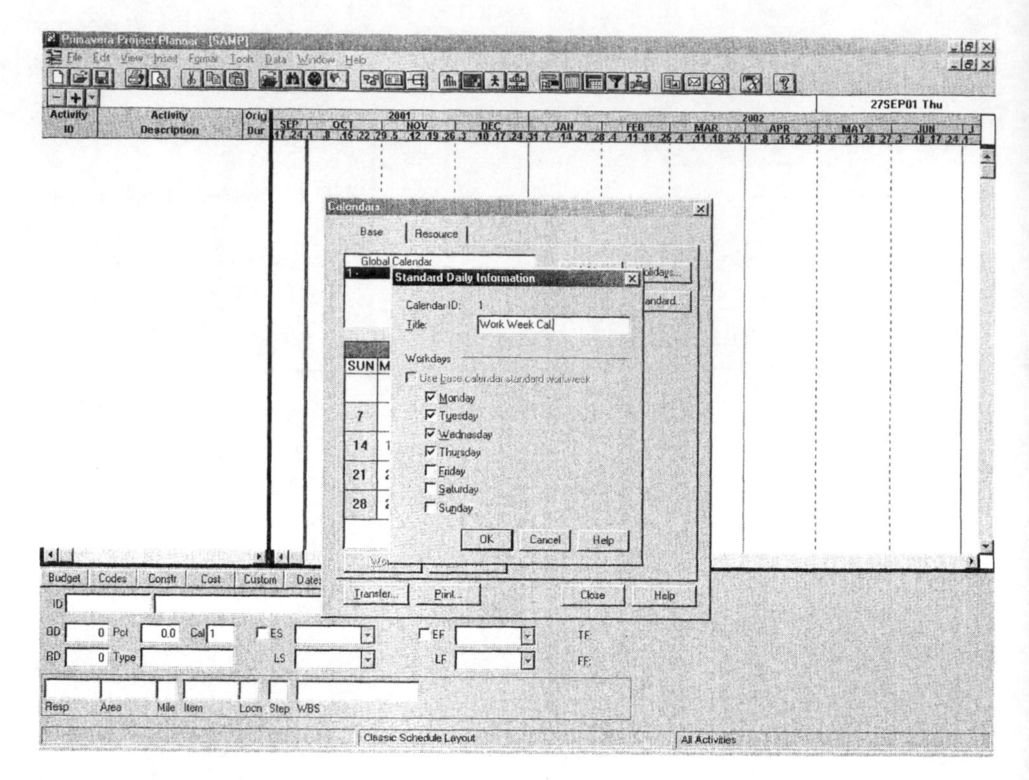

Screen Image 6-3
Standard Daily Information

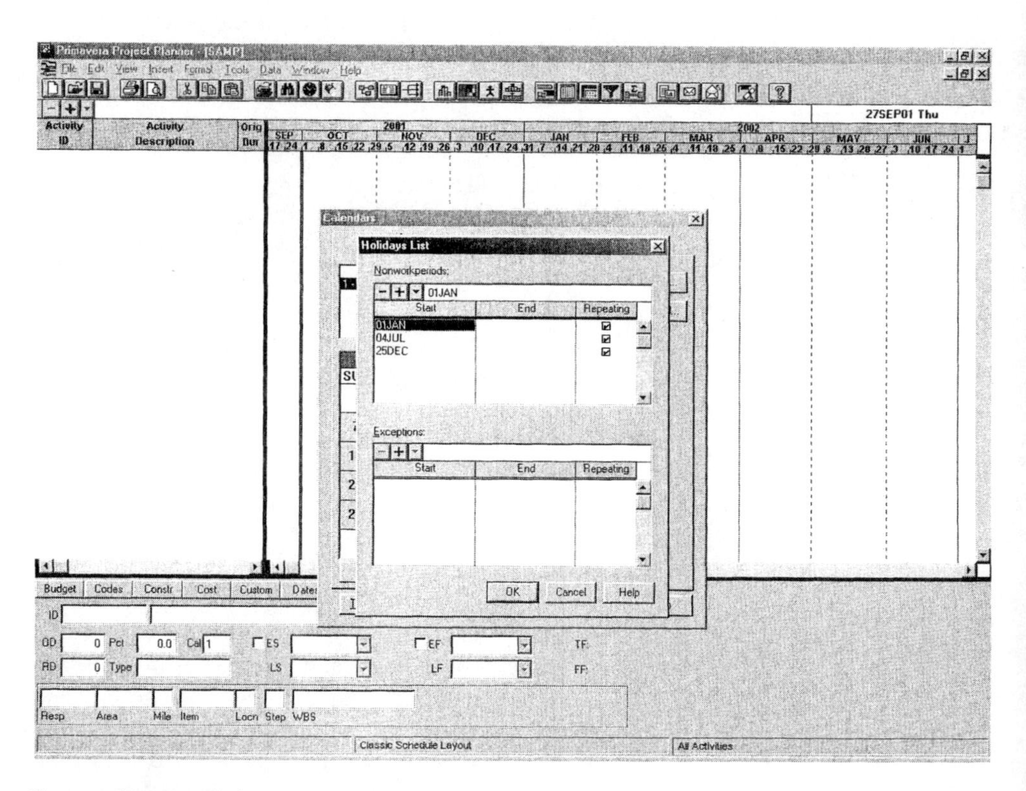

Screen Image 6-4
Holidays List

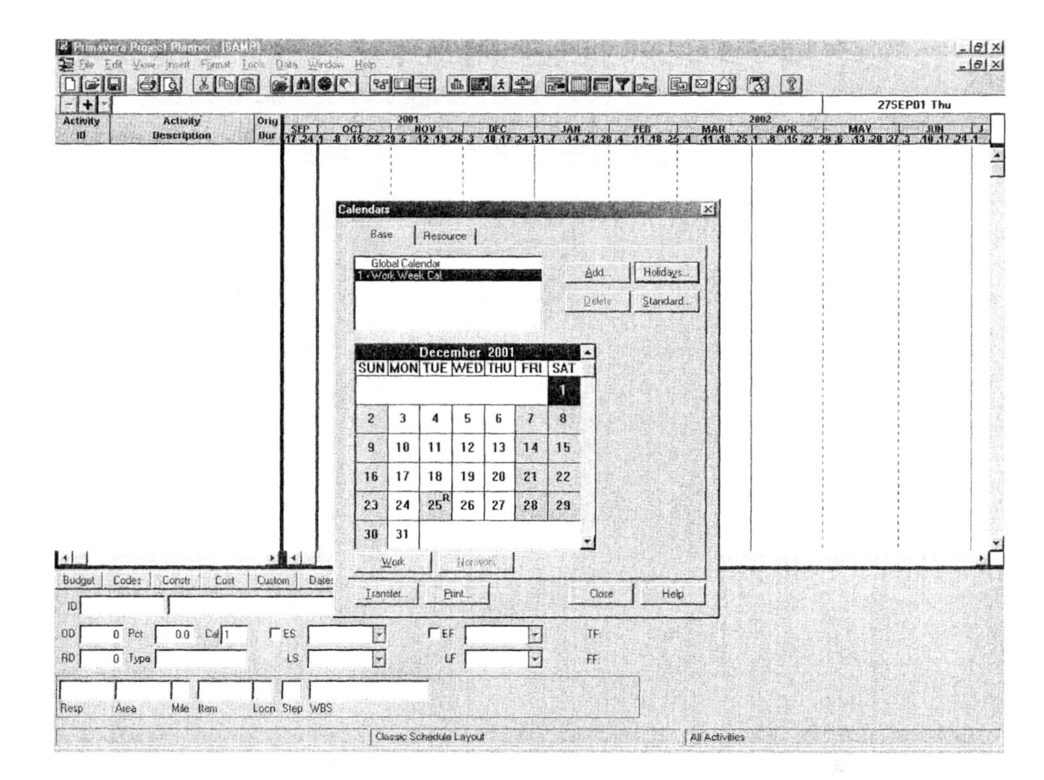

Screen Image 6-5
Repeating Holidays

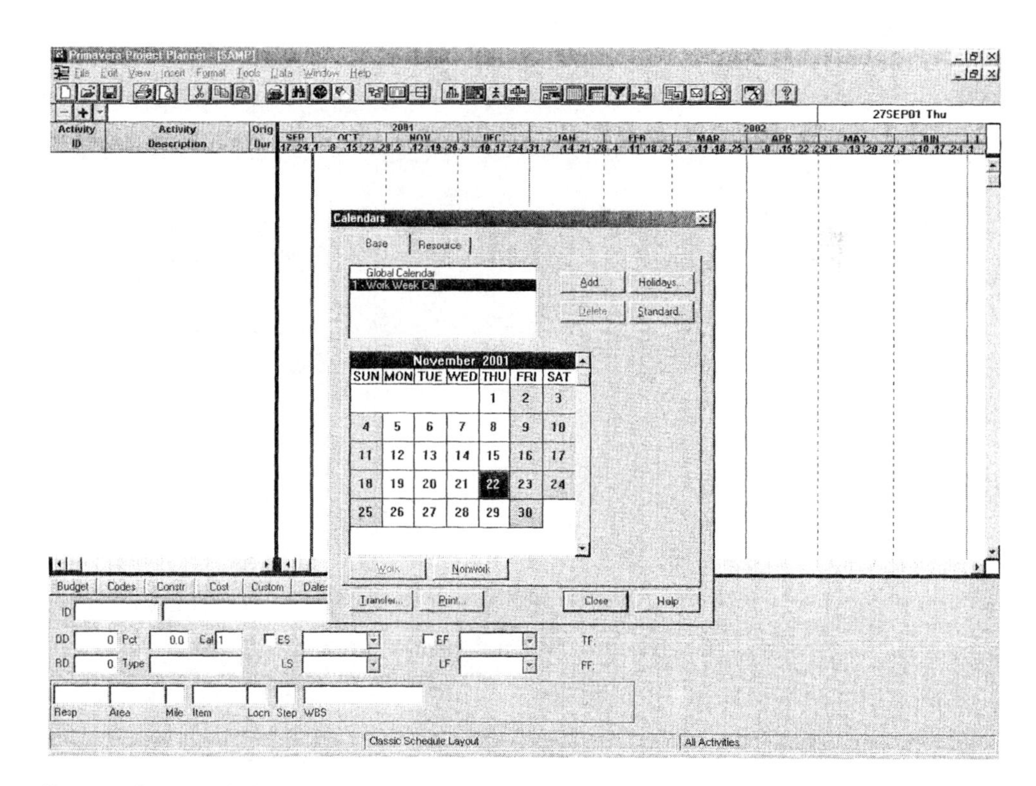

Screen Image 6-6
Selecting Nonwork Period

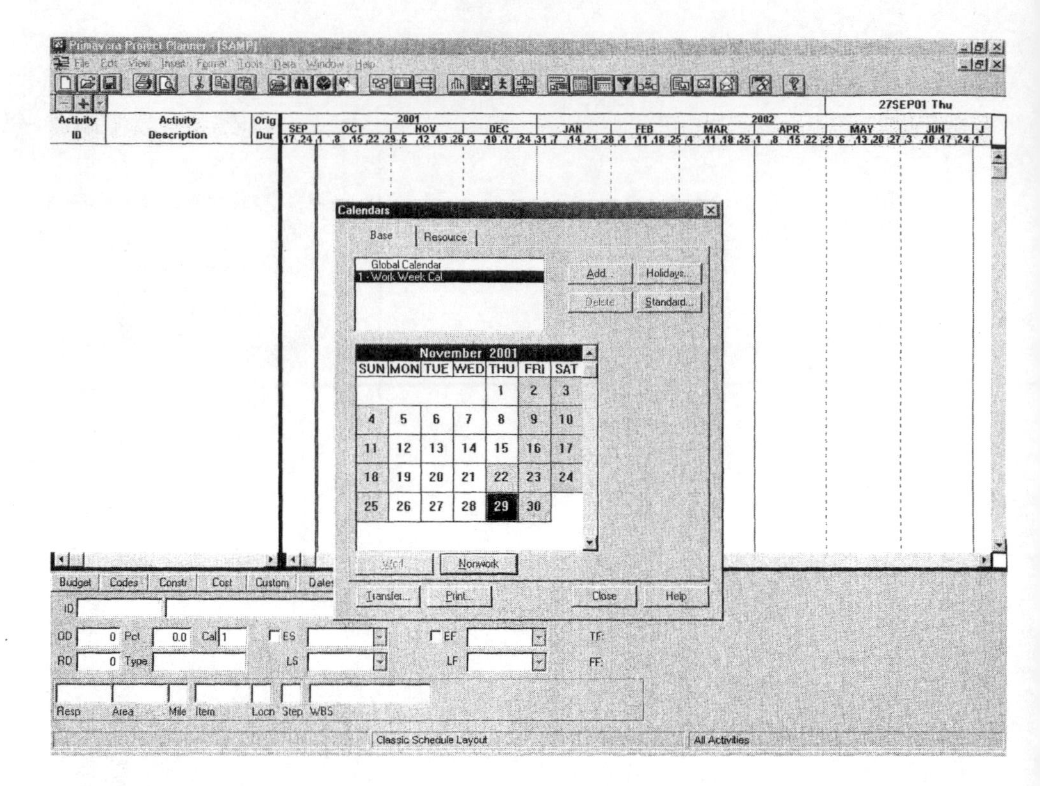

Screen Image 6-7
Nonwork Period

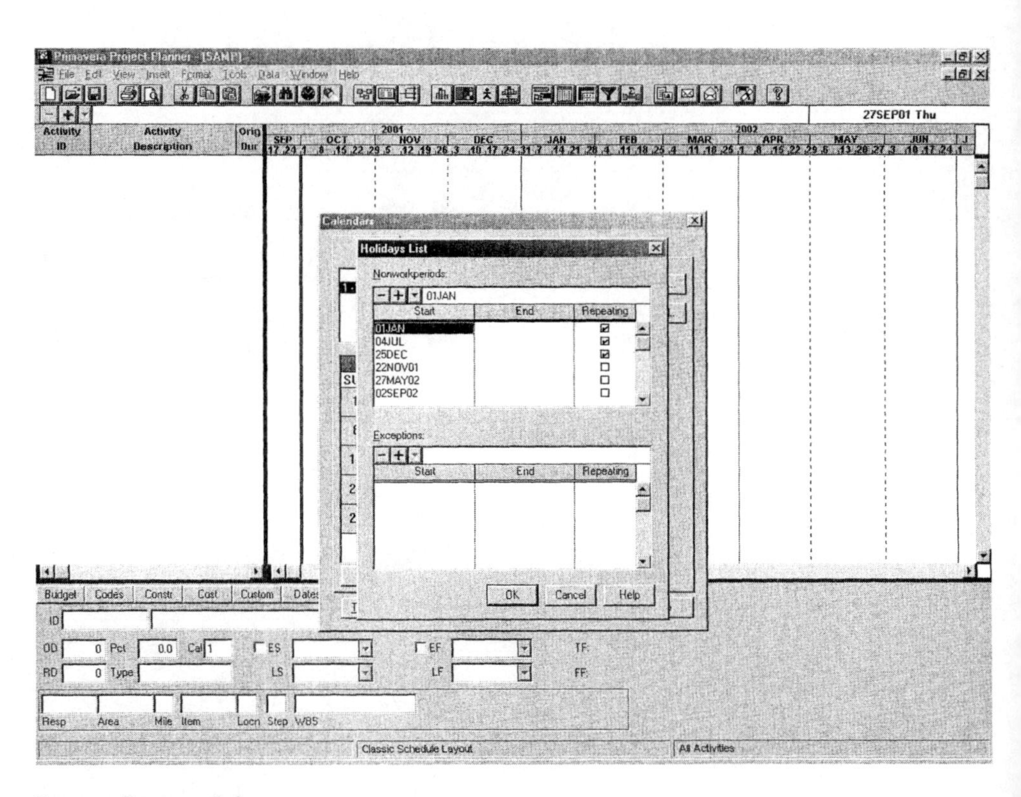

Screen Image 6-8
Complete Holiday List

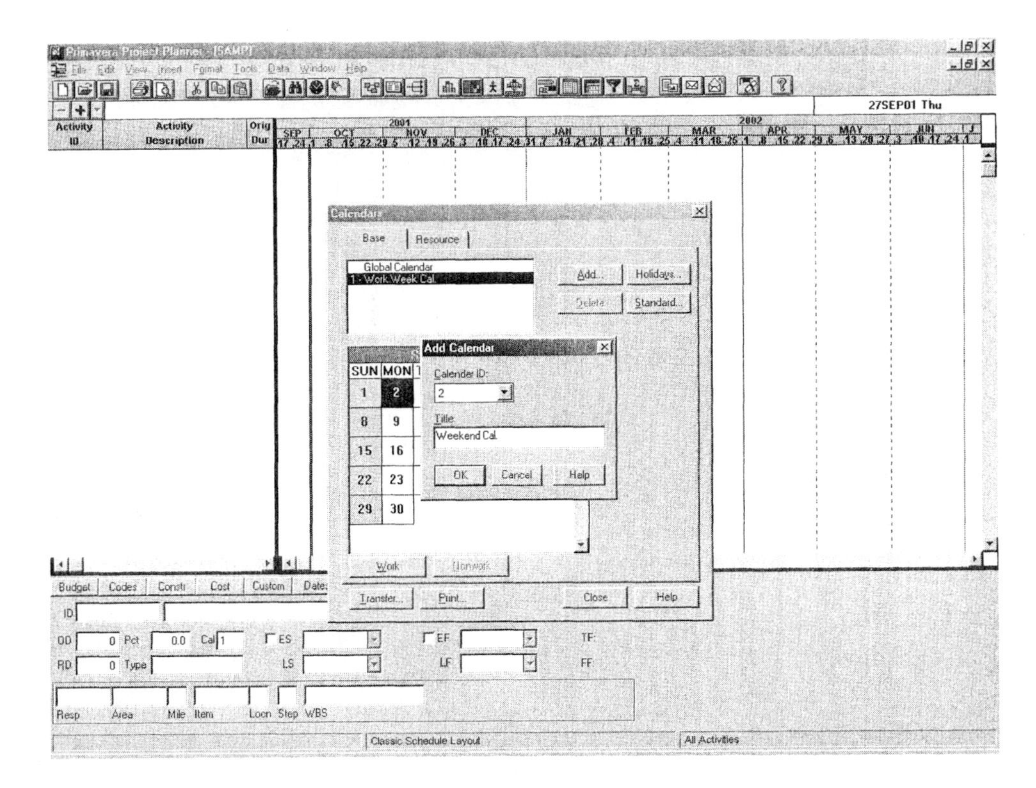

Screen Image 6-9
Add Calendar Window

Primavera Project Planner supports multiple calendars for the same project. This feature is helpful if some of the project activities need to be performed using different length workweeks or workdays. For example, two calendars would be needed if some of the project activities were to be done on a 4-day workweek and the remainder done on the weekends. The activities that would be performed on the 4-day workweek would be attached to one calendar and the weekend activities would be attached to another calendar.

Additional calendars can be added by clicking the *Add* button on the *Calendars* window. This action will open the *Add Calendar* window as shown in Screen Image 6-9. In that Screen Image the *Calendar ID* and *Title* are assigned. When the *OK* button is clicked the program will return to the *Calendars* window as shown in Screen Image 6-10.

In that window the list of calendars is available and the user can shift between calendars simply by clicking on the desired calendar. In Screen Image 6-10 the *Weekend Calendar* is active and all dates are specified as workdays. This occurred since there has been no specific information entered pertaining to this calendar. The quickest way to get information into the new calendar is to copy it from an existing calendar. By clicking on the *Transfer* button on the *Calendars* window the *Calendar Transfer* window can be opened as shown in Screen Image 6-11. In that window the user can specify the *From* and *To* calendar. In Screen Image 6-11 when the *Transfer* button is clicked Calendar 2 and Calendar 1 will be identical. Screen Image 6-12 details Calendar 2 after Calendar 1 has been copied. The only task remaining is to specify the workdays for the second calendar. This is done by clicking the *Standard* button and opening the *Standard Daily Information* window as shown in Screen Image 6-13. In that window Monday through Thursday have been unchecked and Friday, Saturday and Sunday have been checked. Screen Image 6-14 shows the modified Calendar 2, which has only Friday, Saturday, and Sunday specified as workdays.

Once all of the calendar information has been entered, clicking the *Print* button on the *Calendars* window will generate a hard copy of the project calendar. This command will open the *Calendar Print* window, as shown in Screen Image 6-15. The *Detailed* option produces a concise graphical calendar that shows the planned workdays and their

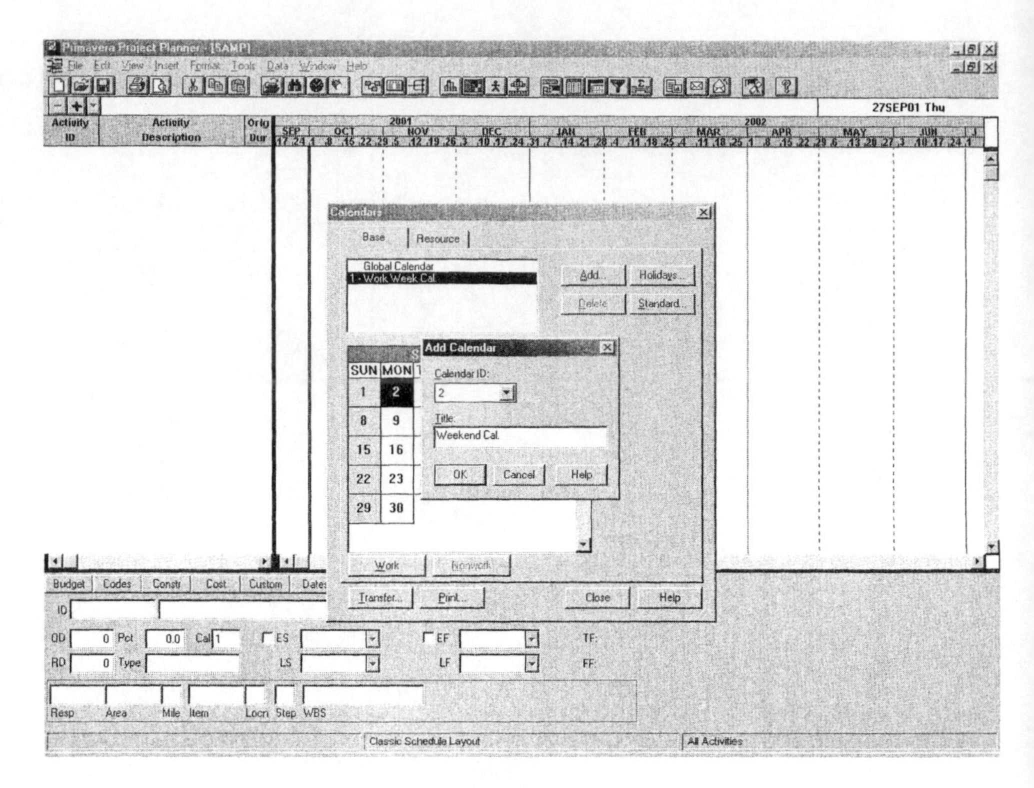

Screen Image 6-10
Initial Second Calendar

Screen Image 6-11
Calendar Transfer Window

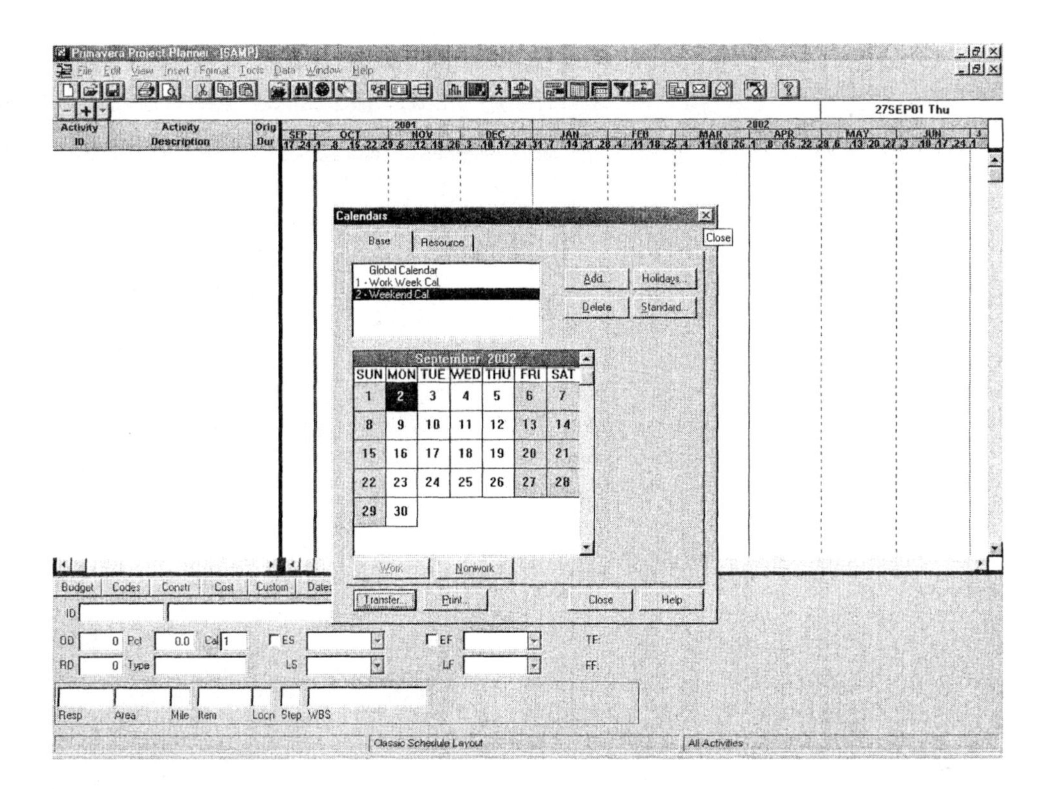

Screen Image 6-12
Transferred Calendar

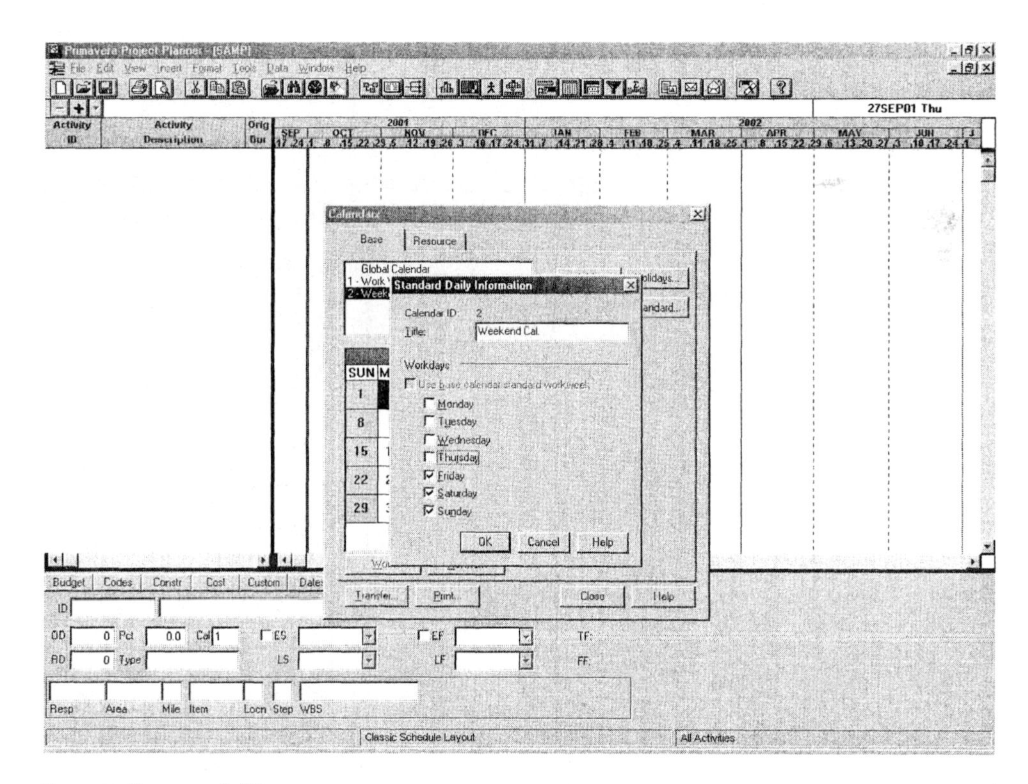

Screen Image 6-13
Specifying the Workdays for Calendar 2

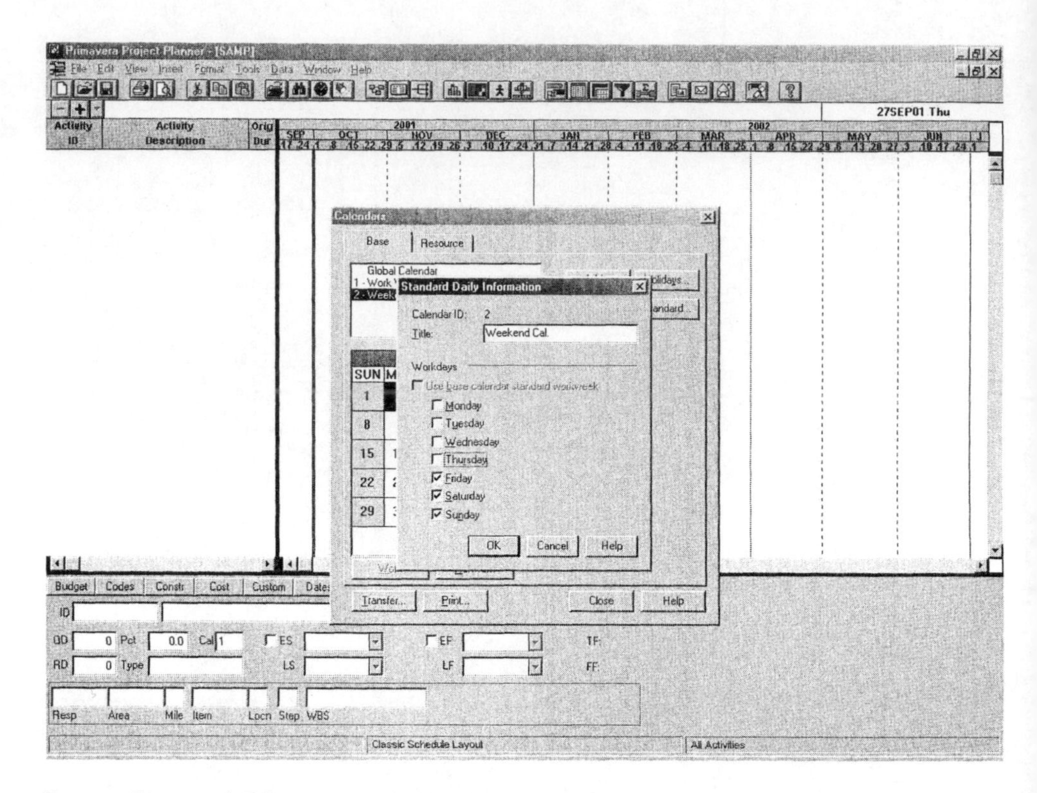

Screen Image 6-14
Calendar 2

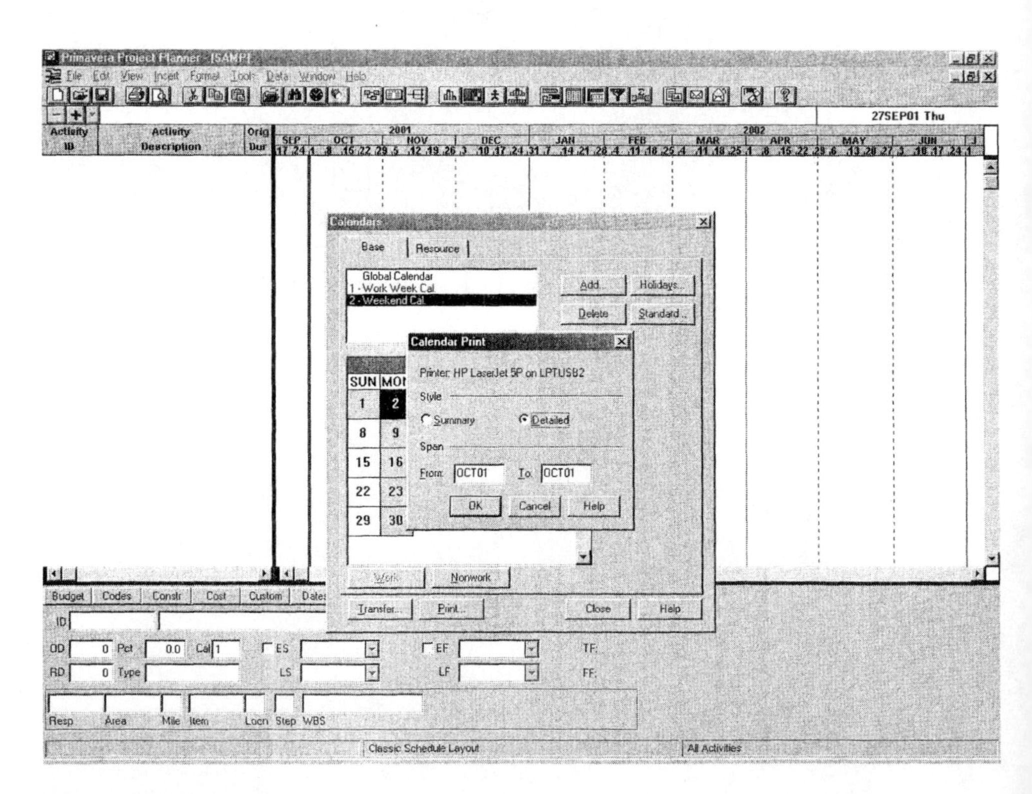

Screen Image 6-15
Calendar Print Window

REPORT DATE 16FEB00 CAL-1 Work Week Cal. PAGE NO. 1

OCTOBER 2001

SUN	MON	TUE	WED	THUR	FRI	SAT
	1 WP=1	2 WP=2	3 WP=3	4 WP=4	5	6
7	8 WP=5	9 WP=6	10 WP=7	11 WP=8	12	13
14	15 WP=9	16 WP=10	17 WP=11	18 WP=12	19	20
21	22 WP=13	23 WP=14	24 WP=15	25 WP=16	26	27
28	29 WP=17	30 WP=18	31 WP=19			

NOVEMBER 2001

SUN	MON	TUE	WED	THUR	FRI	SAT
				1 WP=20	2	3
4	5 WP=21	6 WP=22	7 WP=23	8 WP=24	9	10
11	12 WP=25	13 WP=26	14 WP=27	15 WP=28	16	17
18	19 WP=29	20 WP=30	21 WP=31	22	23	24
25	26 WP=32	27 WP=33	28 WP=34	29 WP=35	30	

DECEMBER 2001

SUN	MON	TUE	WED	THUR	FRI	SAT
						1
2	3 WP=36	4 WP=37	5 WP=38	6 WP=39	7	8
9	10 WP=40	11 WP=41	12 WP=42	13 WP=43	14	15
16	17 WP=44	18 WP=45	19 WP=46	20 WP=47	21	22
23	24 WP=48	25 R	26 WP=49	27 WP=50	28	29
30	31 WP=51					

JANUARY 2002

SUN	MON	TUE	WED	THUR	FRI	SAT
		1 R	2 WP=52	3 WP=53	4	5
6	7 WP=54	8 WP=55	9 WP=56	10 WP=57	11	12
13	14 WP=58	15 WP=59	16 WP=60	17 WP=61	18	19
20	21 WP=62	22 WP=63	23 WP=64	24 WP=65	25	26
27	28 WP=66	29 WP=67	30 WP=68	31 WP=69		

FEBRUARY 2002

SUN	MON	TUE	WED	THUR	FRI	SAT
					1	2
3	4 WP=70	5 WP=71	6 WP=72	7 WP=73	8	9
10	11 WP=74	12 WP=75	13 WP=76	14 WP=77	15	16
17	18 WP=78	19 WP=79	20 WP=80	21 WP=81	22	23
24	25 WP=82	26 WP=83	27 WP=84	28 WP=85		

MARCH 2002

SUN	MON	TUE	WED	THUR	FRI	SAT
					1	2
3	4 WP=86	5 WP=87	6 WP=88	7 WP=89	8	9
10	11 WP=90	12 WP=91	13 WP=92	14 WP=93	15	16
17	18 WP=94	19 WP=95	20 WP=96	21 WP=97	22	23
24	25 WP=98	26 WP=99	27 WP=100	28 WP=101	29	30
31						

Output 6-1
4-Day Calendar

chronological number. Output 6-1 is an example of the detailed calendar. The summary calendar produces a tabular list of the workdays. That report, in spite of its name, typically generates more pages of output than does the detailed report. The span option allows the user to designate a start and stop date for the calendar output.

Activity Codes

In addition to the customary activity data, Primavera Project Planner allows for the creation of user defined data fields. These custom fields are referred to as *activity codes* and can be created and modified at any time. The activity code fields can be sorted and selected, allowing the user to generate output that meets specific needs. For example, if the project covers a large geographic area it may be advantageous, for management purposes, to divide

REPORT DATE 16FEB00 CAL-2 Weekend Cal. PAGE NO. 4

OCTOBER 2001

SUN	MON	TUE	WED	THUR	FRI	SAT
	1	2	3	4	5 WP=1	6 WP=2
7 WP=3	8	9	10	11	12 WP=4	13 WP=5
14 WP=6	15	16	17	18	19 WP=7	20 WP=8
21 WP=9	22	23	24	25	26 WP=10	27 WP=11
28 WP=12	29	30	31			

NOVEMBER 2001

SUN	MON	TUE	WED	THUR	FRI	SAT
				1	2 WP=13	3 WP=14
4 WP=15	5	6	7	8	9 WP=16	10 WP=17
11 WP=18	12	13	14	15	16 WP=19	17 WP=20
18 WP=21	19	20	21	22	23	24 WP=22
25 WP=23	26	27	28	29	30 WP=24	

DECEMBER 2001

SUN	MON	TUE	WED	THUR	FRI	SAT
						1 WP=25
2 WP=26	3	4	5	6	7 WP=27	8 WP=28
9 WP=29	10	11	12	13	14 WP=30	15 WP=31
16 WP=32	17	18	19	20	21 WP=33	22 WP=34
23	24	25 R	26	27	28 WP=35	29 WP=36
30	31					

JANUARY 2002

SUN	MON	TUE	WED	THUR	FRI	SAT
		1 R	2	3	4 WP=37	5 WP=38
6 WP=39	7	8	9	10	11 WP=40	12 WP=41
13 WP=42	14	15	16	17	18 WP=43	19 WP=44
20 WP=45	21	22	23	24	25 WP=46	26 WP=47
27 WP=48	28	29	30	31		

FEBRUARY 2002

SUN	MON	TUE	WED	THUR	FRI	SAT
					1 WP=49	2 WP=50
3 WP=51	4	5	6	7	8 WP=52	9 WP=53
10 WP=54	11	12	13	14	15 WP=55	16 WP=56
17 WP=57	18	19	20	21	22 WP=58	23 WP=59
24 WP=60	25	26	27	28		

MARCH 2002

SUN	MON	TUE	WED	THUR	FRI	SAT
					1 WP=61	2 WP=62
3 WP=63	4	5	6	7	8 WP=64	9 WP=65
10 WP=66	11	12	13	14	15 WP=67	16 WP=68
17 WP=69	18	19	20	21	22 WP=70	23 WP=71
24 WP=72	25	26	27	28	29 WP=73	30 WP=74
31 WP=75						

Output 6-1, continued
3-Day Calendar

the project into smaller geographic areas. This option would allow the project planner to track the project by specific geographic areas as well as generate information about the entire project. The same holds true for supervision; it is desirable to have specific individuals responsible for specific activities or project elements.

Through the use of activity codes, individual superintendents/foremen can be assigned to specific activities. This concept enhances accountability by making known who is responsible for which aspects of the project. By removing ambiguity, clear lines of responsibility and communications are developed, as well as ownership in the project. Activity codes can be added at any time; however, it is recommended that they be initialized prior to any input of schedule data. At the bottom of Screen Image 6-16 are the *Resp, Area, Mile, Item, Locn*, and *Step* activity codes. There can be any number of activity code fields

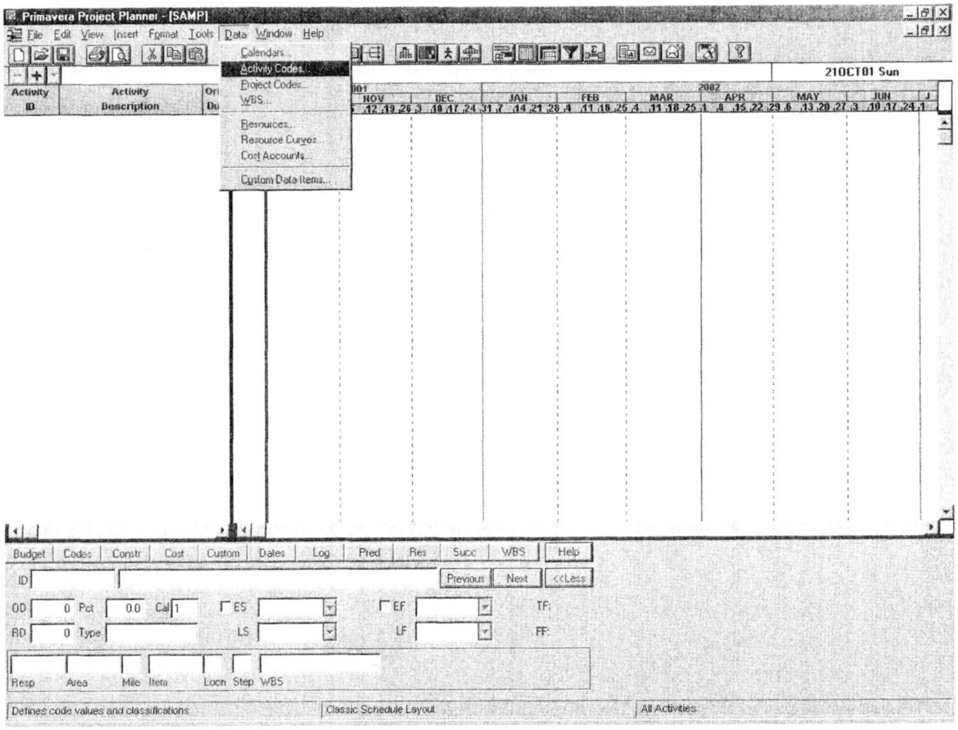

Screen Image 6-16
Opening the Activity Codes Dictionary

so long as the total number of characters for all of the fields added together does not exceed 64.

To define or change the activity codes fields, the *Activity Codes* window needs to be opened. This is done by selecting the *Data* command followed by the *Activity Codes* option as shown in Screen Image 6-16. Screen Image 6-17 is an example of the default settings for the *Activity Codes*.

The activity codes shown in Screen Image 6-17 match those shown on the bottom portion of Screen Image 6-16. The *Name* column is where the activity code field descriptors are assigned. These descriptors can be any five unique alphanumeric characters. The *Length* column defines how many characters of input will be allowed in that field. Ten characters is the maximum field length for any single activity code.

Finally, the *Description* column is used to more fully describe the function of the field descriptor. In order for the pull-down menus to work properly and be understood by subsequent users, good descriptions are a necessity. Clicking on any of the field descriptors will highlight that activity code, allowing it to be deleted by clicking on the minus button. Conversely, clicking on the plus button will facilitate adding a new activity code.

The bottom half of the *Activity Codes* window is used to input all of the possible entries for each individual activity code. Clicking on the individual activity codes activates that portion of the window.

In Screen Image 6-18, all of the default activity codes have been deleted and the *SUPR* activity code has been added. If there have been modifications, the user will be prompted to save or discard the edits when closing this window. Screen Image 6-18 shows the prompt to save or delete the activity code edits. Screen Image 6-19 shows the modified *Activity Codes* window with the names of all of the superintendents who will be working on the project. Adding all of the possible values will facilitate assigning superintendents to the activities. If a check is entered in the *When new code values occur in project, validate against dictionary*, all activity code entries to the activities will be checked for accuracy.

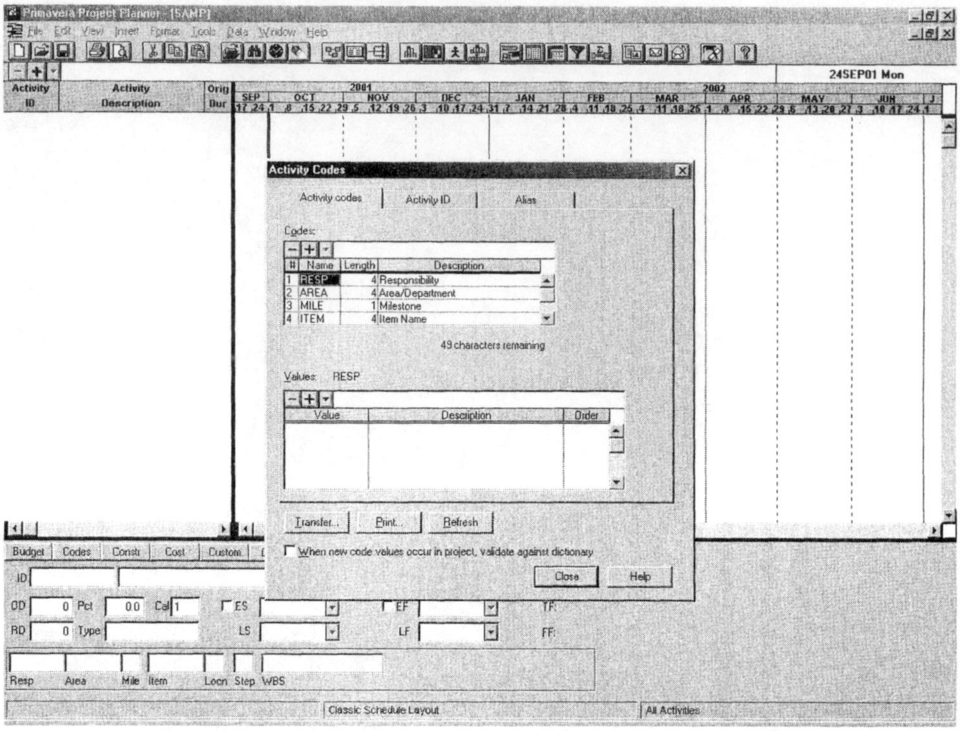

Primavera Project Planner - [SAMP]
File Edit View Insert Format Tools Data Window Help
24SEP01 Mon
Activity ID | Activity Description | Orig Dur
2001 SEP OCT NOV DEC 2002 JAN FEB MAR APR MAY JUN
Activity Codes
Activity codes | Activity ID | Alias
Codes:

#	Name	Length	Description
1	RESP	4	Responsibility
2	AREA	4	Area/Department
3	MILE	1	Milestone
4	ITEM	4	Item Name

49 characters remaining
Values: RESP

Value	Description	Order

Transfer... Print... Refresh
When new code values occur in project, validate against dictionary
Close Help
Budget Codes Constr Cost Custom
ID
OD 0 Pct 0.0 Cal 1 ES EF TF
RD 0 Type LS LF FF
Resp Area Mile Item Locn Step WBS
Classic Schedule Layout | All Activities

Screen Image 6-17
Activity Codes Window

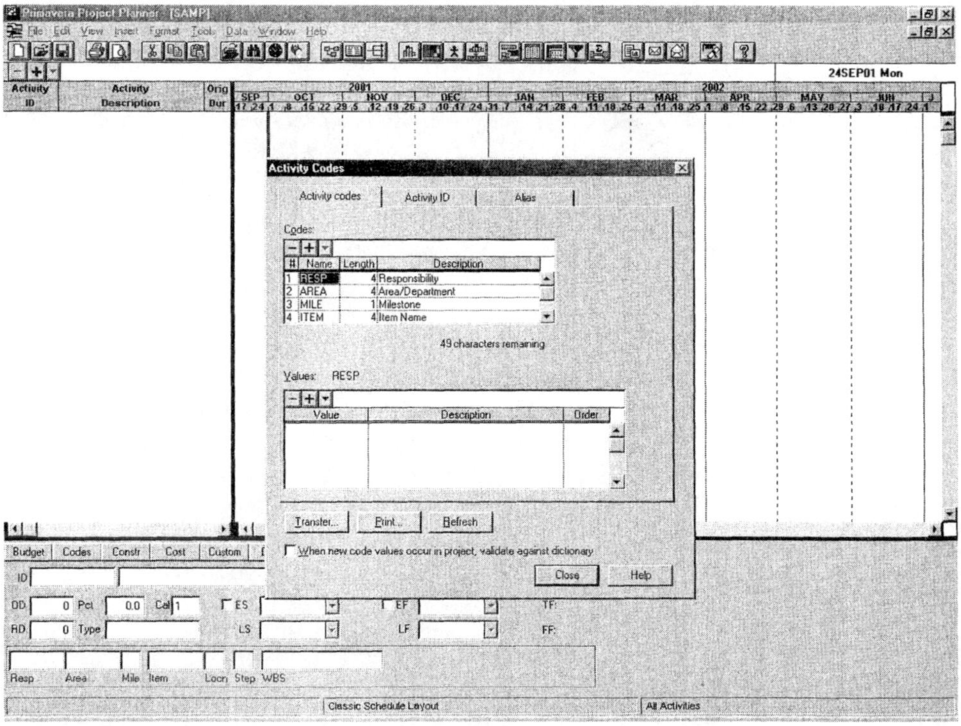

Primavera Project Planner - [SAMP]
File Edit View Insert Format Tools Data Window Help
24SEP01 Mon
Activity ID | Activity Description | Orig Dur
2001 SEP OCT NOV DEC 2002 JAN FEB MAR APR MAY JUN
Activity Codes
Activity codes | Activity ID | Alias
Codes:

#	Name	Length	Description
1	RESP	4	Responsibility
2	AREA	4	Area/Department
3	MILE	1	Milestone
4	ITEM	4	Item Name

49 characters remaining
Values: RESP

Value	Description	Order

Transfer... Print... Refresh
When new code values occur in project, validate against dictionary
Close Help
Budget Codes Constr Cost Custom
ID
OD 0 Pct 0.0 Cal 1 ES EF TF
RD 0 Type LS LF FF
Resp Area Mile Item Locn Step WBS
Classic Schedule Layout | All Activities

Screen Image 6-18
Editing the Activity Codes Dictionary

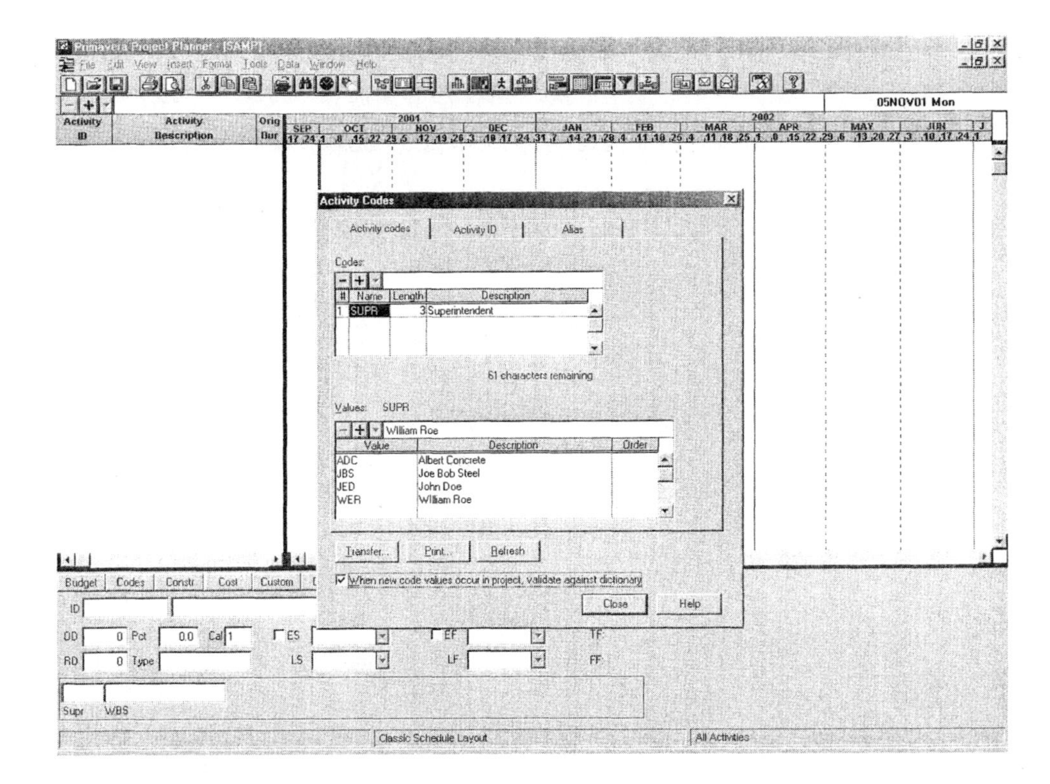

Screen Image 6-19
Completed Activity Codes Dictionary

The default setting for the *Activity ID* field is set at ten characters. However, that field can be subdivided into additional fields. This is accomplished by clicking the *Activity ID* tab and entering the descriptors and values just as with the activity codes (Screen Image 6-20). The *Alias* tab, as shown in Screen Image 6-21, allows the user to combine activity code fields into a single descriptor.

Once all of the activity codes have been defined and their allowable values established, this information can be printed out by clicking on the *Print* button. This will open the *Print Options* window (Screen Image 6-22), which allows the user to direct the output to a specific location or device. Output 6-2 is an example of the activity code output. When all of the activity codes have been entered, click the *Close* button to return to the main input screen.

Project Codes

Project codes are a set of fixed fields that allow the user to enter project specific descriptive information. The Project Codes window is opened by selecting the *Project Codes* option from within the *Data* command as shown in Screen Image 6-23. The *Project Codes Definition* tab provides a list of project descriptive information and an area to input all of the possible values. Screen Image 6-24 shows some of the available descriptive information. These descriptive fields are limited to Project Manager, Status, Project Location, Reason, Industry, Division, Plan Prepared by, Key Contact, Priority, and Currency. The *Values for* tab is where the user attaches the list from the *Project Code Definition* to a specific project. Screen Image 6-25 is an example of the completed *Values for* tab.

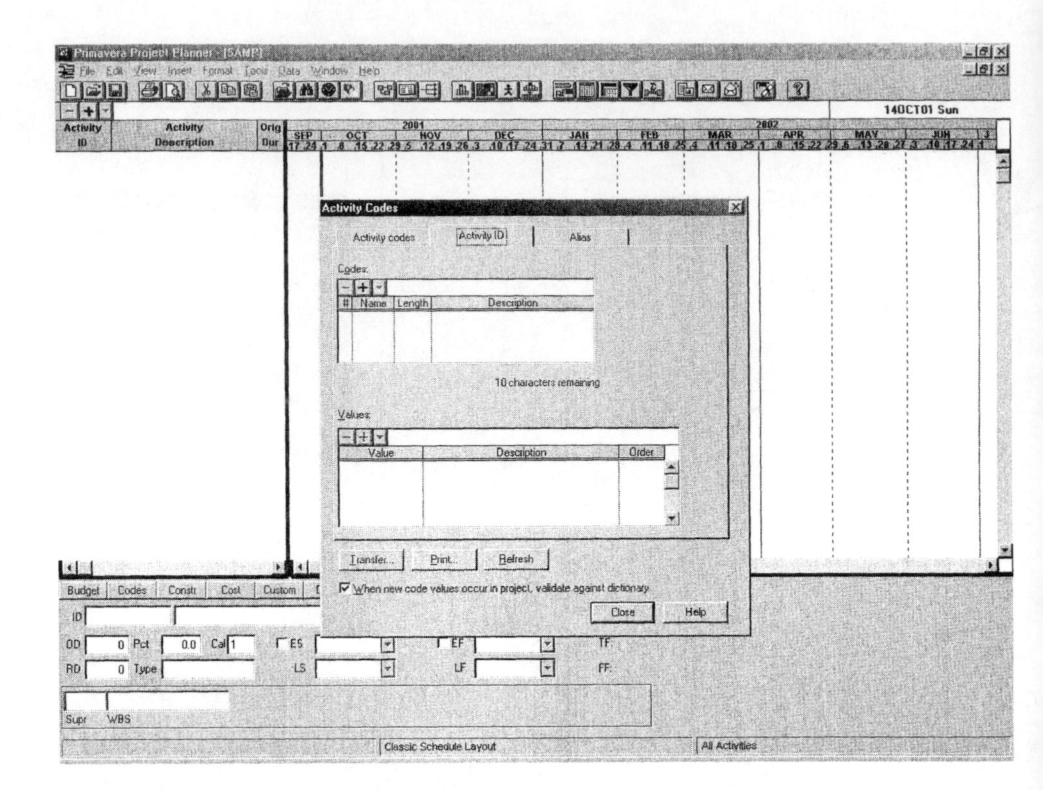

Screen Image 6-20
Editing Activity ID's

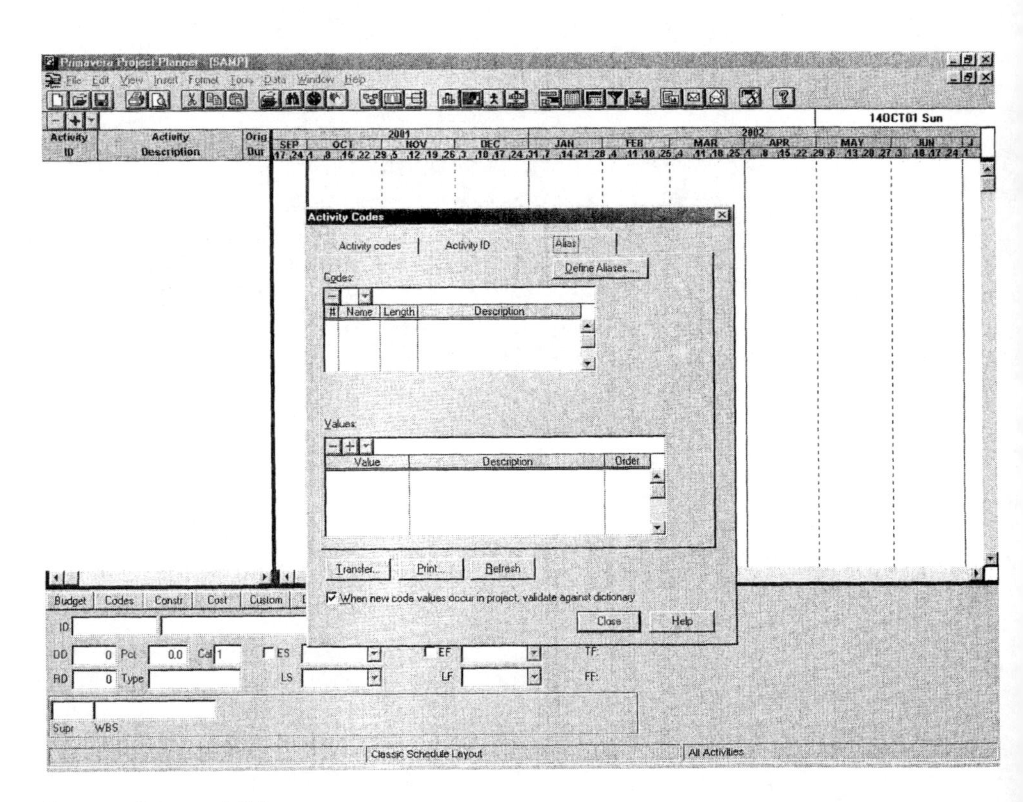

Screen Image 6-21
Editing Using the Alias Tab

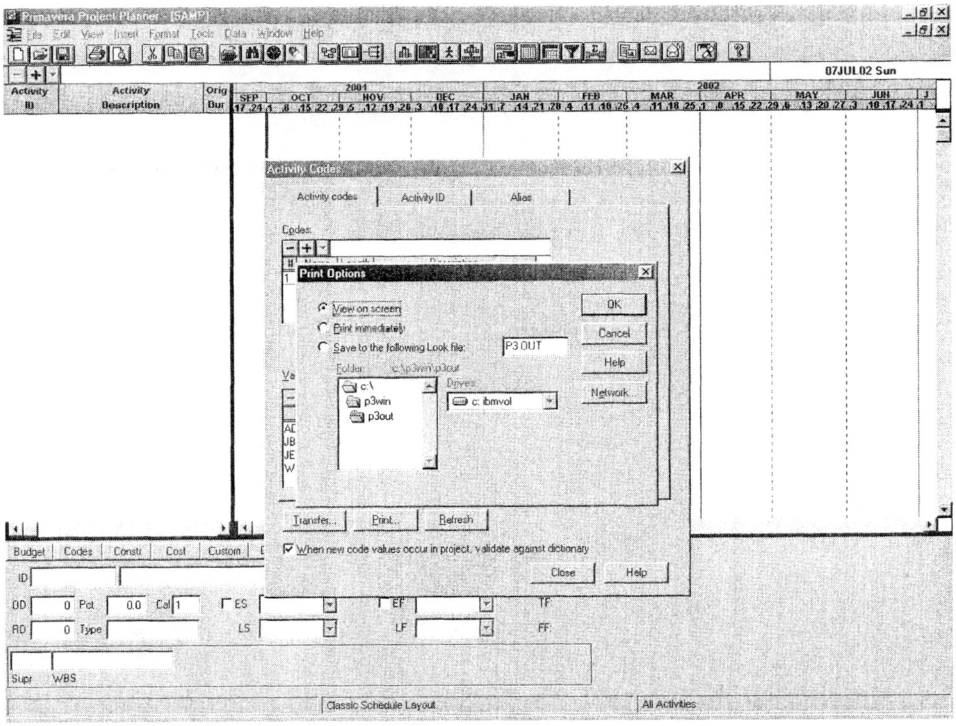

Screen Image 6-22
Print Options Windows

```
------------------------------------------------------------------------------
                          PRIMAVERA PROJECT PLANNER

Date 26FEB00          -----ACTIVITY CODES DICTIONARY-----              Page    1

SAMP - Sample Construction Project
------------------------------------------------------------------------------
   CODE    VALUE                       TITLE                         SEQUENCE
------------------------------------------------------------------------------

Activity Codes:

  SUPR  Superintendent
           ADC         Albert Concrete
           JBS         Joe Bob Steel
           JED         John Doe
           WER         WIlliam Roe

Activity ID Codes:

Alias Codes:
```

Output 6-2
Activity Codes Dictionary Report

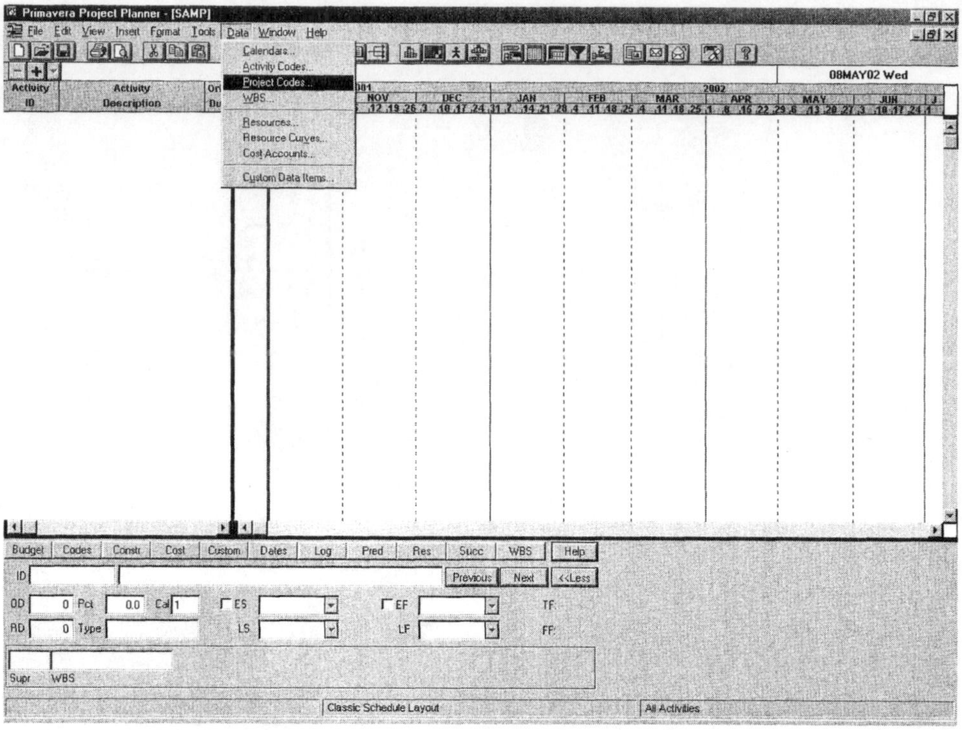

Primavera Project Planner - [SAMP]

File Edit View Insert Format Tools Data Window Help

Calendars...
Activity Codes...
Project Codes...
WBS...
Resources...
Resource Curves...
Cost Accounts...
Custom Data Items...

08MAY02 Wed

Activity ID | Activity Description | Orig Dur

2001 NOV DEC | 2002 JAN FEB MAR APR MAY JUN

Budget | Codes | Constr | Cost | Custom | Dates | Log | Pred | Res | Succ | WBS | Help

ID | Previous | Next | <<Less

OD 0 Pct 0.0 Cal 1 | ES | EF | TF

RD 0 Type | LS | LF | FF

Supr WBS

Classic Schedule Layout | All Activities

Screen Image 6-23
Opening Project Codes

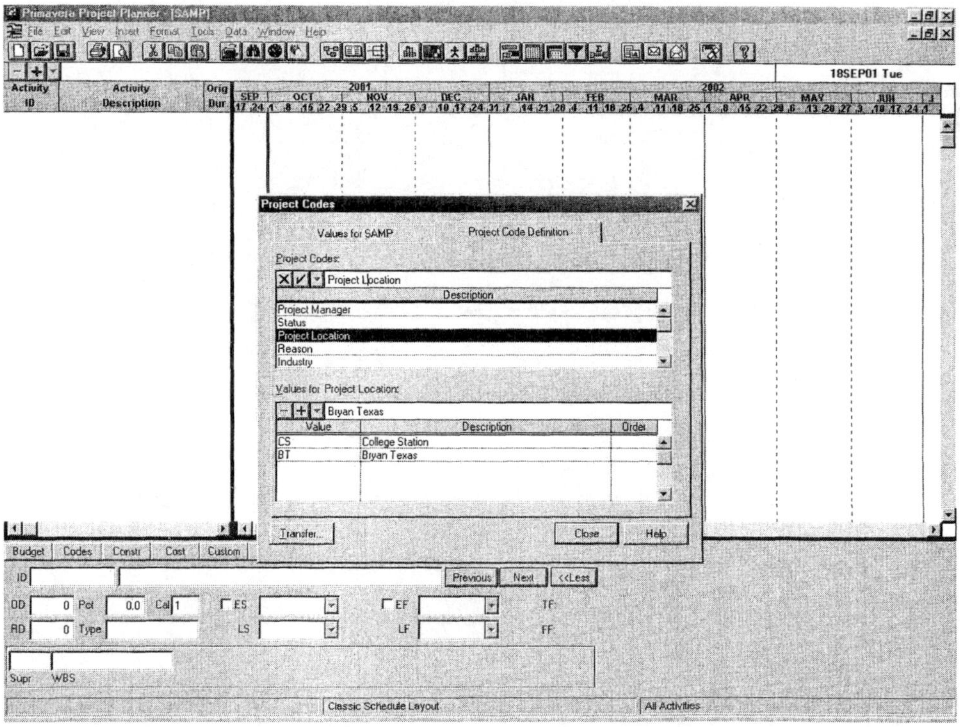

Primavera Project Planner - [SAMP]

File Edit View Insert Format Tools Data Window Help

18SEP01 Tue

Activity ID | Activity Description | Orig Dur

2001 SEP OCT NOV DEC | 2002 JAN FEB MAR APR MAY JUN

Project Codes

Values for SAMP | Project Code Definition

Project Codes:

Project Location

Description
Project Manager
Status
Project Location
Reason
Industry

Values for Project Location:

Bryan Texas

Value	Description	Order
CS	College Station	
BT	Bryan Texas	

Transfer... | Close | Help

Budget | Codes | Constr | Cost | Custom

ID | Previous | Next | <<Less

OD 0 Pct 0.0 Cal 1 | ES | EF | TF

RD 0 Type | LS | LF | FF

Supr WBS

Classic Schedule Layout | All Activities

Screen Image 6-24
Project Codes Window

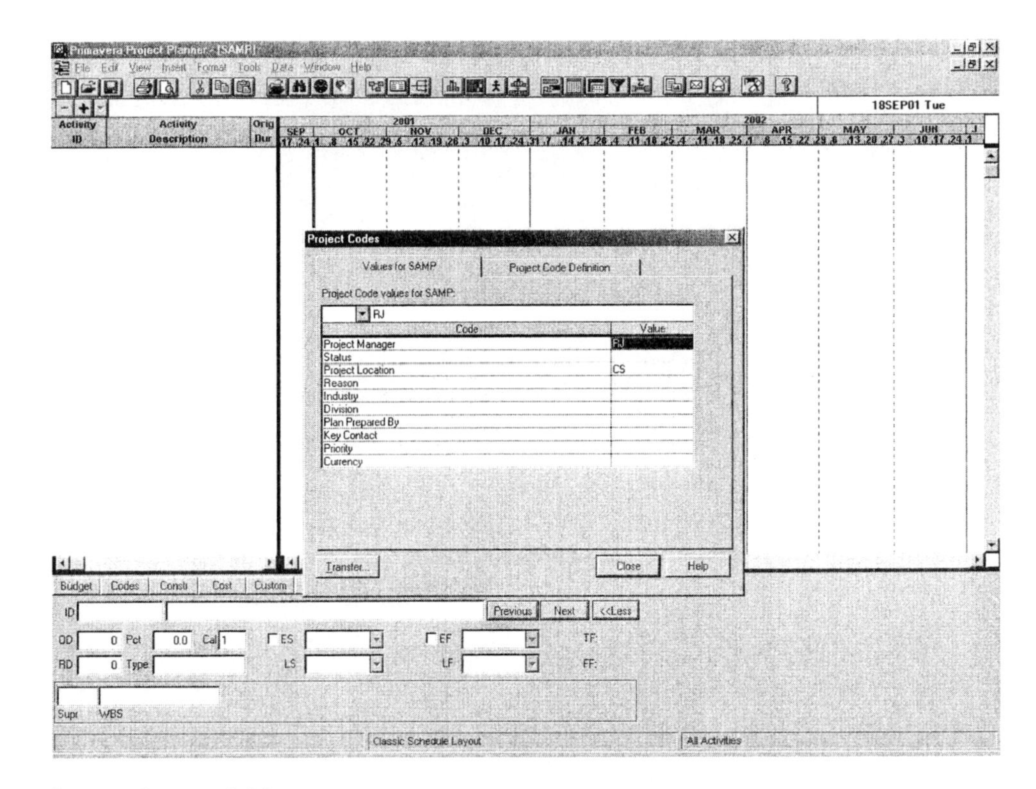

Screen Image 6-25
Project Codes Values

Conclusion

When setting up a new project, the first step is to modify the project calendar so that it matches the construction project work schedule. The second step is to identify how the schedule will be divided so that easily controllable portions can be extracted and managed. This is done by developing an activity code scheme that will allow the user to select portions of the project schedule for review and evaluation.

Suggested Exercises

1. Create a project work calendar with repeating and nonrepeating holidays. At a minimum the holidays should be New Year's Day, Martin Luther King Jr.'s Birthday, Memorial Day, Independence Day, Labor Day, Thanksgiving, and Christmas. When this is completed, print out a one-year detailed calendar.
2. Assume five superintendents will be working on your project. Set up an activity code and a name for each of these superintendents and print out a copy of the activity codes.
3. Set up required project codes (for example, superintendent).

Loading the Schedule Logic

Icon	Description	Icon	Description
	Format Columns		Organize
	View PERT		View Bar Chart (PERT)
	Add Activity (PERT)		Delete Activity (PERT)
	Extract Activity (PERT)		Select All (PERT)
	Activity Box Configuration (PERT)		Activity Box Ends & Colors (PERT)
	Format Relationship (PERT)		Organize (PERT)

Activity Input–Activity Form

The loading of the schedule logic consists of two operations: input of the activities, and tying the activities together with relationship arrows. A number of different options and processes are available to accomplish these tasks. The activity-input process begins with the Primavera Project Planner main input screen, as shown in Screen Image 7-1. If the activity form at the bottom of the screen is not present, it can be turned on and off by pressing the F7 key. This portion of the screen can also be turned on and off by selecting the *View* command followed by the *Activity Form* option. It is essential that the activity form be present since the majority of the activity input will be performed from that portion of the screen.

The activity-input process begins by clicking the plus button immediately beneath the row of icons. This will activate the adjacent field, which is where the first activity identifier *(Activity ID)* or activity number can be entered. This is the only information that is entered at the top of the screen. The remaining information about the activity is entered through the activity form.

Primavera Project Planner offers the option of either having the user assign the activity identifiers or having them automatically assigned based upon the first entry. Whichever method is selected, it is important for the activity identifiers to be incremented in steps of 5 or 10. This will allow for additional activities to be inserted at a later date. To verify, set, or remove the automatic numbering process, select the *Tools* command followed by the *Options* and the *Activity Inserting* option as shown in Screen Image 7-2. This will open the *Activity Inserting Options* window as shown in Screen Image 7-3. There are two options associated with this window. The first deals with turning on or off the automatic numbering process. If this option is turned on, then the increment amount needs to be set. As previously mentioned, this should be set to 5 or 10. The final option deals with having Primavera Project Planner automatically turn on the activity form when adding an activity.

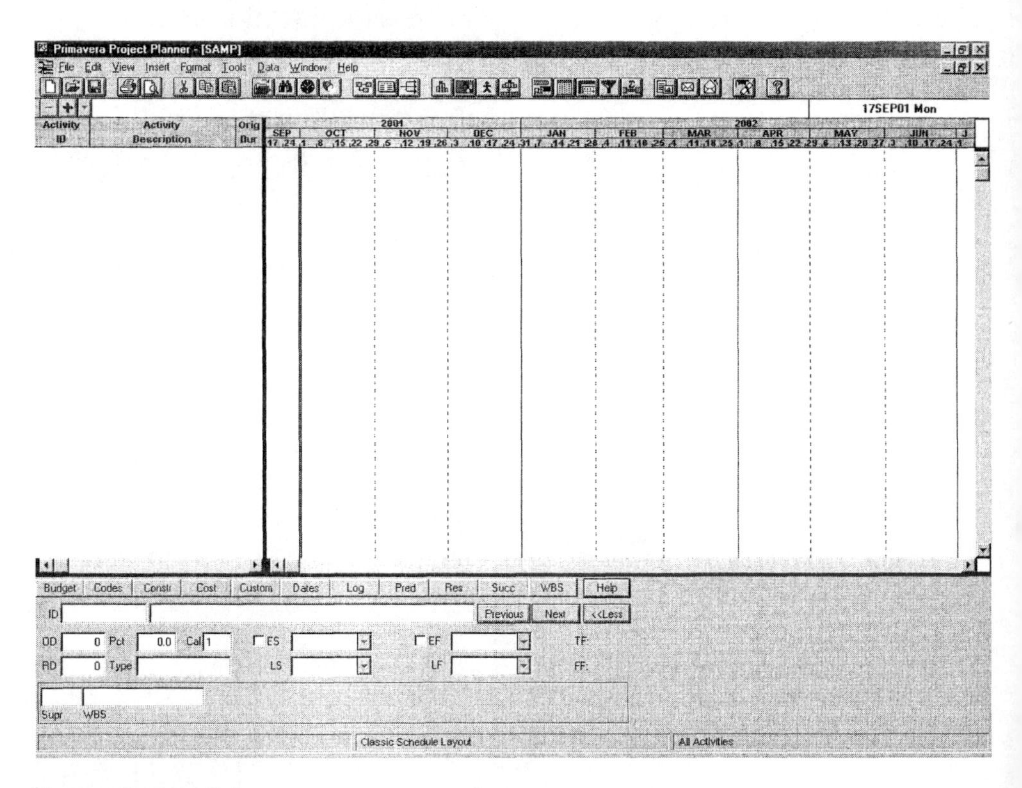

Screen Image 7-1
Main Input Screen

Screen Image 7-2
Activity Incrementing

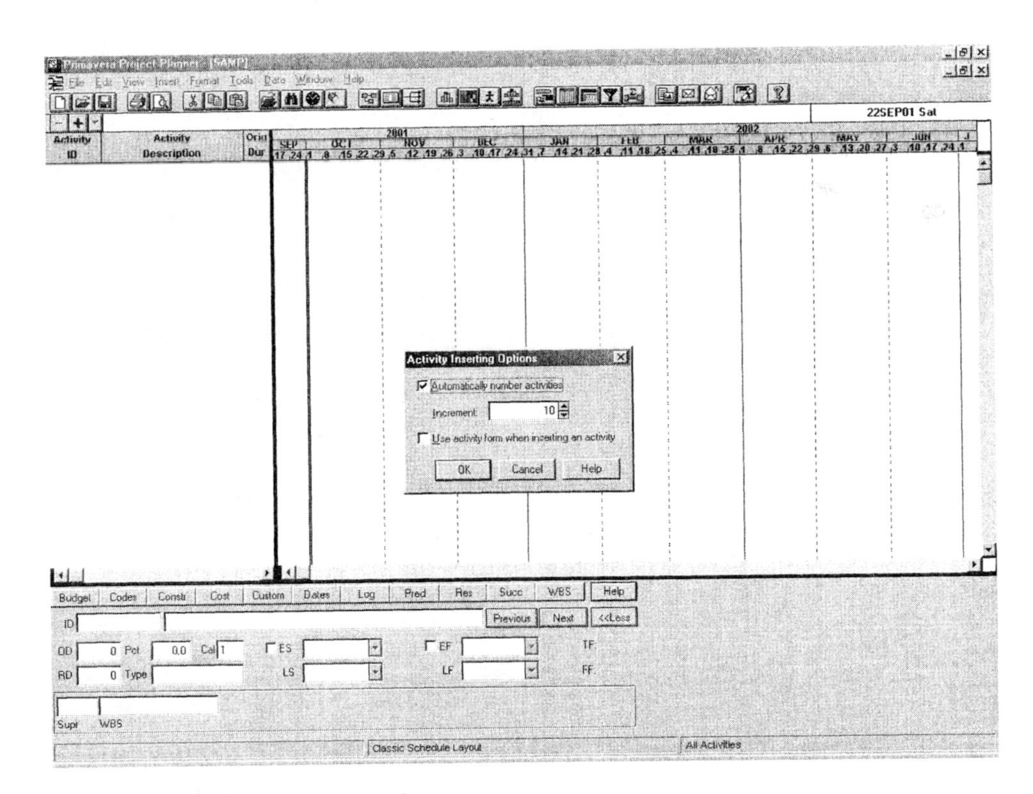

Screen Image 7-3
Activity Inserting Options Window

It is a good practice to have this set to turn on automatically, since all of the specifics of an activity must be entered from that screen area. If that option is selected, the activity form will automatically appear when the plus button is clicked.

Regardless of whether the auto incrementing is turned on or off, the first activity is initialized by clicking the plus button directly below the command line and entering the first activity number. When this is completed, press the *Enter* key to activate the *Activity Form* portion of the screen. Once the activity identifier has been initialized, the mouse or *Tab* key can be used to navigate the *Activity Form*. The first step should be to click on the long field that is adjacent to the activity *ID* field. This field, with no descriptor, is for the activity description. In order to keep the schedule compact the activity description should be limited to between 30 and 35 characters. By limiting the length of the activity description, more concise graphic output can be generated. After the activity description has been entered, the cursor can be moved from field to field on the activity form by pressing the *Tab* key. The next field on the activity form is for the original duration (*OD*). This is where the planned activity duration is entered. This duration must correspond with the workweek used in the estimate and with the project calendar. If a 4-day workweek calendar is being used for that activity, then the calculated duration must be based on the number of hours in that standard workday, which is typically 10. Similarly, if a 5-day calendar is being used, the duration for that activity typically would be calculated using an 8-hour workday. The *RD* field, or remaining duration, should be the same as the OD until actual work begins on the activity. If this field is tabbed over, the remaining duration will be calculated by Primavera Project Planner. The *Pct* field is used for inputting the percent complete for the individual activities. Since the project has not started, this field should be tabbed over. The *Cal* field is where the activity is tied to a specific calendar. If only one calendar is used, Calendar 1—the default calendar—should be entered in this field. However, if the activity is associated with one of the alternate calendars, then the appropriate calendar identifiers should be entered in that field. If good descriptions of the calendars were used to set up the calendars, then the pull-down menu will prove helpful. There are a number of different activity types that are used in Primavera Project Planner; however, *Task* is the most commonly used when inputting construction activities. Table 7-1 gives a brief description of these types of activities.

In later chapters, the other types of activities will be discussed and demonstrated. The *ES*, *LS*, *EF*, and *LF* fields correspond to early start, late start, early finish, and late finish dates.

Table 7-1
Activity Types

Activity Type	Description
Task	Schedules activities according to the base calendar assigned to it.
Independent	Works according to its own resource calendar.
Meeting	Schedules a meeting activity when all resources attached to the activity are available.
Start Milestone	Marks the start of a significant event in the project.
Finish Milestone	Marks the end of a significant event in the project.
Start Flag	Marks the start of a string of activities.
Finish Flag	Marks the end of a string of activities.
Hammock	An activity that spans a group of activities.
WBS	Rolls up or summarizes activities that have common prefixes.

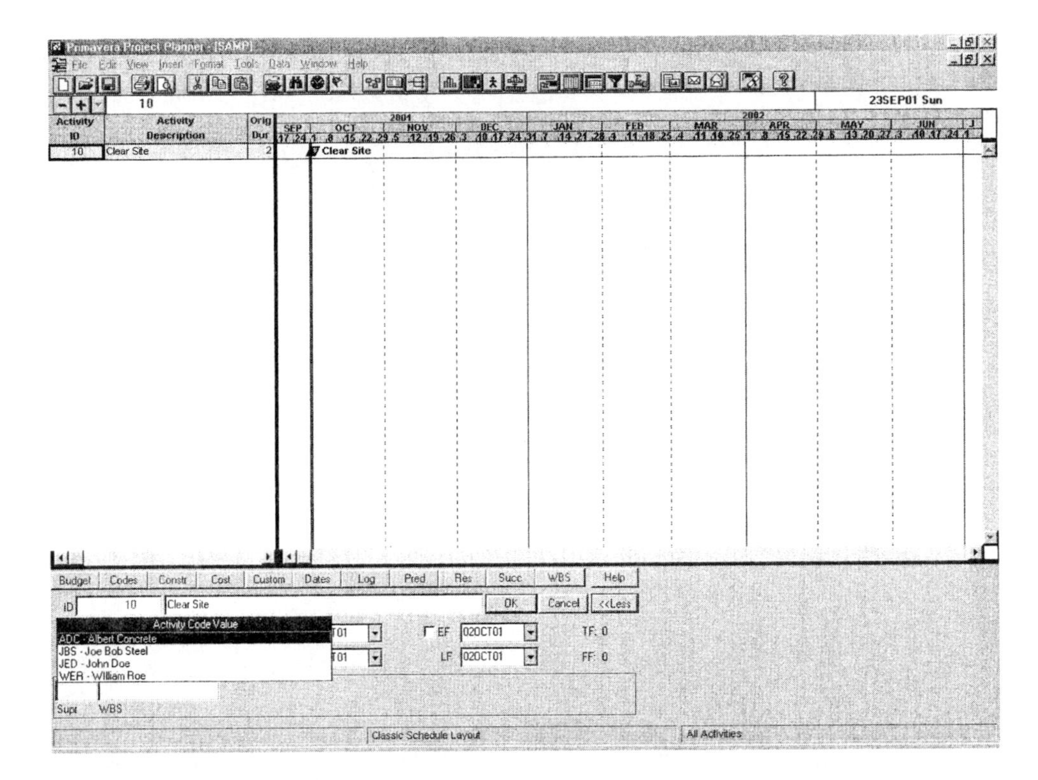

Screen Image 7-4
Activity Code Values

Since the initial objective is to have the software calculate these dates, they should be tabbed over. The *Supr* field, shown in Screen Image 7-4, is the activity code field that was created in the example from Chapter 6. By tabbing to the activity code field, a pull-down menu can be used that will list all initialized and available entries for that specific activity code field. Screen Image 7-4 shows the pull-down menu options for the *SUPR* activity code field. Clicking the desired item on that list will automatically enter it in the appropriate activity code field. The user can also enter items in the activity code field that are not currently listed in the activity codes dictionary. However, the user may be prompted to verify that the input is correct before it is added to that activity and to the activity codes dictionary. This prompting is an option that can be changed within the *Activity Codes* window. When all of the information about a specific activity has been input, clicking on the *OK* button will save all of the data. Clicking on the plus button will save the current data and start the process for adding the next activity.

Screen Image 7-5 is an example of the input for the first activity, while Screen Image 7-6 is an example showing multiple activities that have been added. From these examples it becomes clear that Primavera Project Planner creates a bar for each of the activities and that the length of the bar is equal to the specified duration of the activity. In the later example (Screen Image 7-6), all of the activities have the same start date. This is because there are no relationship arrows. Until the relationship arrows have been added, it is assumed that all activities qualify as the first activity, since they have no predecessor.

Importing Activity Data

If the activity list or estimate was prepared using a spreadsheet program such as Excel® or Lotus®, that data can be imported directly into Primavera Project Planner. A variety of information can be imported from spreadsheet or database programs. The primary advantage

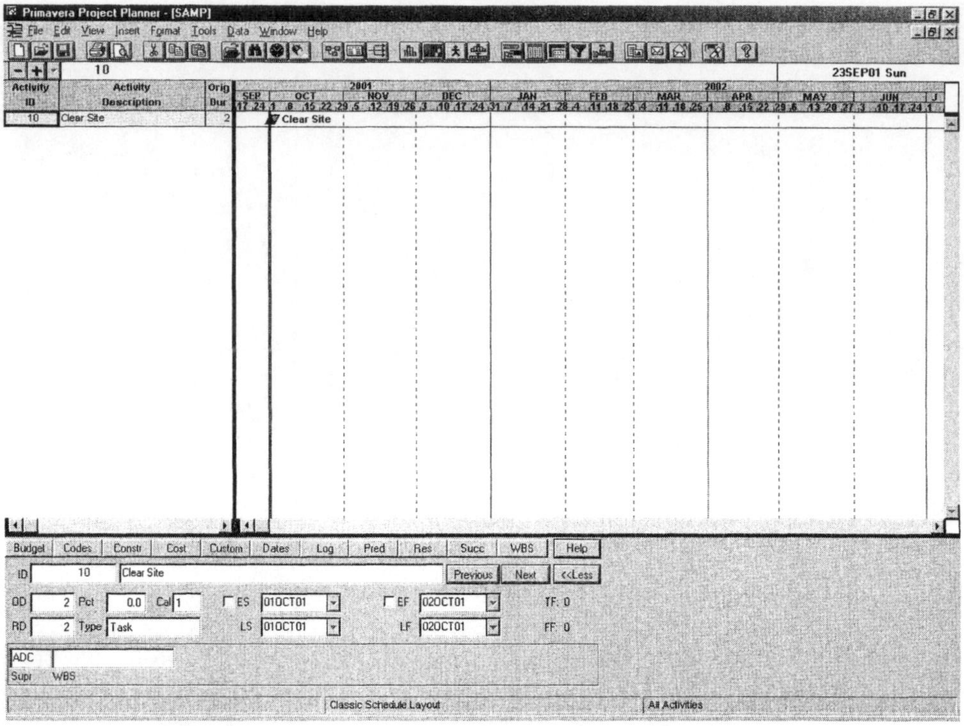

Primavera Project Planner - [SAMP]
File Edit View Insert Format Tools Data Window Help
10 | 23SEP01 Sun

Activity ID	Activity Description	Orig Dur
10	Clear Site	2

2001 SEP OCT NOV DEC 2002 JAN FEB MAR APR MAY JUN J
Clear Site

Budget Codes Constr Cost Custom Dates Log Pred Res Succ WBS Help
ID 10 Clear Site Previous Next <<Less
OD 2 Pct 0.0 Cal 1 ES 01OCT01 EF 02OCT01 TF: 0
RD 2 Type Task LS 01OCT01 LF 02OCT01 FF: 0
ADC
Supr WBS
Classic Schedule Layout | All Activities

Screen Image 7-5
First Activity Added

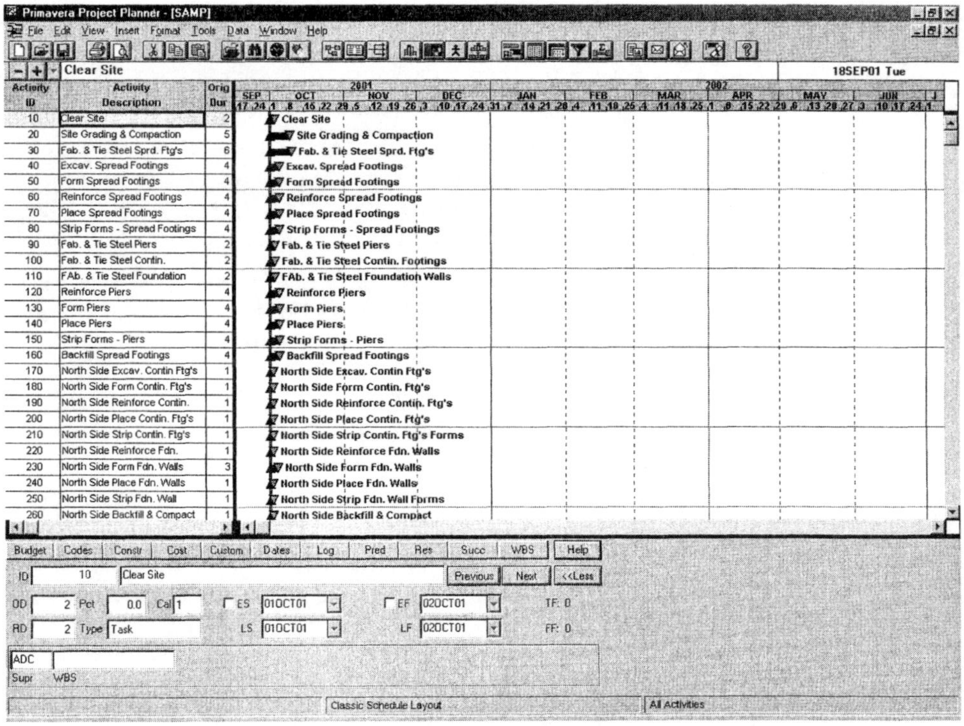

Primavera Project Planner - [SAMP]
File Edit View Insert Format Tools Data Window Help
Clear Site | 18SEP01 Tue

Activity ID	Activity Description	Orig Dur
10	Clear Site	2
20	Site Grading & Compaction	5
30	Fab. & Tie Steel Sprd. Ftg's	6
40	Excav. Spread Footings	4
50	Form Spread Footings	4
60	Reinforce Spread Footings	4
70	Place Spread Footings	4
80	Strip Forms - Spread Footings	4
90	Fab. & Tie Steel Piers	2
100	Fab. & Tie Steel Contin.	2
110	FAb. & Tie Steel Foundation	2
120	Reinforce Piers	4
130	Form Piers	4
140	Place Piers	4
150	Strip Forms - Piers	4
160	Backfill Spread Footings	4
170	North Side Excav. Contin Ftg's	1
180	North Side Form Contin. Ftg's	1
190	North Side Reinforce Contin.	1
200	North Side Place Contin. Ftg's	1
210	North Side Strip Contin. Ftg's	1
220	North Side Reinforce Fdn.	1
230	North Side Form Fdn. Walls	3
240	North Side Place Fdn. Walls	1
250	North Side Strip Fdn. Wall	1
260	North Side Backfill & Compact	1

2001 SEP OCT NOV DEC 2002 JAN FEB MAR APR MAY JUN J
Clear Site
Site Grading & Compaction
Fab. & Tie Steel Sprd. Ftg's
Excav. Spread Footings
Form Spread Footings
Reinforce Spread Footings
Place Spread Footings
Strip Forms - Spread Footings
Fab. & Tie Steel Piers
Fab. & Tie Steel Contin. Footings
FAb. & Tie Steel Foundation Walls
Reinforce Piers
Form Piers
Place Piers
Strip Forms - Piers
Backfill Spread Footings
North Side Excav. Contin Ftg's
North Side Form Contin. Ftg's
North Side Reinforce Contin. Ftg's
North Side Place Contin. Ftg's
North Side Strip Contin. Ftg's Forms
North Side Reinforce Fdn. Walls
North Side Form Fdn. Walls
North Side Place Fdn. Walls
North Side Strip Fdn. Wall Forms
North Side Backfill & Compact

Budget Codes Constr Cost Custom Dates Log Pred Res Succ WBS Help
ID 10 Clear Site Previous Next <<Less
OD 2 Pct 0.0 Cal 1 ES 01OCT01 EF 02OCT01 TF: 0
RD 2 Type Task LS 01OCT01 LF 02OCT01 FF: 0
ADC
Supr WBS
Classic Schedule Layout | All Activities

Screen Image 7-6
Multiple Activities Added

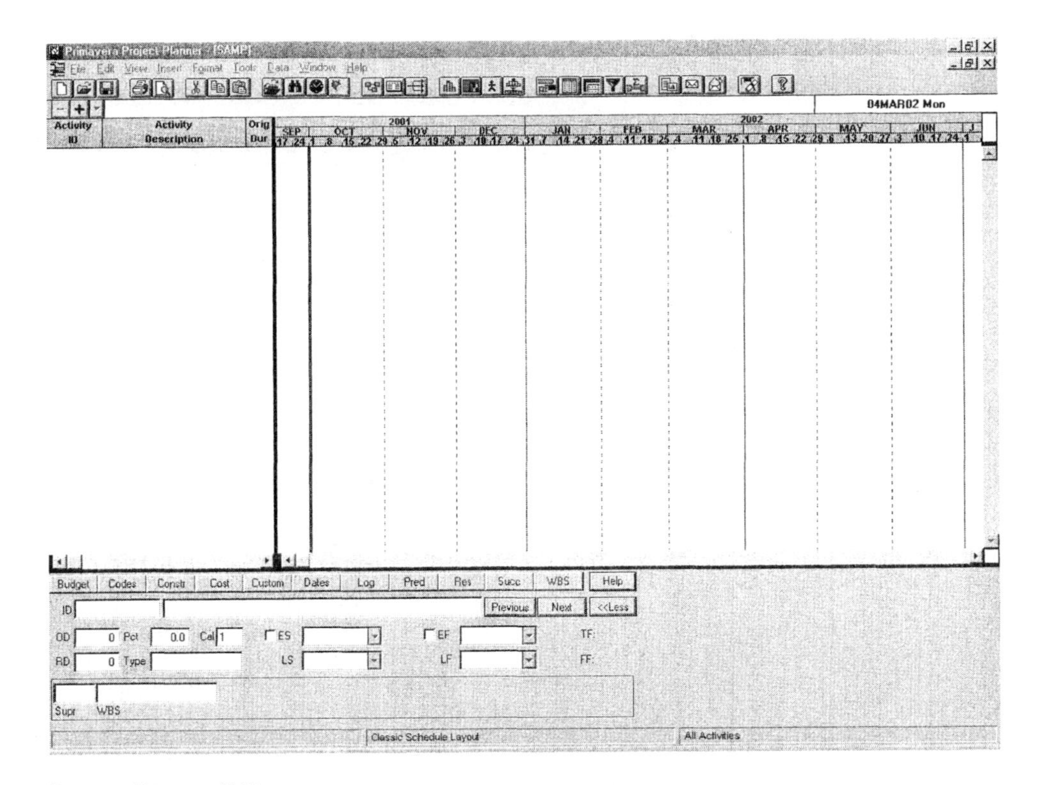

Screen Image 7-7
New Project

of using this feature is that it prevents the user from having to key in activity information more than once. To effectively utilize this feature, a record layout has to be created that will show what information can be imported and how it should be organized on the spreadsheet. Starting with a newly created project, the screen should look like Screen Image 7-7. On this screen add an activity and description. (The specifics of this activity are not important, as it will soon be deleted.) This activity is needed only to create a sample record layout. In Screen Image 7-8, Activity 10—Sample Activity—has been added to the project.

The next step in importing data is to create a record layout that can be used by the spreadsheet program. This is done by opening the *Export* window. This window can be opened by selecting the *Tools* command followed by *Project Utilities* and finally the *Export* option. Screen Image 7-9 details the sequence of these commands and options. The *Export* window as shown in Screen Image 7-10 lists any number of previously established export instructions. From this window click the *Add* button which will open the *Add a New Report* window as shown in Screen Image 7-11. This window will automatically assign a *Report ID*. The number that is returned is simply a sequential count and should be accepted by clicking the *OK* button.

The *Export* window has three separate import tabs, *Content, Format*, and *Selection*. The *Content* import area is perhaps the most important as it will be used to determine what information will be included in the record layout. The *Title* field holds the description for this set of instructions. This description is the one that will be shown later on the list of already prepared export instructions. If this option is used on a regular basis it is important that a good description be used so that it can easily be identified for future use. In Screen Image 7-12 three fields have been identified for export, *Activity ID, Activity description*, and *Original duration*. While there are a number of other options, these three fields represent the information that will be imported from the estimate into Primavera Project Planner.

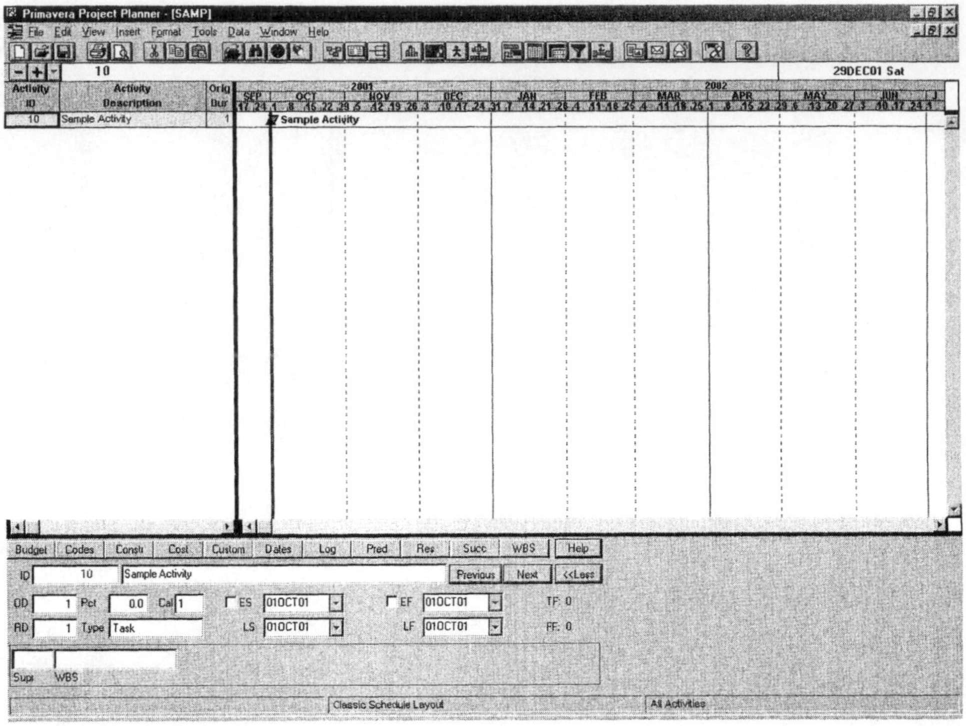

Primavera Project Planner - [SAMP]
File Edit View Insert Format Tools Data Window Help
10 29DEC01 Sat
Activity ID | Activity Description | Orig Dur | 2001 SEP OCT NOV DEC 2002 JAN FEB MAR APR MAY JUN
10 | Sample Activity | 1 | Sample Activity
Budget Codes Constr Cost Custom Dates Log Pred Res Succ WBS Help
ID 10 Sample Activity Previous Next <<Less
OD 1 Pct 0.0 Cal 1 ES 01OCT01 EF 01OCT01 TF: 0
RD 1 Type Task LS 01OCT01 LF 01OCT01 FF: 0
Supr WBS
Classic Schedule Layout All Activities

Screen Image 7-8
Temporary Activity Added

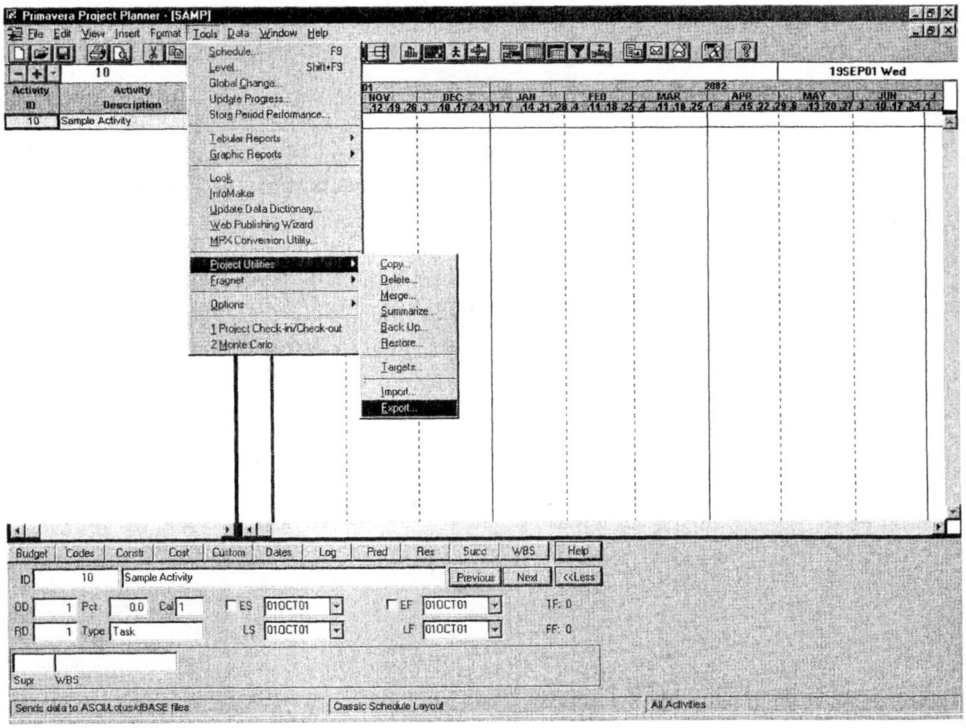

Primavera Project Planner - [SAMP]
File Edit View Insert Format Tools Data Window Help
10 19SEP01 Wed
Activity ID | Activity Description
10 | Sample Activity
Schedule... F9
Level... Shift+F9
Global Change...
Update Progress...
Store Period Performance...
Tabular Reports
Graphic Reports
Look
InfoMaker
Update Data Dictionary...
Web Publishing Wizard
MPX Conversion Utility...
Project Utilities
Fragnet
Options
1 Project Check-in/Check-out
2 Monte Carlo
Copy...
Delete...
Merge...
Summarize...
Back Up...
Restore...
Targets...
Import...
Export...
Budget Codes Constr Cost Custom Dates Log Pred Res Succ WBS Help
ID 10 Sample Activity Previous Next <<Less
OD 1 Pct 0.0 Cal 1 ES 01OCT01 EF 01OCT01 TF: 0
RD 1 Type Task LS 01OCT01 LF 01OCT01 FF: 0
Supr WBS
Sends data to ASCII/Lotus/dBASE files Classic Schedule Layout All Activities

Screen Image 7-9
Opening the Export Window

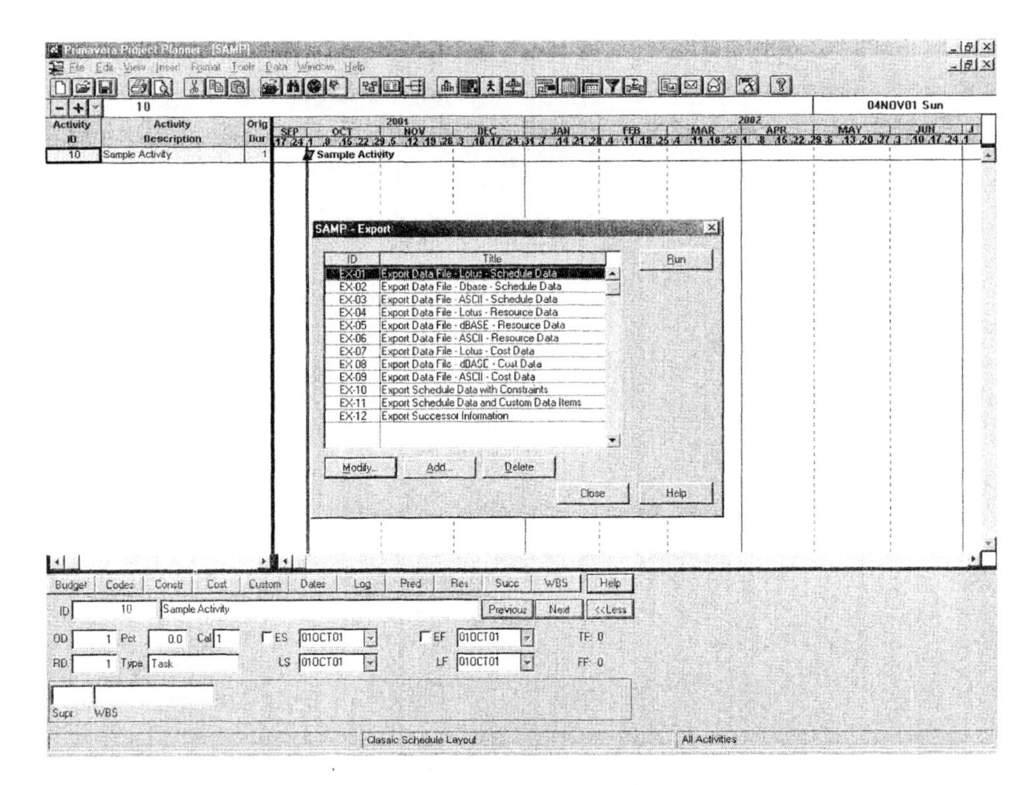

Screen Image 7-10
Export Window

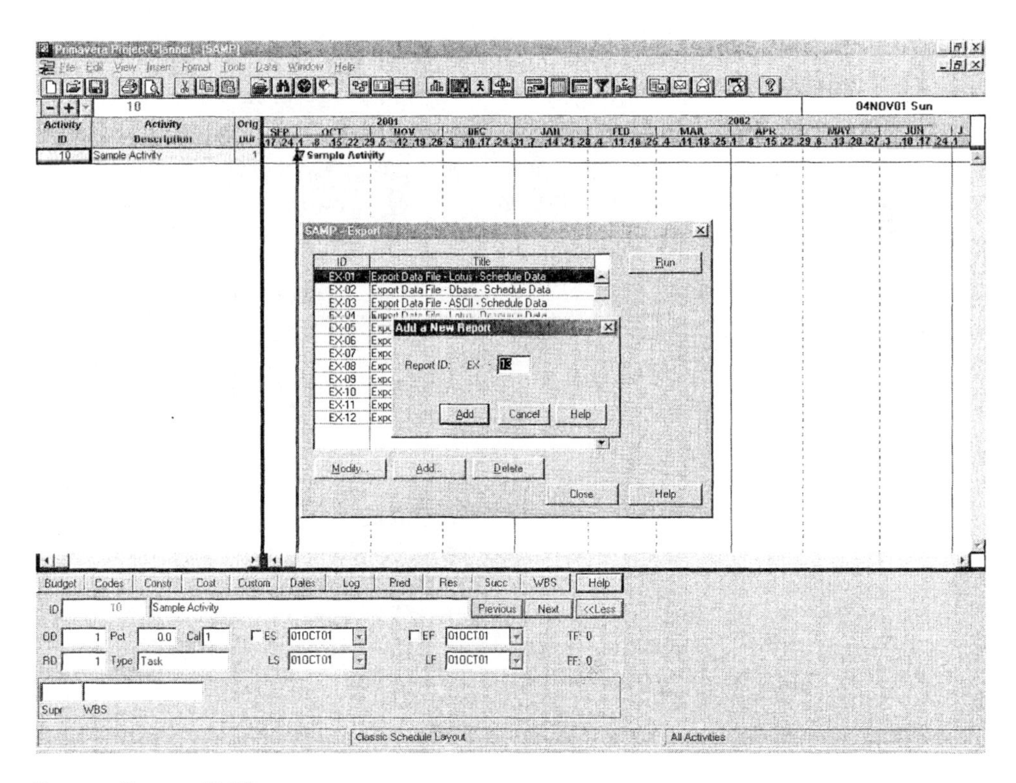

Screen Image 7-11
Add a New Report Window

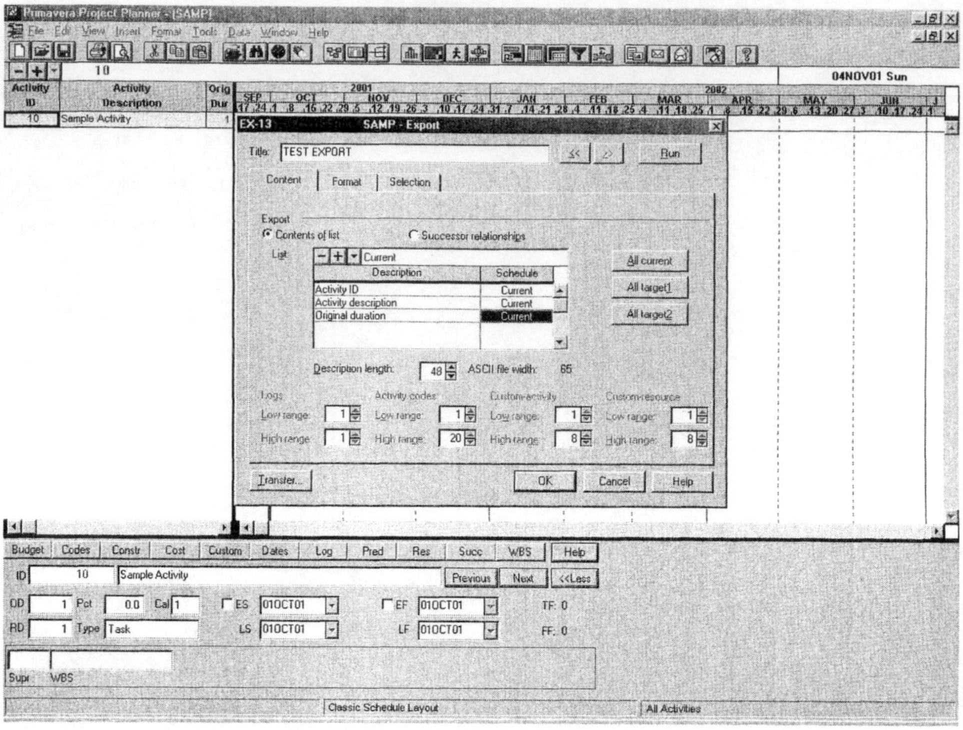

Screen Image 7-12
Export Content Window

There are two important items on the *Format* import area, the *Name* and *Type* field. In Screen Image 7-13 the *Name* field is set to *C:\P3WIN\P3OUT\P3*. This field specifies the location where the output file will be written and its name. In this screen image the file name will be *P3.WK1* and will be written in the *P3OUT* subdirectory of the *P3WIN* folder. This setting is the default setting, and the user should pay particular close attention to the location to avoid having to find the file when it is needed for subsequent operations. The *Type* field is where the user can specify what type of output file will be generated. There are a number of options that are available, however one of the Lotus (WKS or WK1) file formats will be the most universal and can be read by a number of different spreadsheet programs.

The *Selection* input area window as shown in Screen Image 7-14 is used to filter in or out specific activities for the file download. Since there is only one activity in the current schedule, no input selection criteria are required. If the *Selection Criteria* table is left blank, all activities within the project will be exported. When the three input areas have been completed, the *Run* button can be clicked. This will open the *Output Options* window as shown in Screen Image 7-15. This window allows the user to route the printed output. For the purpose of this example, the routing is not important since the volume of the output is so small. Output 7-1 is a copy of the output that would be generated; all it shows is the path where the exported file has been written.

Once the export has been completed, the user should go back and delete the initial activity from Primavera Project Planner. When this is done the project will exist but will contain no data. The newly created export file can now be opened using a spreadsheet program. In Screen Image 7-16 the exported file has been opened using Excel. That screen shows the layout that will be used to import the activity information. By using a cut and paste operation the information that needs to be imported can be added. In Screen Image 7-17 the exported record has been replaced with the information that will be imported into Primavera Project Planner. When the required information has been pasted onto the spreadsheet, it

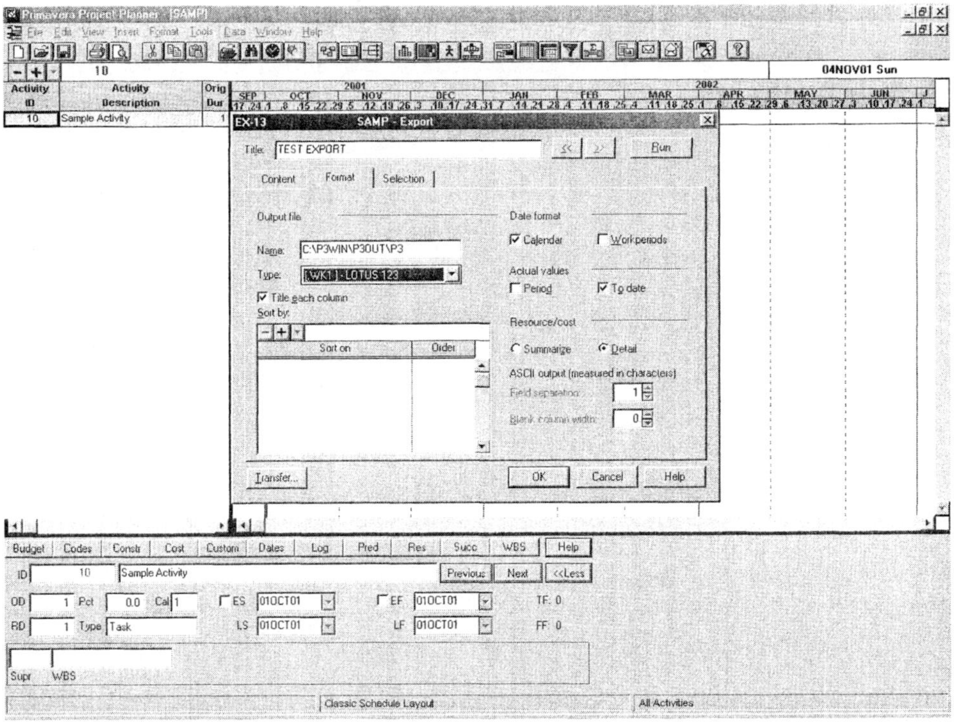

Primavera Project Planner - [SAMP]
File Edit View Insert Format Tools Data Window Help
10
04NOV01 Sun
Activity ID | Activity Description | Orig Dur
10 | Sample Activity | 1
EX-13 SAMP - Export
Title: TEST EXPORT
Run
Content | Format | Selection
Output file
Name: C:\P3WIN\P3OUT\P3
Type: WK1 - LOTUS 123
Title each column
Sort by:
Sort on | Order
Date format
Calendar
Workperiods
Actual values
Period
To date
Resource/cost
Summarize
Detail
ASCII output (measured in characters)
Field separation: 1
Blank column width: 0
Transfer...
OK
Cancel
Help
Budget | Codes | Constr | Cost | Custom | Dates | Log | Pred | Res | Succ | WBS | Help
ID 10 Sample Activity
Previous | Next | <<Less
OD 1 Pct 0.0 Cal 1
ES 01OCT01
EF 01OCT01
TF: 0
RD 1 Type Task
LS 01OCT01
LF 01OCT01
FF: 0
Supr WBS
Classic Schedule Layout
All Activities

Screen Image 7-13
Export Format Window

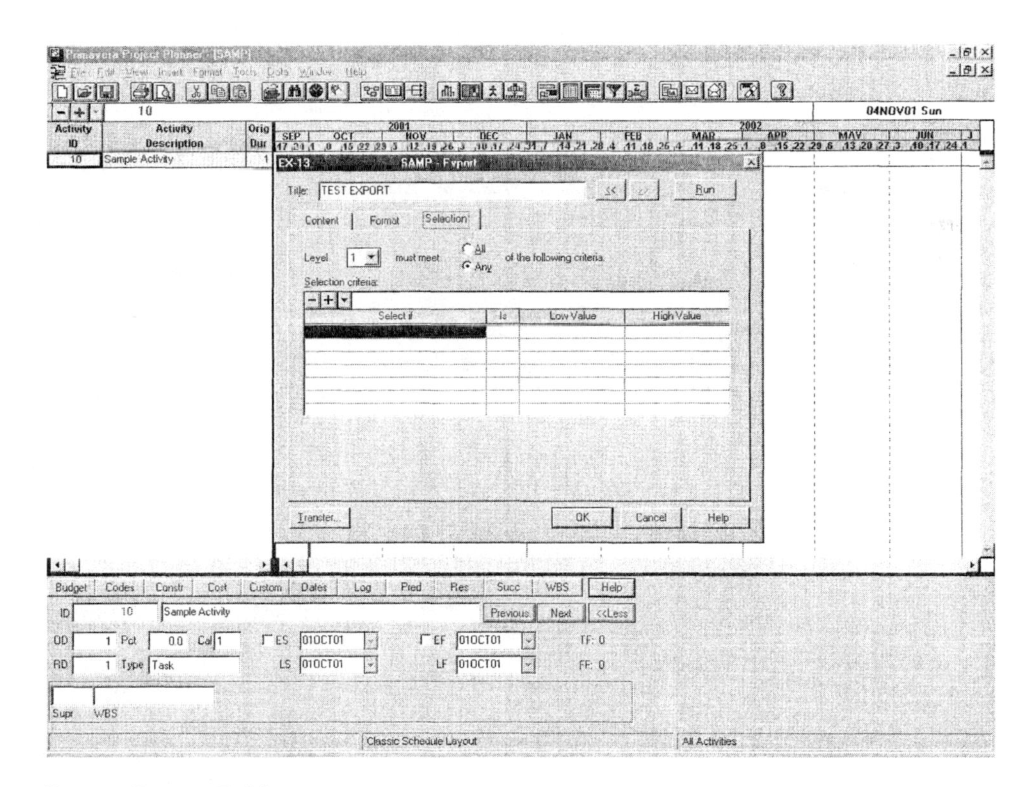

Screen Image 7-14
Export Selection Window

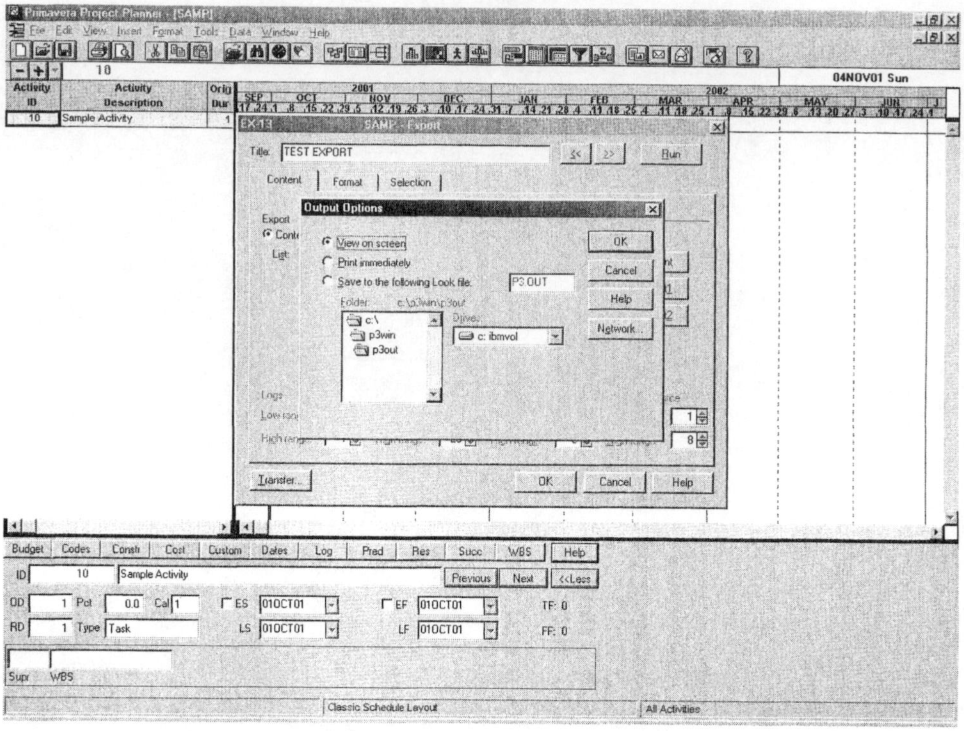

Screen Image 7-15
Export Output Options

```
---------- EXPORT DATA CRITERIA
 ACT    10 DES    48 OD      5


LOTUS file generated into P3.WK1
```

Output 7-1
Export Report

needs to be saved. However, Primavera Project Planner is extremely particular about the file format that can be imported. The most universally available file format is the Lotus WKS format. This is the original Lotus file format and is compatible with Primavera Project Planner. The user should note the exact location where this file is being saved and its name. In Screen Image 7-18 the file name is ALIST.WKS. After the file has been saved the spreadsheet program can be exited and the user should return to Primavera Project Planner.

In Screen Image 7-19 the importing operation is begun by selecting the *Tools* and *Project Utilities* command followed by the *Import* option. These commands will open the *Import* window as shown in Screen Image 7-20. This window, just as with the *Export* window, lists instructions that have previously been used. Select the *Add* option, which will open the *Add a New Report* window as shown in Screen Image 7-21. Just as with the *Export* options, this is simply a sequential counter and should be accepted.

By accepting the report ID, the *Import* window will be opened as shown in Screen Image 7-22. In addition to a name, the only other required input is the path and name of the file that is to be imported. This must be entered perfectly in order for this operation to work. After clicking the *Run* button, the *Output Options* window as shown in Screen Image 7-23 will appear. This window allows the user to specify where the output

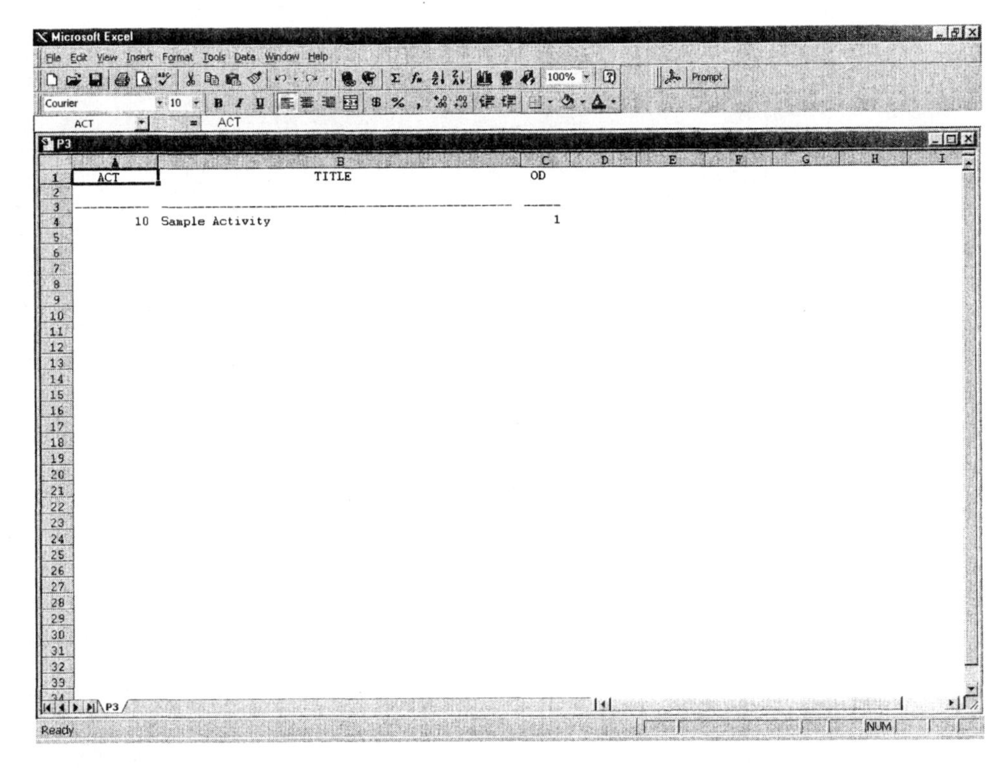

ACT	TITLE	OD
10	Sample Activity	1

Screen Image 7-16
Export File Opened in Excel

ACT	TITLE	OD
10	Clear Site	2
20	Site Grading & Compaction	5
30	Fab. & Tie Steel Spread Ftg's	6
40	Excav. Spread Footings	4
50	Form Spread Footings	4
60	Reinforce Spread Footings	4
70	Place Spread Footings	4
80	Strip Forms - Spread Footings	4
90	Fab. & Tie Steel Piers	2
100	Fab. & Tie Steel Contin. Footings	2
110	Fab. & Tie Steel Foundation Walls	2
120	Reinforce Piers	4
130	Form Piers	4
140	Place Piers	4
150	Strip Forms - Piers	4
160	Backfill Spread Footings	4
170	North Side Excav. Contin Ftg's	1
180	North Side Form Contin. Ftg's	1
190	North Side Reinforce Contin. Ftg's	1
200	North Side Place Contin. Ftg's	1
210	North Side Strip Contin. Ftg's Forms	1
220	North Side Reinforce Fdn. Walls	1
230	North Side Form Fdn. Walls	3
240	North Side Place Fdn. Walls	1
250	North Side Strip Fdn. Wall Forms	1
260	North Side Backfill & Compact	1
270	East Side Excav. Contin Ftg's	1
280	East Side Form Contin. Ftg's	1
290	East Side Reinforce Contin. Ftg's	1
300	East Side Place Contin. Ftg's	1
310	East Side Strip Contin. Ftg's Forms	1
320	East Side Reinforce Fdn. Walls	1

Screen Image 7-17
File with Data to be Imported

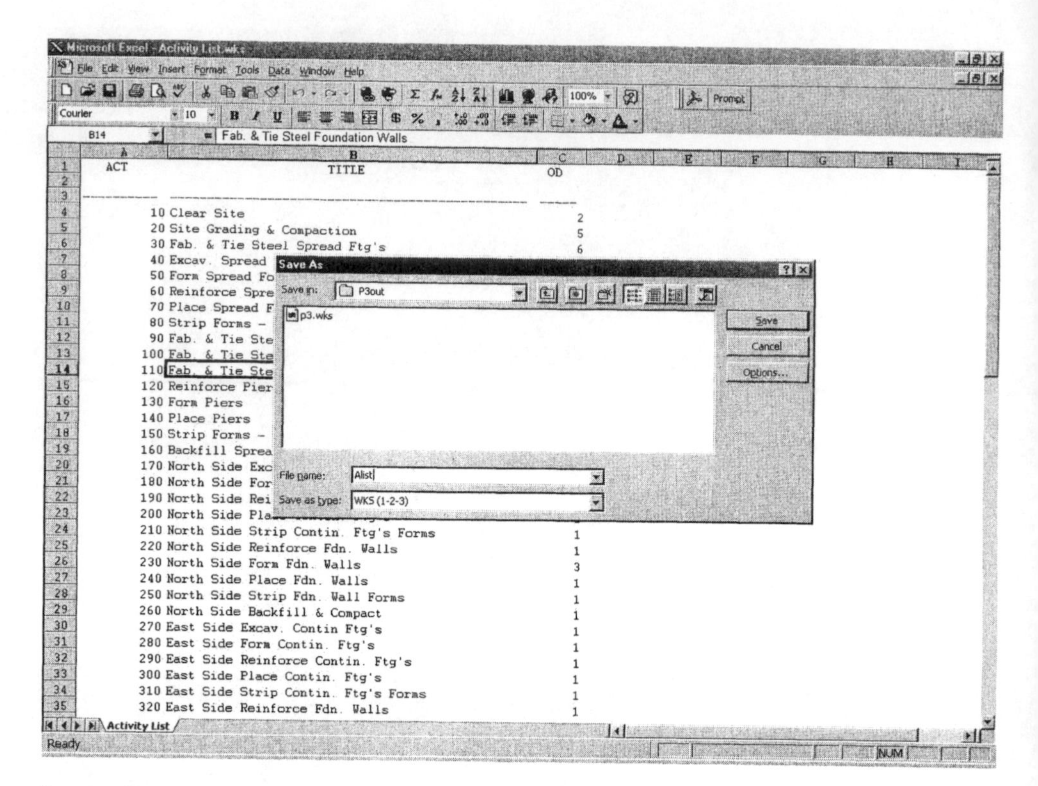

Screen Image 7-18
Saving File to Import

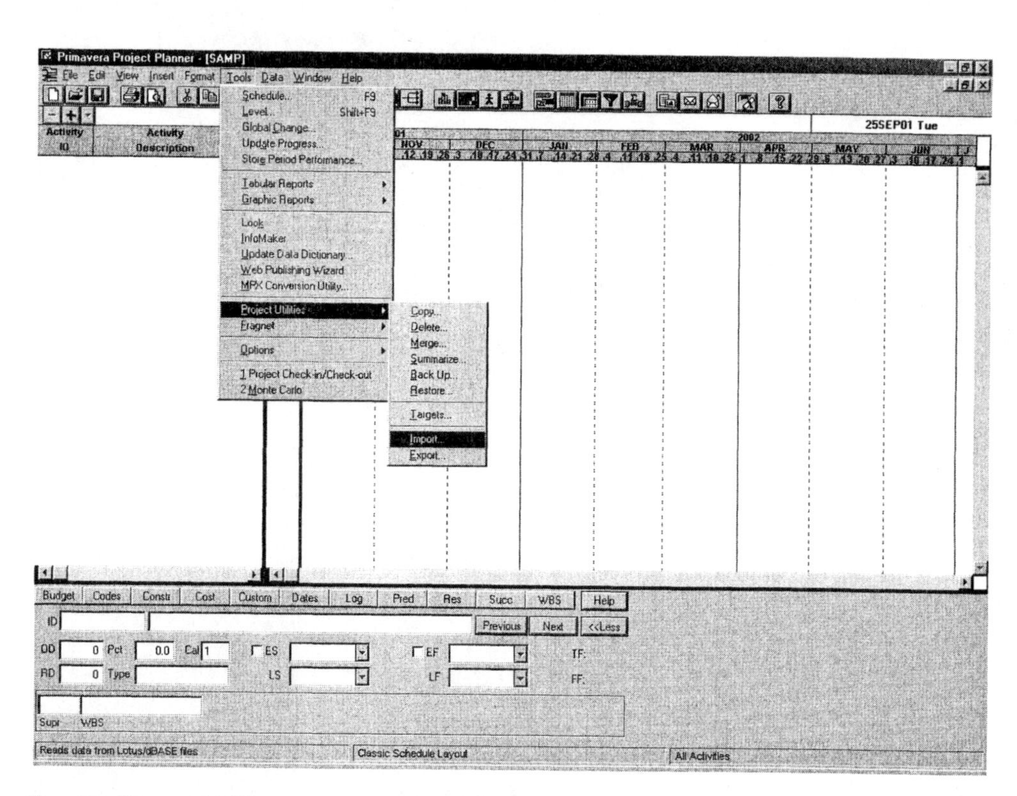

Screen Image 7-19
Opening the Import Window

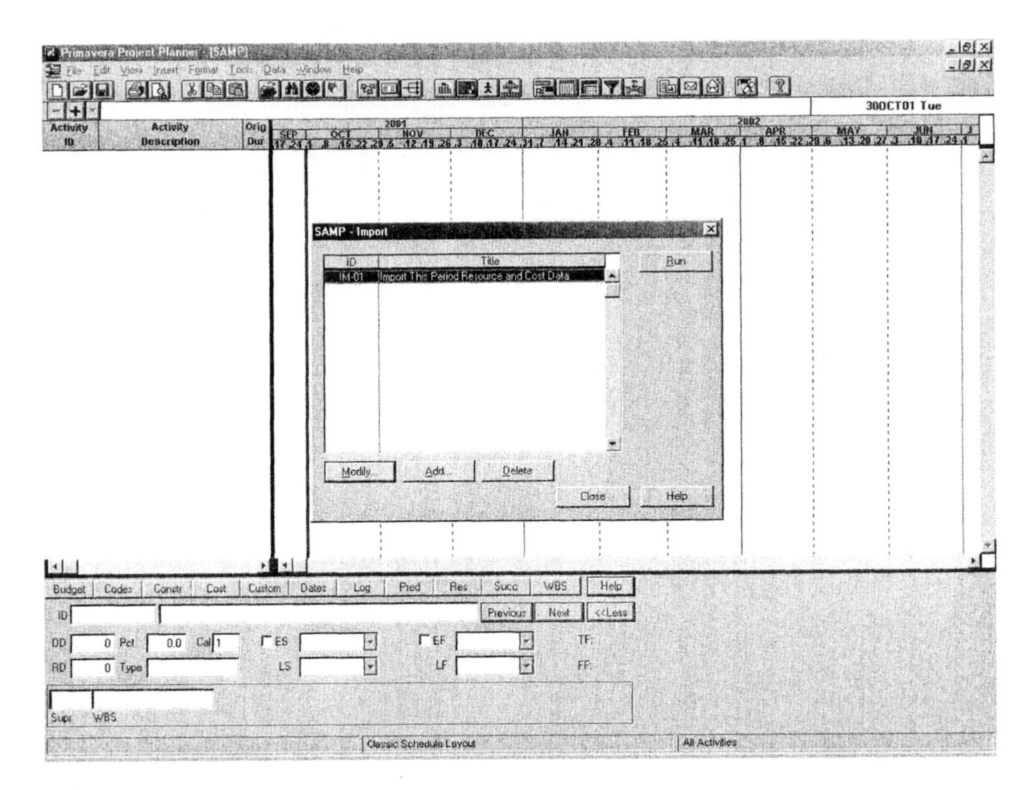

Screen Image 7-20
Previous Import Instructions

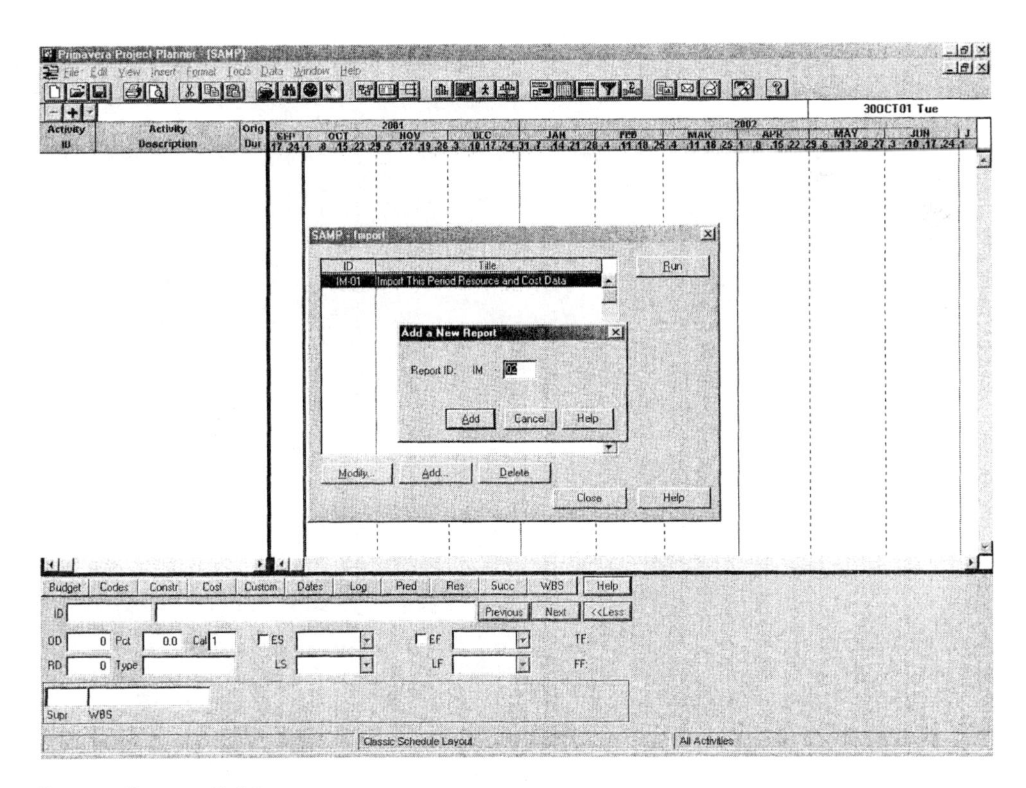

Screen Image 7-21
Add a New Report Window

Screen Image 7-22
Import Content Window

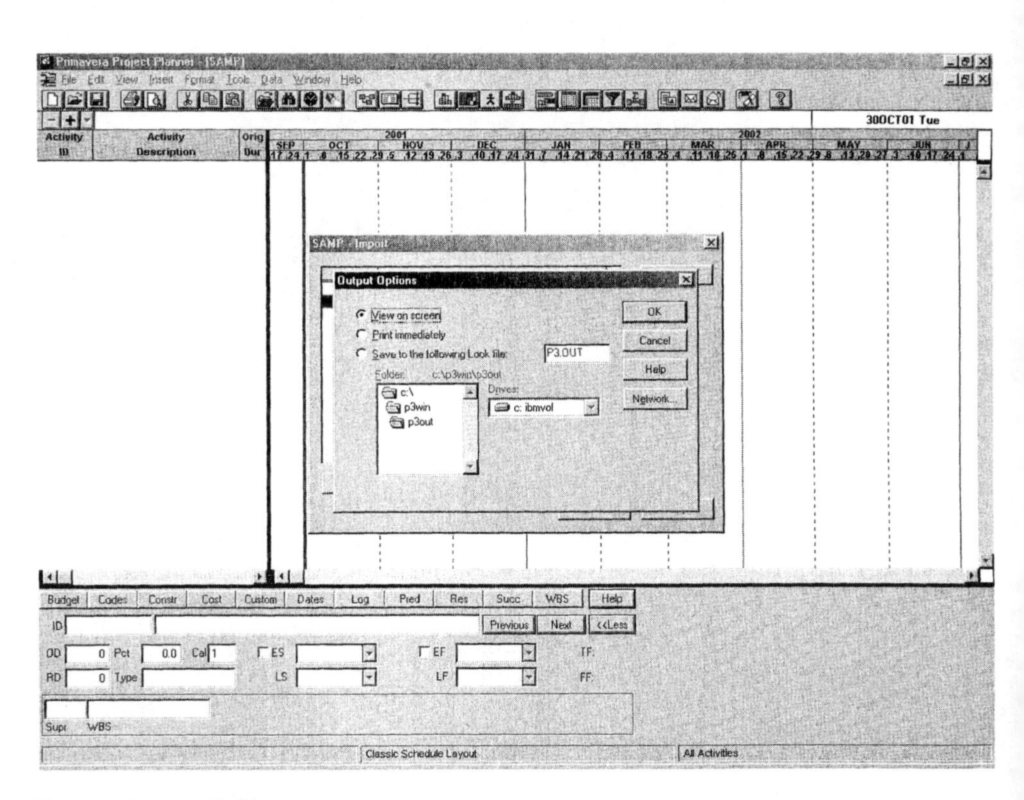

Screen Image 7-23
Output Options Window

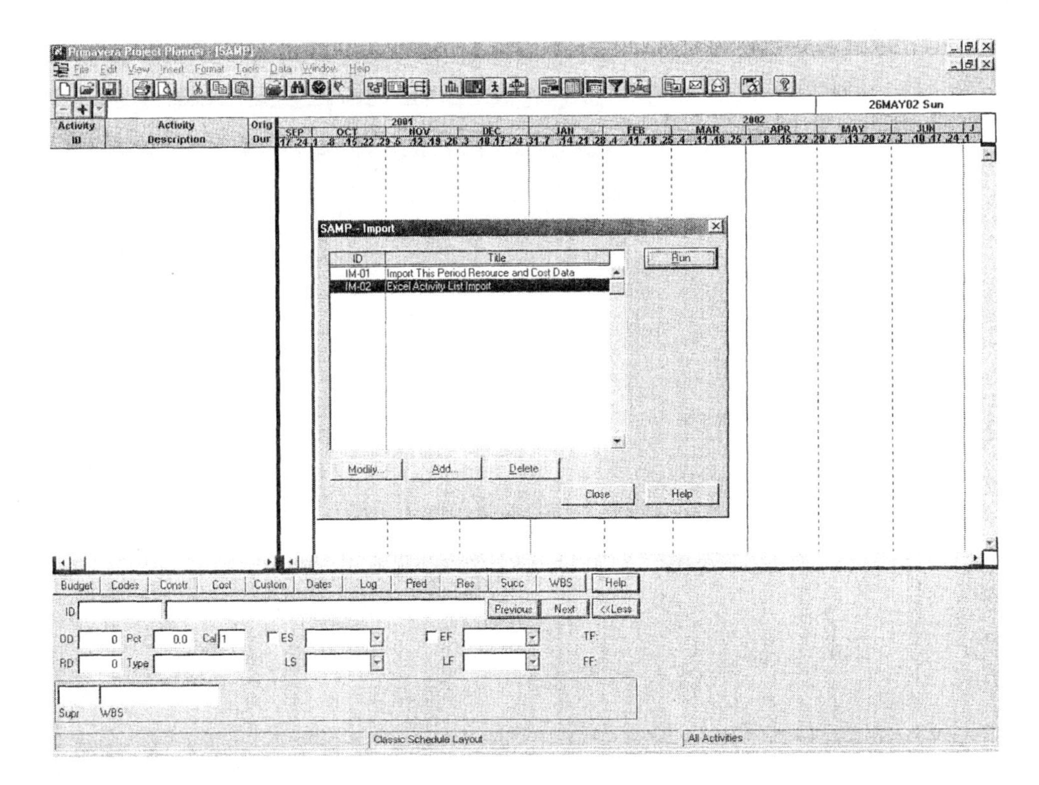

Screen Image 7-24
Import Window

report is to be routed. This report makes a trial run on the import and generates a report detailing what information will be entered in what fields. If this report shows that the instructions and data are compatible, click the *OK* button. This will return the program to the initial import window as shown in Screen Image 7-24. To complete the importing process, click the *Close* button, which will open the *Rerun current filter* option window as shown in Screen Image 7-25.

If the *Yes* option is selected, the data from the spreadsheet program will be imported. In Screen Image 7-26 data from the activity list that was in Excel is now imported into Primavera Project Planner. These activities now need to have their activity codes added by typing the appropriate information on the activity form for each of the activities. When this has been completed it is impossible to discern which activities were imported and which were manually entered.

The importing process can be the mechanism that links data from other systems to Primavera Project Planner. For example, if the accounting system can be downloaded into a spreadsheet program, then that information can ultimately be imported with a minimal amount of effort.

Relationship Input

The relationship arrows can be added through keyboard input or through a combination of mouse and keyboard operations. The keyboard only method works by describing the successor to each activity. To begin this operation, click the *Succ* button at the top of the activity form window which will open the *Successors* window. Screen Image 7-27 is an example of the main input screen with the *Successors* window opened. In that screen image

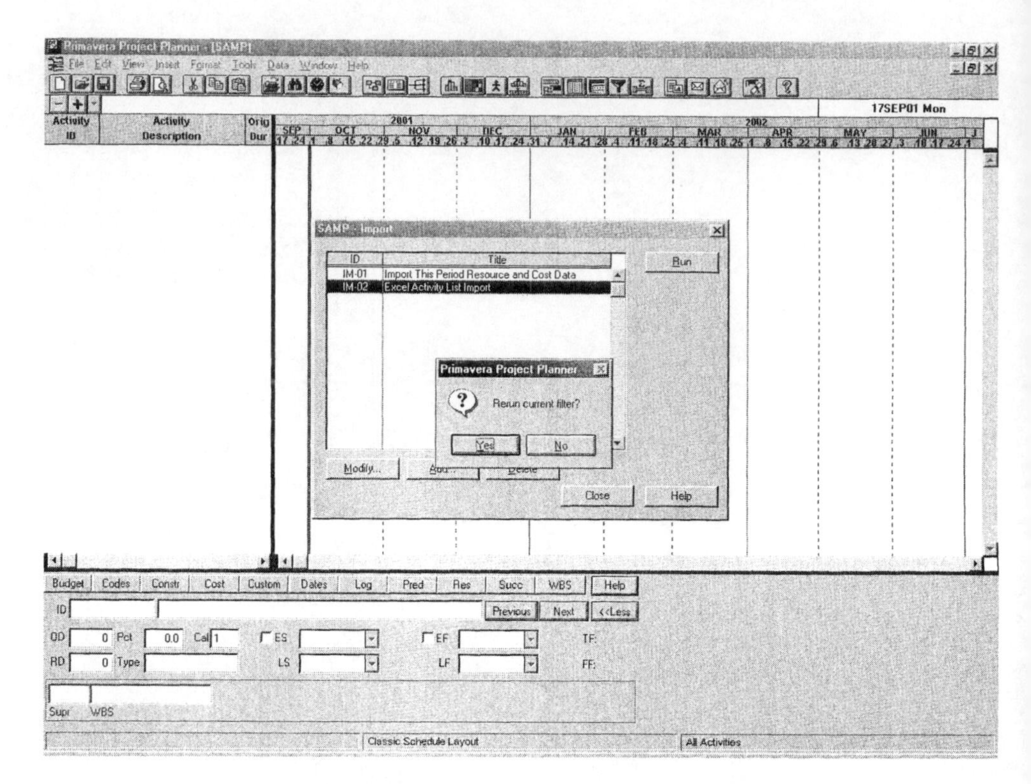

Screen Image 7-25
Rerun Current Filter

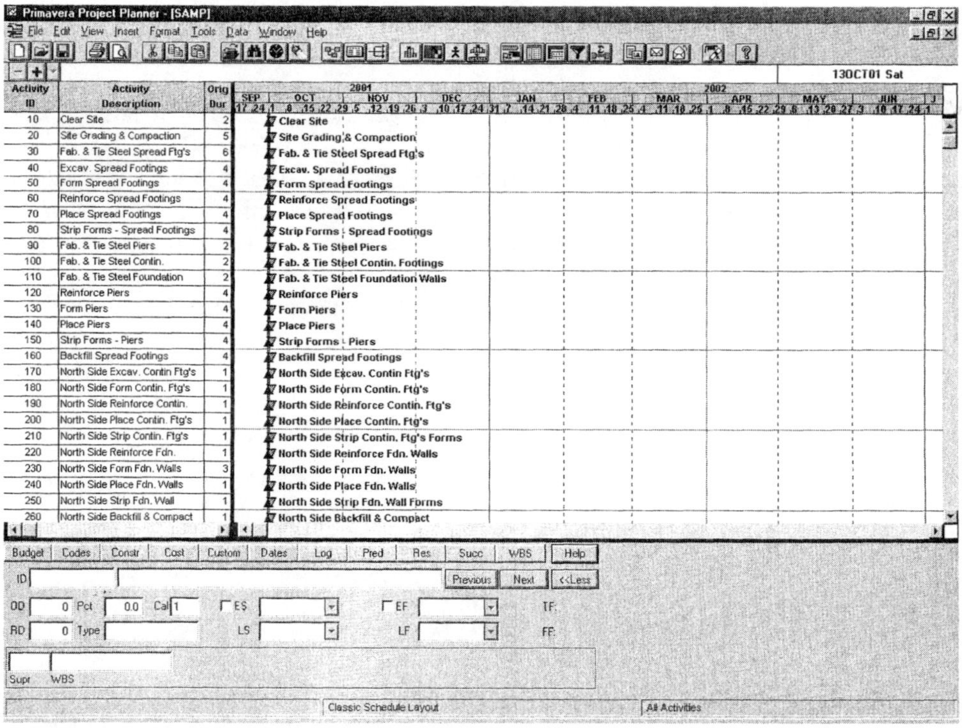

Primavera Project Planner - [SAMP]

File Edit View Insert Format Tools Data Window Help

13OCT01 Sat

Activity ID	Activity Description	Orig Dur	2001–2002 (SEP OCT NOV DEC JAN FEB MAR APR MAY JUN)
10	Clear Site	2	Clear Site
20	Site Grading & Compaction	5	Site Grading & Compaction
30	Fab. & Tie Steel Spread Ftg's	6	Fab. & Tie Steel Spread Ftg's
40	Excav. Spread Footings	4	Excav. Spread Footings
50	Form Spread Footings	4	Form Spread Footings
60	Reinforce Spread Footings	4	Reinforce Spread Footings
70	Place Spread Footings	4	Place Spread Footings
80	Strip Forms - Spread Footings	4	Strip Forms - Spread Footings
90	Fab. & Tie Steel Piers	2	Fab. & Tie Steel Piers
100	Fab. & Tie Steel Contin.	2	Fab. & Tie Steel Contin. Footings
110	Fab. & Tie Steel Foundation	2	Fab. & Tie Steel Foundation Walls
120	Reinforce Piers	4	Reinforce Piers
130	Form Piers	4	Form Piers
140	Place Piers	4	Place Piers
150	Strip Forms - Piers	4	Strip Forms - Piers
160	Backfill Spread Footings	4	Backfill Spread Footings
170	North Side Excav. Contin Ftg's	1	North Side Excav. Contin Ftg's
180	North Side Form Contin. Ftg's	1	North Side Form Contin. Ftg's
190	North Side Reinforce Contin.	1	North Side Reinforce Contin. Ftg's
200	North Side Place Contin. Ftg's	1	North Side Place Contin. Ftg's
210	North Side Strip Contin. Ftg's	1	North Side Strip Contin. Ftg's Forms
220	North Side Reinforce Fdn.	1	North Side Reinforce Fdn. Walls
230	North Side Form Fdn. Walls	3	North Side Form Fdn. Walls
240	North Side Place Fdn. Walls	1	North Side Place Fdn. Walls
250	North Side Strip Fdn. Wall	1	North Side Strip Fdn. Wall Forms
260	North Side Backfill & Compact	1	North Side Backfill & Compact

Budget | Codes | Constr | Cost | Custom | Dates | Log | Pred | Res | Succ | WBS | Help

ID | Previous | Next | <<Less

OD 0 Pct 0.0 Cal 1 | ES | EF | TF:

RD 0 Type | LS | LF | FF:

Supr WBS

Classic Schedule Layout | All Activities

Screen Image 7-26
Data Imported into Primavera Project Planner

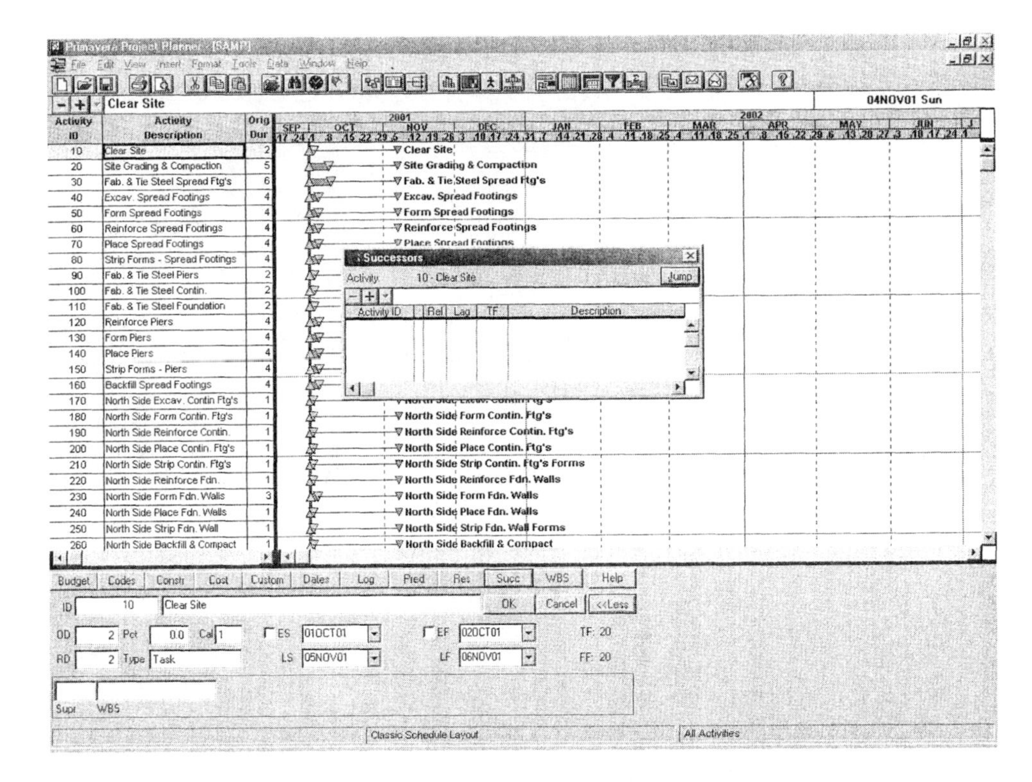

Screen Image 7-27
Successors Activity Window

Activity 10 is highlighted, meaning that any activities entered in the *Successors* window would be for the immediate successors to Activity 10. Using the schedule introduced in Chapter 3 as an example, Activity 20 is the only successor to Activity 10, and they have a finish-to-start relationship. To enter this information in the *Successors* window, click on the plus button in that window and enter 20 as the *Activity ID*, enter *FS* as the relationship type *(Rel)*, and enter a 0 for the lag. By clicking on any of the fields in the successor window, the pull-down menu next to the plus button can be used to identify and select the entry for that field. Screen Image 7-28 shows the *Successors* window using the pull-down menu for selecting a successor *Activity ID*. Simply clicking on the desired activity will add that activity as a successor. The same process can be followed to specify the relationship and lag. Then by clicking on the *OK* button on the activity form or on another activity on the main input screen or by pressing the *Jump* button this information will be written to the schedule. The *Jump* button will move the pointer to the activity ID that is highlighted in the *Successors* window. As the successors are added to the schedule they should automatically be drawn on the main input screen. If this is not happening, then most likely that option has been turned off. The relationship lines are turned on and off by clicking on the relationship icon.

Screen Image 7-29 is an example of the *Successors* window after the relationship between Activity 40 and Activity 50 has been entered. This pair of activities has both a start-to-start and a finish-to-finish relationship. When activities have multiple relationships, these relationships are shown with multiple entries in the *Successors* window.

The relationship arrows between activities can also be added by using the mouse to draw them on the main input screen. However, this method requires some modifications to the main input screen. Since this method requires using the mouse to draw the relationship arrows, the current configuration, with small fonts and closely spaced activity bars, makes this operation difficult. To remedy this situation, the spacing between the activity bars needs to be increased. This is done by clicking on the *Format* command and selecting

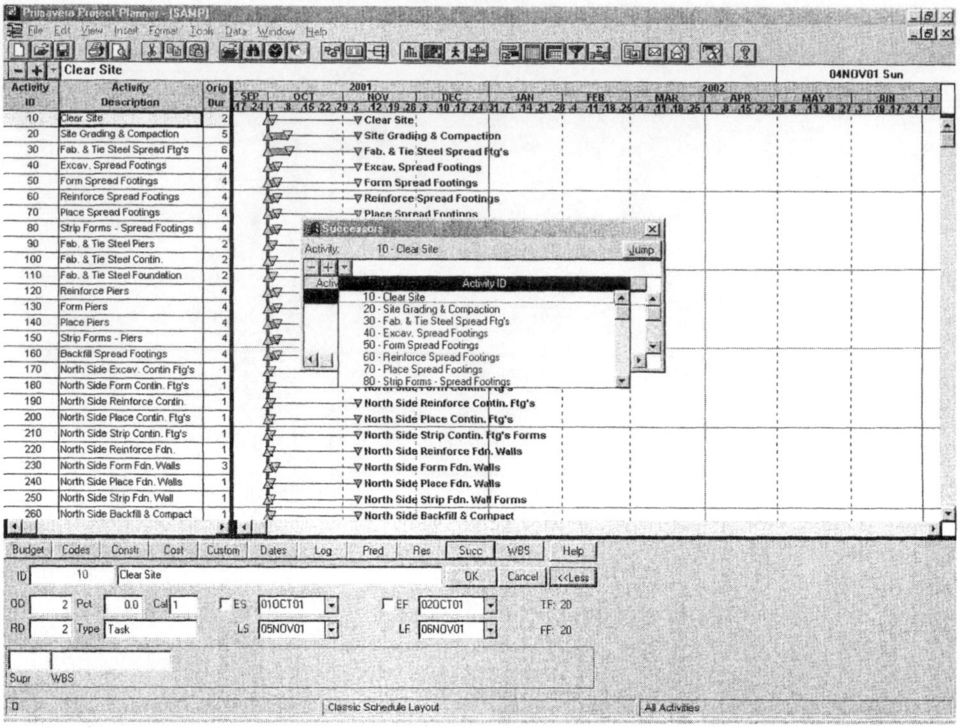

Primavera Project Planner - [SAMP]

File Edit View Insert Format Tools Data Window Help

Clear Site — 04NOV01 Sun

Activity ID	Activity Description	Orig Dur
10	Clear Site	2
20	Site Grading & Compaction	5
30	Fab. & Tie Steel Spread Ftg's	6
40	Excav. Spread Footings	4
50	Form Spread Footings	4
60	Reinforce Spread Footings	4
70	Place Spread Footings	4
80	Strip Forms - Spread Footings	4
90	Fab. & Tie Steel Piers	2
100	Fab. & Tie Steel Contin.	2
110	Fab. & Tie Steel Foundation	2
120	Reinforce Piers	4
130	Form Piers	4
140	Place Piers	4
150	Strip Forms - Piers	4
160	Backfill Spread Footings	4
170	North Side Excav. Contin Ftg's	1
180	North Side Form Contin. Ftg's	1
190	North Side Reinforce Contin.	1
200	North Side Place Contin. Ftg's	1
210	North Side Strip Contin. Ftg's	1
220	North Side Reinforce Fdn.	1
230	North Side Form Fdn. Walls	3
240	North Side Place Fdn. Walls	1
250	North Side Strip Fdn. Wall	1
260	North Side Backfill & Compact	1

2001: SEP OCT NOV DEC — 2002: JAN FEB MAR APR MAY JUN

Successors

Activity: 10 - Clear Site — Jump

Activity ID
10 - Clear Site
20 - Site Grading & Compaction
30 - Fab. & Tie Steel Spread Ftg's
40 - Excav. Spread Footings
50 - Form Spread Footings
60 - Reinforce Spread Footings
70 - Place Spread Footings
80 - Strip Forms - Spread Footings

Budget | Codes | Constr | Cost | Custom | Dates | Log | Pred | Res | Succ | WBS | Help

ID 10 Clear Site — OK — Cancel — <<Less

OD 2 Pct 0.0 Cal 1 — ES 01OCT01 — EF 02OCT01 — TF: 20

RD 2 Type Task — LS 05NOV01 — LF 06NOV01 — FF: 20

Supr WBS

0 — Classic Schedule Layout — All Activities

Screen Image 7-28
Successors Window with Activity ID List

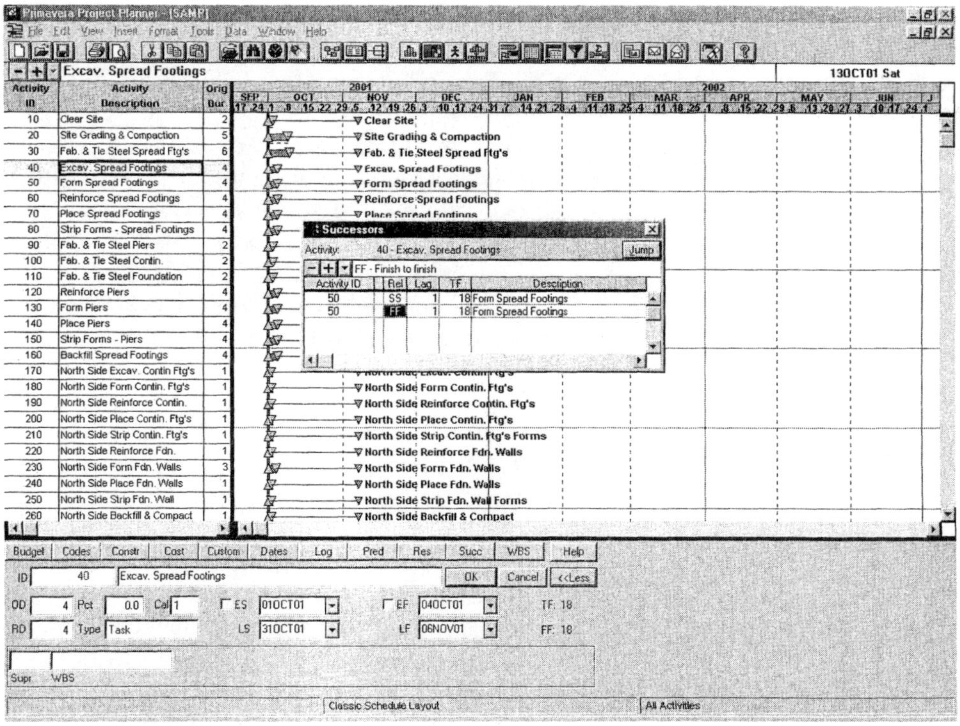

Primavera Project Planner - [SAMP]

File Edit View Insert Format Tools Data Window Help

Excav. Spread Footings — 13OCT01 Sat

(Activity list as in Screen Image 7-28, activities 10 through 260.)

Successors

Activity: 40 - Excav. Spread Footings — Jump

FF - Finish to finish

Activity ID	Rel	Lag	TF	Description
50	SS	1	18	Form Spread Footings
50	FF	1	18	Form Spread Footings

Budget | Codes | Constr | Cost | Custom | Dates | Log | Pred | Res | Succ | WBS | Help

ID 40 Excav. Spread Footings — OK — Cancel — <<Less

OD 4 Pct 0.0 Cal 1 — ES 01OCT01 — EF 04OCT01 — TF: 18

RD 4 Type Task — LS 31OCT01 — LF 06NOV01 — FF: 18

Supr WBS

Classic Schedule Layout — All Activities

Screen Image 7-29
Activity Pair 10–20 Entered

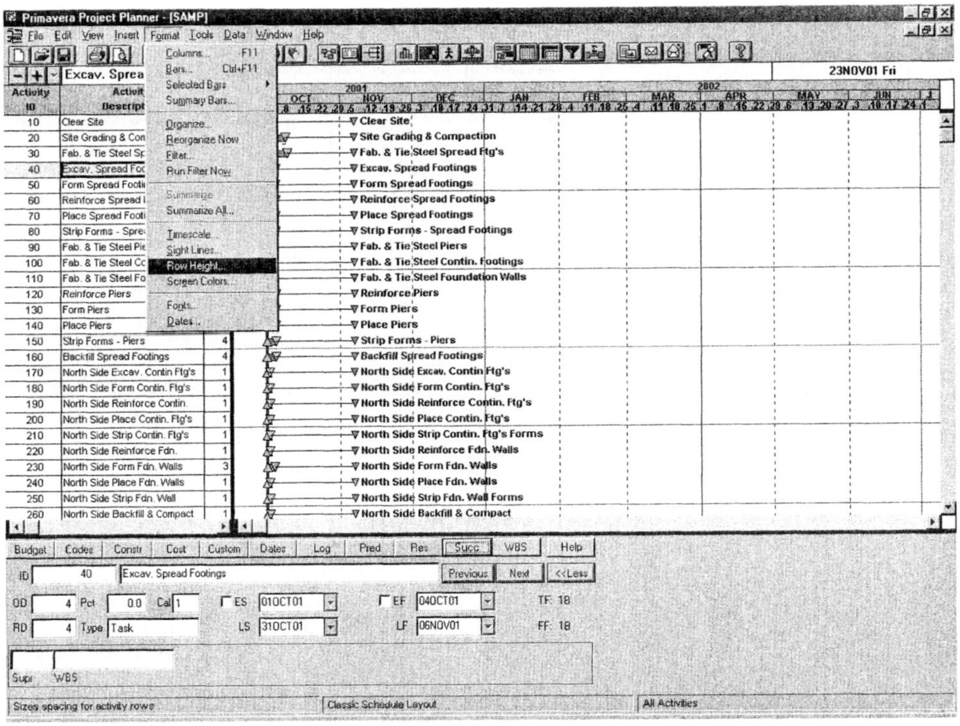

Screen Image 7-30
Opening the Row Height Window

the *Row Height* option as shown in Screen Image 7-30. This will activate the *Row Height* window, as seen in Screen Image 7-31.

The default setting allows the row height to be automatically sized. To change this setting, remove the check from the *Automatic size* box and enter the desired row height. Twenty-five should be considered a minimum value. The next step is to select the *Apply to all activities* option. If this is not done, then only the activities that are highlighted on the main input screen will be affected. Screen Image 7-32 is an example of the *Row Height* window after the necessary modifications have been entered. When the desired changes have been made, the *OK* button should be clicked. Screen Image 7-33 illustrates the main input screen with the increased row height.

Once the row heights have been enlarged, the mouse can be used to draw the relationship arrows. When the pointer is placed near the beginning or end of the activity, it will change shape to that of the symbol on the relationship icon. If this does not happen, the relationship lines have been turned off and must be turned on. When the pointer changes shape, depress the left mouse button and hold it down while moving the mouse to that activity's successor. If a start-to-start relationship is to be drawn, the pointer should be placed on the left end of the activity bar and drug down to the left end of the successor activity bar. When activities are of short duration, the user needs to be very careful to get the relationship arrow appropriately placed.

Once the relationship lines have been drawn, the lags need to be added. To add the lags, point the mouse at the relationship arrow and click the left button. This will activate the *Edit Relationship* window. From this window, the relationship type and lag can be modified. This process can be used to modify the relationships regardless of whether they were input from the keyboard or drawn. Screen Image 7-34 shows how the relationship between Activity 40 and Activity 50 could be edited. After the needed modifications to this window have been made, the user can click on the next relationship line or click the OK button to close this window.

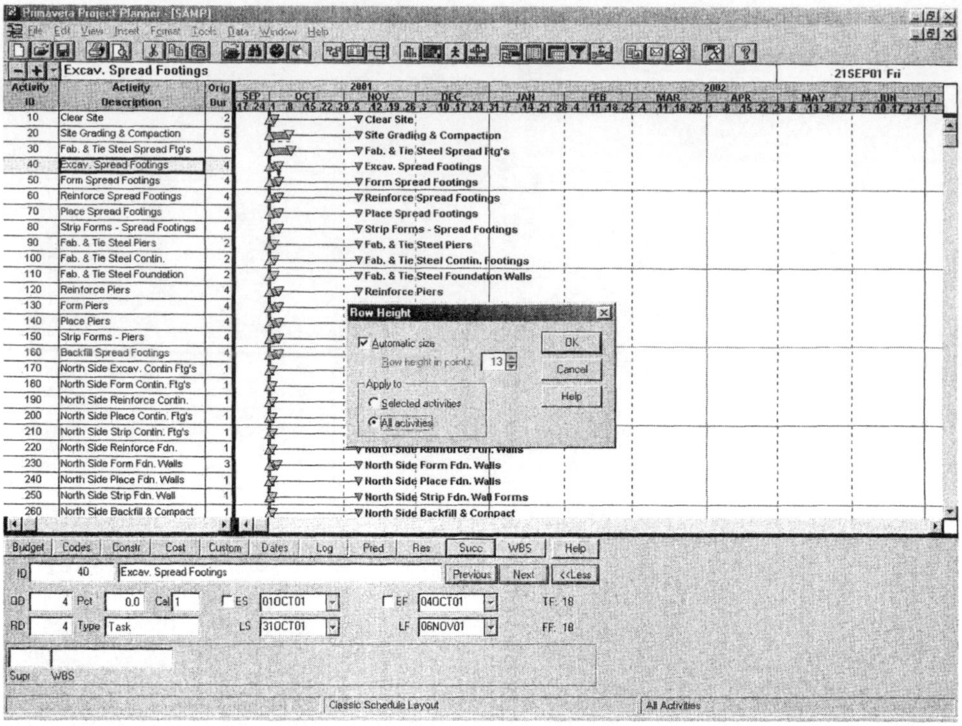

Primavera Project Planner - [SAMP]
File Edit View Insert Format Tools Data Window Help
Excav. Spread Footings 21SEP01 Fri

Activity ID	Activity Description	Orig Dur
10	Clear Site	2
20	Site Grading & Compaction	5
30	Fab. & Tie Steel Spread Ftg's	6
40	Excav. Spread Footings	4
50	Form Spread Footings	4
60	Reinforce Spread Footings	4
70	Place Spread Footings	4
80	Strip Forms - Spread Footings	4
90	Fab. & Tie Steel Piers	2
100	Fab. & Tie Steel Contin.	2
110	Fab. & Tie Steel Foundation	2
120	Reinforce Piers	4
130	Form Piers	4
140	Place Piers	4
150	Strip Forms - Piers	4
160	Backfill Spread Footings	4
170	North Side Excav. Contin Ftg's	1
180	North Side Form Contin. Ftg's	1
190	North Side Reinforce Contin.	1
200	North Side Place Contin. Ftg's	1
210	North Side Strip Contin. Ftg's	1
220	North Side Reinforce Fdn.	1
230	North Side Form Fdn. Walls	3
240	North Side Place Fdn. Walls	1
250	North Side Strip Fdn. Wall	1
260	North Side Backfill & Compact	1

2001 SEP OCT NOV DEC JAN 2002 FEB MAR APR MAY JUN

Clear Site
Site Grading & Compaction
Fab. & Tie Steel Spread Ftg's
Excav. Spread Footings
Form Spread Footings
Reinforce Spread Footings
Place Spread Footings
Strip Forms - Spread Footings
Fab. & Tie Steel Piers
Fab. & Tie Steel Contin. Footings
Fab. & Tie Steel Foundation Walls
Reinforce Piers
North Side Reinforce Fdn. Walls
North Side Form Fdn. Walls
North Side Place Fdn. Walls
North Side Strip Fdn. Wall Forms
North Side Backfill & Compact

Row Height
Automatic size
Row height in points: 13
Apply to
Selected activities
All activities
OK
Cancel
Help

Budget Codes Constr Cost Custom Dates Log Pred Res Succ WBS Help
ID 40 Excav. Spread Footings Previous Next <<Less
OD 4 Pct 0.0 Cal 1 ES 01OCT01 EF 04OCT01 TF: 18
RD 4 Type Task LS 31OCT01 LF 06NOV01 FF: 18
Supr WBS
Classic Schedule Layout All Activities

Screen Image 7-31
The Row Height Window

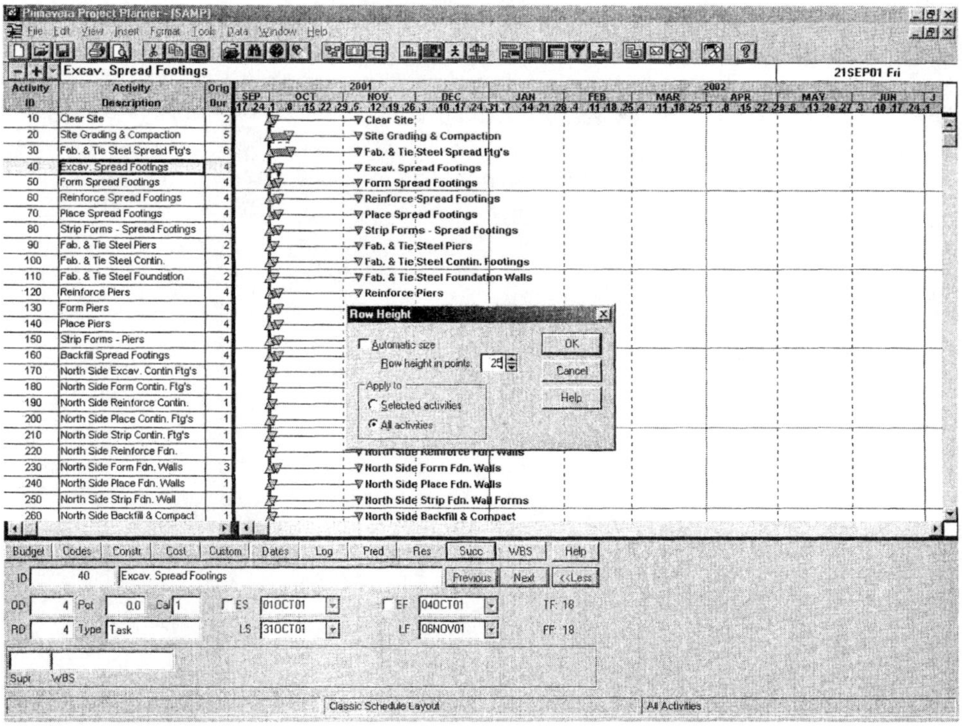

Primavera Project Planner - [SAMP]
File Edit View Insert Format Tools Data Window Help
Excav. Spread Footings 21SEP01 Fri

Activity ID	Activity Description	Orig Dur
10	Clear Site	2
20	Site Grading & Compaction	5
30	Fab. & Tie Steel Spread Ftg's	6
40	Excav. Spread Footings	4
50	Form Spread Footings	4
60	Reinforce Spread Footings	4
70	Place Spread Footings	4
80	Strip Forms - Spread Footings	4
90	Fab. & Tie Steel Piers	2
100	Fab. & Tie Steel Contin.	2
110	Fab. & Tie Steel Foundation	2
120	Reinforce Piers	4
130	Form Piers	4
140	Place Piers	4
150	Strip Forms - Piers	4
160	Backfill Spread Footings	4
170	North Side Excav. Contin Ftg's	1
180	North Side Form Contin. Ftg's	1
190	North Side Reinforce Contin.	1
200	North Side Place Contin. Ftg's	1
210	North Side Strip Contin. Ftg's	1
220	North Side Reinforce Fdn.	1
230	North Side Form Fdn. Walls	3
240	North Side Place Fdn. Walls	1
250	North Side Strip Fdn. Wall	1
260	North Side Backfill & Compact	1

2001 SEP OCT NOV DEC JAN 2002 FEB MAR APR MAY JUN

Clear Site
Site Grading & Compaction
Fab. & Tie Steel Spread Ftg's
Excav. Spread Footings
Form Spread Footings
Reinforce Spread Footings
Place Spread Footings
Strip Forms - Spread Footings
Fab. & Tie Steel Piers
Fab. & Tie Steel Contin. Footings
Fab. & Tie Steel Foundation Walls
Reinforce Piers
North Side Reinforce Fdn. Walls
North Side Form Fdn. Walls
North Side Place Fdn. Walls
North Side Strip Fdn. Wall Forms
North Side Backfill & Compact

Row Height
Automatic size
Row height in points: 24
Apply to
Selected activities
All activities
OK
Cancel
Help

Budget Codes Constr Cost Custom Dates Log Pred Res Succ WBS Help
ID 40 Excav. Spread Footings Previous Next <<Less
OD 4 Pct 0.0 Cal 1 ES 01OCT01 EF 04OCT01 TF: 18
RD 4 Type Task LS 31OCT01 LF 06NOV01 FF: 18
Supr WBS
Classic Schedule Layout All Activities

Screen Image 7-32
Default Row Height Settings

Primavera Project Planner - [SAMP]

File Edit View Insert Format Tools Data Window Help

Excav. Spread Footings 27JAN02 Sun

Activity ID	Activity Description	Orig Dur
10	Clear Site	2
20	Site Grading & Compaction	5
30	Fab. & Tie Steel Spread Ftg's	6
40	Excav. Spread Footings	4
50	Form Spread Footings	4
60	Reinforce Spread Footings	4
70	Place Spread Footings	4
80	Strip Forms - Spread Footings	4
90	Fab. & Tie Steel Piers	2
100	Fab. & Tie Steel Contin. Footings	2
110	Fab. & Tie Steel Foundation Walls	2
120	Reinforce Piers	4
130	Form Piers	4

Budget Codes Constr Cost Custom Dates Log Pred Res Succ WBS Help

ID 40 Excav. Spread Footings Previous Next <<Less

OD 4 Pct 0.0 Cal 1 ES 01OCT01 EF 04OCT01 TF 18

RD 4 Type Task LS 31OCT01 LF 06NOV01 FF 18

Supr WBS

Classic Schedule Layout All Activities

Screen Image 7-33
Modified Row Height Window

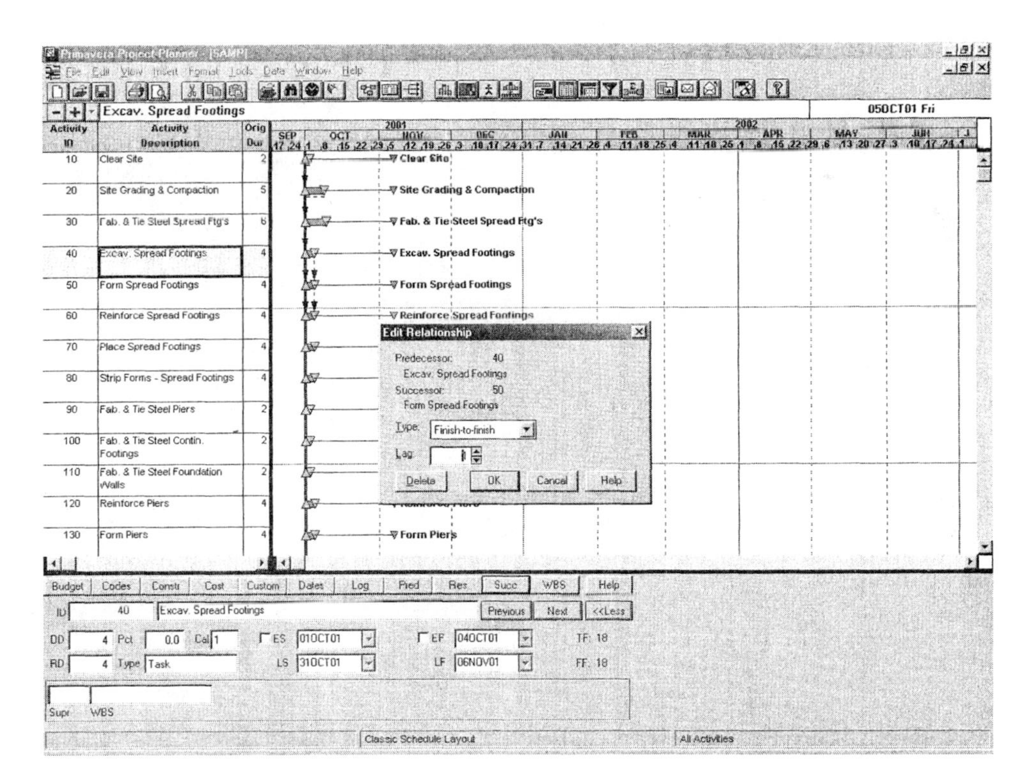

Screen Image 7-34
Edit Relationship Window

PERT Activity Input

In addition to inputting activities from the main input screen, Primavera Project Planner allows the user to physically draw the logic diagram. Once the diagram has been drawn Primavera Project Planner will automatically convert that information into the bar chart data as previously discussed. Primavera Project Planner will also convert the bar chart data to a logic diagram; however, the activity placement is often end to end and very difficult to read. Even with this poor activity placement, the user can reposition the activities by using the mouse to pick up and actually move an activity to a new physical location on the page.

To change from the bar chart mode to the drawing mode select the *View* command and the *PERT* option as shown in Screen Image 7-35 or click on the view PERT icon. This will open a window similar to the one shown in Screen Image 7-36.

At this point the user basically has a blank page of paper on which a construction logic diagram can be drawn. To add activities to this diagram, place the mouse pointer at the location for the new activity and double click. This will add a blank activity box similar to the one shown in Screen Image 7-37. Just as with the bar chart entry, the activity information is added via the *Activity Form*. Once the specifics of an activity have been added, the *OK* button can be clicked on the activity form to save that activity information. Screen Image 7-38 shows how activities 10 and 20 would have been added to the schedule. Just as with the bar chart schedule, the relationships need to be added. This is done by moving the mouse to the left side of the activity box, if the start of the activity requires a relationship, or the right side of the box if a finish is required. As the mouse pointer moves near the end of the activity boxes it changes shape from an arrow to the relationship tool. When this happens click and hold the mouse and drag the relationship arrow to the desired location and release

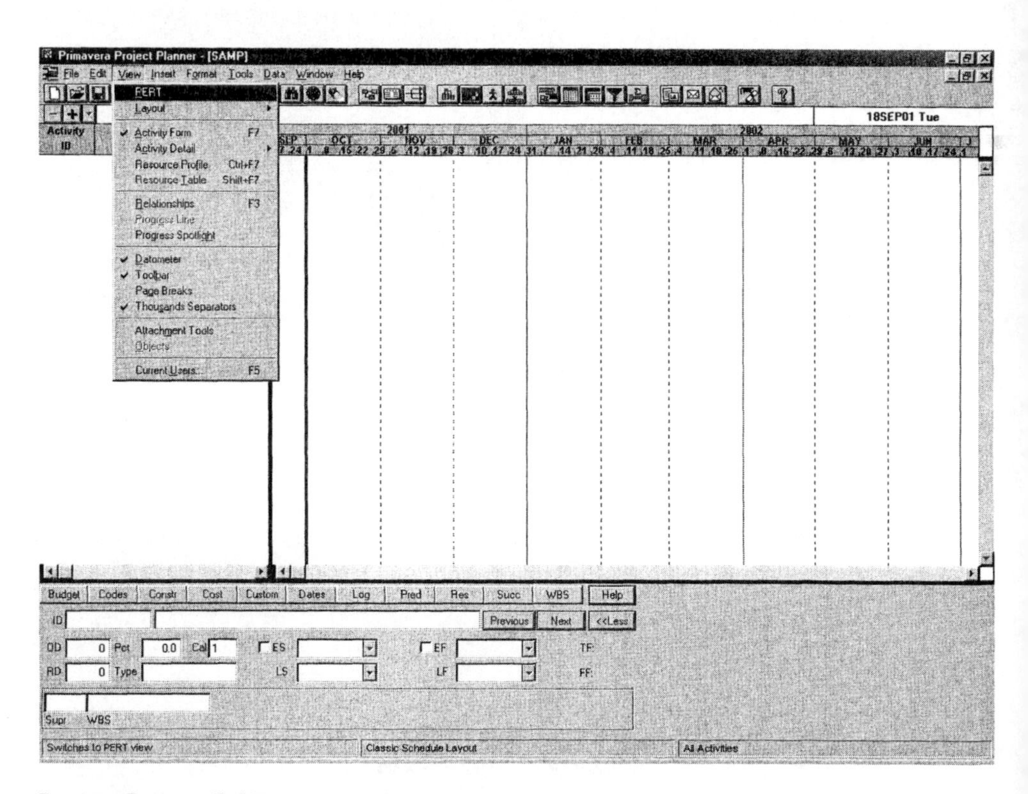

Screen Image 7-35
PERT Option

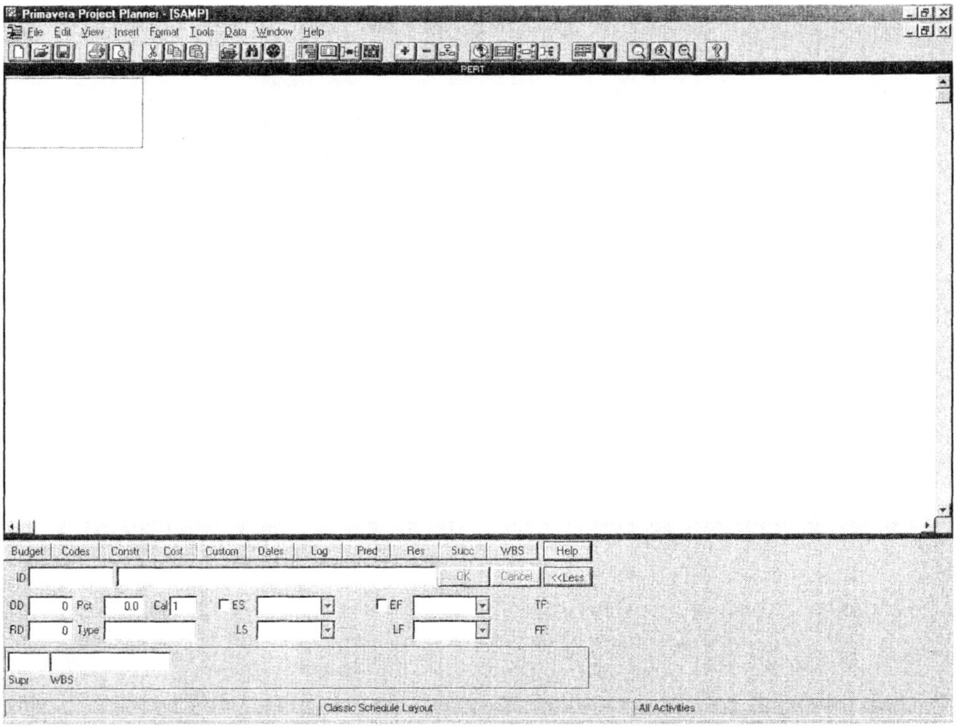

Primavera Project Planner - [SAMP]
File Edit View Insert Format Tools Data Window Help
PERT
Budget Codes Constr Cost Custom Dates Log Pred Res Succ WBS Help
ID OK Cancel <<Less
OD 0 Pct 0.0 Cal 1 ES EF TF:
RD 0 Type LS LF FF:
Supr WBS
Classic Schedule Layout All Activities

Screen Image 7-36
PERT Window

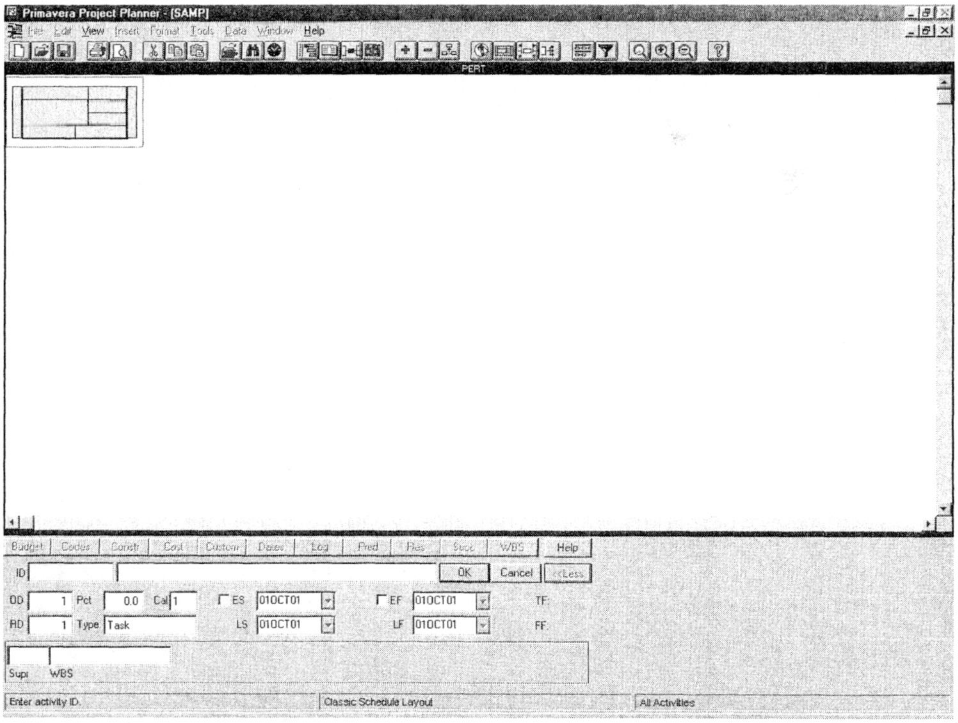

Primavera Project Planner - [SAMP]
File Edit View Insert Format Tools Data Window Help
PERT
Budget Codes Constr Cost Custom Dates Log Pred Res Succ WBS Help
ID OK Cancel <<Less
OD 1 Pct 0.0 Cal 1 ES 01OCT01 EF 01OCT01 TF:
RD 1 Type Task LS 01OCT01 LF 01OCT01 FF:
Supr WBS
Enter activity ID. Classic Schedule Layout All Activities

Screen Image 7-37
Adding an Activity

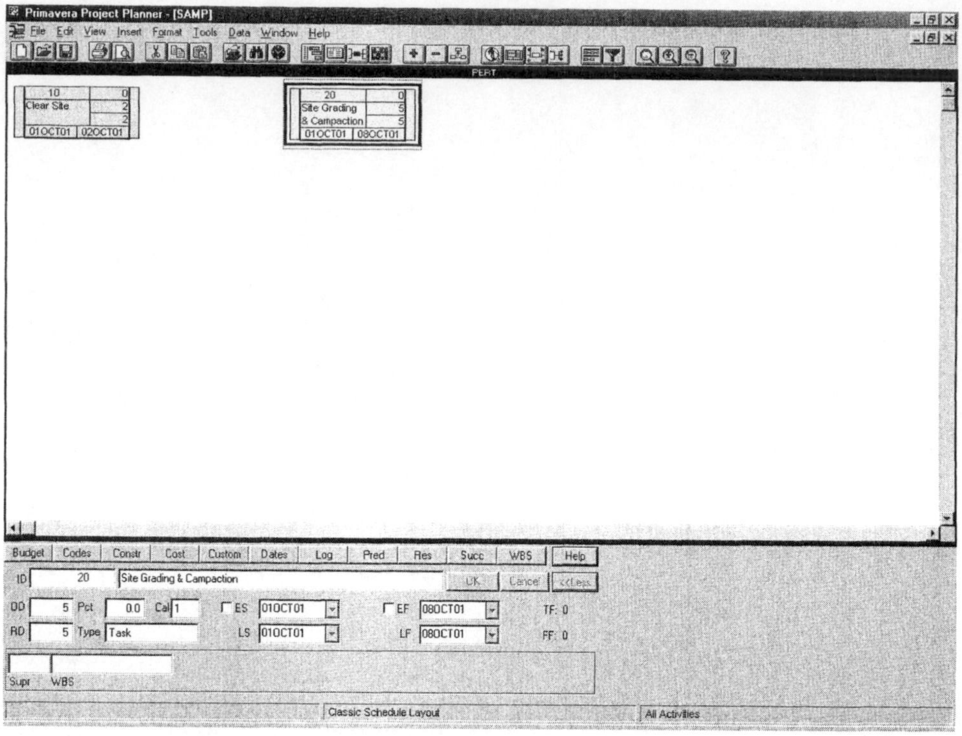

Screen Image 7-38
Adding Activity 20 via PERT Window

the mouse key at the desired activity. This drag and click operation will add the desired relationship. To specify the lag for that relationship simply click on the arrow to activate the *Edit Relationship* window, as shown in Screen Image 7-39. To view the entire network, click on the print preview icon. Screen Image 7-40 is a view of a portion of the schedule from the print preview mode. By pressing the *Close* button the original logic diagram reappears. To return to the bar chart display, click on the bar chart view icon.

Calculating the Schedule

Even after all of the relationships have been input, the bar chart schedule still shows all of the activities beginning on the same date. This is easily corrected by calculating the schedule. There are numerous ways in which Primavera Project Planner can be instructed to calculate the schedule. One is through the use of the F9 key. Another is with the schedule calculate icon. The last is through the *Tools/Schedule* command as shown in Screen Image 7-41. When the schedule is being calculated for the first time, the *Tools/Schedule* command should be used. This method allows the user to modify the rules for calculating the schedule and to generate some needed output. When the schedule is being calculated for the first time, these reports will be helpful in debugging and finding any loops.

The *Schedule* window as shown in Screen Image 7-42 has a number of important features. The *Options* button will activate the *Schedule/Level Calculation Options* window as shown in Screen Image 7-43. While there are many options on this screen, two are of interest at the present time. The first is the *Schedule durations* as *Contiguous* or *Interruptible*. If the activities cannot be halted once they are started, then the contiguous

Screen Image 7-39
Edit Relationship Window

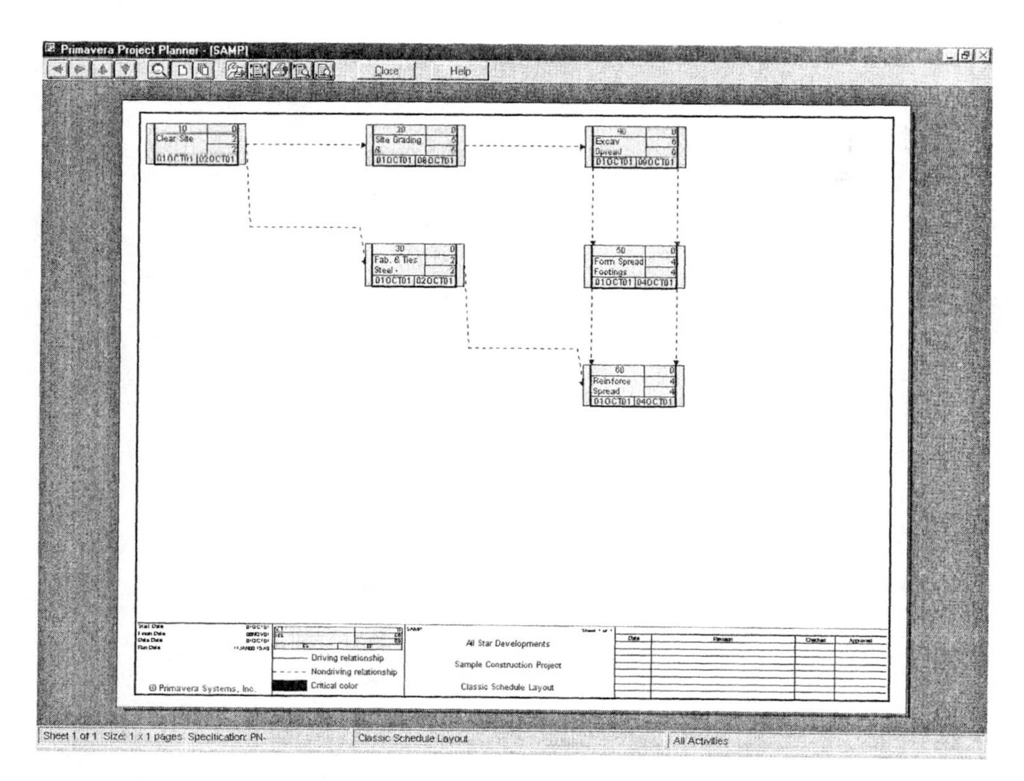

Screen Image 7-40
PERT Print Preview

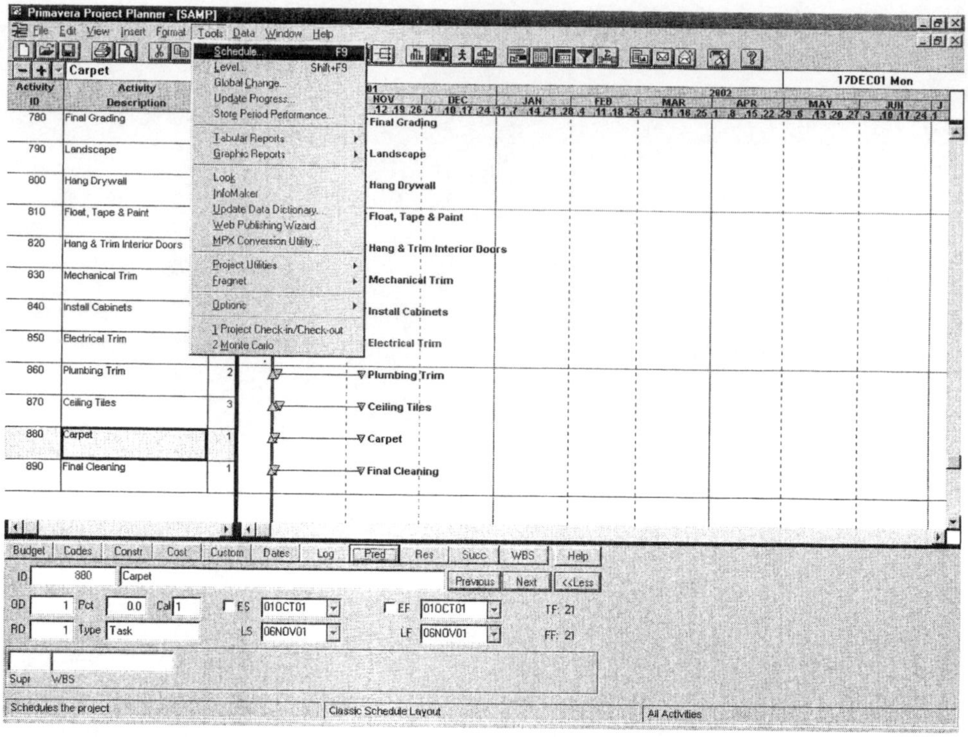

Primavera Project Planner - [SAMP]
File Edit View Insert Format Tools Data Window Help
Carpet
17DEC01 Mon

Schedule... F9
Level... Shift+F9
Global Change...
Update Progress...
Store Period Performance...
Tabular Reports
Graphic Reports
Look
InfoMaker
Update Data Dictionary...
Web Publishing Wizard
MPX Conversion Utility...
Project Utilities
Fragnet
Options
1 Project Check-in/Check-out
2 Monte Carlo

Activity ID	Activity Description	Orig Dur
780	Final Grading	
790	Landscape	
800	Hang Drywall	
810	Float, Tape & Paint	
820	Hang & Trim Interior Doors	
830	Mechanical Trim	
840	Install Cabinets	
850	Electrical Trim	
860	Plumbing Trim	2
870	Ceiling Tiles	3
880	Carpet	1
890	Final Cleaning	1

Budget Codes Constr Cost Custom Dates Log Pred Res Succ WBS Help
ID 880 Carpet Previous Next <<Less
OD 1 Pct 0.0 Cal 1 ES 01OCT01 EF 01OCT01 TF: 21
RD 1 Type Task LS 06NOV01 LF 06NOV01 FF: 21
Supr WBS
Schedules the project Classic Schedule Layout All Activities

Screen Image 7-41
Calculating the Schedule

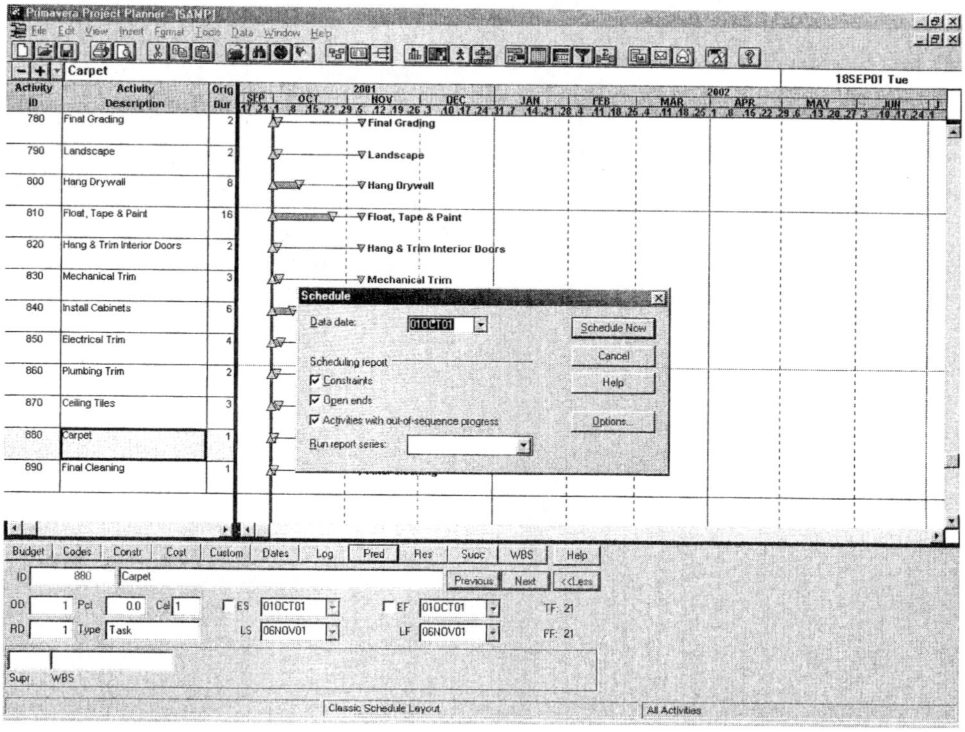

Primavera Project Planner - [SAMP]
File Edit View Insert Format Tools Data Window Help
Carpet
18SEP01 Tue

Activity ID	Activity Description	Orig Dur
780	Final Grading	2
790	Landscape	2
800	Hang Drywall	8
810	Float, Tape & Paint	16
820	Hang & Trim Interior Doors	2
830	Mechanical Trim	3
840	Install Cabinets	6
850	Electrical Trim	4
860	Plumbing Trim	2
870	Ceiling Tiles	3
880	Carpet	1
890	Final Cleaning	1

Schedule
Data date: 01OCT01
Schedule Now
Cancel
Help
Scheduling report
Constraints
Open ends
Activities with out-of-sequence progress
Options...
Run report series:

Budget Codes Constr Cost Custom Dates Log Pred Res Succ WBS Help
ID 880 Carpet Previous Next <<Less
OD 1 Pct 0.0 Cal 1 ES 01OCT01 EF 01OCT01 TF: 21
RD 1 Type Task LS 06NOV01 LF 06NOV01 FF: 21
Supr WBS
Classic Schedule Layout All Activities

Screen Image 7-42
Schedule Calculation

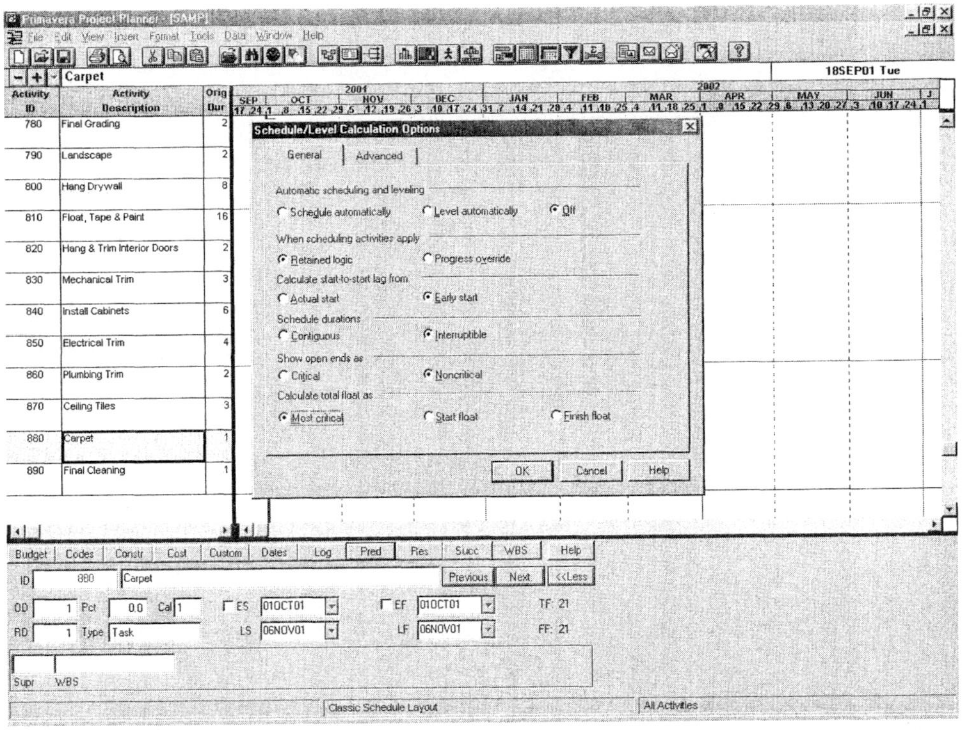

Screen Image 7-43
Schedule Calculation—Recommended Settings

option should be selected. However, if the activities can be started, stopped, and restarted and stopped until they are complete, the interruptible option should be selected. The manual schedule calculations that were demonstrated in the previous chapters assumed that the activities were interruptible. This assumption allows for greater flexibility when scheduling crews and reflects how work is actually performed. Since contiguous activities are typically exceptions, it is recommended that the interruptible option be selected. This will allow the project manager greater flexibility when deciding when to actually start a specific activity.

The second important option deals with how the float will be calculated. Four types of floats were discussed in Chapter 4. However, Primavera Project Planner only recognizes one type of float, referred to as *total float*. The user may specify that the total float is the *Start float*, *Finish float*, or the *Most critical*. The *Most critical* option, while not consistent, will state the total float as the smaller of the start float and finish float. This option provides the most accurate picture of the float in the schedule. Screen Image 7-43 shows the recommended settings.

Once the calculation options have been selected, clicking the *OK* button will save the changes and close the window. Then click on the *Schedule Now* button on the *Schedule* window to initiate the calculation process. During the calculation process, the user will be prompted on where to route the output. Since this is typically diagnostic output, having the output return to the screen will save both time and paper.

Screen Image 7-44 is an example of the main input screen after the schedule has been calculated. Notice that the starts of the activities are no longer all on one date but are correctly distributed.

Output 7-2 is an example of the output that was generated when the schedule was calculated. Typically there are at least three pages of this output. The first page is just header information, while the second page lists all of those activities without predecessors or successors. It is important to remember that activities without predecessors could be the first activities on the project, whereas those without successors could be the last activities

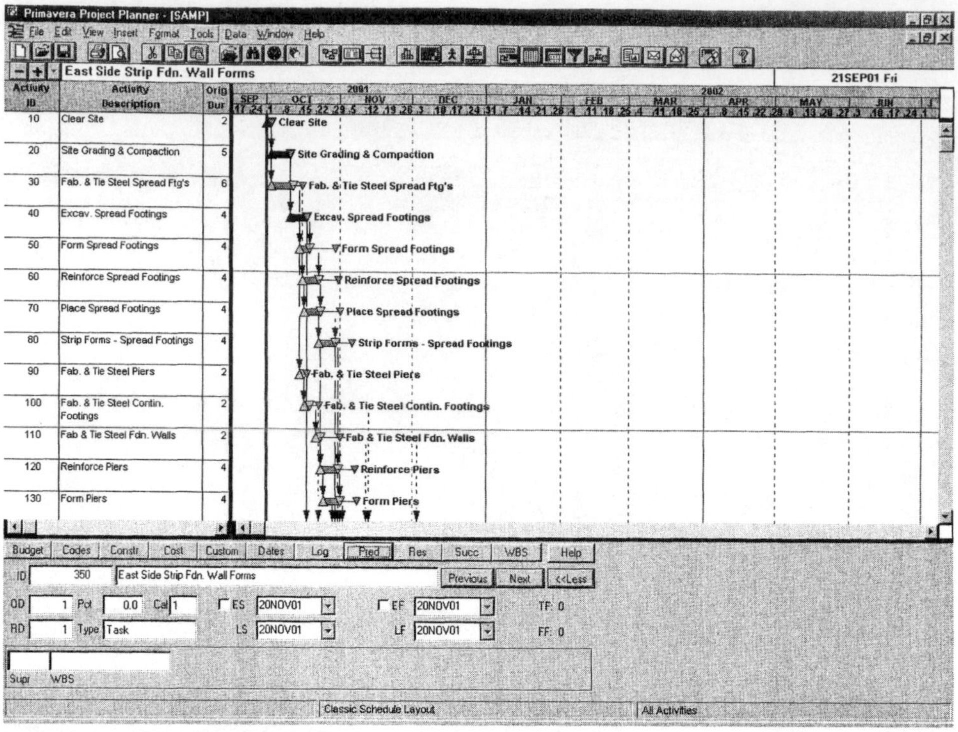

Primavera Project Planner - [SAMP]

East Side Strip Fdn. Wall Forms — 21SEP01 Fri

Activity ID	Activity Description	Orig Dur
10	Clear Site	2
20	Site Grading & Compaction	5
30	Fab. & Tie Steel Spread Ftg's	6
40	Excav. Spread Footings	4
50	Form Spread Footings	4
60	Reinforce Spread Footings	4
70	Place Spread Footings	4
80	Strip Forms - Spread Footings	4
90	Fab. & Tie Steel Piers	2
100	Fab. & Tie Steel Contin. Footings	2
110	Fab & Tie Steel Fdn. Walls	2
120	Reinforce Piers	4
130	Form Piers	4

ID 350 East Side Strip Fdn. Wall Forms

OD 1 Pct 0.0 Cal 1 ES 20NOV01 EF 20NOV01 TF: 0

RD 1 Type Task LS 20NOV01 LF 20NOV01 FF: 0

Classic Schedule Layout — All Activities

Screen Image 7-44

Schedule After Calculation

```
Scheduling and Leveling Calculations -- Scheduling Report Page: 1



 This Primavera software is registered to TAMU.
 Start of schedule for project SAMP.
 Serial number...19554609

 User name LESLIE  .
```

```
 Open end listing -- Scheduling Report Page: 2
 ----------------
 Activity         10  has no predecessors
                                          Activity       770 has no successors
                                          Activity       790 has no successors
                                          Activity       890 has no successors
```

```
 Scheduling Statistics for Project SAMP:
 Schedule calculation mode - Retained logic
 Schedule calculation mode - Interruptible activities
 Float calculation mode     - Use more critical float from start or finish dates
 SS relationships           - Use early start of predecessor


        Schedule run on Sat Feb 26 16:28:53 2000
                  Run Number  8.

      Number of activities.................       89
      Number of activities in longest path..      50
      Started activities...................        0
      Completed activities.................        0
      Number of relationships..............      130
      Percent complete.....................      0.0

      Data date............................  01OCT01
      Start date...........................  01OCT01
      Imposed finish date..................
      Latest calculated early finish.......  25JUN02
```

Output 7-2

Schedule Calculation Report

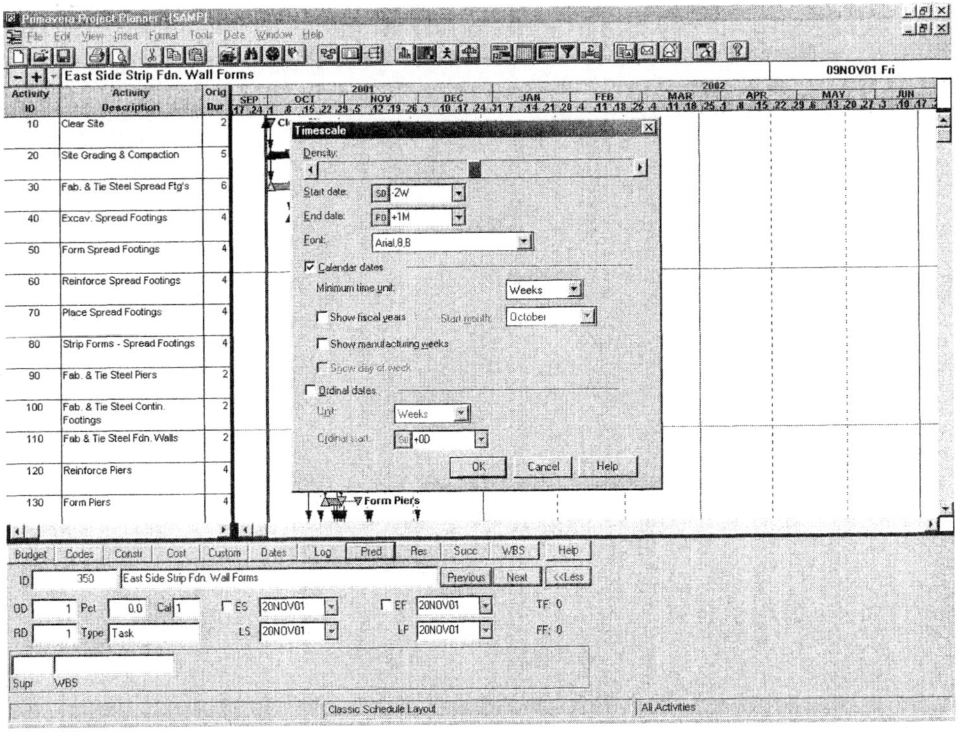

Screen Image 7-45
Time Scale Option

on the project. The activities listed on this report should be checked to see if they meet that criterion. The last page gives specific information about the number of activities and relationships found on the schedule.

If the *F9* or schedule icon is used to initiate a schedule, the user will be prompted to verify the data date. This is the date on which the actual progress in the field was measured. Since the project has not begun, the project start date should be in this field.

Changing the Time Scale

The time scale located at the top of the main input screen can be changed to show more or less of the schedule. The changing of the date density impacts the readability and accuracy of the schedule. To change the time scale, double click on that area of the main input screen. That will open the *Timescale* window as shown in Screen Image 7-45. By sliding the density bar to the left, the spacing between the dates will be decreased, allowing for more dates to be shown. Conversely, sliding the *Density* bar to the right will increase spacing, decreasing the number of dates shown. In Screen Image 7-46 the density bar has been moved to spread the dates apart, so as to make the schedule easier to read.

Reorganizing the Schedule

It is often helpful to change the order of the activities on the main input screen. A common example may be to change the order from activity ID to early start. All of the changes to the organization of the main input screen are accomplished through the *Organize* window. This window is opened by selecting the *Format* command followed by the *Organize* option as shown in Screen Image 7-47, or by clicking the organize icon. Either of these methods

Screen Image 7-46
Less Dense Time Scale

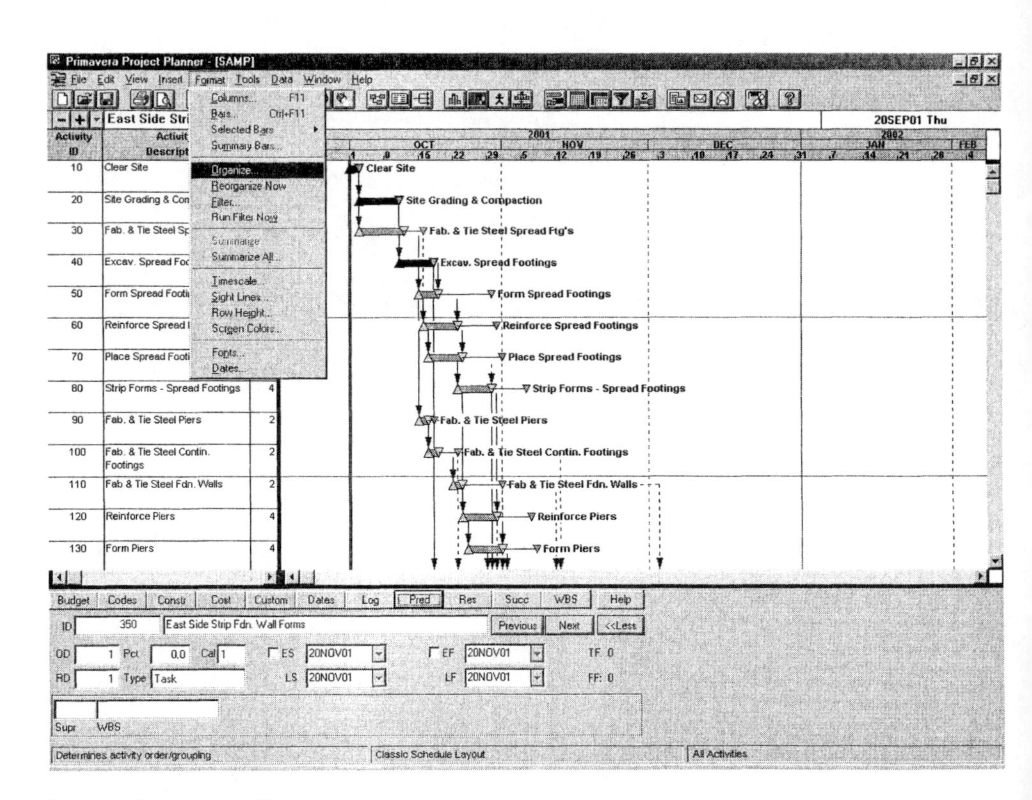

Screen Image 7-47
Beginning Schedule Organization

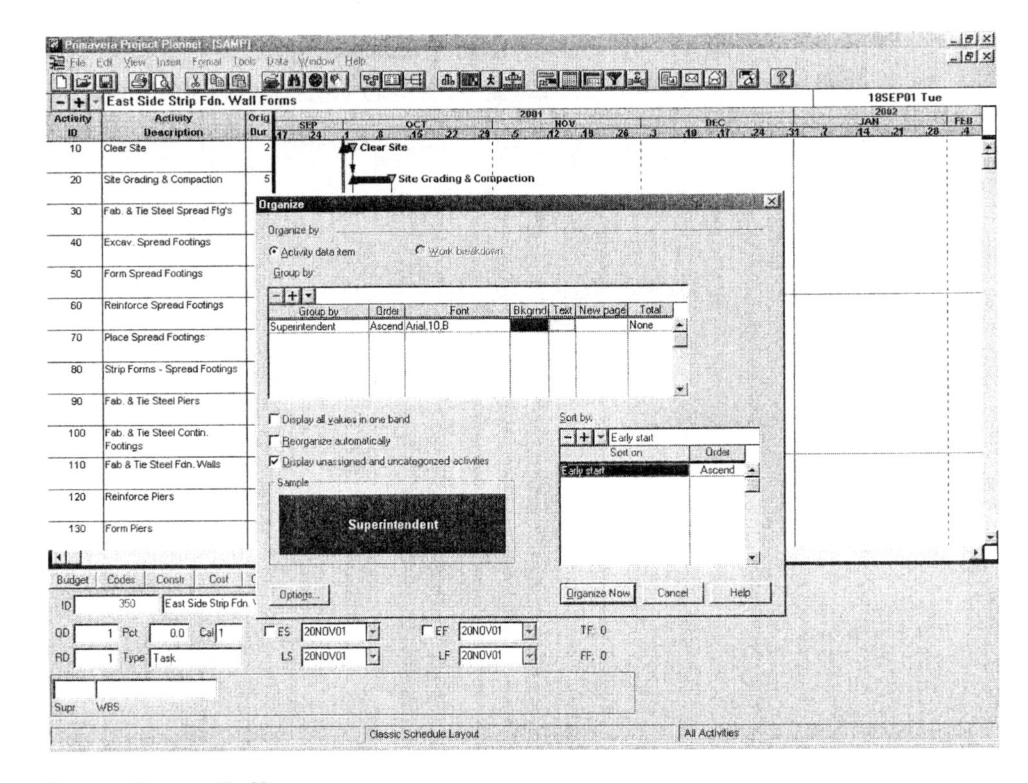

Screen Image 7-48
Organize Window

will open the *Organize* window as shown in Screen Image 7-48. By selecting the *Sort by* option, the user dictates the order in which the activities will be presented on the main input screen. However, for this example the *Group by* code is changed from blank to activity code *Superintendent*, and the *Sort by* option is by *Activity ID*. This combination of options groups all of the activities together that are the responsibility of a single supervisor or superintendent and then lists all of those activities by *Early start*. When the *Organize Now* button is pressed, the main input screen has a changed appearance as shown in Screen Image 7-49. Screen Image 7-50 is the same schedule but with the relationships turned off. This is done by pressing the relationship icon. By turning off the relationship lines the information presented, although not as concise, is more understandable.

Printing Out the Main Input Screen Bar Chart

Generating output with Primavera Project Planner is similar to any Windows based program. The first step is to select the *File* command followed by the *Print Setup* option. This opens the standard windows dialogue box for specifying the printer. The proper printer, orientation, paper size, and graphic settings need to be specified. While the print setup is fairly standard, the page setup is unique to Primavera Project Planner. The *Page Setup* window (Screen Image 7-52) is opened by selecting the *File* command followed by the *Page Setup* option as shown in Screen Image 7-51. The default values for most of the fields on this window are acceptable and self-explanatory. However, the user needs to be extremely

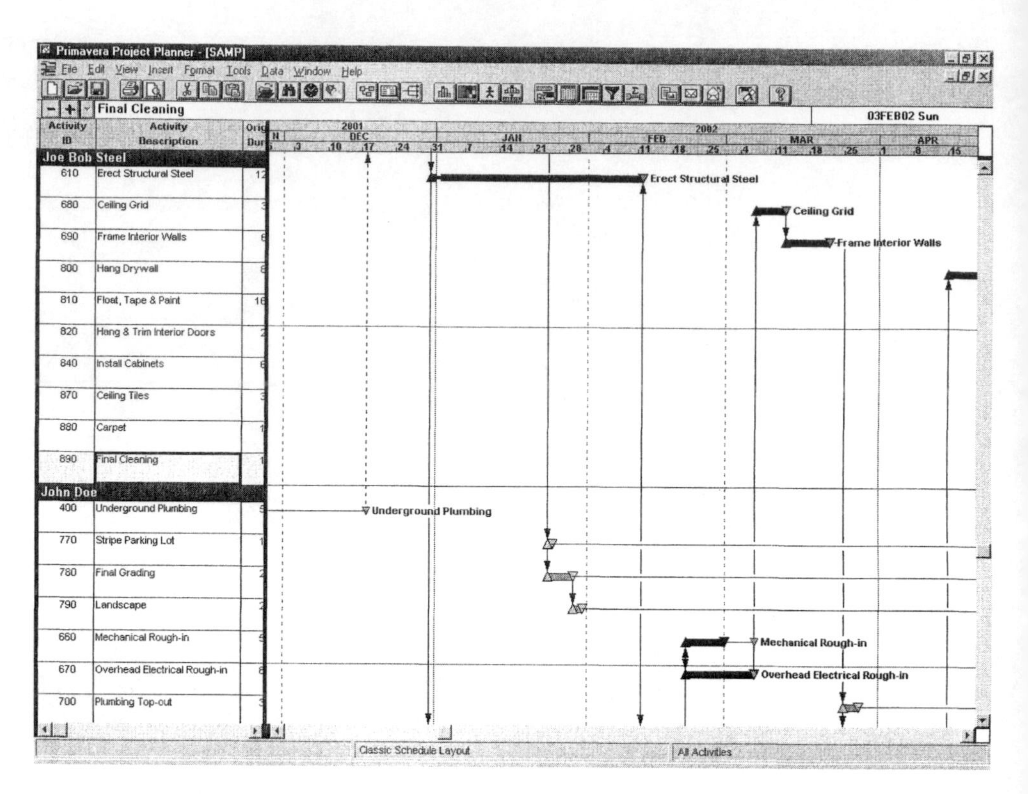

Screen Image 7-49
Organized by Superintendent

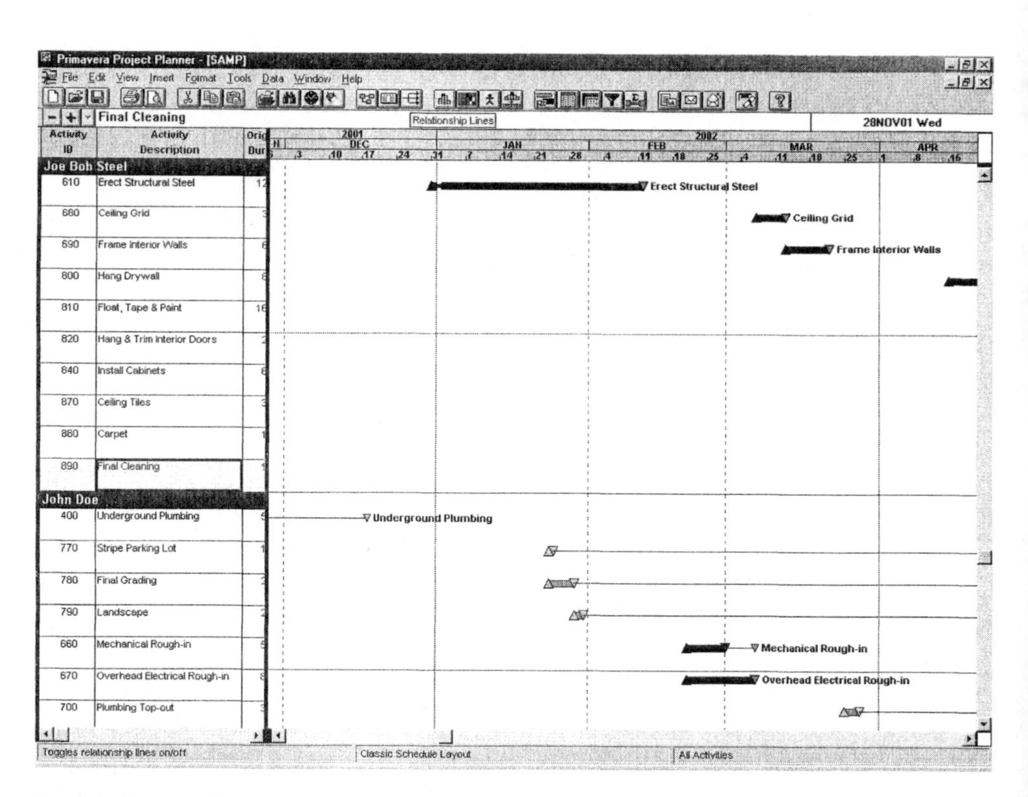

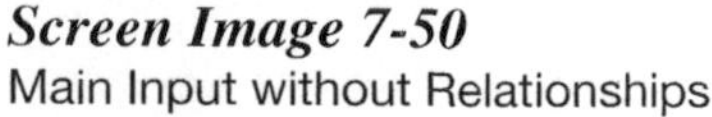

Screen Image 7-50
Main Input without Relationships

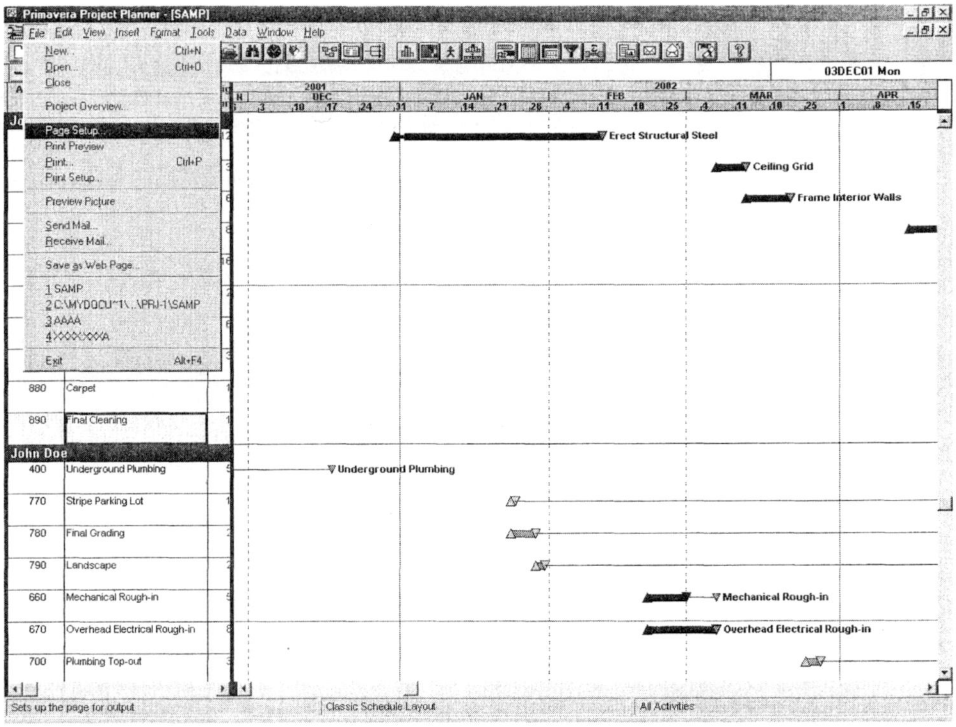

Primavera Project Planner - [SAMP]
File Edit View Insert Format Tools Data Window Help
New... Ctrl+N
Open... Ctrl+O
Close
Project Overview...
Page Setup...
Print Preview
Print... Ctrl+P
Print Setup...
Preview Picture
Send Mail...
Receive Mail...
Save as Web Page...
1 SAMP
2 C:\MYDOCU~1\...\PRJ-1\SAMP
3 AAAA
4 XXXXXXXA
Exit Alt+F4
03DEC01 Mon
2001 DEC 3 10 17 24 31 JAN 7 14 21 28 2002 FEB 4 11 18 25 MAR 4 11 18 25 APR 1 8 15
Erect Structural Steel
Ceiling Grid
Frame Interior Walls

Activity ID	Activity Description
880	Carpet
890	Final Cleaning
John Doe	
400	Underground Plumbing
770	Stripe Parking Lot
780	Final Grading
790	Landscape
660	Mechanical Rough-in
670	Overhead Electrical Rough-in
700	Plumbing Top-out

Underground Plumbing
Mechanical Rough-in
Overhead Electrical Rough-in
Sets up the page for output
Classic Schedule Layout
All Activities

Screen Image 7-51
Opening the Page Setup Window

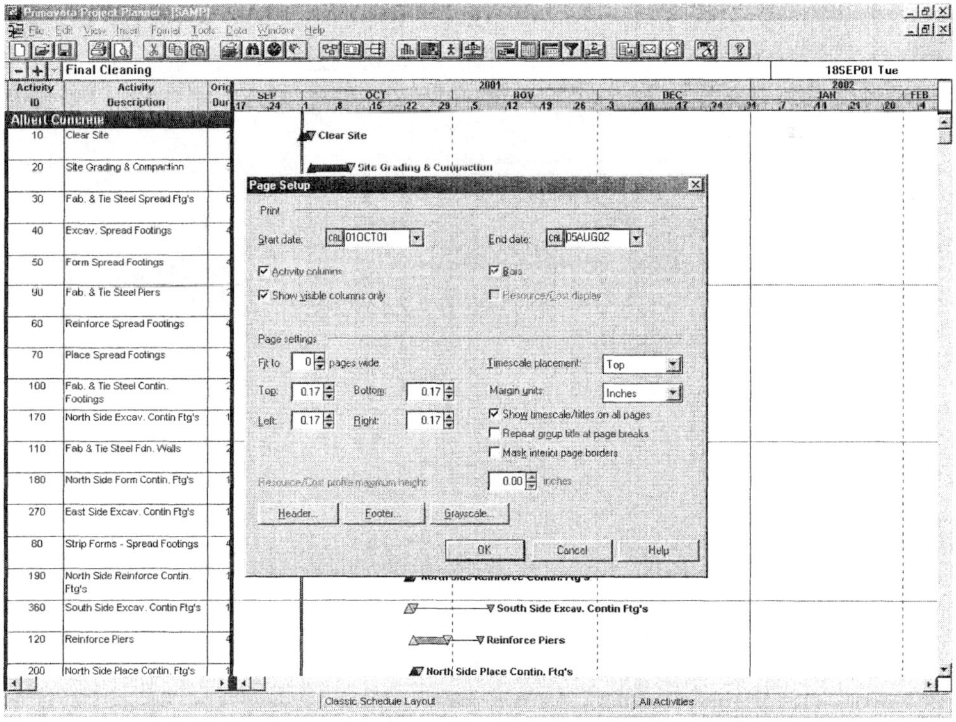

Primavera Project Planner - [SAMP]
File Edit View Insert Format Tools Data Window Help
Final Cleaning
18SEP01 Tue
2001 SEP 17 24 OCT 1 8 15 22 29 NOV 5 12 19 26 DEC 3 10 17 24 31 2002 JAN 7 14 21 28 FEB 4

Activity ID	Activity Description
Albert Concrete	
10	Clear Site
20	Site Grading & Compaction
30	Fab. & Tie Steel Spread Ftg's
40	Excav. Spread Footings
50	Form Spread Footings
90	Fab. & Tie Steel Piers
60	Reinforce Spread Footings
70	Place Spread Footings
100	Fab. & Tie Steel Contin. Footings
170	North Side Excav. Contin Ftg's
110	Fab & Tie Steel Fdn. Walls
180	North Side Form Contin. Ftg's
270	East Side Excav. Contin Ftg's
80	Strip Forms - Spread Footings
190	North Side Reinforce Contin. Ftg's
360	South Side Excav. Contin Ftg's
120	Reinforce Piers
200	North Side Place Contin. Ftg's

Clear Site
Site Grading & Compaction

Page Setup
Print
Start date: CAL 01OCT01
End date: CAL 05AUG02
Activity columns
Bars
Show visible columns only
Resource/Cost display
Page settings
Fit to 0 pages wide
Timescale placement: Top
Top: 0.17 Bottom: 0.17
Margin units: Inches
Left: 0.17 Right: 0.17
Show timescale/titles on all pages
Repeat group title at page breaks
Mask interior page borders
Resource/Cost profile maximum height: 0.00 inches
Header... Footer... Grayscale...
OK Cancel Help

South Side Excav. Contin Ftg's
Reinforce Piers
North Side Place Contin. Ftg's
Classic Schedule Layout
All Activities

Screen Image 7-52
Page Setup Window

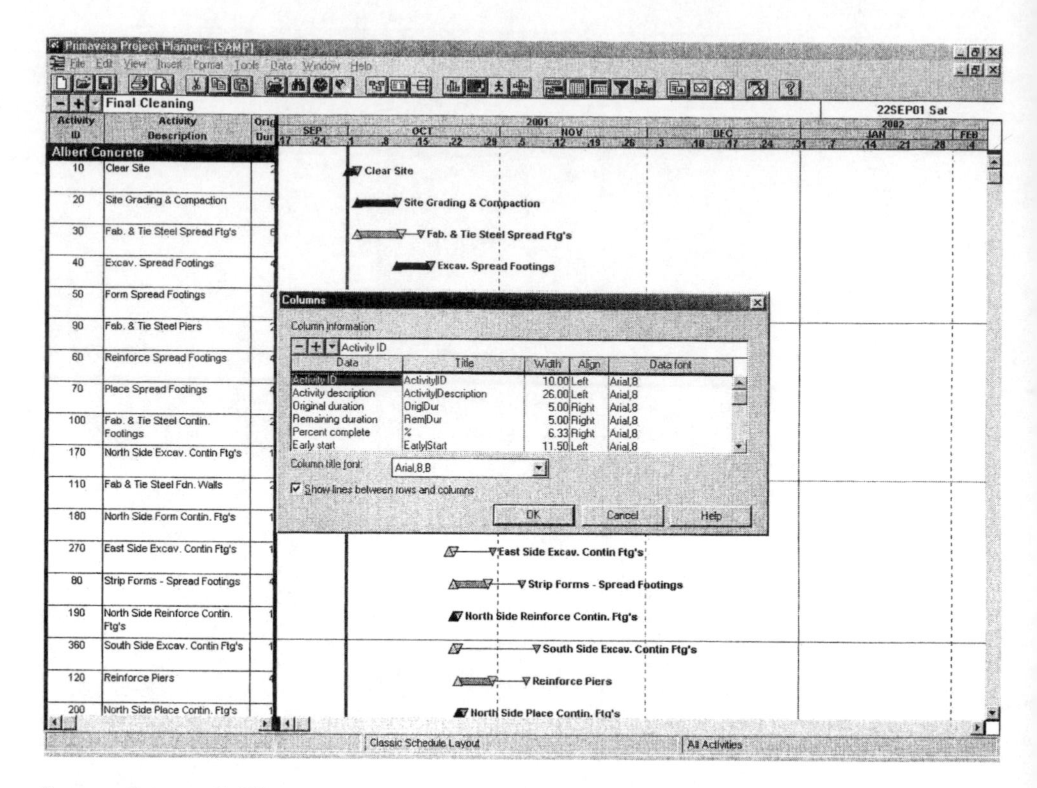

Screen Image 7-53
Columns Window

cautious in selecting the appropriate *Start date* and *End date*. If this is not done, hundreds of useless pages of output could be generated. In Screen Image 7-52, the time scale will range from October 1, 2001 until August 5, 2002. Once this window is modified, the *OK* button needs to be clicked to return to the main input screen.

The next decision about the layout of the printout concerns what columnar data should be presented on the left-hand side of the page. By moving the divider between the bars and the text data, any number of columns containing data can be exposed. Whatever is exposed on the screen is exactly what will be exposed on the printout. Additional columnar information can be added to this area by clicking on the format column icon. This will open the *Columns* window as shown in Screen Image 7-53. From this window the user can define the order of the columns, the font size, and what columnar data is presented. Once all of these issues have been addressed, the print icon can be pressed and output will be generated similar to Output 7-3. However, this output works best when a color printer is available.

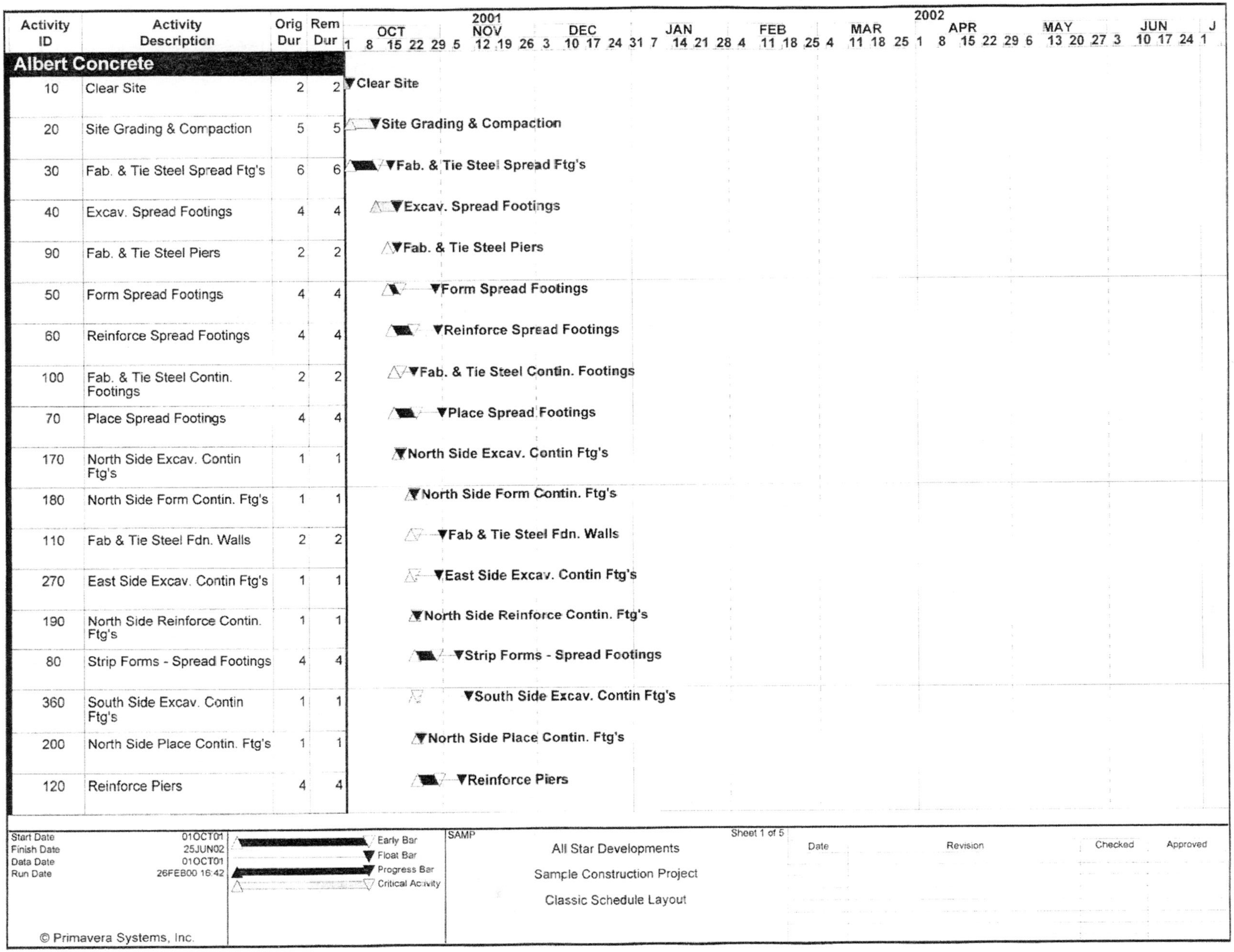
Activity ID
Activity Description
Orig Dur
Rem Dur
2001
OCT
NOV
DEC
2002
JAN
FEB
MAR
APR
MAY
JUN
J
1 8 15 22 29 5 12 19 26 3 10 17 24 31 7 14 21 28 4 11 18 25 4 11 18 25 1 8 15 22 29 6 13 20 27 3 10 17 24 1
Albert Concrete
10 Clear Site 2 2
20 Site Grading & Compaction 5 5
30 Fab. & Tie Steel Spread Ftg's 6 6
40 Excav. Spread Footings 4 4
90 Fab. & Tie Steel Piers 2 2
50 Form Spread Footings 4 4
60 Reinforce Spread Footings 4 4
100 Fab. & Tie Steel Contin. Footings 2 2
70 Place Spread Footings 4 4
170 North Side Excav. Contin Ftg's 1 1
180 North Side Form Contin. Ftg's 1 1
110 Fab & Tie Steel Fdn. Walls 2 2
270 East Side Excav. Contin Ftg's 1 1
190 North Side Reinforce Contin. Ftg's 1 1
80 Strip Forms - Spread Footings 4 4
360 South Side Excav. Contin Ftg's 1 1
200 North Side Place Contin. Ftg's 1 1
120 Reinforce Piers 4 4
Clear Site
Site Grading & Compaction
Fab. & Tie Steel Spread Ftg's
Excav. Spread Footings
Fab. & Tie Steel Piers
Form Spread Footings
Reinforce Spread Footings
Fab. & Tie Steel Contin. Footings
Place Spread Footings
North Side Excav. Contin Ftg's
North Side Form Contin. Ftg's
Fab & Tie Steel Fdn. Walls
East Side Excav. Contin Ftg's
North Side Reinforce Contin. Ftg's
Strip Forms - Spread Footings
South Side Excav. Contin Ftg's
North Side Place Contin. Ftg's
Reinforce Piers
Start Date 01OCT01
Finish Date 25JUN02
Data Date 01OCT01
Run Date 26FEB00 16:42
© Primavera Systems, Inc.
Early Bar
Float Bar
Progress Bar
Critical Activity
SAMP
Sheet 1 of 5
All Star Developments
Sample Construction Project
Classic Schedule Layout
Date
Revision
Checked
Approved

Output 7-3
Main Input Screen Bar Chart

Conclusion

The loading of activity data and relationships can be accomplished in several ways, either keyed in, or imported from an outside database, or drawn in by mouse. Regardless of the method selected, the information can be presented in the bar chart or PERT view. Once all of the data has been entered, the main input screen can be modified to develop a chart that is easy to read. This bar chart schedule can subsequently be printed out and used as a progress tracking mechanism.

Suggested Exercises

1. Using the schedule that was developed in Chapter 4, load all of the activities, relationships, and lags into Primavera Project Planner. This can be done with direct keyboard entry or by using the import feature.
2. Have Primavera Project Planner calculate the schedule and compare the results with the schedule that was manually calculated. If there is a discrepancy between the dates, verify that the calculation logic within Primavera Project Planner is set up with the activities being interruptible. If this does not solve the problem, check the manual schedule for forward and backward pass calculation errors. Finally, verify that the inputted logic is identical to that of the manually calculated schedule.
3. Using the schedule calculation report from Primavera Project Planner, verify that the counts of starting, finishing, and critical activities are correct.
4. Group the activities on the main input screen by activity code and print out that bar chart schedule.

Tabular and Graphic Output

KEY TERMS

Activity List Report A list of the early and late dates for all activities.

Critical Activity List Report A report of the early and late dates for all of the activities on the critical path.

Responsibility Schedule A customized schedule report generated for all of the superintendents on the project, outlining those activities that are their direct responsibility.

Generating Reports

Primavera Project Planner is equipped with a flexible report writer that can produce both tabular and graphical schedule reports. Both of these types of reports are developed by generating a set of report specifications. Tabular reports, as their name implies, have limited graphic capabilities and are intended for line printers, while graphical reports are visual presentations of information requiring a color plotter or printer. For the purpose of explaining how this report writer works, a series of practical reports will be developed in this chapter.

Early Start Report

The early start activity list report is a tabular report that lists all of the activities in early start order. This report details the basic order of the construction tasks. The first step in generating this or any of the tabular reports is to open the *Schedule Reports* window. This is accomplished by selecting the *Tools* command followed by the *Tabular Reports/Schedule* options as shown in Screen Image 8-1. The *Schedule Reports* window as shown in Screen Image 8-2 lists all of the existing tabular report specifications.

These specifications are not project specific. Once a set of report specifications has been developed, these specs can be used on any project, as long as the project is set up to handle specific activity numbering patterns and activity codes. This feature provides consistency in the management of projects with a minimal amount of effort. However, the sets of report specifications that are provided by the software publisher are of little use unless the current project is set up using the same activity numbering pattern and activity codes. If you do not know the activity numbering and activity code configuration, it is often simpler to create new report specifications rather than modify existing ones. Once the user achieves a better understanding of how Primavera Project Planner works, standardized project setup procedures can be developed that will allow the report specifications to be shared between projects. This does not preclude using the existing report specifications, as they can be modified by clicking on the desired report, highlighting it, and then clicking on the *Modify* button. The process for modifying and creating new reports is identical.

For simplicity the examples presented here will deal with developing only new report specifications. To add a new set of report specifications, click the *Add* button from the *Schedule Reports* window to open the *Add a New Report* window, as shown on Screen Image 8-3.

The *Add a New Report* window prompts the user to input or accept a *Report ID*. In Screen Image 8-3, the program has assigned *SR-22* as the identifier for this new set of report specifications. The *SR* stands for schedule report, and the 22 signifies that this report is the twenty-second to be developed.

Screen Image 8-1
Opening the Schedule Reports Window

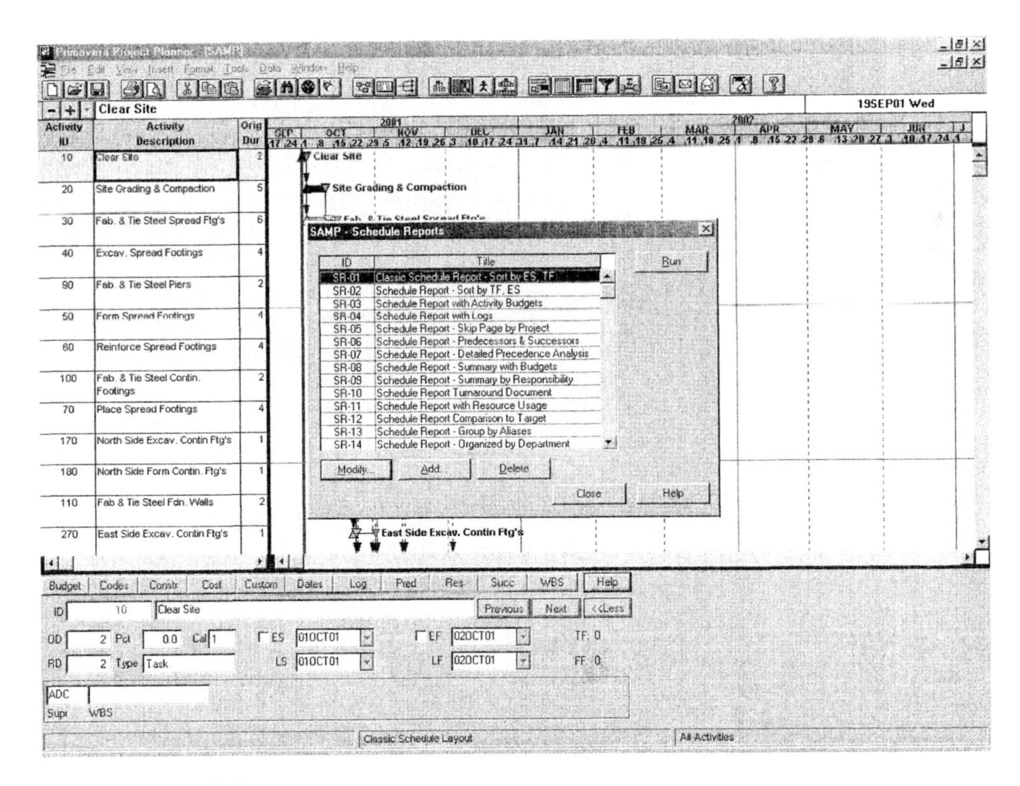

Screen Image 8-2
Schedule Reports Window

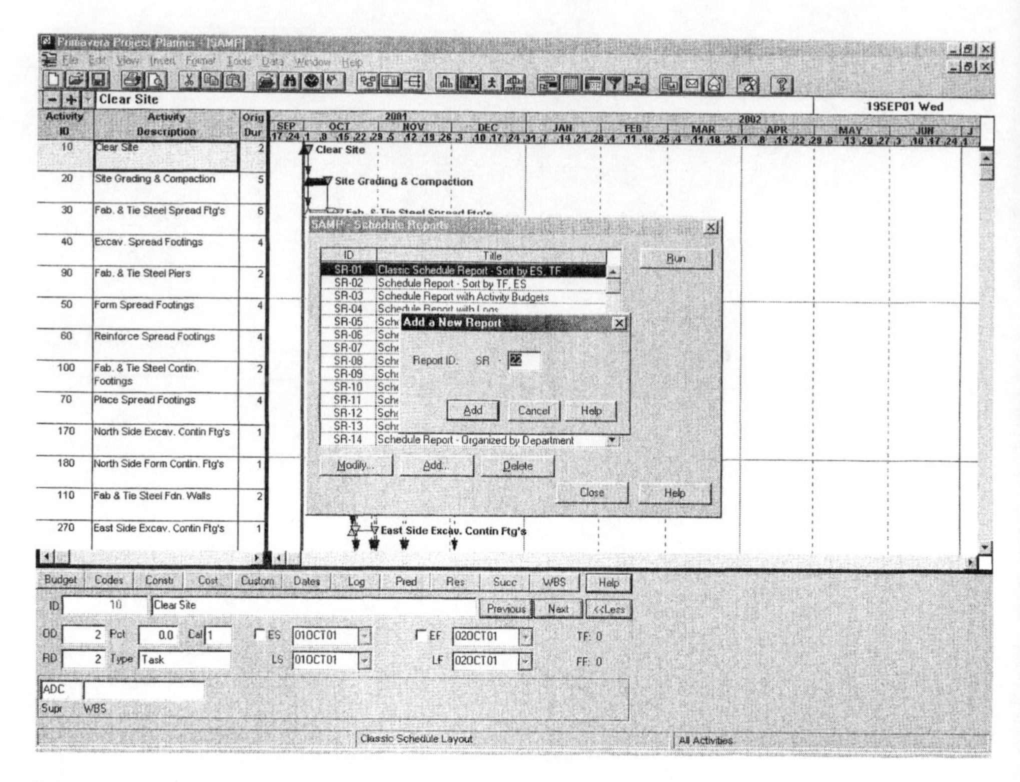

Screen Image 8-3
Add a New Report Window

In almost all circumstances, it is recommended that the identification issued by the program be accepted. To accept this report identification number, click the *Add* button. This will activate the *Schedule Reports* window shown in Screen Image 8-4. The first item to correct is the report title, which is done by typing over the report ID and replacing it with a more descriptive title. In Screen Image 8-5 the title has been changed from *SR-22* to *Early Start Schedule*. This is the title that will now be listed on the *Schedule Reports* window. Immediately under the Title field there are three tabs that when clicked provide a series of report options. The *Content* tab allows the user to designate what descriptive information should be included in the schedule report.

The first item within the *Content* tab is the *Include the following data* choice; this is where the user specifies what data outside of the early dates, late dates, calendar, activity codes, and float will be included in the report. The *Activity code line* is required for all reports. This entry adds the activity ID, activity description, original duration, remaining duration, and percent complete to the report. There are any number of additional items that can be added such as successor activity, predecessor activity, or budget information. However, for the *Early Start Schedule* report only the *Activity code line* will be used. The second item is the *Show these activity codes on activity code line* option; this command directs the report writer as to which activity codes to integrate into the report. The sample project currently has only one activity code, *Superintendent*, and that code has been selected. Screen Image 8-5 shows the *Content* tab after the previously discussed modifications.

The *Format* tab (Screen Image 8-6) is where the order and presentation of the data is specified. The first option is the *Skip line on* field. This field instructs the report writer to add a blank line to the report whenever the value in the specified field changes. For example, this would be used to add a blank line whenever the assigned superintendent changes. The only valid entry in this field is one of the activity codes or *<none>*. In the current example skipping a line whenever the superintendent changes would destroy the coherence of the report. Therefore, this field is left with its default *<none>* entry. The *Skip page on* field is similar to the *Skip line* command, except this option instructs the report writer to

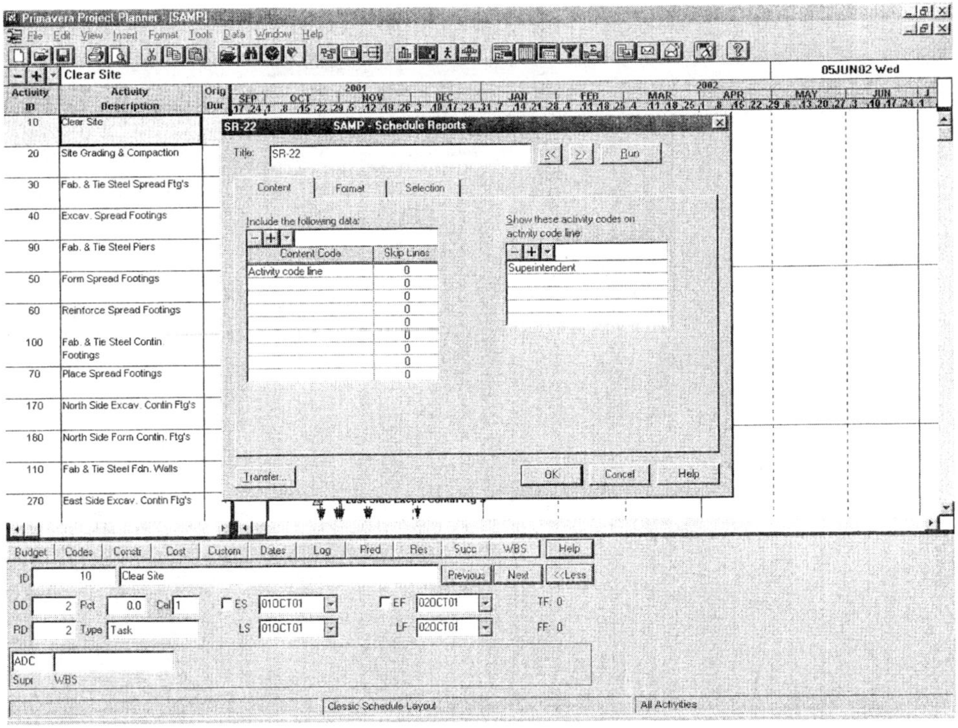

Primavera Project Planner - [SAMP]
File Edit View Insert Format Tools Data Window Help
Clear Site
05JUN02 Wed

Activity ID	Activity Description
10	Clear Site
20	Site Grading & Compaction
30	Fab. & Tie Steel Spread Ftg's
40	Excav. Spread Footings
90	Fab. & Tie Steel Piers
50	Form Spread Footings
60	Reinforce Spread Footings
100	Fab. & Tie Steel Contin. Footings
70	Place Spread Footings
170	North Side Excav. Contin Ftg's
180	North Side Form Contin. Ftg's
110	Fab & Tie Steel Fdn. Walls
270	East Side Excav. Contin Ftg's

Orig Dur
2001 SEP OCT NOV DEC
2002 JAN FEB MAR APR MAY JUN

SR-22 SAMP - Schedule Reports
Title: SR-22 << >> Run
Content | Format | Selection
Include the following data:

Content Code	Skip Lines
Activity code line	0
	0
	0
	0
	0
	0
	0
	0
	0

Show these activity codes on activity code line:
Superintendent
Transfer... OK Cancel Help

Budget Codes Constr Cost Custom Dates Log Pred Res Succ WBS Help
ID 10 Clear Site Previous Next <<Less
OD 2 Pct 0.0 Cal 1 ES 01OCT01 EF 02OCT01 TF: 0
RD 2 Type Task LS 01OCT01 LF 02OCT01 FF: 0
ADC
Supr WBS
Classic Schedule Layout All Activities

Screen Image 8-4
Schedule Reports Content Window

Primavera Project Planner - [SAMP]
File Edit View Insert Format Tools Data Window Help
Clear Site
05JUN02 Wed

Activity ID	Activity Description
10	Clear Site
20	Site Grading & Compaction
30	Fab. & Tie Steel Spread Ftg's
40	Excav. Spread Footings
90	Fab. & Tie Steel Piers
50	Form Spread Footings
60	Reinforce Spread Footings
100	Fab. & Tie Steel Contin. Footings
70	Place Spread Footings
170	North Side Excav. Contin Ftg's
180	North Side Form Contin. Ftg's
110	Fab & Tie Steel Fdn. Walls
270	East Side Excav. Contin Ftg's

Orig Dur
2001 SEP OCT NOV DEC
2002 JAN FEB MAR APR MAY JUN

SR-22 SAMP - Schedule Reports
Title: Early Start Schedule << >> Run
Content | Format | Selection
Include the following data:

Content Code	Skip Lines
Activity code line	0
	0
	0
	0
	0
	0
	0
	0
	0

Show these activity codes on activity code line:
Superintendent
Transfer... OK Cancel Help

Budget Codes Constr Cost Custom Dates Log Pred Res Succ WBS Help
ID 10 Clear Site Previous Next <<Less
OD 2 Pct 0.0 Cal 1 ES 01OCT01 EF 02OCT01 TF: 0
RD 2 Type Task LS 01OCT01 LF 02OCT01 FF: 0
ADC
Supr WBS
Classic Schedule Layout All Activities

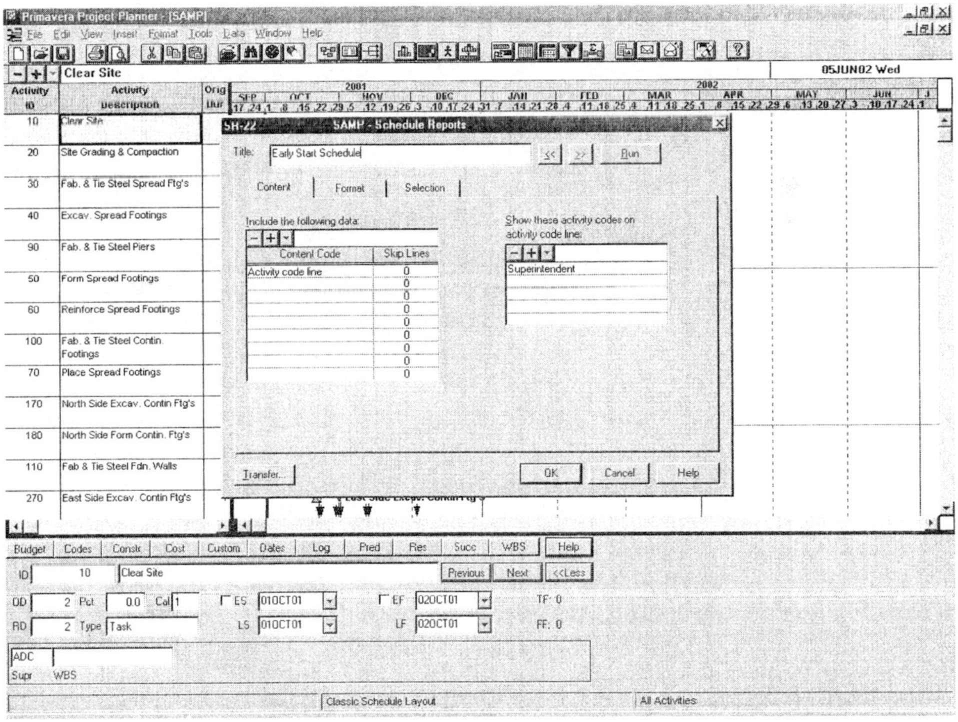

Screen Image 8-5
Schedule Reports Content Window Modified

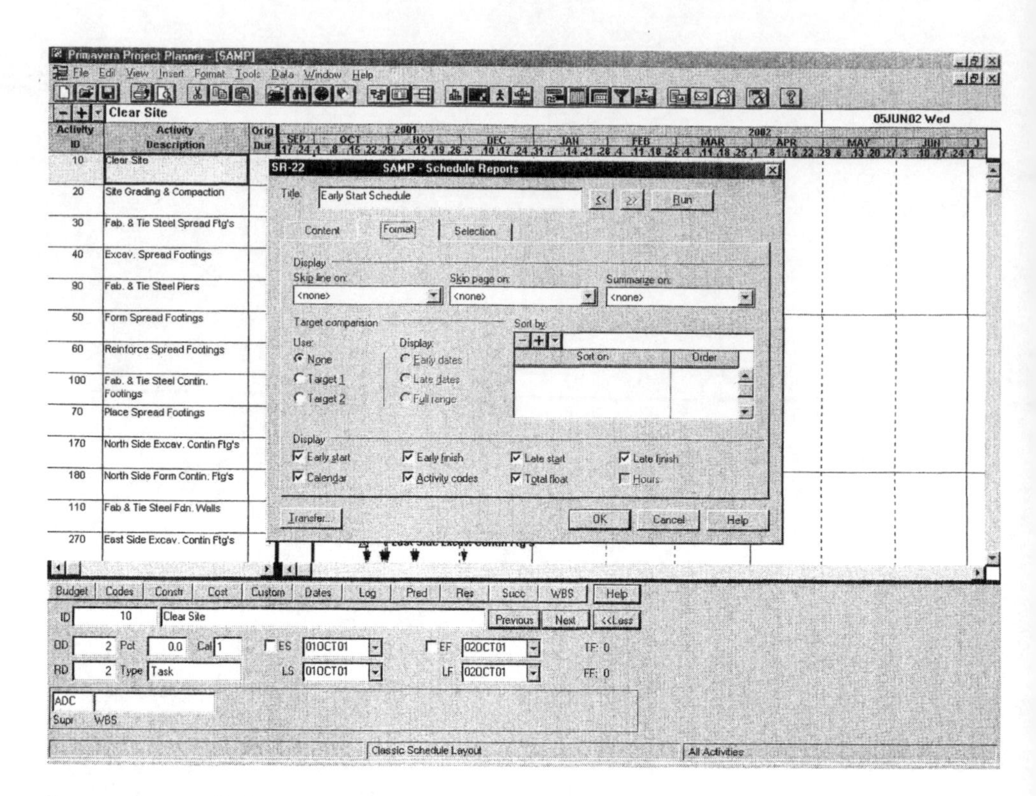

Screen Image 8-6

Schedule Reports Format Window

skip to the top of the next page when the selected activity code changes. Once again the only valid entry is one of the activity codes or *<none>*. For the purpose of this report the *<none>* option will be used. The *Summarize on* field is used when specific activity codes have been added for the purpose of condensing the schedule. For the time being the *<none>* option will be used. The *Target comparison* options address how and which target schedules to display. Since no target schedules have been created, the default *None* setting will be used.

The *Sort by* table allows the user to specify the order in which the data will be presented in the report. In the example shown in Screen Image 8-7, the sort order is *Early start* and *Total float*, which is consistent with the title of the report. With that sort order the primary sort is on *Early start* which will organize all of the activities based on their early start. Since the *Ascend* order is selected, this sort will be from the earliest early start date to the latest early start date. The secondary sort, *Total float*, will be used only when there are activities with the same early start date. With the *Ascend* order selected, the activities that have the same early start will be listed from most critical to least critical.

The sort order is relatively straightforward unless either the *Skip line on* or *Skip page on* option is being used. If one of these options is selected, the user needs to be careful that the sort order is compatible with the desired breaks. Since these commands tell the printer to skip a line or go to the top of the next page when the specified activity code changes, it is important that the sort order match these commands. Finally, the bottom portion of this tab is where the desired columns of data are specified. By clicking on the box adjacent to these items they may be turned on and off. Screen Image 8-7 is an example of the modified *Format* window.

The final step in producing this sample report is to specify which activities to include in the report. This is done via the *Selection* tab as shown in Screen Image 8-8. The first item of importance within this window is the *Any* or *All* option. If only one selection criteria is entered, the *Any* or *All* option is of no significance. These items come into play only if multiple selection criteria are entered on the *Selection criteria* table. If the *Any* option is selected, then if *any one* of the listed criteria is met that activity will be included in the

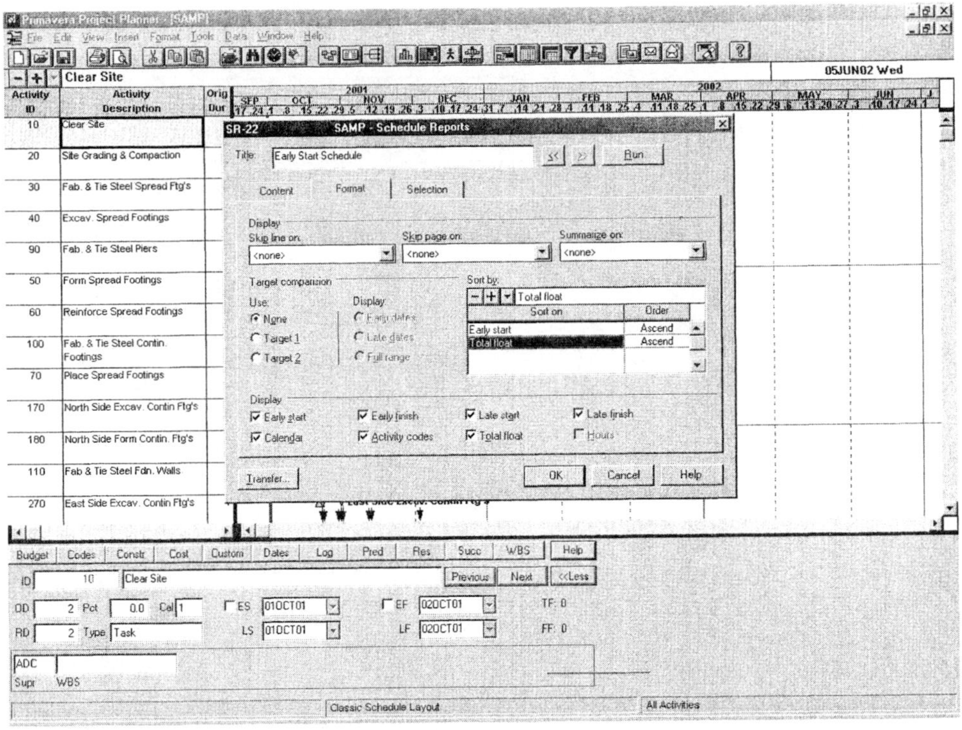

Primavera Project Planner - [SAMP]

File Edit View Insert Format Tools Data Window Help

Clear Site — 05JUN02 Wed

Activity ID	Activity Description	Orig Dur
10	Clear Site	
20	Site Grading & Compaction	
30	Fab. & Tie Steel Spread Ftg's	
40	Excav. Spread Footings	
90	Fab. & Tie Steel Piers	
50	Form Spread Footings	
60	Reinforce Spread Footings	
100	Fab. & Tie Steel Contin. Footings	
70	Place Spread Footings	
170	North Side Excav. Contin Ftg's	
180	North Side Form Contin. Ftg's	
110	Fab & Tie Steel Fdn. Walls	
270	East Side Excav. Contin Ftg's	

2001: SEP OCT NOV DEC — 2002: JAN FEB MAR APR MAY JUN J

SR-22 SAMP - Schedule Reports

Title: Early Start Schedule — Run

Content | Format | Selection

Display — Skip line on: <none> — Skip page on: <none> — Summarize on: <none>

Target comparison — Use: None, Target 1, Target 2 — Display: Early dates, Late dates, Full range

Sort by: Total float

Sort on	Order
Early start	Ascend
Total float	Ascend

Display: Early start, Early finish, Late start, Late finish, Calendar, Activity codes, Total float, Hours

Transfer... OK Cancel Help

Budget | Codes | Constr | Cost | Custom | Dates | Log | Pred | Res | Succ | WBS | Help

ID 10 Clear Site — Previous Next <<Less

OD 2 Pct 0.0 Cal 1 — ES 01OCT01 — EF 02OCT01 — TF: 0

RD 2 Type Task — LS 01OCT01 — LF 02OCT01 — FF: 0

ADC — Supr WBS

Classic Schedule Layout — All Activities

Screen Image 8-7
Modified Schedule Reports Format Window

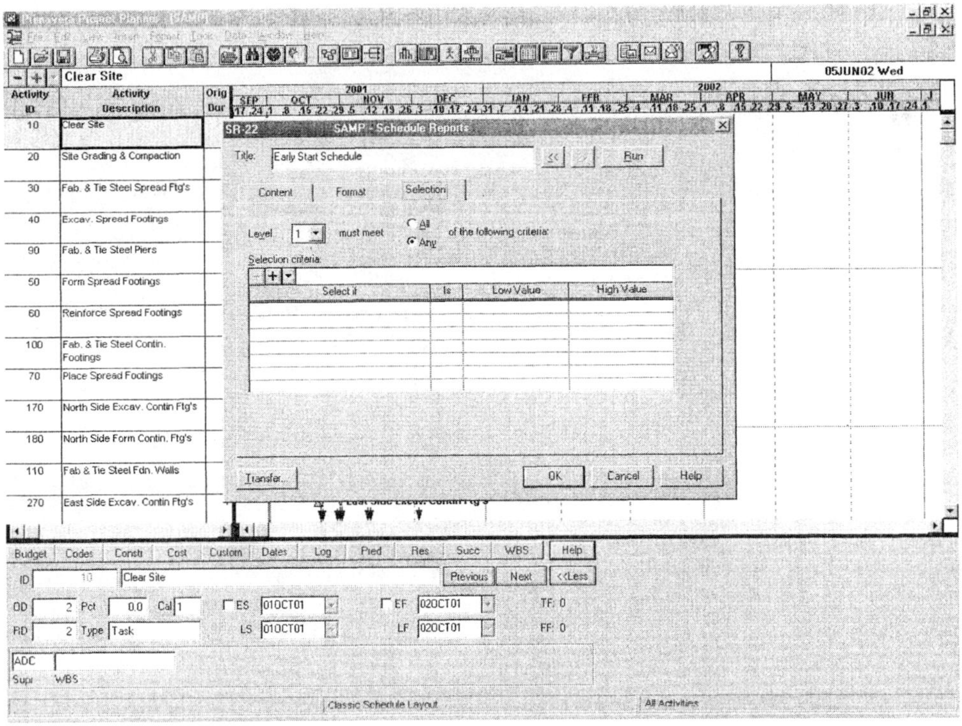

Primavera Project Planner - [SAMP]

Clear Site — 05JUN02 Wed

Activity ID	Activity Description	Orig Dur
10	Clear Site	
20	Site Grading & Compaction	
30	Fab. & Tie Steel Spread Ftg's	
40	Excav. Spread Footings	
90	Fab. & Tie Steel Piers	
50	Form Spread Footings	
60	Reinforce Spread Footings	
100	Fab. & Tie Steel Contin. Footings	
70	Place Spread Footings	
170	North Side Excav. Contin Ftg's	
180	North Side Form Contin. Ftg's	
110	Fab & Tie Steel Fdn. Walls	
270	East Side Excav. Contin Ftg's	

SR-22 SAMP - Schedule Reports

Title: Early Start Schedule — Run

Content | Format | Selection

Level 1 must meet — All / Any — of the following criteria:

Selection criteria:

Select if	Is	Low Value	High Value

Transfer... OK Cancel Help

Budget | Codes | Constr | Cost | Custom | Dates | Log | Pred | Res | Succ | WBS | Help

ID 10 Clear Site — Previous Next <<Less

OD 2 Pct 0.0 Cal 1 — ES 01OCT01 — EF 02OCT01 — TF: 0

RD 2 Type Task — LS 01OCT01 — LF 02OCT01 — FF: 0

ADC — Supr WBS

Classic Schedule Layout — All Activities

Screen Image 8-8
Schedule Reports Selection Window

Table 8-1
Acceptable Operators

Operator Designation	Description
EQ	Equal to
NE	Not equal to
GT	Greater than
LT	Less than
WR	Within range
NR	Not within range
CN	Contains
SN	Does not contain

report. If the *All* option is selected, then only the activities that meet *all* of the criteria are included in the report.

On the *Selection criteria* table the *Select if* field is where the variable's field name is entered. If the precise name and spelling are not known, clicking on the *Select if* box will activate a pull-down menu. The next field on the *Selection criteria* table is the operator, or *Is* field. This field holds the conditional operators. Primavera Project Planner uses a two-character code rather than graphical mathematical symbols. The acceptable codes and their use are listed in Table 8-1 and Screen Image 8-9.

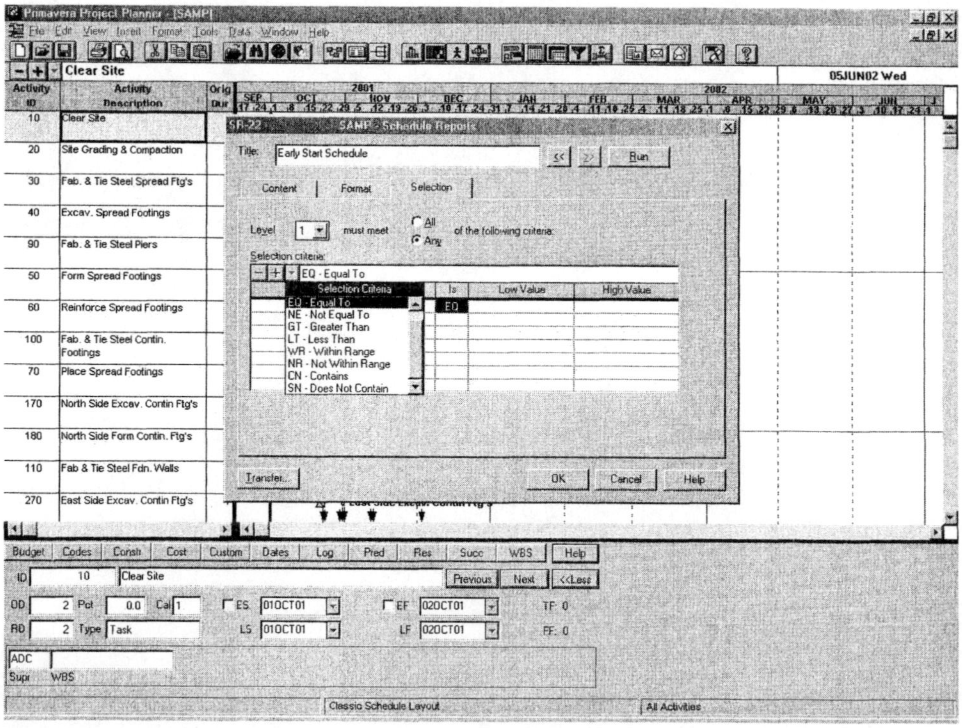

Screen Image 8-9
Selection Criteria

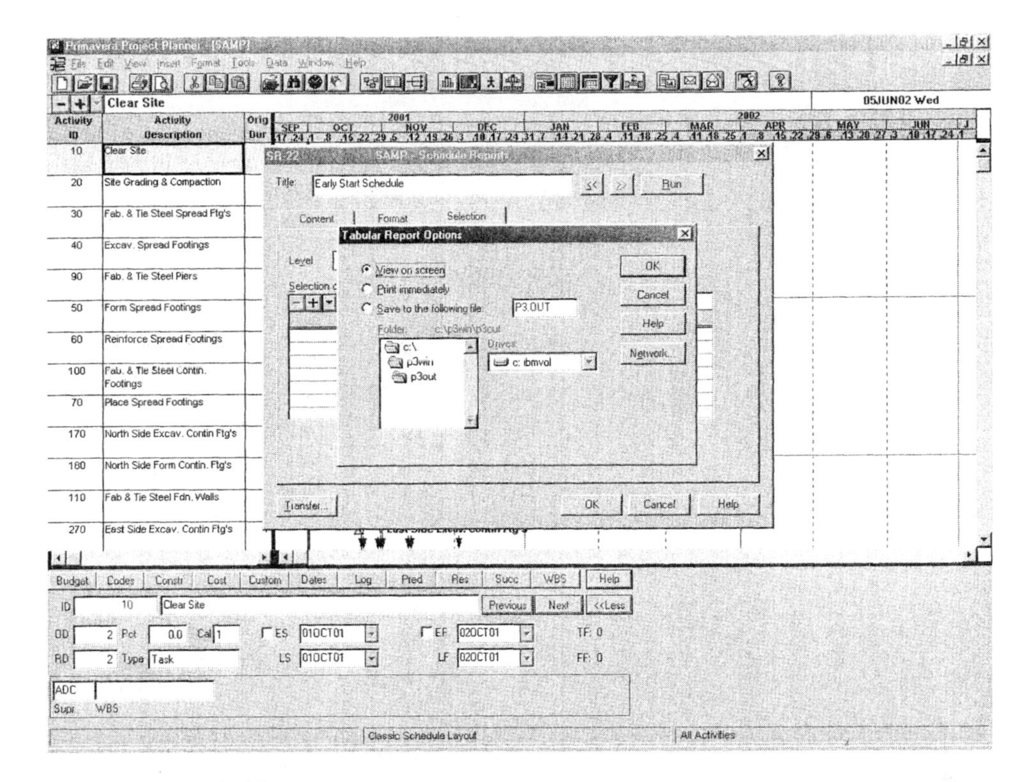

Screen Image 8-10
Tabular Report Options Window

The operators can be selected either by entering the appropriate two-character code or by using the pull-down menu. The *Low Value* and *High Value* fields hold the parameters for the conditional statements. If the conditional operator addresses a single value (EQ, NE, LT, or GT), then that criteria is entered in the *Low Value* column. Conversely, if the selection criteria deals with a range of possible values, then a lower and upper limit needs to be specified. For example, to select all activity ID's from 10 to 100 enter *Activity ID* in the *Select if* field, *WR* in the *Is* field, and set the *Low Value* to 10 and the *High Value* to 100. For the current report all of the activities are to be included; therefore no input on the *Selection* tab is required. If there are no entries on the *Selection criteria* table, all of the activities within the schedule database will be included.

Once the three tabs have been updated, clicking the *Run* button will activate the *Tabular Report Options* window as shown in Screen Image 8-10. This window allows the user to specify where to route the report. The options on this window are to send the output directly to the printer, view it on the computer screen, or save it to a file. If one of the latter two options is selected, the reports can still be printed. If the *Print immediately* option is selected, it is important that the default printer, paper orientation, and size be specified prior to executing the report specifications. The default printer setup is done through the *Windows*© printer configuration or by selecting the *File/Print Setup* option within Primavera Project Planner.

If the *View on screen* option is selected, Primavera Project Planner will initiate a program called *Primavera Look*. This program is designed to allow the user to view the output, annotate the output, and route the output to a printer. Screen Image 8-11 is a sample of the output when viewed through Primavera Look. To route this output to the printer, simply click on the printer icon or select the *File* command followed by the *Print* option as shown in Screen Image 8-12. Either of these actions will open the *Print* window as shown in Screen Image 8-13. From that window the pages, copies, and quality can be manipulated. The printer is the default printer that is specified within the Windows operating system.

To dispose of the output, click on the *File* command and select either the *Exit* or *Close* option as shown in Screen Image 8-14. The *Close* option disposes of the output but

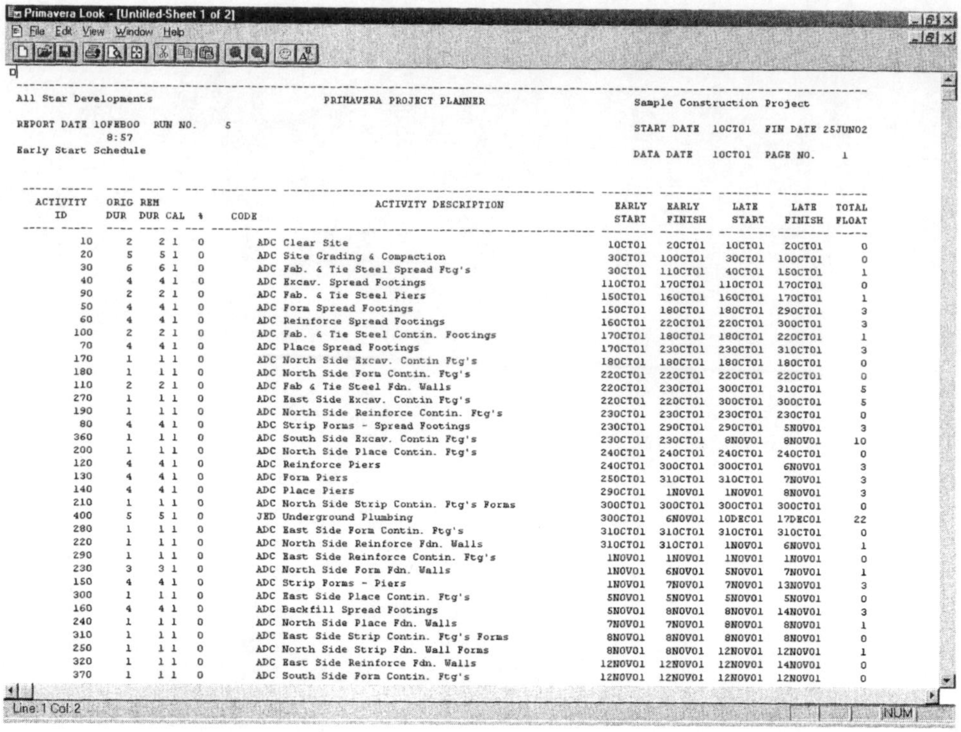

Primavera Look - [Untitled-Sheet 1 of 2]

File Edit View Window Help

All Star Developments PRIMAVERA PROJECT PLANNER Sample Construction Project

REPORT DATE 10FEB00 RUN NO. 5 START DATE 1OCT01 FIN DATE 25JUN02
8:57
Early Start Schedule DATA DATE 1OCT01 PAGE NO. 1

ACTIVITY ID	ORIG DUR	REM DUR	CAL	%	CODE	ACTIVITY DESCRIPTION	EARLY START	EARLY FINISH	LATE START	LATE FINISH	TOTAL FLOAT
10	2	2	1	0	ADC	Clear Site	1OCT01	2OCT01	1OCT01	2OCT01	0
20	5	5	1	0	ADC	Site Grading & Compaction	3OCT01	10OCT01	3OCT01	10OCT01	0
30	6	6	1	0	ADC	Fab. & Tie Steel Spread Ftg's	3OCT01	11OCT01	4OCT01	15OCT01	1
40	4	4	1	0	ADC	Excav. Spread Footings	11OCT01	17OCT01	11OCT01	17OCT01	0
90	2	2	1	0	ADC	Fab. & Tie Steel Piers	15OCT01	16OCT01	16OCT01	17OCT01	1
50	4	4	1	0	ADC	Form Spread Footings	15OCT01	18OCT01	18OCT01	29OCT01	3
60	4	4	1	0	ADC	Reinforce Spread Footings	16OCT01	22OCT01	22OCT01	30OCT01	3
100	2	2	1	0	ADC	Fab. & Tie Steel Contin. Footings	17OCT01	18OCT01	18OCT01	22OCT01	1
70	4	4	1	0	ADC	Place Spread Footings	17OCT01	23OCT01	23OCT01	31OCT01	3
170	1	1	1	0	ADC	North Side Excav. Contin Ftg's	18OCT01	18OCT01	18OCT01	18OCT01	0
180	1	1	1	0	ADC	North Side Form Contin. Ftg's	22OCT01	22OCT01	22OCT01	22OCT01	0
110	2	2	1	0	ADC	Fab & Tie Steel Fdn. Walls	22OCT01	23OCT01	30OCT01	31OCT01	5
270	1	1	1	0	ADC	East Side Excav. Contin Ftg's	22OCT01	22OCT01	30OCT01	30OCT01	5
190	1	1	1	0	ADC	North Side Reinforce Contin. Ftg's	23OCT01	23OCT01	23OCT01	23OCT01	0
80	4	4	1	0	ADC	Strip Forms - Spread Footings	23OCT01	29OCT01	29OCT01	5NOV01	3
360	1	1	1	0	ADC	South Side Excav. Contin Ftg's	23OCT01	23OCT01	8NOV01	8NOV01	10
200	1	1	1	0	ADC	North Side Place Contin. Ftg's	24OCT01	24OCT01	24OCT01	24OCT01	0
120	4	4	1	0	ADC	Reinforce Piers	24OCT01	30OCT01	30OCT01	6NOV01	3
130	4	4	1	0	ADC	Form Piers	25OCT01	31OCT01	31OCT01	7NOV01	3
140	4	4	1	0	ADC	Place Piers	29OCT01	1NOV01	1NOV01	8NOV01	3
210	1	1	1	0	ADC	North Side Strip Contin. Ftg's Forms	30OCT01	30OCT01	30OCT01	30OCT01	0
400	5	5	1	0	JED	Underground Plumbing	30OCT01	6NOV01	10DEC01	17DEC01	22
280	1	1	1	0	ADC	East Side Form Contin. Ftg's	31OCT01	31OCT01	31OCT01	31OCT01	0
220	1	1	1	0	ADC	North Side Reinforce Fdn. Walls	31OCT01	31OCT01	1NOV01	6NOV01	1
290	1	1	1	0	ADC	East Side Reinforce Contin. Ftg's	1NOV01	1NOV01	1NOV01	1NOV01	0
230	3	3	1	0	ADC	North Side Form Fdn. Walls	1NOV01	6NOV01	5NOV01	7NOV01	1
150	4	4	1	0	ADC	Strip Forms - Piers	1NOV01	7NOV01	7NOV01	13NOV01	3
300	1	1	1	0	ADC	East Side Place Contin. Ftg's	5NOV01	5NOV01	5NOV01	5NOV01	0
160	4	4	1	0	ADC	Backfill Spread Footings	5NOV01	8NOV01	8NOV01	14NOV01	3
240	1	1	1	0	ADC	North Side Place Fdn. Walls	7NOV01	7NOV01	8NOV01	8NOV01	1
310	1	1	1	0	ADC	East Side Strip Contin. Ftg's Forms	8NOV01	8NOV01	8NOV01	8NOV01	0
250	1	1	1	0	ADC	North Side Strip Fdn. Wall Forms	8NOV01	8NOV01	12NOV01	12NOV01	1
320	1	1	1	0	ADC	East Side Reinforce Fdn. Walls	12NOV01	12NOV01	12NOV01	14NOV01	0
370	1	1	1	0	ADC	South Side Form Contin. Ftg's	12NOV01	12NOV01	12NOV01	12NOV01	0

Line: 1 Col: 2 NUM

Screen Image 8-11
Primavera Look

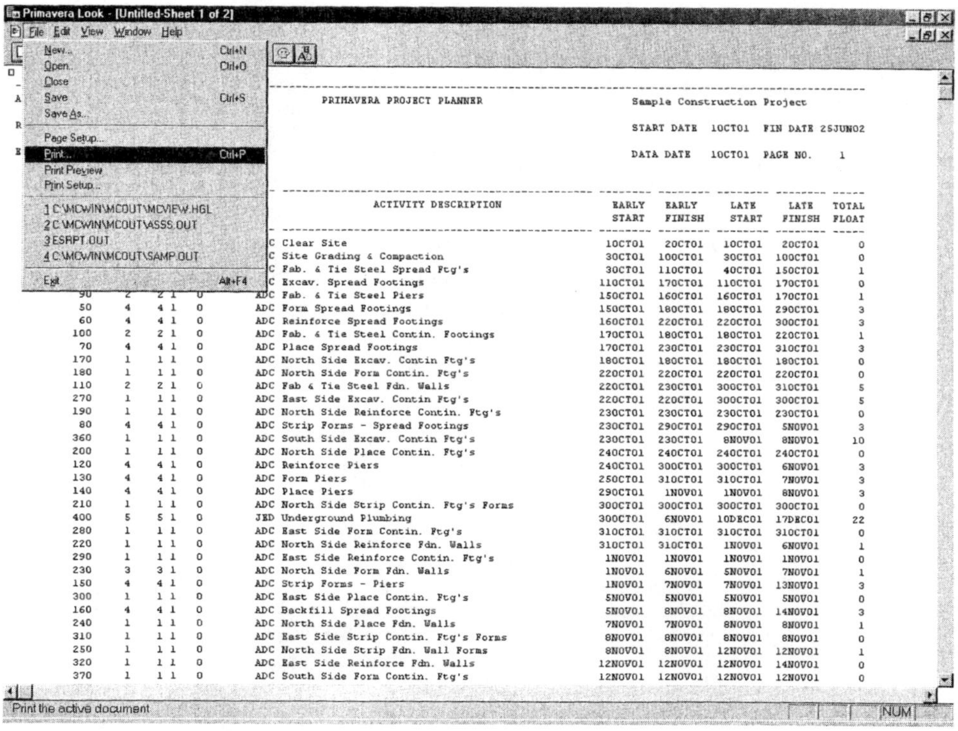

Primavera Look - [Untitled-Sheet 1 of 2]

File Edit View Window Help

New... Ctrl+N
Open... Ctrl+O
Close
Save Ctrl+S
Save As...
Page Setup...
Print... Ctrl+P
Print Preview
Print Setup...
1 C:\MCWIN\MCOUT\MCVIEW.HGL
2 C:\MCWIN\MCOUT\ASSS.OUT
3 ESRPT.OUT
4 C:\MCWIN\MCOUT\SAMP.OUT
Exit Alt+F4

PRIMAVERA PROJECT PLANNER Sample Construction Project

START DATE 1OCT01 FIN DATE 25JUN02

DATA DATE 1OCT01 PAGE NO. 1

ACTIVITY ID	ORIG DUR	REM DUR	CAL	%	CODE	ACTIVITY DESCRIPTION	EARLY START	EARLY FINISH	LATE START	LATE FINISH	TOTAL FLOAT
					C	Clear Site	1OCT01	2OCT01	1OCT01	2OCT01	0
					C	Site Grading & Compaction	3OCT01	10OCT01	3OCT01	10OCT01	0
					C	Fab. & Tie Steel Spread Ftg's	3OCT01	11OCT01	4OCT01	15OCT01	1
					C	Excav. Spread Footings	11OCT01	17OCT01	11OCT01	17OCT01	0
90	2	2	1	0	ADC	Fab. & Tie Steel Piers	15OCT01	16OCT01	16OCT01	17OCT01	1
50	4	4	1	0	ADC	Form Spread Footings	15OCT01	18OCT01	18OCT01	29OCT01	3
60	4	4	1	0	ADC	Reinforce Spread Footings	16OCT01	22OCT01	22OCT01	30OCT01	3
100	2	2	1	0	ADC	Fab. & Tie Steel Contin. Footings	17OCT01	18OCT01	18OCT01	22OCT01	1
70	4	4	1	0	ADC	Place Spread Footings	17OCT01	23OCT01	23OCT01	31OCT01	3
170	1	1	1	0	ADC	North Side Excav. Contin Ftg's	18OCT01	18OCT01	18OCT01	18OCT01	0
180	1	1	1	0	ADC	North Side Form Contin. Ftg's	22OCT01	22OCT01	22OCT01	22OCT01	0
110	2	2	1	0	ADC	Fab & Tie Steel Fdn. Walls	22OCT01	23OCT01	30OCT01	31OCT01	5
270	1	1	1	0	ADC	East Side Excav. Contin Ftg's	22OCT01	22OCT01	30OCT01	30OCT01	5
190	1	1	1	0	ADC	North Side Reinforce Contin. Ftg's	23OCT01	23OCT01	23OCT01	23OCT01	0
80	4	4	1	0	ADC	Strip Forms - Spread Footings	23OCT01	29OCT01	29OCT01	5NOV01	3
360	1	1	1	0	ADC	South Side Excav. Contin Ftg's	23OCT01	23OCT01	8NOV01	8NOV01	10
200	1	1	1	0	ADC	North Side Place Contin. Ftg's	24OCT01	24OCT01	24OCT01	24OCT01	0
120	4	4	1	0	ADC	Reinforce Piers	24OCT01	30OCT01	30OCT01	6NOV01	3
130	4	4	1	0	ADC	Form Piers	25OCT01	31OCT01	31OCT01	7NOV01	3
140	4	4	1	0	ADC	Place Piers	29OCT01	1NOV01	1NOV01	8NOV01	3
210	1	1	1	0	ADC	North Side Strip Contin. Ftg's Forms	30OCT01	30OCT01	30OCT01	30OCT01	0
400	5	5	1	0	JED	Underground Plumbing	30OCT01	6NOV01	10DEC01	17DEC01	22
280	1	1	1	0	ADC	East Side Form Contin. Ftg's	31OCT01	31OCT01	31OCT01	31OCT01	0
220	1	1	1	0	ADC	North Side Reinforce Fdn. Walls	31OCT01	31OCT01	1NOV01	6NOV01	1
290	1	1	1	0	ADC	East Side Reinforce Contin. Ftg's	1NOV01	1NOV01	1NOV01	1NOV01	0
230	3	3	1	0	ADC	North Side Form Fdn. Walls	1NOV01	6NOV01	5NOV01	7NOV01	1
150	4	4	1	0	ADC	Strip Forms - Piers	1NOV01	7NOV01	7NOV01	13NOV01	3
300	1	1	1	0	ADC	East Side Place Contin. Ftg's	5NOV01	5NOV01	5NOV01	5NOV01	0
160	4	4	1	0	ADC	Backfill Spread Footings	5NOV01	8NOV01	8NOV01	14NOV01	3
240	1	1	1	0	ADC	North Side Place Fdn. Walls	7NOV01	7NOV01	8NOV01	8NOV01	1
310	1	1	1	0	ADC	East Side Strip Contin. Ftg's Forms	8NOV01	8NOV01	8NOV01	8NOV01	0
250	1	1	1	0	ADC	North Side Strip Fdn. Wall Forms	8NOV01	8NOV01	12NOV01	12NOV01	1
320	1	1	1	0	ADC	East Side Reinforce Fdn. Walls	12NOV01	12NOV01	12NOV01	14NOV01	0
370	1	1	1	0	ADC	South Side Form Contin. Ftg's	12NOV01	12NOV01	12NOV01	12NOV01	0

Print the active document NUM

Screen Image 8-12
Opening the Print Window

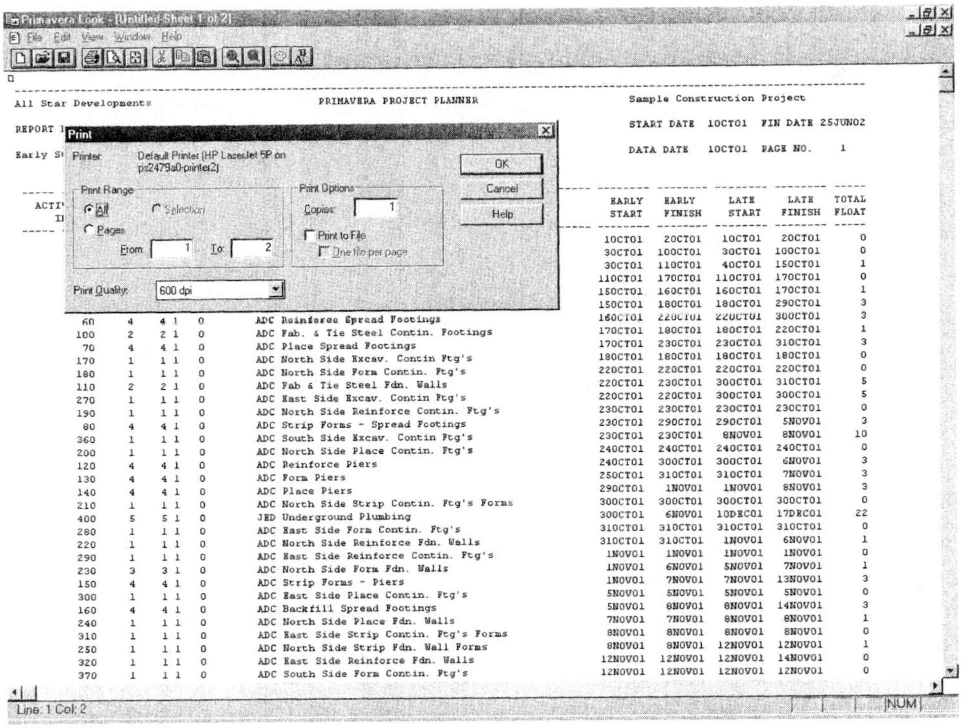

Primavera Look - [Untitled-Sheet 1 of 2]

File Edit View Window Help

All Star Developments PRIMAVERA PROJECT PLANNER Sample Construction Project

REPORT I

START DATE 1OCT01 FIN DATE 25JUN02

Early S

DATA DATE 1OCT01 PAGE NO. 1

ACTI

I

Print

Printer: Default Printer (HP LaserJet 5P on ps2479a0-printer2)

OK

Cancel

Help

Print Range

All

Selection

Pages

From: 1 To: 2

Print Options

Copies: 1

Print to File

One file per page

Print Quality: 600 dpi

						EARLY START	EARLY FINISH	LATE START	LATE FINISH	TOTAL FLOAT
						1OCT01	2OCT01	1OCT01	2OCT01	0
						3OCT01	10OCT01	3OCT01	10OCT01	0
						3OCT01	11OCT01	4OCT01	15OCT01	1
						11OCT01	17OCT01	11OCT01	17OCT01	0
						15OCT01	16OCT01	16OCT01	17OCT01	1
						15OCT01	18OCT01	18OCT01	29OCT01	3
60	4	4	1	0	ADC Reinforce Spread Footings	16OCT01	22OCT01	22OCT01	30OCT01	3
100	2	2	1	0	ADC Fab. & Tie Steel Contin. Footings	17OCT01	18OCT01	18OCT01	22OCT01	1
70	4	4	1	0	ADC Place Spread Footings	17OCT01	23OCT01	23OCT01	31OCT01	3
170	1	1	1	0	ADC North Side Excav. Contin Ftg's	18OCT01	18OCT01	18OCT01	18OCT01	0
180	1	1	1	0	ADC North Side Form Contin. Ftg's	22OCT01	22OCT01	22OCT01	22OCT01	0
110	2	2	1	0	ADC Fab & Tie Steel Fdn. Walls	22OCT01	23OCT01	30OCT01	31OCT01	5
270	1	1	1	0	ADC East Side Excav. Contin Ftg's	22OCT01	22OCT01	30OCT01	30OCT01	5
190	1	1	1	0	ADC North Side Reinforce Contin. Ftg's	23OCT01	23OCT01	23OCT01	23OCT01	0
80	4	4	1	0	ADC Strip Forms - Spread Footings	23OCT01	29OCT01	29OCT01	5NOV01	3
360	1	1	1	0	ADC South Side Excav. Contin Ftg's	23OCT01	23OCT01	8NOV01	8NOV01	10
200	1	1	1	0	ADC North Side Place Contin. Ftg's	24OCT01	24OCT01	24OCT01	24OCT01	0
120	4	4	1	0	ADC Reinforce Piers	24OCT01	30OCT01	30OCT01	6NOV01	3
130	4	4	1	0	ADC Form Piers	25OCT01	31OCT01	31OCT01	7NOV01	3
140	4	4	1	0	ADC Place Piers	29OCT01	1NOV01	1NOV01	8NOV01	3
210	1	1	1	0	ADC North Side Strip Contin. Ftg's Forms	30OCT01	30OCT01	30OCT01	30OCT01	0
400	5	5	1	0	JED Underground Plumbing	30OCT01	6NOV01	10DEC01	17DEC01	22
280	1	1	1	0	ADC East Side Form Contin. Ftg's	31OCT01	31OCT01	31OCT01	31OCT01	0
220	1	1	1	0	ADC North Side Reinforce Fdn. Walls	31OCT01	31OCT01	1NOV01	6NOV01	1
290	1	1	1	0	ADC East Side Reinforce Contin. Ftg's	1NOV01	1NOV01	1NOV01	1NOV01	0
230	3	3	1	0	ADC North Side Form Fdn. Walls	1NOV01	6NOV01	5NOV01	7NOV01	1
150	4	4	1	0	ADC Strip Forms - Piers	1NOV01	7NOV01	7NOV01	13NOV01	3
300	1	1	1	0	ADC East Side Place Contin. Ftg's	5NOV01	5NOV01	5NOV01	5NOV01	0
160	4	4	1	0	ADC Backfill Spread Footings	5NOV01	8NOV01	8NOV01	14NOV01	3
240	1	1	1	0	ADC North Side Place Fdn. Walls	7NOV01	7NOV01	8NOV01	8NOV01	1
310	1	1	1	0	ADC East Side Strip Contin. Ftg's Forms	8NOV01	8NOV01	8NOV01	8NOV01	0
250	1	1	1	0	ADC North Side Strip Fdn. Wall Forms	8NOV01	8NOV01	12NOV01	12NOV01	1
320	1	1	1	0	ADC East Side Reinforce Fdn. Walls	12NOV01	12NOV01	12NOV01	14NOV01	0
370	1	1	1	0	ADC South Side Form Contin. Ftg's	12NOV01	12NOV01	12NOV01	12NOV01	0

Line: 1 Col: 2 NUM

Screen Image 8-13
The Print Window

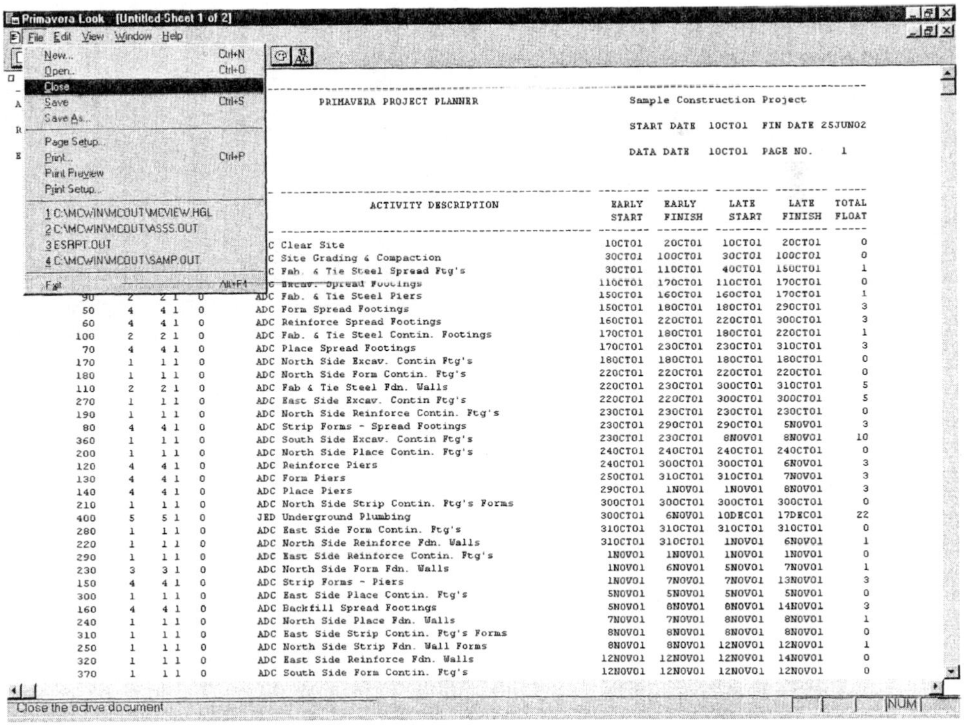

Primavera Look [Untitled-Sheet 1 of 2]

File Edit View Window Help

New... Ctrl+N

Open... Ctrl+O

Close

Save Ctrl+S

Save As...

Page Setup...

Print... Ctrl+P

Print Preview

Print Setup...

1 C:\MCWIN\MCOUT\MCVIEW.HGL

2 C:\MCWIN\MCOUT\ASSS.OUT

3 ESRPT.OUT

4 C:\MCWIN\MCOUT\SAMP.OUT

Exit Alt+F4

PRIMAVERA PROJECT PLANNER Sample Construction Project

START DATE 1OCT01 FIN DATE 25JUN02

DATA DATE 1OCT01 PAGE NO. 1

					ACTIVITY DESCRIPTION	EARLY START	EARLY FINISH	LATE START	LATE FINISH	TOTAL FLOAT
					C Clear Site	1OCT01	2OCT01	1OCT01	2OCT01	0
					C Site Grading & Compaction	3OCT01	10OCT01	3OCT01	10OCT01	0
					C Fab. & Tie Steel Spread Ftg's	3OCT01	11OCT01	4OCT01	15OCT01	1
					C Excav. Spread Footings	11OCT01	17OCT01	11OCT01	17OCT01	0
90	2	2	1	0	ADC Fab. & Tie Steel Piers	15OCT01	16OCT01	16OCT01	17OCT01	1
50	4	4	1	0	ADC Form Spread Footings	15OCT01	18OCT01	18OCT01	29OCT01	3
60	4	4	1	0	ADC Reinforce Spread Footings	16OCT01	22OCT01	22OCT01	30OCT01	3
100	2	2	1	0	ADC Fab. & Tie Steel Contin. Footings	17OCT01	18OCT01	18OCT01	22OCT01	1
70	4	4	1	0	ADC Place Spread Footings	17OCT01	23OCT01	23OCT01	31OCT01	3
170	1	1	1	0	ADC North Side Excav. Contin Ftg's	18OCT01	18OCT01	18OCT01	18OCT01	0
180	1	1	1	0	ADC North Side Form Contin. Ftg's	22OCT01	22OCT01	22OCT01	22OCT01	0
110	2	2	1	0	ADC Fab & Tie Steel Fdn. Walls	22OCT01	23OCT01	30OCT01	31OCT01	5
270	1	1	1	0	ADC East Side Excav. Contin Ftg's	22OCT01	22OCT01	30OCT01	30OCT01	5
190	1	1	1	0	ADC North Side Reinforce Contin. Ftg's	23OCT01	23OCT01	23OCT01	23OCT01	0
80	4	4	1	0	ADC Strip Forms - Spread Footings	23OCT01	29OCT01	29OCT01	5NOV01	3
360	1	1	1	0	ADC South Side Excav. Contin Ftg's	23OCT01	23OCT01	8NOV01	8NOV01	10
200	1	1	1	0	ADC North Side Place Contin. Ftg's	24OCT01	24OCT01	24OCT01	24OCT01	0
120	4	4	1	0	ADC Reinforce Piers	24OCT01	30OCT01	30OCT01	6NOV01	3
130	4	4	1	0	ADC Form Piers	25OCT01	31OCT01	31OCT01	7NOV01	3
140	4	4	1	0	ADC Place Piers	29OCT01	1NOV01	1NOV01	8NOV01	3
210	1	1	1	0	ADC North Side Strip Contin. Ftg's Forms	30OCT01	30OCT01	30OCT01	30OCT01	0
400	5	5	1	0	JED Underground Plumbing	30OCT01	6NOV01	10DEC01	17DEC01	22
280	1	1	1	0	ADC East Side Form Contin. Ftg's	31OCT01	31OCT01	31OCT01	31OCT01	0
220	1	1	1	0	ADC North Side Reinforce Fdn. Walls	31OCT01	31OCT01	1NOV01	6NOV01	1
290	1	1	1	0	ADC East Side Reinforce Contin. Ftg's	1NOV01	1NOV01	1NOV01	1NOV01	0
230	3	3	1	0	ADC North Side Form Fdn. Walls	1NOV01	6NOV01	5NOV01	7NOV01	1
150	4	4	1	0	ADC Strip Forms - Piers	1NOV01	7NOV01	7NOV01	13NOV01	3
300	1	1	1	0	ADC East Side Place Contin. Ftg's	5NOV01	5NOV01	5NOV01	5NOV01	0
160	4	4	1	0	ADC Backfill Spread Footings	5NOV01	8NOV01	8NOV01	14NOV01	3
240	1	1	1	0	ADC North Side Place Fdn. Walls	7NOV01	7NOV01	8NOV01	8NOV01	1
310	1	1	1	0	ADC East Side Strip Contin. Ftg's Forms	8NOV01	8NOV01	8NOV01	8NOV01	0
250	1	1	1	0	ADC North Side Strip Fdn. Wall Forms	8NOV01	8NOV01	12NOV01	12NOV01	1
320	1	1	1	0	ADC East Side Reinforce Fdn. Walls	12NOV01	12NOV01	12NOV01	14NOV01	0
370	1	1	1	0	ADC South Side Form Contin. Ftg's	12NOV01	12NOV01	12NOV01	12NOV01	0

Close the active document NUM

Screen Image 8-14
File Options

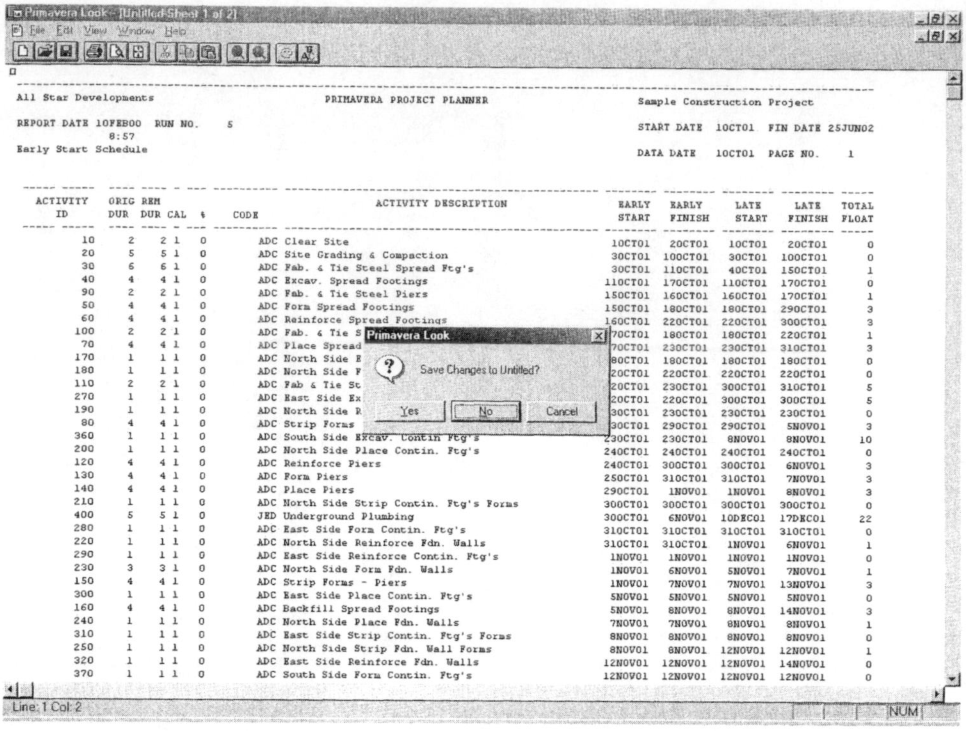

Primavera Look - [Untitled-Sheet 1 of 2]

File Edit View Window Help

All Star Developments PRIMAVERA PROJECT PLANNER Sample Construction Project

REPORT DATE 10FEB00 RUN NO. 5 START DATE 1OCT01 FIN DATE 25JUN02
8:57

Early Start Schedule DATA DATE 1OCT01 PAGE NO. 1

ACTIVITY ID	ORIG DUR	REM DUR	CAL	%	CODE	ACTIVITY DESCRIPTION	EARLY START	EARLY FINISH	LATE START	LATE FINISH	TOTAL FLOAT
10	2	2	1	0	ADC	Clear Site	1OCT01	2OCT01	1OCT01	2OCT01	0
20	5	5	1	0	ADC	Site Grading & Compaction	3OCT01	10OCT01	3OCT01	10OCT01	0
30	6	6	1	0	ADC	Fab. & Tie Steel Spread Ftg's	3OCT01	11OCT01	4OCT01	15OCT01	1
40	4	4	1	0	ADC	Excav. Spread Footings	11OCT01	17OCT01	11OCT01	17OCT01	0
90	2	2	1	0	ADC	Fab. & Tie Steel Piers	15OCT01	16OCT01	16OCT01	17OCT01	1
50	4	4	1	0	ADC	Form Spread Footings	15OCT01	18OCT01	18OCT01	29OCT01	3
60	4	4	1	0	ADC	Reinforce Spread Footings	16OCT01	22OCT01	22OCT01	30OCT01	3
100	2	2	1	0	ADC	Fab. & Tie S	7OCT01	18OCT01	18OCT01	22OCT01	1
70	4	4	1	0	ADC	Place Spread	7OCT01	23OCT01	23OCT01	31OCT01	3
170	1	1	1	0	ADC	North Side E	8OCT01	18OCT01	18OCT01	18OCT01	0
180	1	1	1	0	ADC	North Side F	2OCT01	22OCT01	22OCT01	22OCT01	0
110	2	2	1	0	ADC	Fab & Tie St	2OCT01	23OCT01	30OCT01	31OCT01	5
270	1	1	1	0	ADC	East Side Ex	2OCT01	22OCT01	30OCT01	30OCT01	5
190	1	1	1	0	ADC	North Side R	3OCT01	23OCT01	23OCT01	23OCT01	0
80	4	4	1	0	ADC	Strip Forms	3OCT01	29OCT01	29OCT01	5NOV01	3
360	1	1	1	0	ADC	South Side Excav. Contin Ftg's	23OCT01	23OCT01	8NOV01	8NOV01	10
200	1	1	1	0	ADC	North Side Place Contin. Ftg's	24OCT01	24OCT01	24OCT01	24OCT01	0
120	4	4	1	0	ADC	Reinforce Piers	24OCT01	30OCT01	30OCT01	6NOV01	3
130	4	4	1	0	ADC	Form Piers	25OCT01	31OCT01	31OCT01	7NOV01	3
140	4	4	1	0	ADC	Place Piers	29OCT01	1NOV01	1NOV01	8NOV01	3
210	1	1	1	0	ADC	North Side Strip Contin. Ftg's Forms	30OCT01	30OCT01	30OCT01	30OCT01	0
400	5	5	1	0	JED	Underground Plumbing	30OCT01	6NOV01	10DEC01	17DEC01	22
280	1	1	1	0	ADC	East Side Form Contin. Ftg's	31OCT01	31OCT01	31OCT01	31OCT01	0
220	1	1	1	0	ADC	North Side Reinforce Fdn. Walls	31OCT01	31OCT01	1NOV01	6NOV01	1
290	1	1	1	0	ADC	East Side Reinforce Contin. Ftg's	1NOV01	1NOV01	1NOV01	1NOV01	0
230	3	3	1	0	ADC	North Side Form Fdn. Walls	1NOV01	6NOV01	5NOV01	7NOV01	1
150	4	4	1	0	ADC	Strip Forms - Piers	1NOV01	7NOV01	7NOV01	13NOV01	3
300	1	1	1	0	ADC	East Side Place Contin. Ftg's	5NOV01	5NOV01	5NOV01	5NOV01	0
160	4	4	1	0	ADC	Backfill Spread Footings	5NOV01	8NOV01	8NOV01	14NOV01	3
240	1	1	1	0	ADC	North Side Place Fdn. Walls	7NOV01	7NOV01	8NOV01	8NOV01	1
310	1	1	1	0	ADC	East Side Strip Contin. Ftg's Forms	8NOV01	8NOV01	8NOV01	8NOV01	0
250	1	1	1	0	ADC	North Side Strip Fdn. Wall Forms	8NOV01	8NOV01	12NOV01	12NOV01	1
320	1	1	1	0	ADC	East Side Reinforce Fdn. Walls	12NOV01	12NOV01	12NOV01	14NOV01	0
370	1	1	1	0	ADC	South Side Form Contin. Ftg's	12NOV01	12NOV01	12NOV01	12NOV01	0

Primavera Look

Save Changes to Untitled?

Yes No Cancel

Line: 1 Col: 2 NUM

Screen Image 8-15
Primavera Look File Options

leaves Primavera Look running, while the *Exit* command disposes of the output and closes Primavera Look. Regardless of whether the *Close* or *Exit* option is selected, the user will be prompted to verify any wish to dispose of the output. Screen Image 8-15 is the prompt window that is opened when disposing of the output. By clicking on the *No* button the output is disposed of, whereas if the *Yes* button is clicked, the *Save As* window shown in Screen Image 8-16 will appear. From this window the user can specify the desired location for saving the output file. Once the output file is saved, it can be opened later and manipulated through Primavera Look or a word processor. Output 8-1 is the actual output generated from the Early Start report specifications.

Once this or any report is generated, it is important for the scheduler to review the output. Some of the items that should be critically evaluated are the order of activities and whether the start dates appear reasonable. Furthermore, verify that no extraneous information or activities are presented within the report. This and all output should be critically reviewed prior to publishing or sharing with other individuals.

Critical Activity List Report

Another important report is a list of the activities on the critical path. This report, in terms of report specifications, is very similar to the previous example. The first difference between the two, however, is the sort order. The sort order in the previous report was early start and then total float; in this report the sort order needs to be by early start only. This sort in conjunction with selecting all activities with zero float will create a report which contains only critical activities listed in the order in which they must be executed. By opening the *Format* tab on the *Schedule Reports* window, the sort option of *Early Start* can be added, as shown in Screen Image 8-17. Then the *Selection* tab needs to be clicked and the zero float criteria entered. Screen Image 8-18 shows the selection criteria for including only the critical activities on the report. Once these specifications are entered, the *Run* button can be clicked and the desired output option selected. Output 8-2 is an example of the report that would be generated from these report specifications.

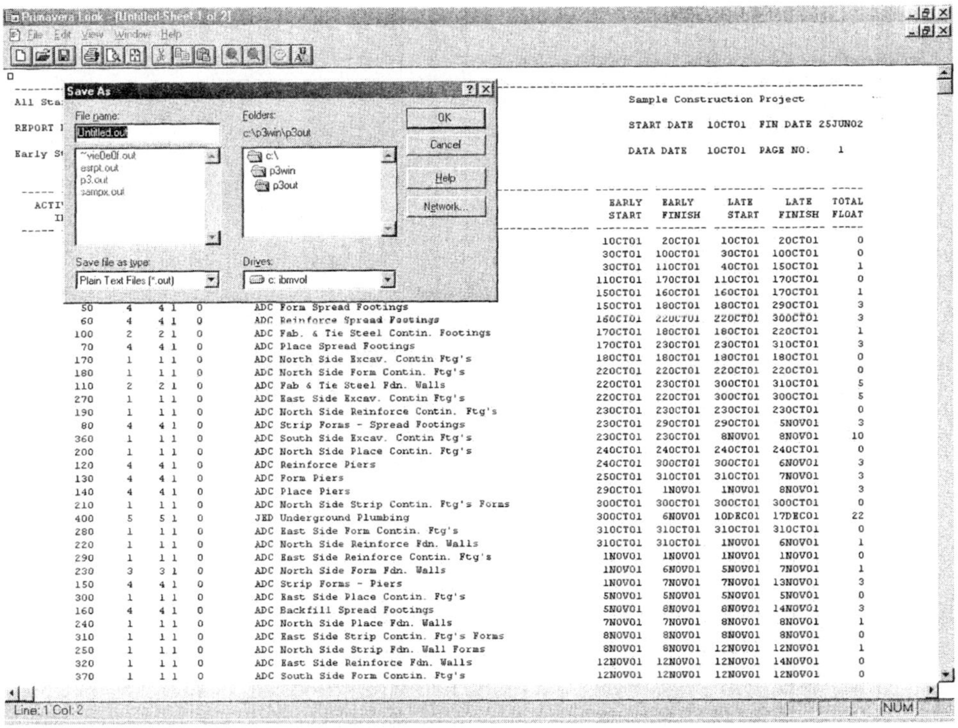

Primavera Look - [Untitled-Sheet 1 of 2]
File Edit View Window Help

Save As
File name: Untitled.out
Folders: c:\p3win\p3out
~vie0e0f.out
esrpt.out
p3.out
sampx.out
c:\
p3win
p3out
OK
Cancel
Help
Network...
Save file as type: Plain Text Files (*.out)
Drives: c: ibmvol

All Sta
REPORT I
Early S

Sample Construction Project

START DATE 1OCT01 FIN DATE 25JUN02

DATA DATE 1OCT01 PAGE NO. 1

ACTI… I…						EARLY START	EARLY FINISH	LATE START	LATE FINISH	TOTAL FLOAT
						1OCT01	2OCT01	1OCT01	2OCT01	0
						3OCT01	10OCT01	3OCT01	10OCT01	0
						3OCT01	11OCT01	4OCT01	15OCT01	1
						11OCT01	17OCT01	11OCT01	17OCT01	0
						15OCT01	16OCT01	16OCT01	17OCT01	1
50	4	4	1	0	ADC Form Spread Footings	15OCT01	18OCT01	18OCT01	29OCT01	3
60	4	4	1	0	ADC Reinforce Spread Footings	16OCT01	22OCT01	22OCT01	30OCT01	3
100	2	2	1	0	ADC Fab. & Tie Steel Contin. Footings	17OCT01	18OCT01	18OCT01	22OCT01	1
70	4	4	1	0	ADC Place Spread Footings	17OCT01	23OCT01	23OCT01	31OCT01	3
170	1	1	1	0	ADC North Side Excav. Contin Ftg's	18OCT01	18OCT01	18OCT01	18OCT01	0
180	1	1	1	0	ADC North Side Form Contin. Ftg's	22OCT01	22OCT01	22OCT01	22OCT01	0
110	2	2	1	0	ADC Fab & Tie Steel Fdn. Walls	22OCT01	23OCT01	30OCT01	31OCT01	5
270	1	1	1	0	ADC East Side Excav. Contin Ftg's	22OCT01	22OCT01	30OCT01	30OCT01	5
190	1	1	1	0	ADC North Side Reinforce Contin. Ftg's	23OCT01	23OCT01	23OCT01	23OCT01	0
80	4	4	1	0	ADC Strip Forms - Spread Footings	23OCT01	29OCT01	29OCT01	5NOV01	3
360	1	1	1	0	ADC South Side Excav. Contin Ftg's	23OCT01	23OCT01	8NOV01	8NOV01	10
200	1	1	1	0	ADC North Side Place Contin. Ftg's	24OCT01	24OCT01	24OCT01	24OCT01	0
120	4	4	1	0	ADC Reinforce Piers	24OCT01	30OCT01	30OCT01	6NOV01	3
130	4	4	1	0	ADC Form Piers	25OCT01	31OCT01	31OCT01	7NOV01	3
140	4	4	1	0	ADC Place Piers	29OCT01	1NOV01	1NOV01	8NOV01	3
210	1	1	1	0	ADC North Side Strip Contin. Ftg's Forms	30OCT01	30OCT01	30OCT01	30OCT01	0
400	5	5	1	0	JED Underground Plumbing	30OCT01	6NOV01	10DEC01	17DEC01	22
280	1	1	1	0	ADC East Side Form Contin. Ftg's	31OCT01	31OCT01	31OCT01	31OCT01	0
220	1	1	1	0	ADC North Side Reinforce Fdn. Walls	31OCT01	31OCT01	1NOV01	6NOV01	1
290	1	1	1	0	ADC East Side Reinforce Contin. Ftg's	1NOV01	1NOV01	1NOV01	1NOV01	0
230	3	3	1	0	ADC North Side Form Fdn. Walls	1NOV01	6NOV01	5NOV01	7NOV01	1
150	4	4	1	0	ADC Strip Forms - Piers	1NOV01	7NOV01	7NOV01	13NOV01	3
300	1	1	1	0	ADC East Side Place Contin. Ftg's	5NOV01	5NOV01	5NOV01	5NOV01	0
160	4	4	1	0	ADC Backfill Spread Footings	5NOV01	8NOV01	8NOV01	14NOV01	3
240	1	1	1	0	ADC North Side Place Fdn. Walls	7NOV01	7NOV01	8NOV01	8NOV01	1
310	1	1	1	0	ADC East Side Strip Contin. Ftg's Forms	8NOV01	8NOV01	8NOV01	8NOV01	0
250	1	1	1	0	ADC North Side Strip Fdn. Wall Forms	8NOV01	8NOV01	12NOV01	12NOV01	1
320	1	1	1	0	ADC East Side Reinforce Fdn. Walls	12NOV01	12NOV01	12NOV01	14NOV01	0
370	1	1	1	0	ADC South Side Form Contin. Ftg's	12NOV01	12NOV01	12NOV01	12NOV01	0

Line: 1 Col: 2 NUM

Screen Image 8-16
Primavera Look—Saving Output

The inclusion of only critical activities within this report can be verified by observing the total float column within Output 8-2. Since this column contains only zeros, all of these activities are on the critical path. Just as with the previous report, the scheduler needs to review the list to ensure that activities on the critical path are credible. Many construction activities by their nature have flexibility as to when they could be performed; none of these types of activities should be on the critical activity report.

Responsibility Schedule Report

The final sample tabular report that will be discussed is the responsibility schedule report. The purpose of this report is to generate an individualized schedule for each of the supervisors or superintendents. When the sample project was set up, the superintendent's initials were validated in the activity codes dictionary and entered for each of the individual activities. This report will organize the activities into separate schedules for each of the superintendents. Just as with the previous report, the sort order needs to be changed. This is done through the *Format* tab of the *Schedule Reports* window. Within this window the primary sort needs to be by activity code *SUPR* (superintendent). This primary sort will group all of the activities together by their value entered in their *SUPR* activity code field. The secondary sort would be by early start. That sort will take those activities that have the same superintendent and list them by early start order. The third sort is for those occasions where there are multiple activities within the control of one superintendent that have the same early start. In that scenario, the most critical activities will be listed first. The other change required from within this window is the page break. By specifying to skip a page whenever activity code SUPR changes, the report can easily be separated into a number of distinct schedules. Screen Image 8-19 is an example of the *Format* window with the changes required to generate the report found in Output 8-3. This report, since it includes all activities, does not require any special selection criteria.

All Star Developments | PRIMAVERA PROJECT PLANNER | Sample Construction Project

REPORT DATE 10FEB00 RUN NO. 5
8:57

START DATE 1OCT01 FIN DATE 25JUN02

Early Start Schedule

DATA DATE 1OCT01 PAGE NO. 1

ACTIVITY ID	ORIG DUR	REM DUR	CAL	%	CODE	ACTIVITY DESCRIPTION	EARLY START	EARLY FINISH	LATE START	LATE FINISH	TOTAL FLOAT
10	2	2	1	0	ADC	Clear Site	1OCT01	2OCT01	1OCT01	2OCT01	0
20	5	5	1	0	ADC	Site Grading & Compaction	3OCT01	10OCT01	3OCT01	10OCT01	0
30	6	6	1	0	ADC	Fab. & Tie Steel Spread Ftg's	3OCT01	11OCT01	4OCT01	15OCT01	1
40	4	4	1	0	ADC	Excav. Spread Footings	11OCT01	17OCT01	11OCT01	17OCT01	0
90	2	2	1	0	ADC	Fab. & Tie Steel Piers	15OCT01	16OCT01	16OCT01	17OCT01	1
50	4	4	1	0	ADC	Form Spread Footings	15OCT01	18OCT01	18OCT01	29OCT01	3
60	4	4	1	0	ADC	Reinforce Spread Footings	16OCT01	22OCT01	22OCT01	30OCT01	3
100	2	2	1	0	ADC	Fab. & Tie Steel Contin. Footings	17OCT01	18OCT01	18OCT01	22OCT01	1
70	4	4	1	0	ADC	Place Spread Footings	17OCT01	23OCT01	23OCT01	31OCT01	3
170	1	1	1	0	ADC	North Side Excav. Contin Ftg's	18OCT01	18OCT01	18OCT01	18OCT01	0
180	1	1	1	0	ADC	North Side Form Contin. Ftg's	22OCT01	22OCT01	22OCT01	22OCT01	0
110	2	2	1	0	ADC	Fab & Tie Steel Fdn. Walls	22OCT01	23OCT01	30OCT01	31OCT01	5
270	1	1	1	0	ADC	East Side Excav. Contin Ftg's	22OCT01	22OCT01	30OCT01	30OCT01	5
190	1	1	1	0	ADC	North Side Reinforce Contin. Ftg's	23OCT01	23OCT01	23OCT01	23OCT01	0
80	4	4	1	0	ADC	Strip Forms - Spread Footings	23OCT01	29OCT01	29OCT01	5NOV01	3
360	1	1	1	0	ADC	South Side Excav. Contin Ftg's	23OCT01	23OCT01	8NOV01	8NOV01	10
200	1	1	1	0	ADC	North Side Place Contin. Ftg's	24OCT01	24OCT01	24OCT01	24OCT01	0
120	4	4	1	0	ADC	Reinforce Piers	24OCT01	30OCT01	30OCT01	6NOV01	3
130	4	4	1	0	ADC	Form Piers	25OCT01	31OCT01	31OCT01	7NOV01	3
140	4	4	1	0	ADC	Place Piers	29OCT01	1NOV01	1NOV01	8NOV01	3
210	1	1	1	0	ADC	North Side Strip Contin. Ftg's Forms	30OCT01	30OCT01	30OCT01	30OCT01	0
400	5	5	1	0	JED	Underground Plumbing	30OCT01	6NOV01	10DEC01	17DEC01	22
280	1	1	1	0	ADC	East Side Form Contin. Ftg's	31OCT01	31OCT01	31OCT01	31OCT01	0
220	1	1	1	0	ADC	North Side Reinforce Fdn. Walls	31OCT01	31OCT01	1NOV01	6NOV01	1
290	1	1	1	0	ADC	East Side Reinforce Contin. Ftg's	1NOV01	1NOV01	1NOV01	1NOV01	0
230	3	3	1	0	ADC	North Side Form Fdn. Walls	1NOV01	6NOV01	5NOV01	7NOV01	1
150	4	4	1	0	ADC	Strip Forms - Piers	1NOV01	7NOV01	7NOV01	13NOV01	3
300	1	1	1	0	ADC	East Side Place Contin. Ftg's	5NOV01	5NOV01	5NOV01	5NOV01	0
160	4	4	1	0	ADC	Backfill Spread Footings	5NOV01	8NOV01	8NOV01	14NOV01	3
240	1	1	1	0	ADC	North Side Place Fdn. Walls	7NOV01	7NOV01	8NOV01	8NOV01	1
310	1	1	1	0	ADC	East Side Strip Contin. Ftg's Forms	8NOV01	8NOV01	8NOV01	8NOV01	0
250	1	1	1	0	ADC	North Side Strip Fdn. Wall Forms	8NOV01	8NOV01	12NOV01	12NOV01	1
320	1	1	1	0	ADC	East Side Reinforce Fdn. Walls	12NOV01	12NOV01	12NOV01	14NOV01	0
370	1	1	1	0	ADC	South Side Form Contin. Ftg's	12NOV01	12NOV01	12NOV01	12NOV01	0
260	1	1	1	0	ADC	North Side Backfill & Compact	12NOV01	12NOV01	17DEC01	17DEC01	19
330	3	3	1	0	ADC	East Side Form Fdn. Walls	13NOV01	15NOV01	13NOV01	15NOV01	0
380	1	1	1	0	ADC	South Side Reinforce Contin. Ftg's	13NOV01	13NOV01	13NOV01	13NOV01	0
390	1	1	1	0	ADC	West Side Excav. Contin Ftg's	13NOV01	13NOV01	15NOV01	15NOV01	2
420	1	1	1	0	ADC	South Side Place Contin. Ftg's	14NOV01	14NOV01	14NOV01	14NOV01	0
430	1	1	1	0	ADC	South Side Strip Contin. Ftg's Forms	15NOV01	15NOV01	15NOV01	15NOV01	0
340	1	1	1	0	ADC	East Side Place Fdn. Walls	19NOV01	19NOV01	19NOV01	19NOV01	0
470	1	1	1	0	ADC	West Side Form Contin. Ftg's	19NOV01	19NOV01	19NOV01	19NOV01	0
440	1	1	1	0	ADC	South Side Reinforce Fdn. Walls	19NOV01	19NOV01	20NOV01	20NOV01	1
350	1	1	1	0	ADC	East Side Strip Fdn. Wall Forms	20NOV01	20NOV01	20NOV01	20NOV01	0
480	1	1	1	0	ADC	West Side Reinforce Contin. Ftg's	20NOV01	20NOV01	20NOV01	20NOV01	0
450	1	1	1	0	ADC	South Side Form Fdn. Walls	21NOV01	21NOV01	21NOV01	21NOV01	0
510	2	2	1	0	ADC	West Side Place Contin. Ftg's	21NOV01	26NOV01	21NOV01	26NOV01	0
410	1	1	1	0	ADC	East Side Backfill & Compact	21NOV01	21NOV01	17DEC01	17DEC01	13
460	1	1	1	0	ADC	South Side Place Fdn. Walls	26NOV01	26NOV01	26NOV01	26NOV01	0
490	2	2	1	0	ADC	South Side Strip Fdn Wall Forms	29NOV01	3DEC01	29NOV01	3DEC01	0
520	1	1	1	0	ADC	West Side Strip Contin. Ftg's Forms	29NOV01	29NOV01	29NOV01	29NOV01	0
530	1	1	1	0	ADC	West Side Reinforce Fdn. Walls	3DEC01	3DEC01	3DEC01	5DEC01	0
540	3	3	1	0	ADC	West Side Form Fdn. Walls	4DEC01	6DEC01	4DEC01	6DEC01	0
500	1	1	1	0	ADC	South Side Backfill & Compact	4DEC01	4DEC01	17DEC01	17DEC01	7
550	1	1	1	0	ADC	West Side Place Fdn. Walls	10DEC01	10DEC01	10DEC01	10DEC01	0
560	1	1	1	0	ADC	West Side Strip Fdn. Wall Forms	13DEC01	13DEC01	13DEC01	13DEC01	0
570	1	1	1	0	ADC	West Side Backfill & Compact	17DEC01	17DEC01	17DEC01	17DEC01	0
580	3	3	1	0	ADC	Reinforce S.O.G.	18DEC01	20DEC01	18DEC01	20DEC01	0
590	1	1	1	0	ADC	Place & Finish S.O.G.	24DEC01	24DEC01	24DEC01	24DEC01	0
650	5	5	1	0	ADC	Grade Parking Lot	26DEC01	3JAN02	21MAY02	29MAY02	82
600	22	22	1	0	WER	CMU	31DEC01	6FEB02	31DEC01	6FEB02	0
610	12	12	1	0	JBS	Erect Structural Steel	31DEC01	11FEB02	22JAN02	11FEB02	0
720	3	3	1	0	ADC	Form Parking Lot	7JAN02	9JAN02	30MAY02	6JUN02	82
730	5	5	1	0	ADC	Reinforce Parking Lot	8JAN02	15JAN02	3JUN02	10JUN02	82
740	1	1	1	0	ADC	Place & Finish Parking Lot	9JAN02	16JAN02	10JUN02	11JUN02	82
750	1	1	1	0	ADC	Strip Forms - Parking Lot	15JAN02	21JAN02	13JUN02	13JUN02	82
760	2	2	1	0	ADC	Place Curbs	22JAN02	23JAN02	17JUN02	18JUN02	82
780	2	2	1	0	JED	Final Grading	24JAN02	28JAN02	19JUN02	20JUN02	82
770	1	1	1	0	JED	Stripe Parking Lot	24JAN02	24JAN02	25JUN02	25JUN02	85
790	2	2	1	0	JED	Landscape	29JAN02	30JAN02	24JUN02	25JUN02	82
620	6	6	1	0	WER	Roofing / Flashing	12FEB02	20FEB02	12FEB02	20FEB02	0
640	3	3	1	0	WER	Overhead Doors	12FEB02	14FEB02	18FEB02	20FEB02	3
630	1	1	1	0	WER	Exterior Glass & Glazing	12FEB02	12FEB02	20FEB02	20FEB02	5
660	5	5	1	0	JED	Mechanical Rough-in	21FEB02	28FEB02	21FEB02	6MAR02	0
670	8	8	1	0	JED	Overhead Electrical Rough-in	21FEB02	6MAR02	21FEB02	6MAR02	0
680	3	3	1	0	JBS	Ceiling Grid	7MAR02	12MAR02	7MAR02	12MAR02	0
690	6	6	1	0	JBS	Frame Interior Walls	13MAR02	21MAR02	13MAR02	21MAR02	0
710	12	12	1	0	JED	Electrical Rough-in	25MAR02	11APR02	25MAR02	11APR02	0
700	3	3	1	0	JED	Plumbing Top-out	25MAR02	27MAR02	12JUN02	17JUN02	45
800	8	8	1	0	JBS	Hang Drywall	15APR02	25APR02	15APR02	25APR02	0
810	16	16	1	0	JBS	Float, Tape & Paint	29APR02	23MAY02	29APR02	23MAY02	0
840	6	6	1	0	JBS	Install Cabinets	28MAY02	5JUN02	28MAY02	5JUN02	0
830	3	3	1	0	JED	Mechanical Trim	28MAY02	30MAY02	10JUN02	12JUN02	7
820	2	2	1	0	JBS	Hang & Trim Interior Doors	28MAY02	29MAY02	13JUN02	17JUN02	10
850	4	4	1	0	JED	Electrical Trim	6JUN02	12JUN02	6JUN02	12JUN02	0
860	2	2	1	0	JED	Plumbing Trim	13JUN02	17JUN02	13JUN02	17JUN02	0
870	3	3	1	0	JBS	Ceiling Tiles	18JUN02	20JUN02	18JUN02	20JUN02	0
880	1	1	1	0	JBS	Carpet	24JUN02	24JUN02	24JUN02	24JUN02	0
890	1	1	1	0	JBS	Final Cleaning	25JUN02	25JUN02	25JUN02	25JUN02	0

Output 8-1
Early Start Schedule Report

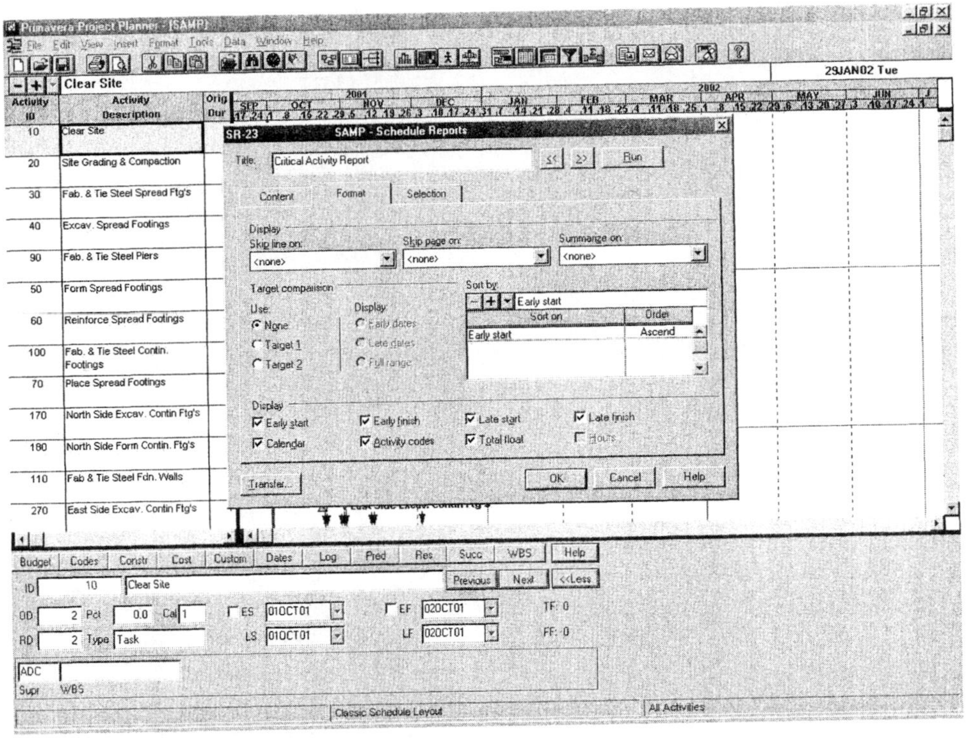

Screen Image 8-17
Critical Activity Report Format Window

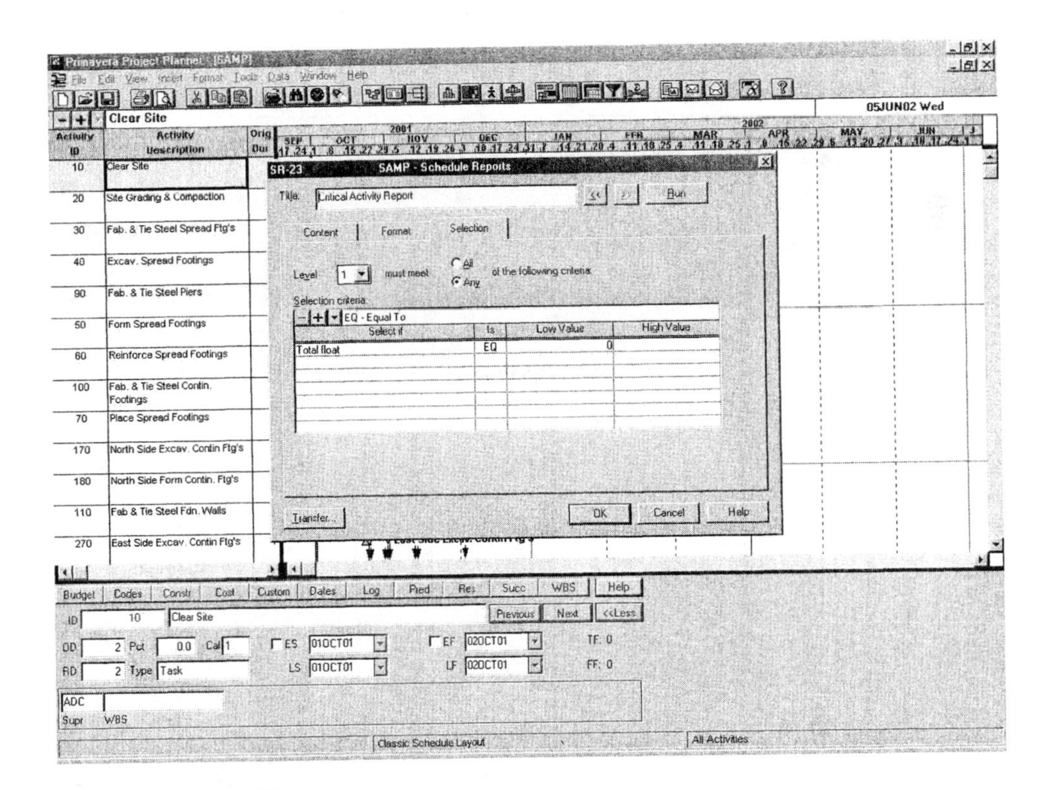

Screen Image 8-18
Selection Criteria for Critical Activities

.opments PRIMAVERA PROJECT PLANNER Sample Construction Project

)FEB00 RUN NO. 6 START DATE 1OCT01 FIN DATE 25JUN02
9:01
'ity Report DATA DATE 1OCT01 PAGE NO. 1

ORIG DUR	REM DUR	CAL	%	CODE	ACTIVITY DESCRIPTION	EARLY START	EARLY FINISH	LATE START	LATE FINISH	TOTAL FLOAT
2	2	1	0	ADC	Clear Site	1OCT01	2OCT01	1OCT01	2OCT01	0
5	5	1	0	ADC	Site Grading & Compaction	3OCT01	10OCT01	3OCT01	10OCT01	0
4	4	1	0	ADC	Excav. Spread Footings	11OCT01	17OCT01	11OCT01	17OCT01	0
1	1	1	0	ADC	North Side Excav. Contin Ftg's	18OCT01	18OCT01	18OCT01	18OCT01	0
1	1	1	0	ADC	North Side Form Contin. Ftg's	22OCT01	22OCT01	22OCT01	22OCT01	0
1	1	1	0	ADC	North Side Reinforce Contin. Ftg's	23OCT01	23OCT01	23OCT01	23OCT01	0
1	1	1	0	ADC	North Side Place Contin. Ftg's	24OCT01	24OCT01	24OCT01	24OCT01	0
1	1	1	0	ADC	North Side Strip Contin. Ftg's Forms	30OCT01	30OCT01	30OCT01	30OCT01	0
1	1	1	0	ADC	East Side Form Contin. Ftg's	31OCT01	31OCT01	31OCT01	31OCT01	0
1	1	1	0	ADC	East Side Reinforce Contin. Ftg's	1NOV01	1NOV01	1NOV01	1NOV01	0
1	1	1	0	ADC	East Side Place Contin. Ftg's	5NOV01	5NOV01	5NOV01	5NOV01	0
1	1	1	0	ADC	East Side Strip Contin. Ftg's Forms	8NOV01	8NOV01	8NOV01	8NOV01	0
1	1	1	0	ADC	East Side Reinforce Fdn. Walls	12NOV01	12NOV01	12NOV01	14NOV01	0
1	1	1	0	ADC	South Side Form Contin. Ftg's	12NOV01	12NOV01	12NOV01	12NOV01	0
3	3	1	0	ADC	East Side Form Fdn. Walls	13NOV01	15NOV01	13NOV01	15NOV01	0
1	1	1	0	ADC	South Side Reinforce Contin. Ftg's	13NOV01	13NOV01	13NOV01	13NOV01	0
1	1	1	0	ADC	South Side Place Contin. Ftg's	14NOV01	14NOV01	14NOV01	14NOV01	0
1	1	1	0	ADC	South Side Strip Contin. Ftg's Forms	15NOV01	15NOV01	15NOV01	15NOV01	0
1	1	1	0	ADC	East Side Place Fdn. Walls	19NOV01	19NOV01	19NOV01	19NOV01	0
1	1	1	0	ADC	West Side Form Contin. Ftg's	19NOV01	19NOV01	19NOV01	19NOV01	0
1	1	1	0	ADC	East Side Strip Fdn. Wall Forms	20NOV01	20NOV01	20NOV01	20NOV01	0
1	1	1	0	ADC	West Side Reinforce Contin. Ftg's	20NOV01	20NOV01	20NOV01	20NOV01	0
1	1	1	0	ADC	South Side Form Fdn. Walls	21NOV01	21NOV01	21NOV01	21NOV01	0
2	2	1	0	ADC	West Side Place Contin. Ftg's	21NOV01	26NOV01	21NOV01	26NOV01	0
1	1	1	0	ADC	South Side Place Fdn. Walls	26NOV01	26NOV01	26NOV01	26NOV01	0
2	2	1	0	ADC	South Side Strip Fdn Wall Forms	29NOV01	3DEC01	29NOV01	3DEC01	0
1	1	1	0	ADC	West Side Strip Contin. Ftg's Forms	29NOV01	29NOV01	29NOV01	29NOV01	0
1	1	1	0	ADC	West Side Reinforce Fdn. Walls	3DEC01	3DEC01	3DEC01	5DEC01	0
3	3	1	0	ADC	West Side Form Fdn. Walls	4DEC01	6DEC01	4DEC01	6DEC01	0
1	1	1	0	ADC	West Side Place Fdn. Walls	10DEC01	10DEC01	10DEC01	10DEC01	0
1	1	1	0	ADC	West Side Strip Fdn. Wall Forms	13DEC01	13DEC01	13DEC01	13DEC01	0
1	1	1	0	ADC	West Side Backfill & Compact	17DEC01	17DEC01	17DEC01	17DEC01	0
3	3	1	0	ADC	Reinforce S.O.G.	18DEC01	20DEC01	18DEC01	20DEC01	0
1	1	1	0	ADC	Place & Finish S.O.G.	24DEC01	24DEC01	24DEC01	24DEC01	0
22	22	1	0	WER	CMU	31DEC01	6FEB02	31DEC01	6FEB02	0
12	12	1	0	JBS	Erect Structural Steel	31DEC01	11FEB02	22JAN02	11FEB02	0
6	6	1	0	WER	Roofing / Flashing	12FEB02	20FEB02	12FEB02	20FEB02	0
5	5	1	0	JED	Mechanical Rough-in	21FEB02	28FEB02	21FEB02	6MAR02	0
8	8	1	0	JED	Overhead Electrical Rough-in	21FEB02	6MAR02	21FEB02	6MAR02	0
3	3	1	0	JBS	Ceiling Grid	7MAR02	12MAR02	7MAR02	12MAR02	0
6	6	1	0	JBS	Frame Interior Walls	13MAR02	21MAR02	13MAR02	21MAR02	0
12	12	1	0	JED	Electrical Rough-in	25MAR02	11APR02	25MAR02	11APR02	0
8	8	1	0	JBS	Hang Drywall	15APR02	25APR02	15APR02	25APR02	0
16	16	1	0	JBS	Float, Tape & Paint	29APR02	23MAY02	29APR02	23MAY02	0
6	6	1	0	JBS	Install Cabinets	28MAY02	5JUN02	28MAY02	5JUN02	0
4	4	1	0	JED	Electrical Trim	6JUN02	12JUN02	6JUN02	12JUN02	0
2	2	1	0	JED	Plumbing Trim	13JUN02	17JUN02	13JUN02	17JUN02	0
3	3	1	0	JBS	Ceiling Tiles	18JUN02	20JUN02	18JUN02	20JUN02	0
1	1	1	0	JBS	Carpet	24JUN02	24JUN02	24JUN02	24JUN02	0
1	1	1	0	JBS	Final Cleaning	25JUN02	25JUN02	25JUN02	25JUN02	0

2

ivity Report

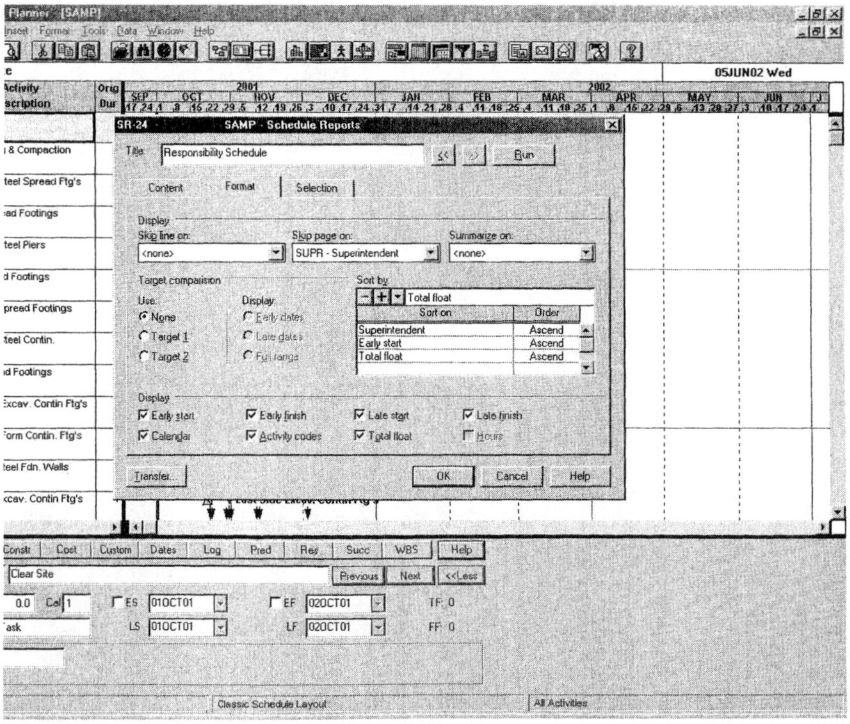

Planner - [SAMP]
Insert Format Tools Data Window Help
05JUN02 Wed
Activity Description | Orig Dur | 2001 SEP OCT NOV DEC | 2002 JAN FEB MAR APR MAY JUN J
& Compaction
teel Spread Ftg's
ad Footings
teel Piers
d Footings
pread Footings
teel Contin.
d Footings
Excav. Contin Ftg's
Form Contin. Ftg's
teel Fdn. Walls
xcav. Contin Ftg's

SR-24 SAMP - Schedule Reports
Title: Responsibility Schedule | << | >> | Run
Content | Format | Selection
Display
Skip line on: <none> | Skip page on: SUPR - Superintendent | Summarize on: <none>
Target comparison
Use: None / Target 1 / Target 2
Display: Early dates / Late dates / Full range
Sort by: Total float

Sort on	Order
Superintendent	Ascend
Early start	Ascend
Total float	Ascend

Display
Early start | Early finish | Late start | Late finish
Calendar | Activity codes | Total float | Hours
Transfer... | OK | Cancel | Help

Constr | Cost | Custom | Dates | Log | Pred | Res | Succ | WBS | Help
Clear Site | Previous | Next | <<Less
0.0 Cal 1 | ES 01OCT01 | EF 02OCT01 | TF 0
Task | LS 01OCT01 | LF 02OCT01 | FF 0
Classic Schedule Layout | All Activities

(I)mage 8-19

for Responsibility Report

All Star Developments PRIMAVERA PROJECT PLANNER Sample Construction Project

REPORT DATE 10FEB00 RUN NO. 7 START DATE 1OCT01 FIN DATE 25JUN02
9:03
Responsibility Schedule DATA DATE 1OCT01 PAGE NO. 1

Albert Concrete

ACTIVITY ID	ORIG DUR	REM DUR	CAL	%	CODE	ACTIVITY DESCRIPTION	EARLY START	EARLY FINISH	LATE START	LATE FINISH	TOTAL FLOAT
10	2	2	1	0	ADC	Clear Site	1OCT01	2OCT01	1OCT01	2OCT01	0
20	5	5	1	0	ADC	Site Grading & Compaction	3OCT01	10OCT01	3OCT01	10OCT01	0
30	6	6	1	0	ADC	Fab. & Tie Steel Spread Ftg's	3OCT01	11OCT01	4OCT01	15OCT01	1
40	4	4	1	0	ADC	Excav. Spread Footings	11OCT01	17OCT01	11OCT01	17OCT01	0
90	2	2	1	0	ADC	Fab. & Tie Steel Piers	15OCT01	16OCT01	16OCT01	17OCT01	1
50	4	4	1	0	ADC	Form Spread Footings	15OCT01	18OCT01	18OCT01	29OCT01	3
60	4	4	1	0	ADC	Reinforce Spread Footings	16OCT01	22OCT01	22OCT01	30OCT01	3
100	2	2	1	0	ADC	Fab. & Tie Steel Contin. Footings	17OCT01	18OCT01	18OCT01	22OCT01	1
70	4	4	1	0	ADC	Place Spread Footings	17OCT01	23OCT01	23OCT01	31OCT01	3
170	1	1	1	0	ADC	North Side Excav. Contin Ftg's	18OCT01	18OCT01	18OCT01	18OCT01	0
180	1	1	1	0	ADC	North Side Form Contin. Ftg's	22OCT01	22OCT01	22OCT01	22OCT01	0
110	2	2	1	0	ADC	Fab & Tie Steel Fdn. Walls	22OCT01	23OCT01	30OCT01	31OCT01	5
270	1	1	1	0	ADC	East Side Excav. Contin Ftg's	22OCT01	22OCT01	30OCT01	30OCT01	5
190	1	1	1	0	ADC	North Side Reinforce Contin. Ftg's	23OCT01	23OCT01	23OCT01	23OCT01	0
80	4	4	1	0	ADC	Strip Forms - Spread Footings	23OCT01	29OCT01	29OCT01	5NOV01	3
360	1	1	1	0	ADC	South Side Excav. Contin Ftg's	23OCT01	23OCT01	8NOV01	8NOV01	10
200	1	1	1	0	ADC	North Side Place Contin. Ftg's	24OCT01	24OCT01	24OCT01	24OCT01	0
120	4	4	1	0	ADC	Reinforce Piers	24OCT01	30OCT01	30OCT01	6NOV01	3
130	4	4	1	0	ADC	Form Piers	25OCT01	31OCT01	31OCT01	7NOV01	3
140	4	4	1	0	ADC	Place Piers	29OCT01	1NOV01	1NOV01	8NOV01	3
210	1	1	1	0	ADC	North Side Strip Contin. Ftg's Forms	30OCT01	30OCT01	30OCT01	30OCT01	0
280	1	1	1	0	ADC	East Side Form Contin. Ftg's	31OCT01	31OCT01	31OCT01	31OCT01	0
220	1	1	1	0	ADC	North Side Reinforce Fdn. Walls	31OCT01	31OCT01	1NOV01	6NOV01	1
290	1	1	1	0	ADC	East Side Reinforce Contin. Ftg's	1NOV01	1NOV01	1NOV01	1NOV01	0
230	3	3	1	0	ADC	North Side Form Fdn. Walls	1NOV01	6NOV01	5NOV01	7NOV01	1
150	4	4	1	0	ADC	Strip Forms - Piers	1NOV01	7NOV01	7NOV01	13NOV01	3
300	1	1	1	0	ADC	East Side Place Contin. Ftg's	5NOV01	5NOV01	5NOV01	5NOV01	0
160	4	4	1	0	ADC	Backfill Spread Footings	5NOV01	8NOV01	8NOV01	14NOV01	3
240	1	1	1	0	ADC	North Side Place Fdn. Walls	7NOV01	7NOV01	8NOV01	8NOV01	1
310	1	1	1	0	ADC	East Side Strip Contin. Ftg's Forms	8NOV01	8NOV01	8NOV01	8NOV01	0
250	1	1	1	0	ADC	North Side Strip Fdn. Wall Forms	8NOV01	8NOV01	12NOV01	12NOV01	1
320	1	1	1	0	ADC	East Side Reinforce Fdn. Walls	12NOV01	12NOV01	12NOV01	14NOV01	0
370	1	1	1	0	ADC	South Side Form Contin. Ftg's	12NOV01	12NOV01	12NOV01	12NOV01	0
260	1	1	1	0	ADC	North Side Backfill & Compact	12NOV01	12NOV01	17DEC01	17DEC01	19
330	3	3	1	0	ADC	East Side Form Fdn. Walls	13NOV01	15NOV01	13NOV01	15NOV01	0
380	1	1	1	0	ADC	South Side Reinforce Contin. Ftg's	13NOV01	13NOV01	13NOV01	13NOV01	0
390	1	1	1	0	ADC	West Side Excav. Contin Ftg's	13NOV01	13NOV01	15NOV01	15NOV01	2
420	1	1	1	0	ADC	South Side Place Contin. Ftg's	14NOV01	14NOV01	14NOV01	14NOV01	0
430	1	1	1	0	ADC	South Side Strip Contin. Ftg's Forms	15NOV01	15NOV01	15NOV01	15NOV01	0
340	1	1	1	0	ADC	East Side Place Fdn. Walls	19NOV01	19NOV01	19NOV01	19NOV01	0
470	1	1	1	0	ADC	West Side Form Contin. Ftg's	19NOV01	19NOV01	19NOV01	19NOV01	0
440	1	1	1	0	ADC	South Side Reinforce Fdn. Walls	19NOV01	19NOV01	20NOV01	20NOV01	1
350	1	1	1	0	ADC	East Side Strip Fdn. Wall Forms	20NOV01	20NOV01	20NOV01	20NOV01	0
480	1	1	1	0	ADC	West Side Reinforce Contin. Ftg's	20NOV01	20NOV01	20NOV01	20NOV01	0
450	1	1	1	0	ADC	South Side Form Fdn. Walls	21NOV01	21NOV01	21NOV01	21NOV01	0
510	2	2	1	0	ADC	West Side Place Contin. Ftg's	21NOV01	26NOV01	21NOV01	26NOV01	0
410	1	1	1	0	ADC	East Side Backfill & Compact	21NOV01	21NOV01	17DEC01	17DEC01	13
460	1	1	1	0	ADC	South Side Place Fdn. Walls	26NOV01	26NOV01	26NOV01	26NOV01	0
490	2	2	1	0	ADC	South Side Strip Fdn Wall Forms	29NOV01	3DEC01	29NOV01	3DEC01	0
520	1	1	1	0	ADC	West Side Strip Contin. Ftg's Forms	29NOV01	29NOV01	29NOV01	29NOV01	0
530	1	1	1	0	ADC	West Side Reinforce Fdn. Walls	3DEC01	3DEC01	3DEC01	5DEC01	0
540	3	3	1	0	ADC	West Side Form Fdn. Walls	4DEC01	6DEC01	4DEC01	6DEC01	0
500	1	1	1	0	ADC	South Side Backfill & Compact	4DEC01	4DEC01	17DEC01	17DEC01	7
550	1	1	1	0	ADC	West Side Place Fdn. Walls	10DEC01	10DEC01	10DEC01	10DEC01	0
560	1	1	1	0	ADC	West Side Strip Fdn. Wall Forms	13DEC01	13DEC01	13DEC01	13DEC01	0
570	1	1	1	0	ADC	West Side Backfill & Compact	17DEC01	17DEC01	17DEC01	17DEC01	0
580	3	3	1	0	ADC	Reinforce S.O.G.	18DEC01	20DEC01	18DEC01	20DEC01	0
590	1	1	1	0	ADC	Place & Finish S.O.G.	24DEC01	24DEC01	24DEC01	24DEC01	0
650	5	5	1	0	ADC	Grade Parking Lot	26DEC01	3JAN02	21MAY02	29MAY02	82
720	3	3	1	0	ADC	Form Parking Lot	7JAN02	9JAN02	30MAY02	6JUN02	82
730	5	5	1	0	ADC	Reinforce Parking Lot	8JAN02	15JAN02	3JUN02	10JUN02	82
740	1	1	1	0	ADC	Place & Finish Parking Lot	9JAN02	16JAN02	10JUN02	11JUN02	82
750	1	1	1	0	ADC	Strip Forms - Parking Lot	15JAN02	21JAN02	13JUN02	13JUN02	82
760	2	2	1	0	ADC	Place Curbs	22JAN02	23JAN02	17JUN02	18JUN02	82

Output 8-3
Responsibility Schedule (page 1)

All Star Developments — PRIMAVERA PROJECT PLANNER — Sample Construction Project

REPORT DATE 10FEB00 RUN NO. 7 — START DATE 1OCT01 FIN DATE 25JUN02
9:03
Responsibility Schedule — DATA DATE 1OCT01 PAGE NO. 2

Joe Bob Steel

ACTIVITY ID	ORIG DUR	REM DUR	CAL	%	CODE	ACTIVITY DESCRIPTION	EARLY START	EARLY FINISH	LATE START	LATE FINISH	TOTAL FLOAT
610	12	12	1	0	JBS	Erect Structural Steel	31DEC01	11FEB02	22JAN02	11FEB02	0
680	3	3	1	0	JBS	Ceiling Grid	7MAR02	12MAR02	7MAR02	12MAR02	0
690	6	6	1	0	JBS	Frame Interior Walls	13MAR02	21MAR02	13MAR02	21MAR02	0
800	8	8	1	0	JBS	Hang Drywall	15APR02	25APR02	15APR02	25APR02	0
810	16	16	1	0	JBS	Float, Tape & Paint	29APR02	23MAY02	29APR02	23MAY02	0
840	6	6	1	0	JBS	Install Cabinets	28MAY02	5JUN02	28MAY02	5JUN02	0
820	2	2	1	0	JBS	Hang & Trim Interior Doors	28MAY02	29MAY02	13JUN02	17JUN02	10
870	3	3	1	0	JBS	Ceiling Tiles	18JUN02	20JUN02	18JUN02	20JUN02	0
880	1	1	1	0	JBS	Carpet	24JUN02	24JUN02	24JUN02	24JUN02	0
890	1	1	1	0	JBS	Final Cleaning	25JUN02	25JUN02	25JUN02	25JUN02	0

Output 8-3
Responsibility Schedule (page 2)

All Star Developments — PRIMAVERA PROJECT PLANNER — Sample Construction Project

REPORT DATE 10FEB00 RUN NO. 7 — START DATE 1OCT01 FIN DATE 25JUN02
9:03
Responsibility Schedule — DATA DATE 1OCT01 PAGE NO. 3

John Doe

ACTIVITY ID	ORIG DUR	REM DUR	CAL	%	CODE	ACTIVITY DESCRIPTION	EARLY START	EARLY FINISH	LATE START	LATE FINISH	TOTAL FLOAT
400	5	5	1	0	JED	Underground Plumbing	30OCT01	6NOV01	10DEC01	17DEC01	22
780	2	2	1	0	JED	Final Grading	24JAN02	28JAN02	19JUN02	20JUN02	82
770	1	1	1	0	JED	Stripe Parking Lot	24JAN02	24JAN02	25JUN02	25JUN02	85
790	2	2	1	0	JED	Landscape	29JAN02	30JAN02	24JUN02	25JUN02	82
660	5	5	1	0	JED	Mechanical Rough-in	21FEB02	28FEB02	21FEB02	6MAR02	0
670	8	8	1	0	JED	Overhead Electrical Rough-in	21FEB02	6MAR02	21FEB02	6MAR02	0
710	12	12	1	0	JED	Electrical Rough-in	25MAR02	11APR02	25MAR02	11APR02	0
700	3	3	1	0	JED	Plumbing Top-out	25MAR02	27MAR02	12JUN02	17JUN02	45
830	3	3	1	0	JED	Mechanical Trim	28MAY02	30MAY02	10JUN02	12JUN02	7
850	4	4	1	0	JED	Electrical Trim	6JUN02	12JUN02	6JUN02	12JUN02	0
860	2	2	1	0	JED	Plumbing Trim	13JUN02	17JUN02	13JUN02	17JUN02	0

Output 8-3
Responsibility Schedule (page 3)

All Star Developments — PRIMAVERA PROJECT PLANNER — Sample Construction Project

REPORT DATE 10FEB00 RUN NO. 7 — START DATE 1OCT01 FIN DATE 25JUN02
9:03
Responsibility Schedule — DATA DATE 1OCT01 PAGE NO. 4

WIlliam Roe

ACTIVITY ID	ORIG DUR	REM DUR	CAL	%	CODE	ACTIVITY DESCRIPTION	EARLY START	EARLY FINISH	LATE START	LATE FINISH	TOTAL FLOAT
600	22	22	1	0	WER	CMU	31DEC01	6FEB02	31DEC01	6FEB02	0
620	6	6	1	0	WER	Roofing / Flashing	12FEB02	20FEB02	12FEB02	20FEB02	0
640	3	3	1	0	WER	Overhead Doors	12FEB02	14FEB02	18FEB02	20FEB02	3
630	1	1	1	0	WER	Exterior Glass & Glazing	12FEB02	12FEB02	20FEB02	20FEB02	5

Output 8-3
Responsibility Schedule (page 4)

Graphical Reports

Primavera Project Planner's graphical report writer is very similar to the tabular report writer. They have similar options, but graphic output devices have special requirements—such as pen configuration and line types—that need to be addressed. The graphic output can be routed to virtually any type of printer or plotter. The main limitation of printer-type output devices is the physical size of the available output. Since most printers only accept paper widths that range from $8^1/_2$ to 11 inches, their output capabilities are limited. These paper widths are often not large enough to produce useable graphic output with the necessary level of detail. Therefore, plotters or specialized inkjet printers, which will support C and D size paper, are often used to generate this type of output.

Output Device Setup

When generating graphic output it is critical that the output device be properly configured before developing any report specifications. The output device setup begins from the main input screen by executing the *File* command followed by the *Print Setup* option as shown in Screen Image 8-20. These commands will open the *Print Setup* window, as shown in Screen Image 8-21. This window is typical of the print setup option used by other Windows applications. To change the printer, click on the pull-down arrow below the *Specific Printer*. In Screen Image 8-21 the default printer is a HP LaserJet 5p. Since this printer does not support wide carriage output, it needs to be changed.

By using the pull-down menu of *Specific Printers*, a list of all available output devices can be found. If a specific plotter/printer is attached to the computer or network, then that device should be selected. Another option is to select *Primavera Plot* as the output device. This device driver, which was installed along with the Primavera Project Planner software, allows the user to emulate plotter output to a printer. After the appropriate output device has been selected, the orientation and paper size needs to be specified. From within the *Paper* box the *Size* can be selected by using the pull-down menu to select the size from those that are compatible with the selected output device. Screen Image 8-22 shows the *Print Setup* window using the *Primavera Plot* driver with D size paper (24″ × 36″).

The paper size on the *Print Setup* window may differ from the exact paper size by a couple of inches. The size shown in the paper box is often the available plot area rather than the exact paper size. Once the needed information has been input, this window can be closed by clicking on the *OK* button.

Bar Chart Schedule

An excellent use of Primavera Project Planner's graphic output options is to generate bar chart schedules. This type of schedule is easily understood, and its graphic nature makes it well suited for this output. The first step in the process is to select the *Tools* command followed by *Graphic Reports* and *Bar* options as shown in Screen Image 8-23. These commands open the *Bar Charts* window that contains a list of all of the graphic bar chart specifications. In Screen Image 8-24 there is a list of previously developed graphic reports specifications. By clicking the *Add* button, the *Add a New Report* window will open. In Screen Image 8-25 the user is being prompted to *Add* or *Cancel* report identifier *BC-24*. This is simply the twenty-fourth set of graphic bar chart specifications that have been developed. If this identifier is acceptable, click the *Add* button, which will open the window shown in Screen Image 8-26. At the top of this window is the default report title. Typing over this assigned title will permanently change it and add the new title to the list of reports. Immediately below the *Title* are eight tabs that orchestrate how information will be presented on the page. The best methodology to use in developing a report is to work through each of these tabs.

Screen Image 8-20
Opening the Print Setup Window

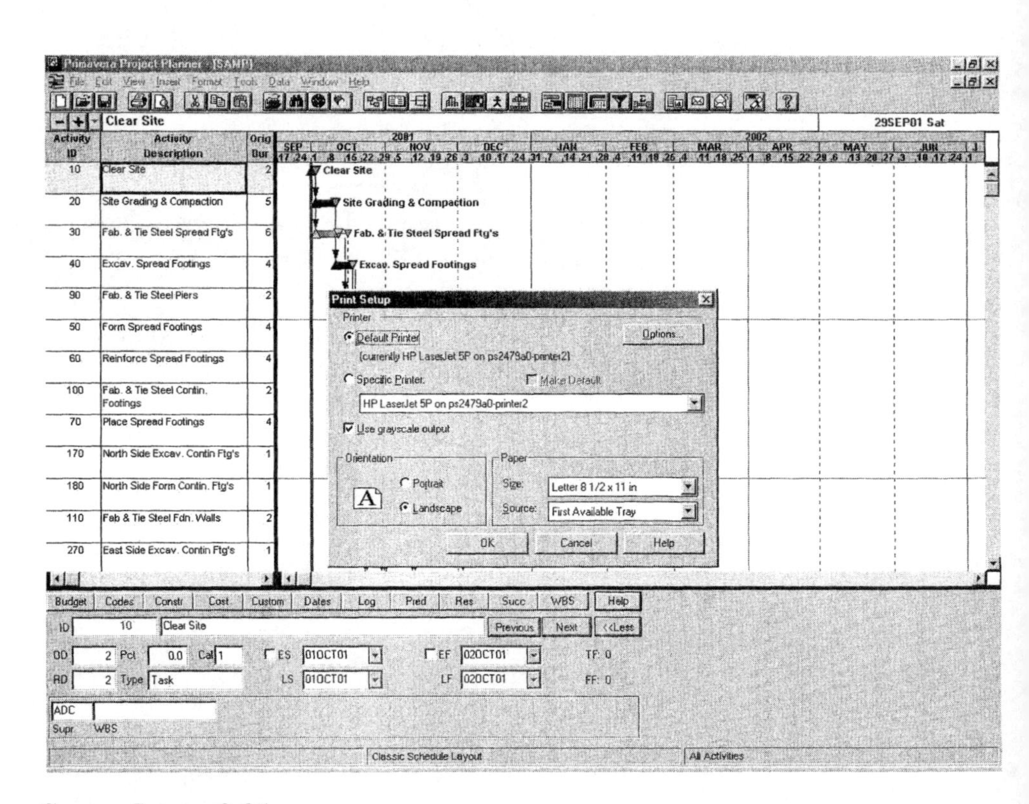

Screen Image 8-21
Print Setup Window

Primavera Project Planner - [SAMP]
File Edit View Insert Format Tools Data Window Help
Clear Site
29SEP01 Sat

Activity ID	Activity Description	Orig Dur
10	Clear Site	2
20	Site Grading & Compaction	5
30	Fab. & Tie Steel Spread Ftg's	6
40	Excav. Spread Footings	4
90	Fab. & Tie Steel Piers	2
50	Form Spread Footings	4
60	Reinforce Spread Footings	4
100	Fab. & Tie Steel Contin. Footings	2
70	Place Spread Footings	4
170	North Side Excav. Contin Ftg's	1
180	North Side Form Contin. Ftg's	1
110	Fab & Tie Steel Fdn. Walls	2
270	East Side Excav. Contin Ftg's	1

Clear Site
Site Grading & Compaction
Fab. & Tie Steel Spread Ftg's
Excav. Spread Footings

Print Setup
Printer
Default Printer
(currently HP LaserJet 5P on ps2479a0-printer2)
Specific Printer:
Make Default
Primavera Plot on COM1:
Options...
Use grayscale output
Orientation
Portrait
Landscape
Paper
Size: ANSI D (22 x 34 in)
Source:
OK
Cancel
Help

Budget Codes Constr Cost Custom Dates Log Pred Res Succ WBS Help
ID 10 Clear Site Previous Next <<Less
OD 2 Pct 0.0 Cal 1 ES 01OCT01 EF 02OCT01 TF: 0
RD 2 Type Task LS 01OCT01 LF 02OCT01 FF: 0
ADC
Supr WBS
Classic Schedule Layout
All Activities

Screen Image 8-22
Revised Print Setup Window

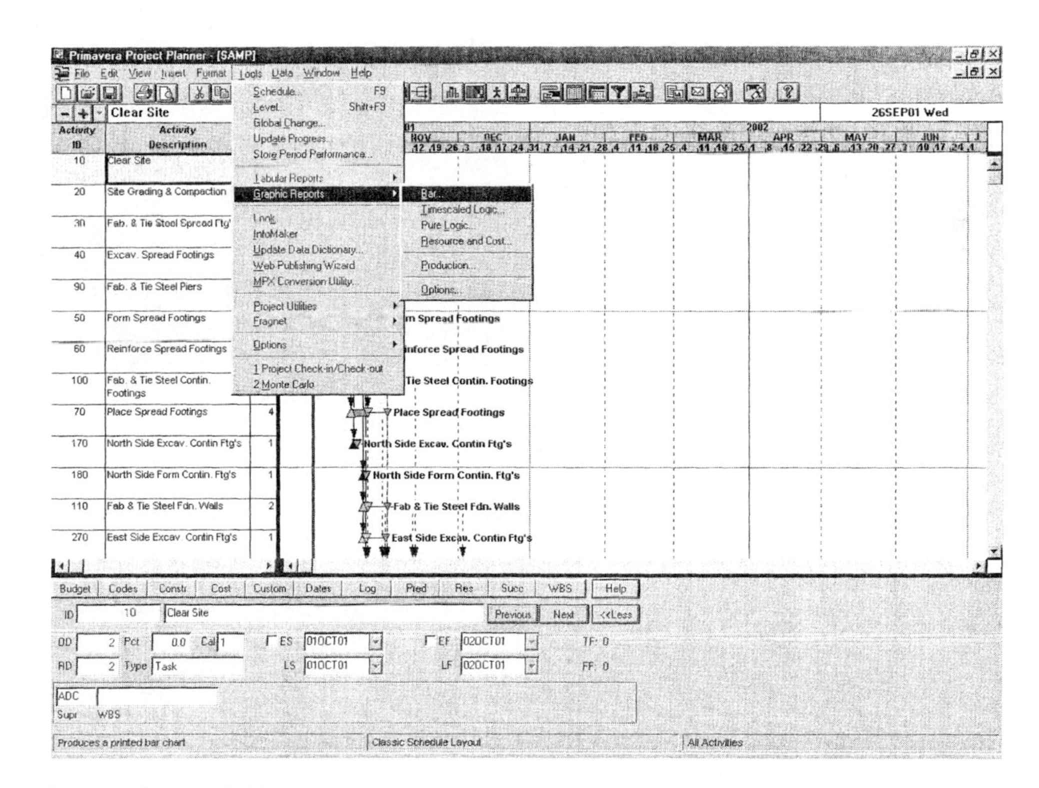

Screen Image 8-23
Opening the Bar Charts Graphic Window

Primavera Project Planner - [SAMP]
File Edit View Insert Format Tools Data Window Help
Clear Site
17SEP01 Mon

Activity ID	Activity Description	Orig Dur
10	Clear Site	2
20	Site Grading & Compaction	5
30	Fab. & Tie Steel Spread Ftg's	6
40	Excav. Spread Footings	4
90	Fab. & Tie Steel Piers	2
50	Form Spread Footings	4
60	Reinforce Spread Footings	4
100	Fab. & Tie Steel Contin. Footings	2
70	Place Spread Footings	4
170	North Side Excav. Contin Ftg's	1
180	North Side Form Contin. Ftg's	1
110	Fab & Tie Steel Fdn. Walls	2
270	East Side Excav. Contin Ftg's	1

2001 SEP OCT NOV DEC 2002 JAN FEB MAR APR MAY JUN
Clear Site
Site Grading & Compaction
East Side Excav. Contin Ftg's

SAMP - Bar Charts

ID	Title
BC-01	Construction Summary by Phase
BC-02	Bar Chart with Activity Logs
BC-03	Group Bars by Phase - Foundation
BC-04	Group Bars by Phase - Structural
BC-05	Group Bars by Phase - Rough-In/Close
BC-06	Group Bars by Phase - Finishes
BC-07	Procurement Summary - Discrete Bars
BC-08	Robotics Automation Summary by Phase
BC-09	Management Summary
BC-10	Group Bars by Phase - Automation Sys
BC-11	Project Summary with Costs
BC-12	Highlighting Responsibilities
BC-13	Activity Schedule/Resource Loading
BC-14	Single Line Activity Data Column

Run Modify... Add... Delete Close Help

Budget Codes Constr Cost Custom Dates Log Pred Res Succ WBS Help
ID 10 Clear Site Previous Next <<Less
OD 2 Pct 0.0 Cal 1 ES 01OCT01 EF 02OCT01 TF: 0
RD 2 Type Task LS 01OCT01 LF 02OCT01 FF: 0
ADC
Supr WBS
Classic Schedule Layout All Activities

Screen Image 8-24
Bar Charts Window

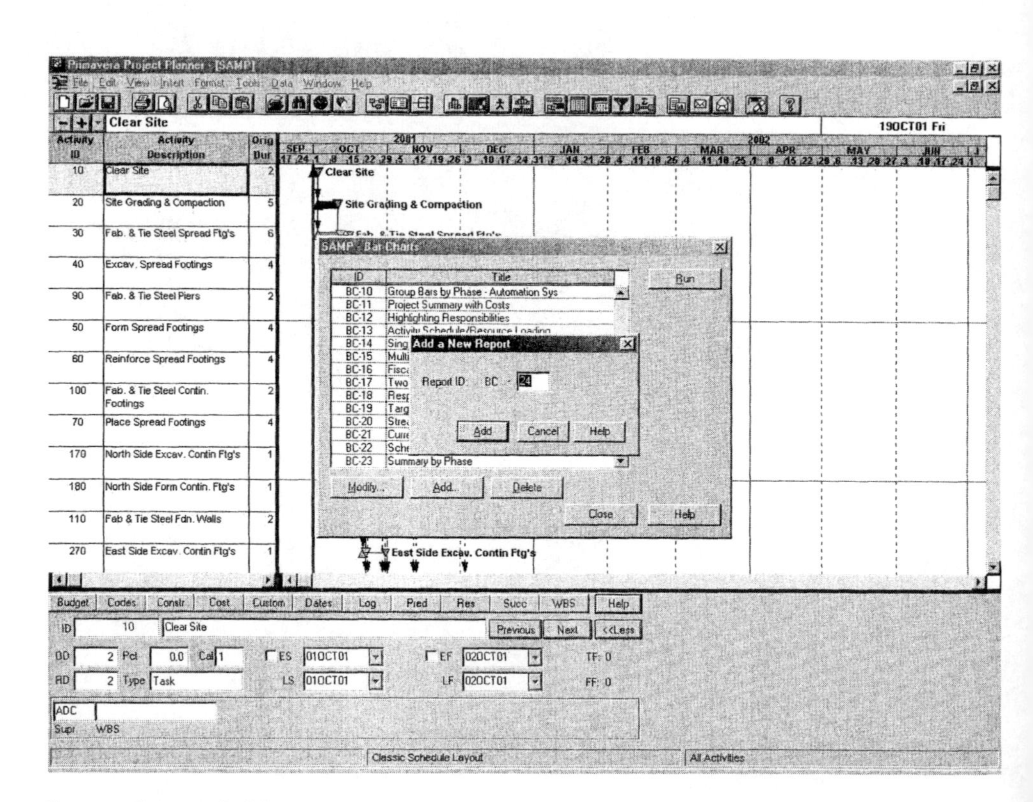

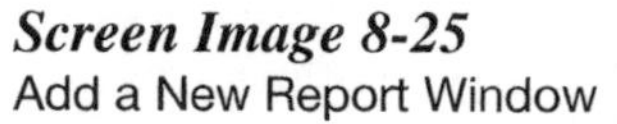

Screen Image 8-25
Add a New Report Window

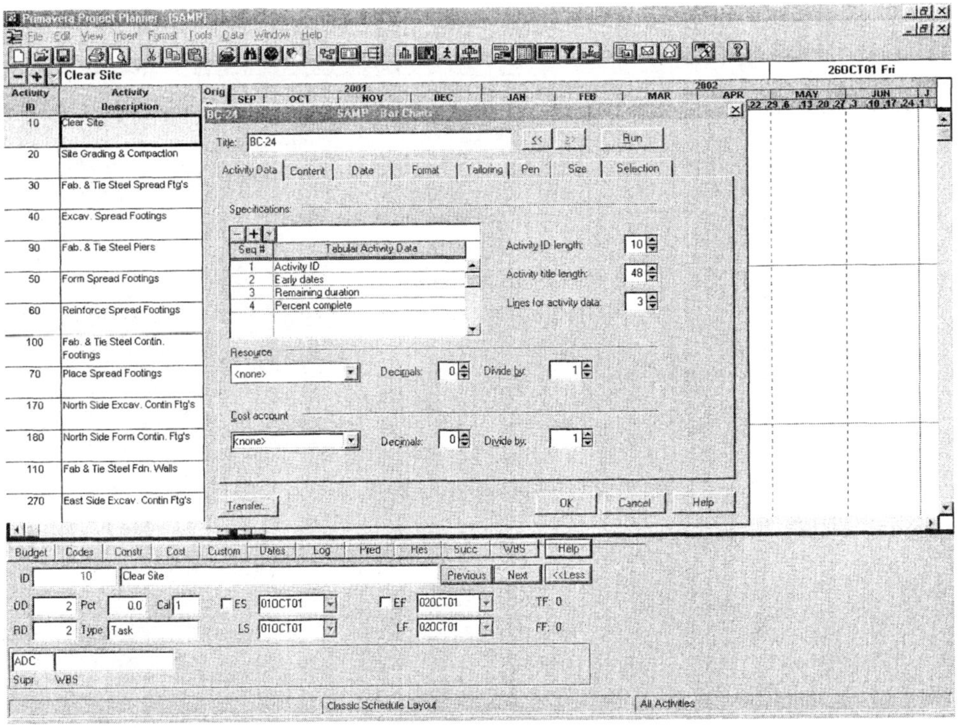

Screen Image 8-26
Initial Activity Window

The *Activity Data* tab allows the user to specify what text data is to be included and how it should be presented. Primavera Project Planner graphic schedules have two basic areas: the left-hand side of the page, which includes text activity information, and the right side, which includes the activity bars and a limited amount of text information. The *Tabular Activity Data* box is used to specify what tabular information to include on the left side of the page. Items can be added and deleted from this list by clicking the plus or minus button. The pull-down menu button, which is adjacent to the plus button, will show a list of all of the possible tabular activity data items. The *Seq#* is the order of the columns of data. The *Activity ID length* specifies how many characters of the activity ID field will be displayed. If a number less than ten is used, the activity ID must be character rather than numeric. For the activity ID to be character, one of the place holders must be character. Numeric activity ID's are right justified, and this option truncates this field from right to left. For example, if an activity ID is 900 and the length is set to 9, only 90 would print for the activity ID. The *Activity title length* option addresses the number of characters of the activity description that will be printed (maximum of 48). In the previous chapters it was recommended that the activity description be limited to between 30 to 35 characters. If that rule has been followed, then the length can be reduced from the max. The final item in the *Specifications* box is the *Lines for activity data*, which can be set to any number between one and six. This field directly impacts the time scale and number of vertical pages. If this field is set to one, all of the tabular activity information, that was specified in the *Tabular Activity Data* box, will be placed on one line. If many columns of text information are all to be presented on one line, that leaves little room for the activity bars and their data. To accommodate this problem Primavera Project Planner will automatically increase the time scale units to fit the page. However, this reduces the precision of the schedule. To overcome this problem, the tabular information can be placed in two or more rows, which will increase the amount of room that is available for the bars. This in turn will allow for smaller units on the time scale, thereby enhancing the precision of the time scale. There are no hard and fast rules for what is the best number of rows. The best approach is one of trial and error to find the setup that best

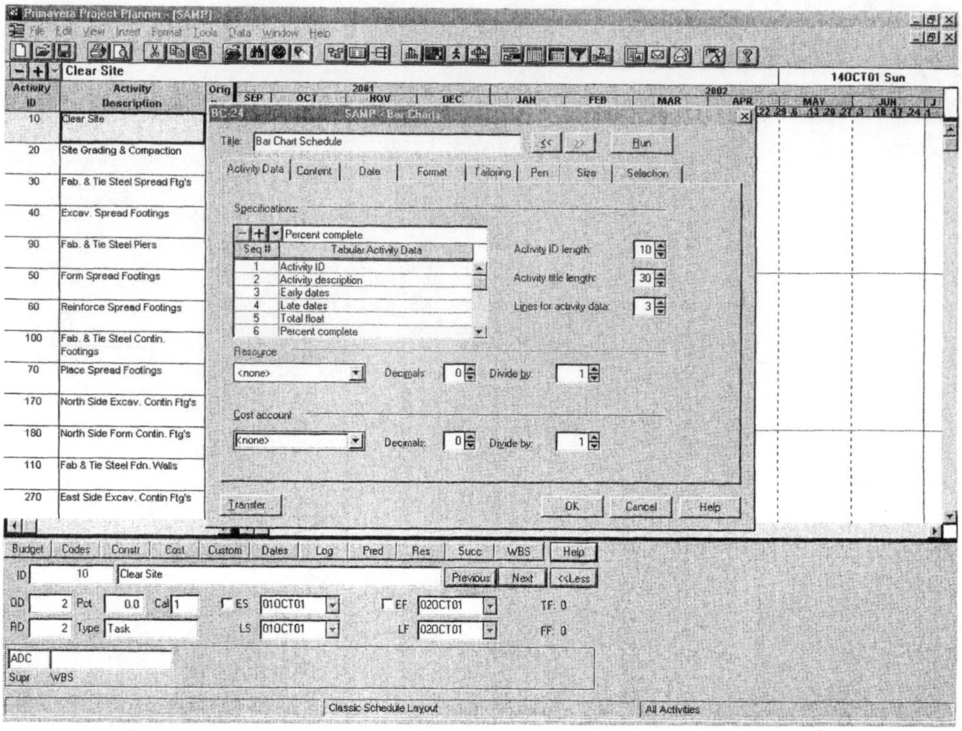

Screen Image 8-27
Revised Activity Data Window

meets the needs of the users. This can easily be done by routing the output to the screen and reviewing it until an acceptable compromise is reached between the precision of the time scale and the number of pages. The *Resource* and *Cost account* options deal with items that will be presented later. In Screen Image 8-27 the *Activity ID, Activity description, Early dates, Late dates, Total float*, and *Percent complete* will be presented in tabular format. In addition, the activity description will be limited to 30 characters, and all of this tabular information will be presented on three rows. The result of these entries can be seen on the left-hand side of the page on Output 8-4 (page 155).

The *Content* tab is used to manipulate the graphic data. The default settings for the content window are found in Screen Image 8-28. The critical input areas on this window are the *Dates* and *Bar data*. The *Dates* area allows for multiple bars and multiple sources of information. The bars shown in the schedule can be either *Early bars, Late bars*, or *Both*. The early bar begins with the early start of the activity and extends through the early finish of that activity, while the late bar begins with the late start and extends through the late finish. If the *Both* option is selected, then both of the bars are displayed. The source of the dates for the bars can be either the current schedule or one of the two target schedules. The target schedules are a previous iteration of a project schedule that has been "frozen." The *Bar data* information is used to specify what data, if any, should be placed adjacent to the activity bars. The *Activity ID, Title*, and *Log* can be masked or shown. If the data is to be shown, it can be placed *Above, Next* to, or *Below* the activity bar. There is also the option of having start and finish triangles added to each of the activity bars. The *Resource/Cost* graphic information will be discussed later. The title information at the bottom of the window is generic project or company information that will be placed at the bottom of the schedule.

In Screen Image 8-29 both the early and late bars will be shown on the schedule with all of the descriptions masked. On Output 8-4 the early bars are shown as thick bars, while the late bars are shown as dashed lines. The amount that the late bar extends beyond the early finish is the float. Screen Image 8-30 shows the default settings for the *Date* information. The *Start date* and *End date* fields are used to specify the beginning and end of the time scale that is shown at the top of the bar chart schedule. The *Minimum interval* is used to

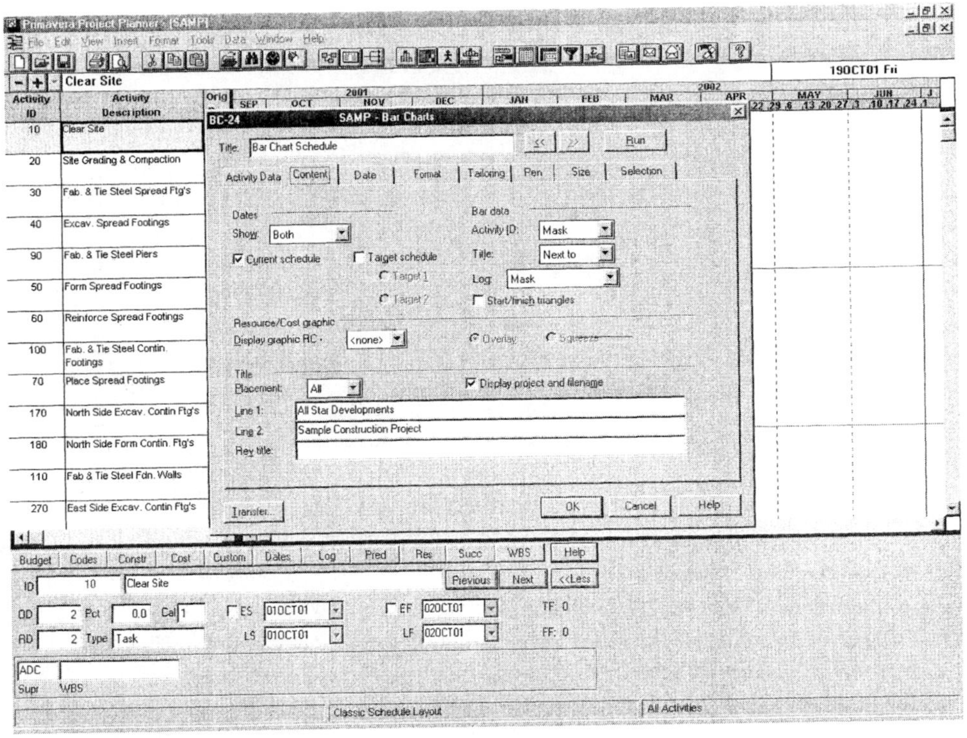

Screen Image 8-28
Initial Content Window

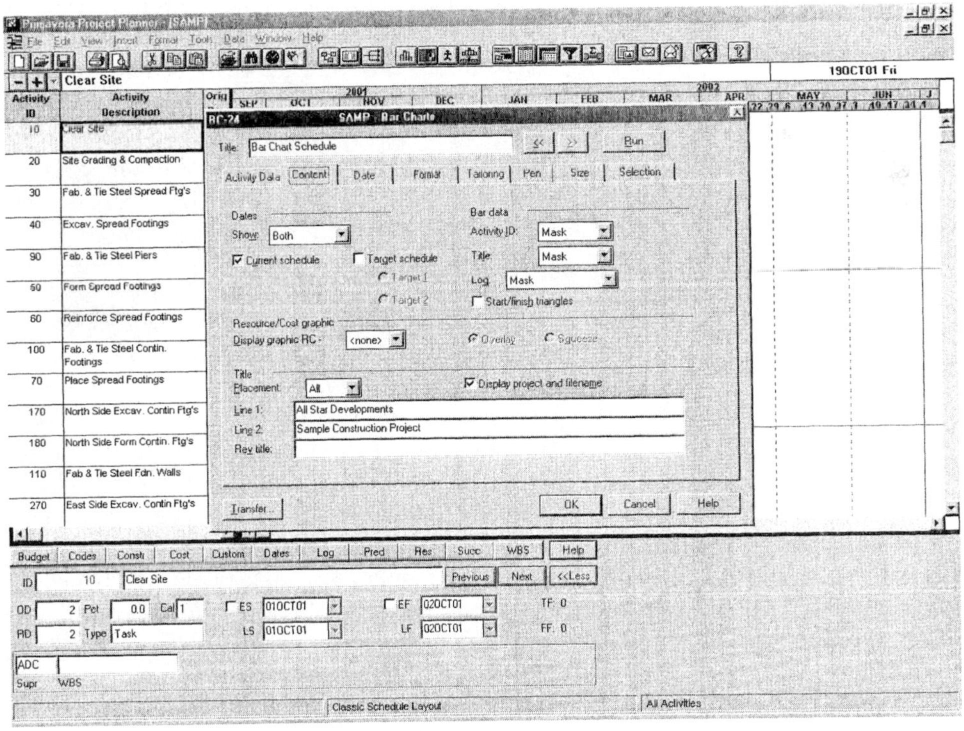

Screen Image 8-29
Revised Content Window

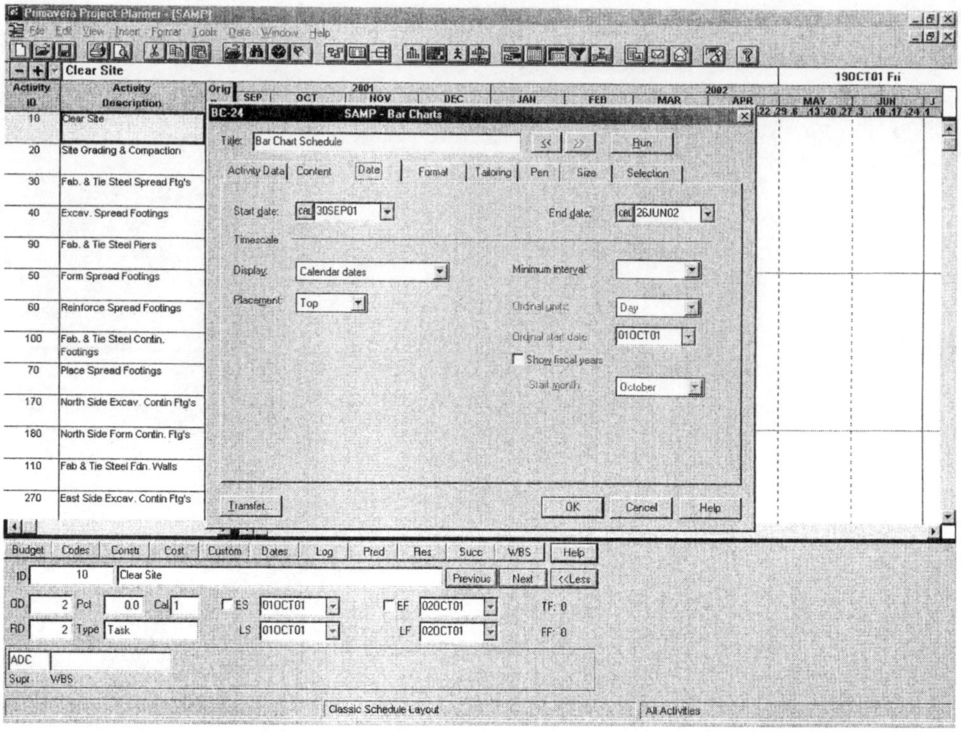

Screen Image 8-30
Date Window

specify a minimum unit of time on the time scale. In most instances the default settings are adequate. The start and end dates should be verified to make sure that they fall within the beginning and end of the project. If activity descriptions are placed next to the activity bars, the end date needs to be extended to allow space for those descriptions.

The *Format* information as shown in Screen Image 8-31 addresses the organization of the information on the page. The *Start new page on* field allows the user to specify when to skip to the top of a new page. The only valid entry in this field is one of the user defined activity codes. Just as with the tabular reports, if this feature is used, the data needs to be sorted correctly. If the sort order is in conflict with this entry, a numerous page report that is of little use will be generated.

The *Group by* field allows the user to specify how the activities are to be grouped on the page. If the field is used, a group description will be printed out everytime that field has a new value. Once again the grouping is limited to one of the activity codes, and the sort must complement this entry. The *Summarize on* box allows for all of the activities to be combined into summary activities for presentation purposes. The summarization is also limited to one of the activity codes. The final box of importance is the *Sort on* in the lower right-hand corner. This box is identical to the one in the tabular reports and allows the user to specify the order in which the data will be presented down the page and if that data should be in ascending or descending order.

Screen Image 8-32 is the *Tailoring* options. This window allows the user to enhance the readability of the schedule by adding vertical and horizontal sight lines. In addition to these features, there is the *Neck for periods of inactivity*. If a bar is necked for a period of inactivity, the thickness of the bar is reduced to show when there is no work being performed on that activity during that period of time. For example, if the bars were necked, they would be thinner on the weekends than on the weekdays. In addition, the user is allowed to define what constitutes a critical activity.

The next window is the *Pen* option, which deals with the logistics of using a pen plotter. Screen Image 8-33 shows the default settings for that window. The *Elements* box is where

Primavera Project Planner - [SAMP]
File Edit View Insert Format Tools Data Window Help
Clear Site
190CT01 Fri

Activity ID	Activity Description
10	Clear Site
20	Site Grading & Compaction
30	Fab. & Tie Steel Spread Ftg's
40	Excav. Spread Footings
90	Fab. & Tie Steel Piers
50	Form Spread Footings
60	Reinforce Spread Footings
100	Fab. & Tie Steel Contin. Footings
70	Place Spread Footings
170	North Side Excav. Contin Ftg's
180	North Side Form Contin. Ftg's
110	Fab & Tie Steel Fdn. Walls
270	East Side Excav. Contin Ftg's

Orig SEP OCT 2001 NOV DEC JAN FEB MAR 2002 APR MAY JUN

BC-24 SAMP - Bar Charts
Title: Bar Chart Schedule << >> Run
Activity Data Content Date Format Tailoring Pen Size Selection
Organize
Start new page on: <none>
Group by: <none>
Summarize on: <none>
Summary graphic
Show individual bars
Place a solid line after each group
Skip description
If less than 0 characters appear
Drop line
Connect bars with 2 lines
Group titles in: Graphic area
Orientation: Landscape
Horizontal sheets: 1
Sort by:

Sort on	Order
Early start	Ascend
	Ascend

Transfer... OK Cancel Help

Budget Codes Constr Cost Custom Dates Log Pred Res Succ WBS Help
ID 10 Clear Site Previous Next <<Less
OD 2 Pct 0.0 Cal 1 ES 01OCT01 EF 02OCT01 TF: 0
RD 2 Type Task LS 01OCT01 LF 02OCT01 FF: 0
ADC
Supr WBS
Classic Schedule Layout All Activities

Screen Image 8-31
Initial Format Window

Primavera Project Planner - [SAMP]
File Edit View Insert Format Tools Data Window Help
Clear Site
190CT01 Fri

Activity ID	Activity Description
10	Clear Site
20	Site Grading & Compaction
30	Fab. & Tie Steel Spread Ftg's
40	Excav. Spread Footings
90	Fab. & Tie Steel Piers
50	Form Spread Footings
60	Reinforce Spread Footings
100	Fab. & Tie Steel Contin. Footings
70	Place Spread Footings
170	North Side Excav. Contin Ftg's
180	North Side Form Contin. Ftg's
110	Fab & Tie Steel Fdn. Walls
270	East Side Excav. Contin Ftg's

Orig SEP OCT 2001 NOV DEC JAN FEB MAR 2002 APR MAY JUN

BC-24 SAMP - Bar Charts
Title: Bar Chart Schedule << >> Run
Activity Data Content Date Format Tailoring Pen Size Selection
Vertical sight lines
Interval: 1
Units: Month
Necking
Neck for periods of inactivity
Expand bar for spanned holidays weekends
For summary bar, use calendar 1
Horizontal sight lines
Interval: 5
Options
Base progress on: Visible percent complete
Critical activities use: Total float
Critical float less than: 1
Page setup
Margin at top (rows): 0
Activity row separation: 0
Begin on sheet: 1
End on sheet:
Confirm preparation
Transfer... OK Cancel Help

Budget Codes Constr Cost Custom Dates Log Pred Res Succ WBS Help
ID 10 Clear Site Previous Next <<Less
OD 2 Pct 0.0 Cal 1 ES 01OCT01 EF 02OCT01 TF: 0
RD 2 Type Task LS 01OCT01 LF 02OCT01 FF: 0

Classic Schedule Layout All Activities

Screen Image 8-32
Tailoring Window

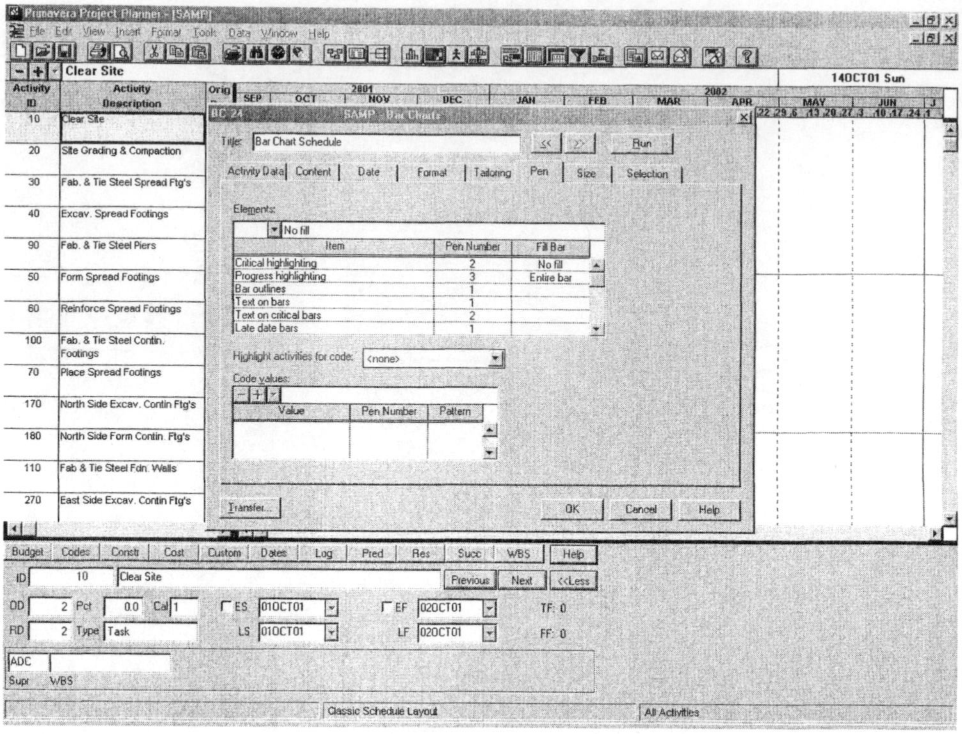

Screen Image 8-33
Pen Window

the user specifies which pens are in which port in the pen carousel. For example, if Pen 1 is black, Pen 2 is red, and Pen 3 is green, the critical activity bars would be outlined in black and filled in red. The *Progress highlighting* entry specifies that as the activities are completed, the bars would be outlined in black and filled with green to show the percent complete. This screen is critical when directing the output to a color-printing device. If the output is being routed to a black-and-white laser printer, most of the input on this screen is not relevant. However, if a black-and-white output device is being used, the user should change the *Critical highlighting* fill from *Entire bar* to *No fill*. With that option the bar would be outlined in black to define the box and outlined within the bar in red to show that it is critical. When printed out on a laser printer the critical highlighting will be a closely spaced double line within the box outline. These extra lines, although very closely spaced, will result in the critical bars being thicker on the output. This can be seen on Activity 40 on Output 8-4.

Screen Image 8-34 shows the default settings for the *Size* window. This window allows the user to set the text and bar size. The best way to set these sizes is through a trial-and-error process. The size of the page and the use of the schedule often dictate what size text will be required.

The *Selection* tab is basically the same window that was used when generating the tabular reports. This is the window where the activities are selected for inclusion or exclusion from the graphic schedule. See Screen Image 8-35.

When the specifications have all been set, the *Run* button can be clicked. This will prompt the user to specify where to route the output, as shown in Screen Image 8-36. Once the destination of the output has been specified, Primavera Project Planner will begin to generate the desired output. However, prior to outputting this data to any destination, the user will be notified as to the number of pages and given the opportunity to proceed or cancel the output. Screen Image 8-37 is an example of the prompt. It is recommended that the output be routed to the screen prior to being routed to any output device. This will speed up the fine-tuning process and save paper. Output 8-4 is an example of the output that the previous report specifications would have produced.

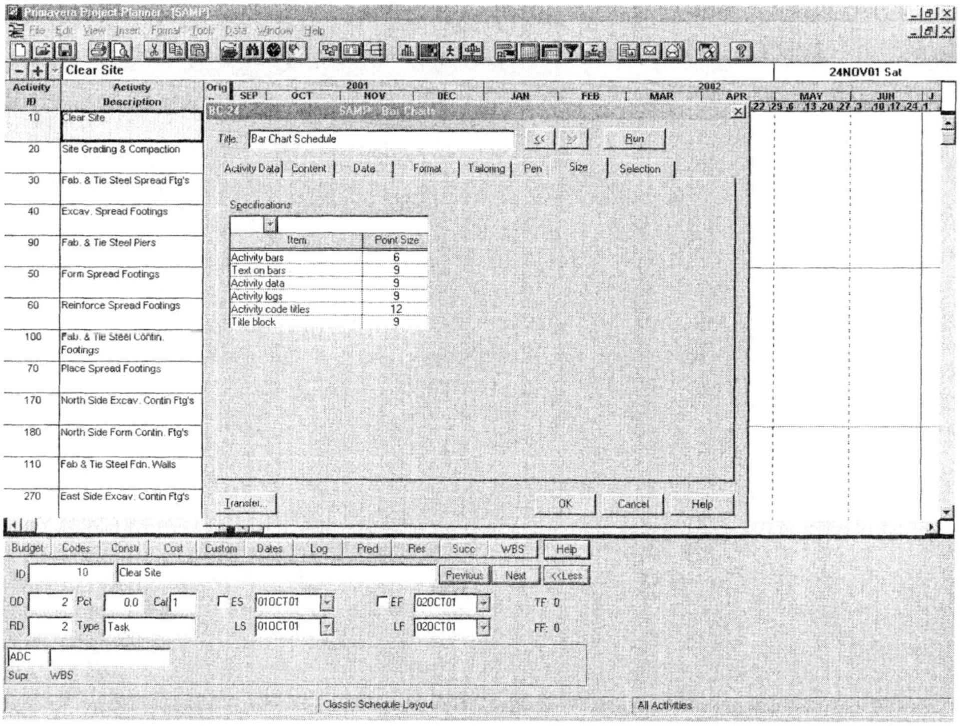

Primavera Project Planner - [SAMP]
File Edit View Insert Format Tools Data Window Help
Clear Site
24NOV01 Sat

Activity ID	Activity Description
10	Clear Site
20	Site Grading & Compaction
30	Fab. & Tie Steel Spread Ftg's
40	Excav. Spread Footings
90	Fab. & Tie Steel Piers
50	Form Spread Footings
60	Reinforce Spread Footings
100	Fab. & Tie Steel Contin. Footings
70	Place Spread Footings
170	North Side Excav. Contin Ftg's
180	North Side Form Contin. Ftg's
110	Fab & Tie Steel Fdn. Walls
270	East Side Excav. Contin Ftg's

2001 SEP OCT NOV DEC 2002 JAN FEB MAR APR MAY JUN

BC-24 SAMP - Bar Charts
Title: Bar Chart Schedule
Run
Activity Data | Content | Date | Format | Tailoring | Pen | Size | Selection
Specifications:

Item	Point Size
Activity bars	6
Text on bars	9
Activity data	9
Activity logs	9
Activity code titles	12
Title block	9

Transfer... OK Cancel Help

Budget Codes Constr Cost Custom Dates Log Pred Res Succ WBS Help
ID 10 Clear Site Previous Next <<Less
OD 2 Pct 0.0 Cal 1 ES 01OCT01 EF 02OCT01 TF 0
RD 2 Type Task LS 01OCT01 LF 02OCT01 FF 0
ADC
Supr WBS
Classic Schedule Layout All Activities

Screen Image 8-34
Size Window

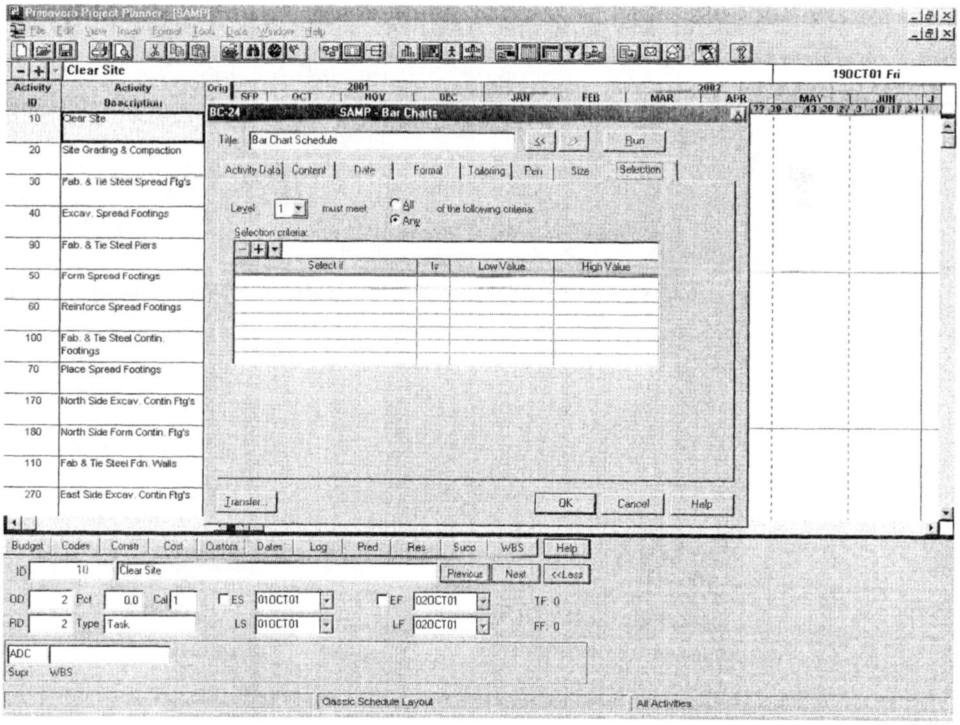

Primavera Project Planner - [SAMP]
File Edit View Insert Format Tools Data Window Help
Clear Site
19OCT01 Fri

Activity ID	Activity Description
10	Clear Site
20	Site Grading & Compaction
30	Fab. & Tie Steel Spread Ftg's
40	Excav. Spread Footings
90	Fab. & Tie Steel Piers
50	Form Spread Footings
60	Reinforce Spread Footings
100	Fab. & Tie Steel Contin. Footings
70	Place Spread Footings
170	North Side Excav. Contin Ftg's
180	North Side Form Contin. Ftg's
110	Fab & Tie Steel Fdn. Walls
270	East Side Excav. Contin Ftg's

2001 SEP OCT NOV DEC 2002 JAN FEB MAR APR MAY JUN

BC-24 SAMP - Bar Charts
Title: Bar Chart Schedule
Run
Activity Data | Content | Date | Format | Tailoring | Pen | Size | Selection
Level 1 must meet All / Any of the following criteria:
Selection criteria:

Select if	Is	Low Value	High Value

Transfer... OK Cancel Help

Budget Codes Constr Cost Custom Dates Log Pred Res Succ WBS Help
ID 10 Clear Site Previous Next <<Less
OD 2 Pct 0.0 Cal 1 ES 01OCT01 EF 02OCT01 TF 0
RD 2 Type Task LS 01OCT01 LF 02OCT01 FF 0
ADC
Supr WBS
Classic Schedule Layout All Activities

Screen Image 8-35
Selection Window

Screen Image 8-36
Run Options

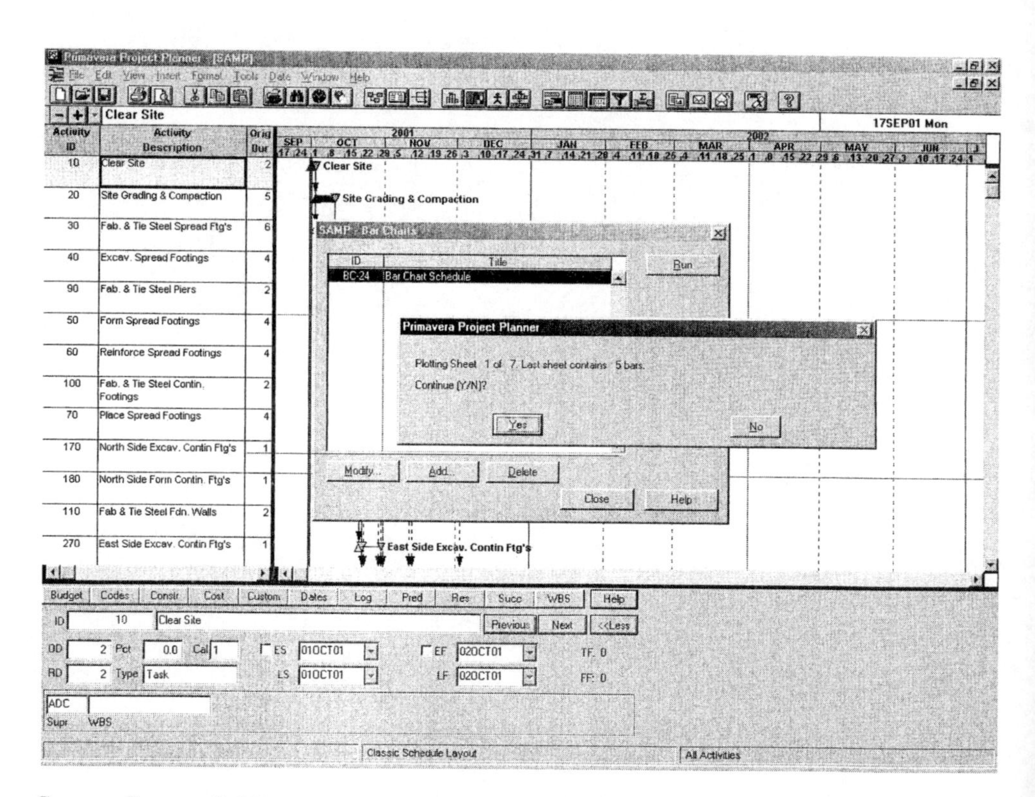

Screen Image 8-37
Output Prompt

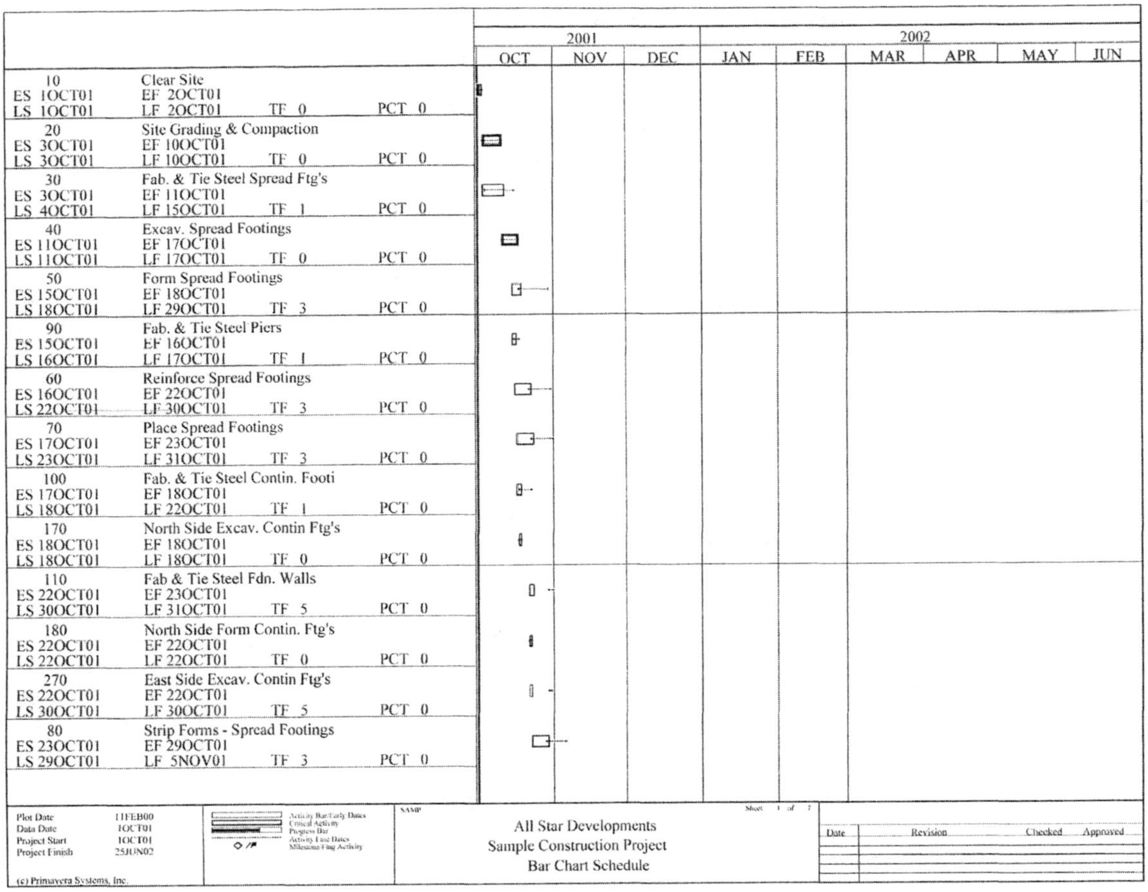

	2001			2002					
	OCT	NOV	DEC	JAN	FEB	MAR	APR	MAY	JUN

Activity	Description	ES	EF	LS	LF	TF	PCT
10	Clear Site	1OCT01	2OCT01	1OCT01	2OCT01	0	0
20	Site Grading & Compaction	3OCT01	10OCT01	3OCT01	10OCT01	0	0
30	Fab. & Tie Steel Spread Ftg's	3OCT01	11OCT01	4OCT01	15OCT01	1	0
40	Excav. Spread Footings	11OCT01	17OCT01	11OCT01	17OCT01	0	0
50	Form Spread Footings	15OCT01	18OCT01	18OCT01	29OCT01	3	0
90	Fab. & Tie Steel Piers	15OCT01	16OCT01	16OCT01	17OCT01	1	0
60	Reinforce Spread Footings	16OCT01	22OCT01	22OCT01	30OCT01	3	0
70	Place Spread Footings	17OCT01	23OCT01	23OCT01	31OCT01	3	0
100	Fab. & Tie Steel Contin. Footi	17OCT01	18OCT01	18OCT01	22OCT01	1	0
170	North Side Excav. Contin Ftg's	18OCT01	18OCT01	18OCT01	18OCT01	0	0
110	Fab & Tie Steel Fdn. Walls	22OCT01	23OCT01	30OCT01	31OCT01	5	0
180	North Side Form Contin. Ftg's	22OCT01	22OCT01	22OCT01	22OCT01	0	0
270	East Side Excav. Contin Ftg's	22OCT01	22OCT01	30OCT01	30OCT01	5	0
80	Strip Forms - Spread Footings	23OCT01	29OCT01	29OCT01	5NOV01	3	0

Plot Date 11FEB00
Data Date 1OCT01
Project Start 1OCT01
Project Finish 25JUN02

Activity Bar/Early Dates
Critical Activity
Progress Bar
Activity Line Dates
Milestone/Flag Activity

SAMP

Sheet 1 of 7

All Star Developments
Sample Construction Project
Bar Chart Schedule

Date	Revision	Checked	Approved

Output 8-4
Graphic Bar Chart Schedule (page 1)

Timescaled Logic Diagrams

A timescaled logic diagram is similar to a bar chart except that the relationships between the bars are drawn. This diagram is produced using a methodology similar to the one employed in producing the bar chart schedule. The first step is to click on the *Tools* command from the main input window followed by the *Graphic reports* and *Timescaled logic* option as shown in Screen Image 8-38. In Screen Image 8-39 there is the list of all of the previously created timescaled logic diagrams.

To create a new report click the *Add* button, which will open a new report window as shown in Screen Image 8-40. Clicking once again on the *Add* button will open the *Timescaled Logic Diagrams* window as shown in Screen Image 8-41. This window has the same tabs as the ones that were seen when creating the bar chart schedule. The *Content* information addresses which bars and data to display. Screen Image 8-41 shows that the *Activity ID, Activity description*, and *Total float* will be displayed along with the early bars. The date information is the same as with the bar chart, and details the beginning and end of the time scale. In Screen Image 8-42, the start date is 30Sep01; the scale extends through 26June2002.

Screen Image 8-43 is for format information. The *Group by* field allows for the activities to be grouped together by activity code. The *Start new page on* specifies any special page breaks. Just as with the tabular reports, the sort order needs to be configured so that all of those activity codes are grouped together. The *Tailoring* option on the *Timescaled Logic Diagrams* page, Screen Image 8-44, is identical in appearance and operation to the respective tab from the bar chart. The font size of the text and bars are changed via the *Size* tab as shown in Screen Image 8-45. The default settings work relatively well for all reports.

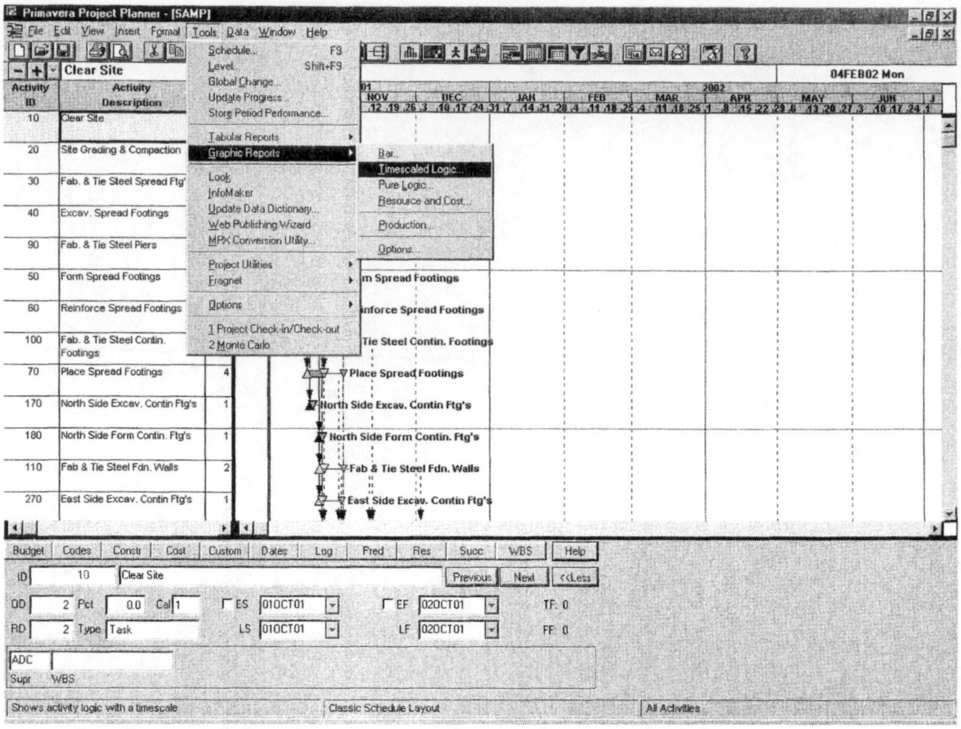

Primavera Project Planner - [SAMP]
File Edit View Insert Format Tools Data Window Help
Clear Site
04FEB02 Mon
Schedule... F9
Level... Shift+F9
Global Change...
Update Progress...
Store Period Performance...
Tabular Reports
Graphic Reports
Bar...
Timescaled Logic...
Pure Logic...
Resource and Cost...
Production...
Options...
Look
InfoMaker
Update Data Dictionary...
Web Publishing Wizard
MPX Conversion Utility...
Project Utilities
Fragnet
Options
1 Project Check-in/Check-out
2 Monte Carlo

Activity ID	Activity Description	
10	Clear Site	
20	Site Grading & Compaction	
30	Fab. & Tie Steel Spread Ftg'	
40	Excav. Spread Footings	
90	Fab. & Tie Steel Piers	
50	Form Spread Footings	
60	Reinforce Spread Footings	
100	Fab. & Tie Steel Contin. Footings	
70	Place Spread Footings	4
170	North Side Excav. Contin Ftg's	1
180	North Side Form Contin. Ftg's	1
110	Fab & Tie Steel Fdn. Walls	2
270	East Side Excav. Contin Ftg's	1

Budget Codes Constr Cost Custom Dates Log Pred Res Succ WBS Help
ID 10 Clear Site Previous Next <<Less
OD 2 Pct 0.0 Cal 1 ES 01OCT01 EF 02OCT01 TF: 0
RD 2 Type Task LS 01OCT01 LF 02OCT01 FF: 0
ADC
Supr WBS
Shows activity logic with a timescale | Classic Schedule Layout | All Activities

Screen Image 8-38
Opening the Timescaled Logic Window

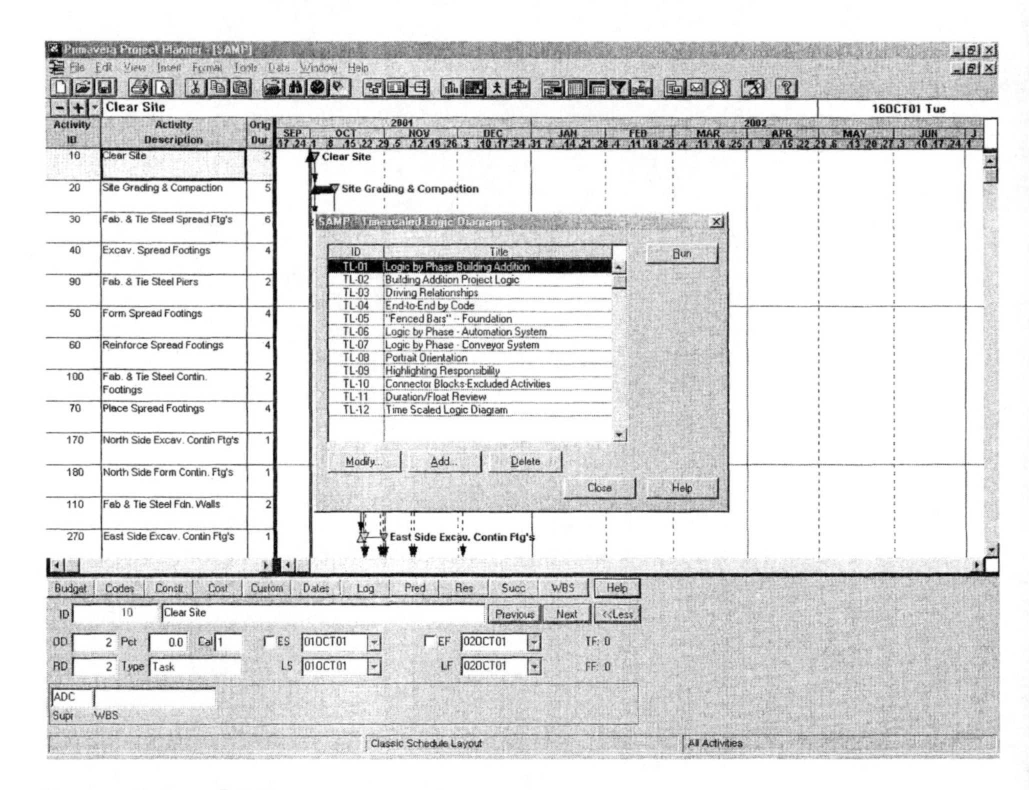

Screen Image 8-39
Previous Timescaled Logic Diagrams

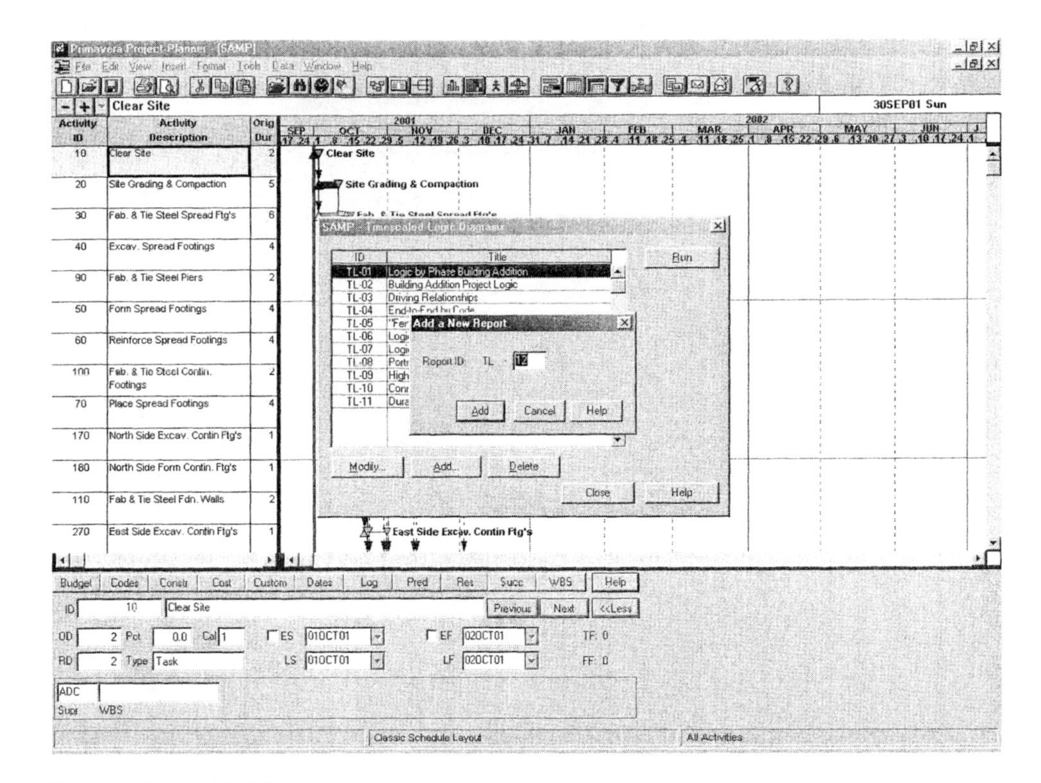

Screen Image 8-40
Opening a New Report

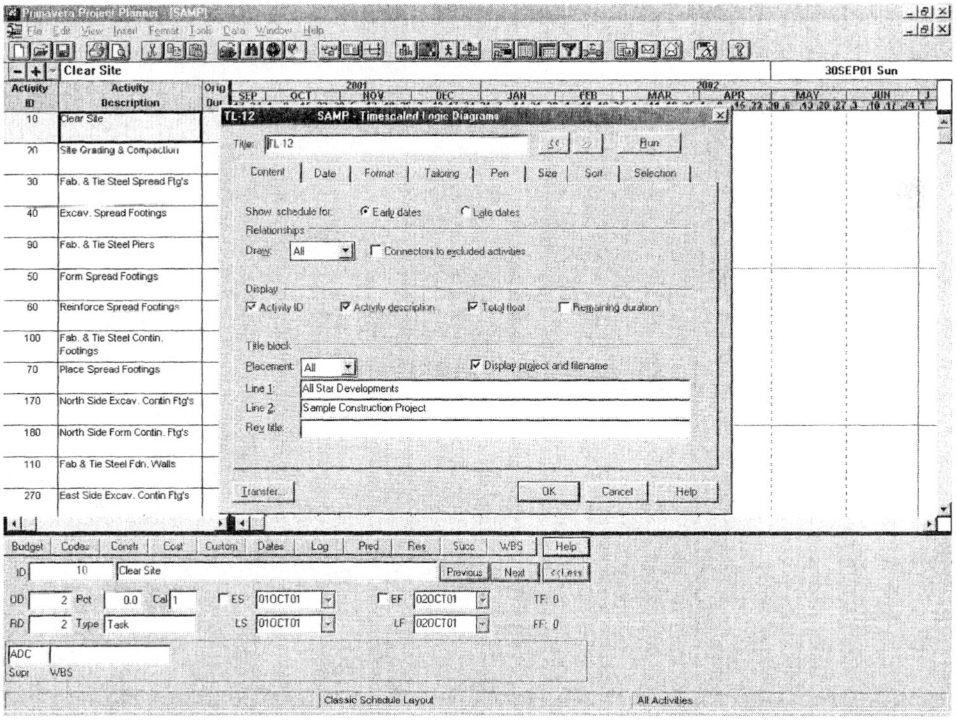

Primavera Project Planner - [SAMP]
File Edit View Insert Format Tools Data Window Help
Clear Site
30SEP01 Sun
Activity ID
Activity Description
Orig Dur
2001
2002
SEP OCT NOV DEC JAN FEB MAR APR MAY JUN J
10 Clear Site
20 Site Grading & Compaction
30 Fab. & Tie Steel Spread Ftg's
40 Excav. Spread Footings
90 Fab. & Tie Steel Piers
50 Form Spread Footings
60 Reinforce Spread Footings
100 Fab. & Tie Steel Contin. Footings
70 Place Spread Footings
170 North Side Excav. Contin Ftg's
180 North Side Form Contin. Ftg's
110 Fab & Tie Steel Fdn. Walls
270 East Side Excav. Contin Ftg's
TL-12 SAMP - Timescaled Logic Diagrams
Title: TL 12
Run
Content Date Format Tailoring Pen Size Sort Selection
Show schedule for: Early dates Late dates
Relationships
Draw: All
Connectors to excluded activities
Display
Activity ID
Activity description
Total float
Remaining duration
Title block
Placement: All
Display project and filename
Line 1: All Star Developments
Line 2: Sample Construction Project
Rev title:
Transfer...
OK
Cancel
Help
Budget Codes Constr Cost Custom Dates Log Pred Res Succ WBS Help
ID 10 Clear Site
Previous Next <<Less
OD 2 Pct 0.0 Cal 1
ES 01OCT01
EF 02OCT01
TF: 0
RD 2 Type Task
LS 01OCT01
LF 02OCT01
FF: 0
ADC
Supr WBS
Classic Schedule Layout
All Activities

Screen Image 8-41
Timescaled Logic Diagrams Window

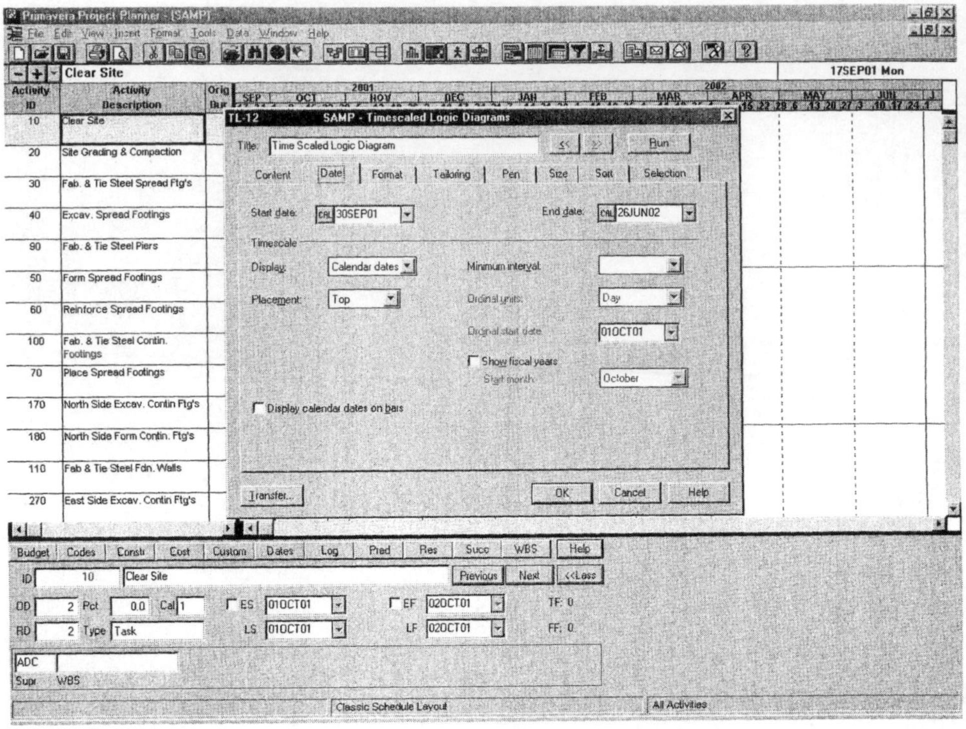

Screen Image 8-42
Date Tab Set

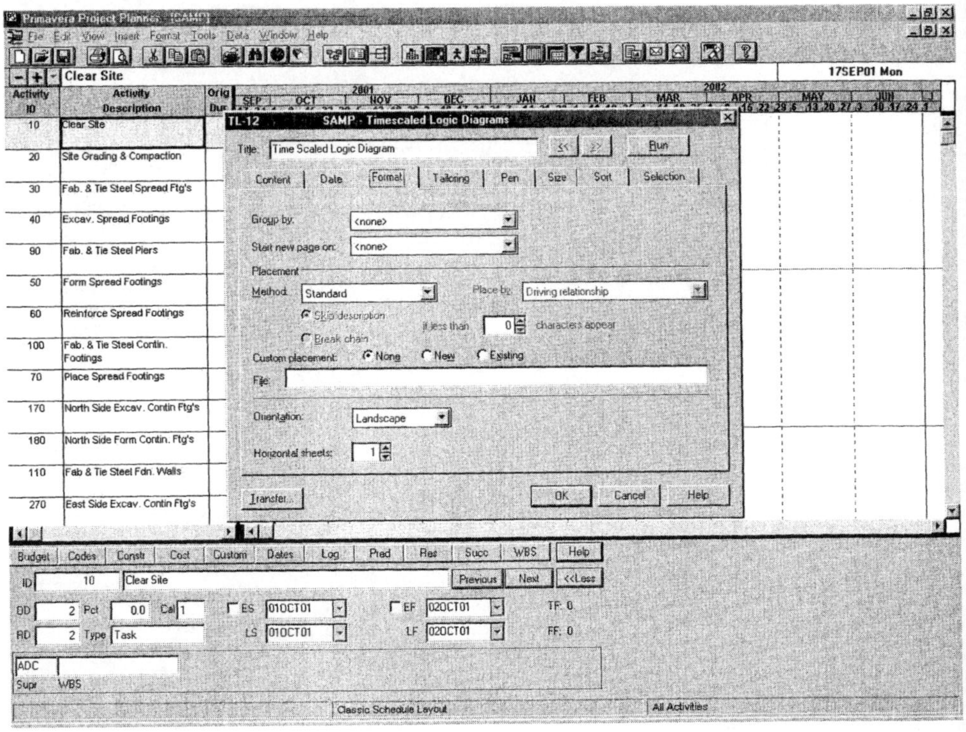

Screen Image 8-43
Format Tab Set

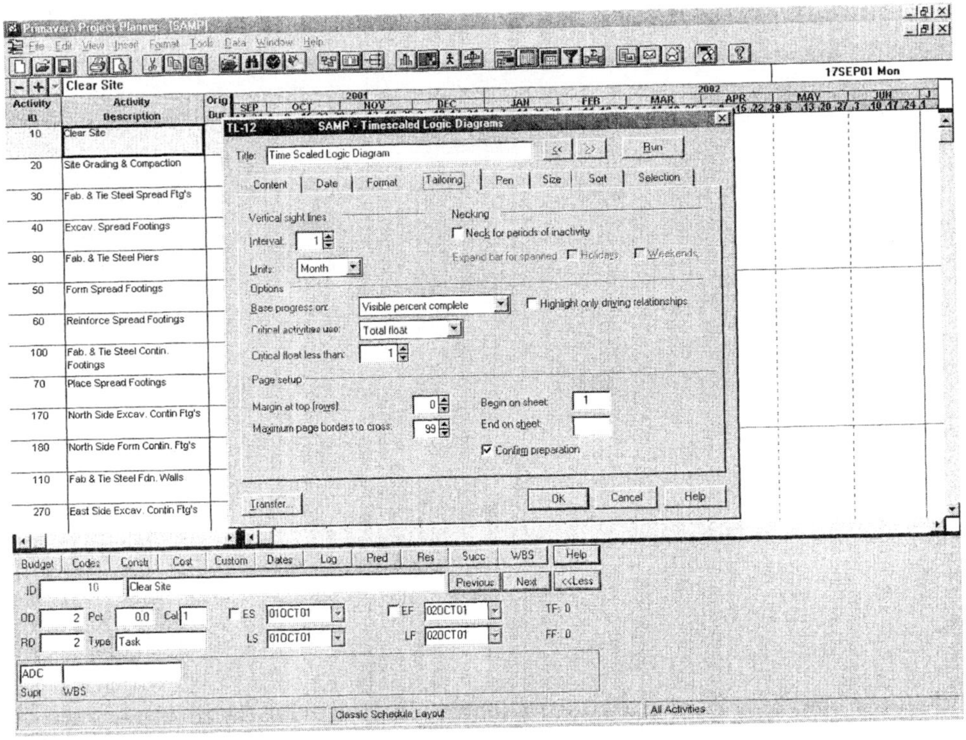

Primavera Project Planner - [SAMP]
File Edit View Insert Format Tools Data Window Help
17SEP01 Mon
Clear Site

Activity ID	Activity Description
10	Clear Site
20	Site Grading & Compaction
30	Fab. & Tie Steel Spread Ftg's
40	Excav. Spread Footings
90	Fab. & Tie Steel Piers
50	Form Spread Footings
60	Reinforce Spread Footings
100	Fab. & Tie Steel Contin. Footings
70	Place Spread Footings
170	North Side Excav. Contin Ftg's
180	North Side Form Contin. Ftg's
110	Fab & Tie Steel Fdn. Walls
270	East Side Excav. Contin Ftg's

TL-12 SAMP - Timescaled Logic Diagrams
Title: Time Scaled Logic Diagram
Run
Content | Date | Format | Tailoring | Pen | Size | Sort | Selection
Vertical sight lines
Interval: 1
Units: Month
Necking
Neck for periods of inactivity
Expand bar for spanned Holidays Weekends
Options
Base progress on: Visible percent complete
Highlight only driving relationships
Critical activities use: Total float
Critical float less than: 1
Page setup
Margin at top (rows): 0
Maximum page borders to cross: 99
Begin on sheet: 1
End on sheet:
Confirm preparation
Transfer... OK Cancel Help

Budget | Codes | Constr | Cost | Custom | Dates | Log | Pred | Res | Succ | WBS | Help
ID 10 Clear Site Previous Next <<Less
OD 2 Pct 0.0 Cal 1 ES 01OCT01 EF 02OCT01 TF: 0
RD 2 Type Task LS 01OCT01 LF 02OCT01 FF: 0
ADC
Supt WBS
Classic Schedule Layout All Activities

Screen Image 8-44
Tailoring Choices

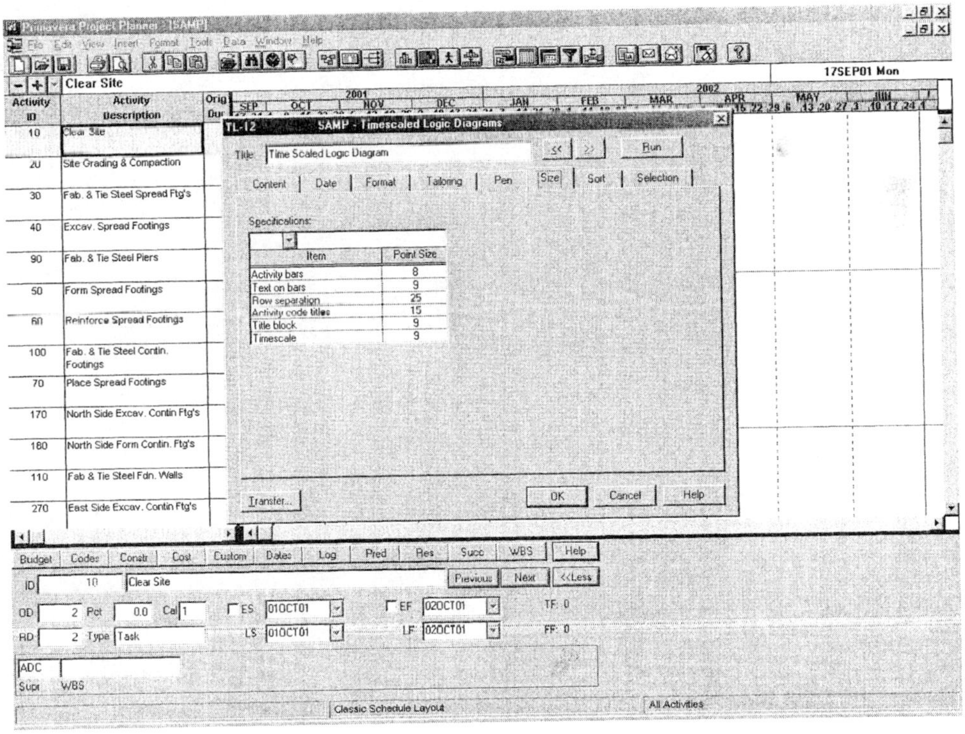

Primavera Project Planner - [SAMP]
File Edit View Insert Format Tools Data Window Help
17SEP01 Mon
Clear Site

Activity ID	Activity Description
10	Clear Site
20	Site Grading & Compaction
30	Fab. & Tie Steel Spread Ftg's
40	Excav. Spread Footings
90	Fab. & Tie Steel Piers
50	Form Spread Footings
60	Reinforce Spread Footings
100	Fab. & Tie Steel Contin. Footings
70	Place Spread Footings
170	North Side Excav. Contin Ftg's
180	North Side Form Contin. Ftg's
110	Fab & Tie Steel Fdn. Walls
270	East Side Excav. Contin Ftg's

TL-12 SAMP - Timescaled Logic Diagrams
Title: Time Scaled Logic Diagram
Run
Content | Date | Format | Tailoring | Pen | Size | Sort | Selection
Specifications:

Item	Point Size
Activity bars	8
Text on bars	9
Row separation	25
Activity code titles	15
Title block	9
Timescale	9

Transfer... OK Cancel Help

Budget | Codes | Constr | Cost | Custom | Dates | Log | Pred | Res | Succ | WBS | Help
ID 10 Clear Site Previous Next <<Less
OD 2 Pct 0.0 Cal 1 ES 01OCT01 EF 02OCT01 TF: 0
RD 2 Type Task LS 01OCT01 LF 02OCT01 FF: 0
ADC
Supt WBS
Classic Schedule Layout All Activities

Screen Image 8-45
Size Choices

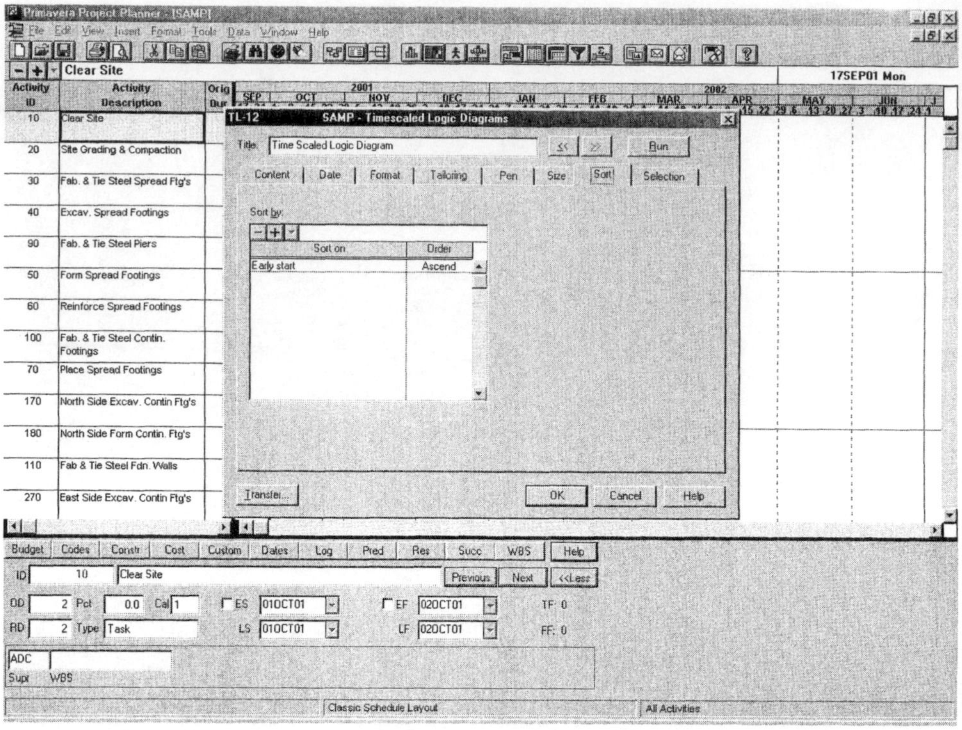

Primavera Project Planner - [SAMP]
File Edit View Insert Format Tools Data Window Help
Clear Site
17SEP01 Mon

Activity ID	Activity Description
10	Clear Site
20	Site Grading & Compaction
30	Fab. & Tie Steel Spread Ftg's
40	Excav. Spread Footings
90	Fab. & Tie Steel Piers
50	Form Spread Footings
60	Reinforce Spread Footings
100	Fab. & Tie Steel Contin. Footings
70	Place Spread Footings
170	North Side Excav. Contin Ftg's
180	North Side Form Contin. Ftg's
110	Fab & Tie Steel Fdn. Walls
270	East Side Excav. Contin Ftg's

Orig Dur
2001 SEP OCT NOV DEC 2002 JAN FEB MAR APR MAY JUN J

TL-12 SAMP - Timescaled Logic Diagrams
Title: Time Scaled Logic Diagram
<< >> Run
Content | Date | Format | Tailoring | Pen | Size | Sort | Selection
Sort by:

Sort on	Order
Early start	Ascend

Transfer... OK Cancel Help

Budget Codes Constr Cost Custom Dates Log Pred Res Succ WBS Help
ID 10 Clear Site Previous Next <<Less
OD 2 Pct 0.0 Cal 1 ES 01OCT01 EF 02OCT01 TF: 0
RD 2 Type Task LS 01OCT01 LF 02OCT01 FF: 0
ADC
Supr WBS
Classic Schedule Layout All Activities

Screen Image 8-46
Sort Selections

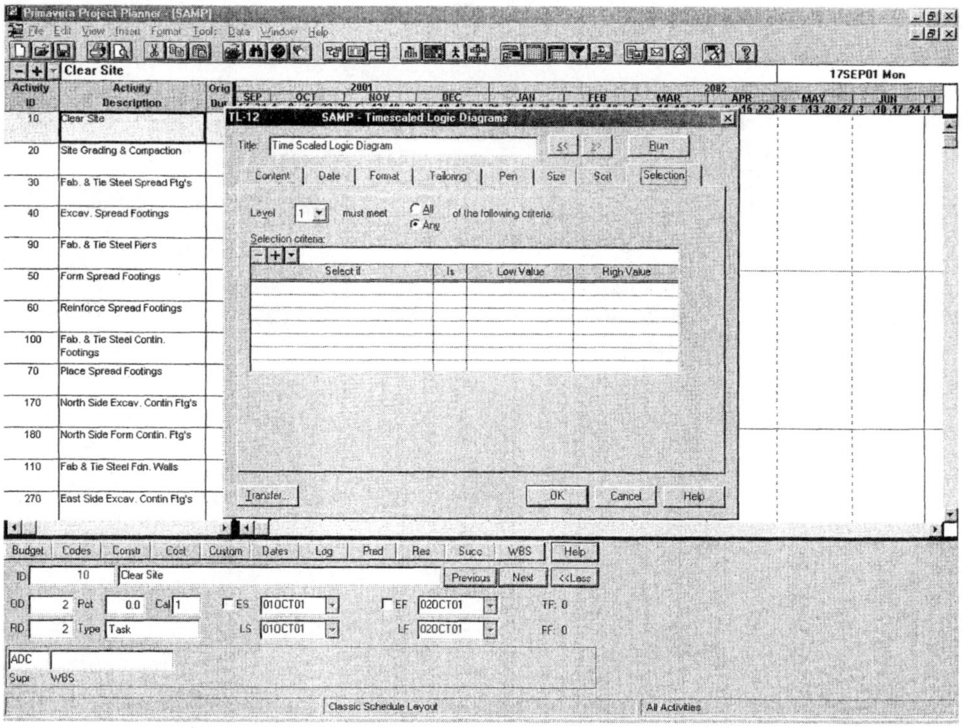

Primavera Project Planner - [SAMP]
File Edit View Insert Format Tools Data Window Help
Clear Site
17SEP01 Mon

Activity ID	Activity Description
10	Clear Site
20	Site Grading & Compaction
30	Fab. & Tie Steel Spread Ftg's
40	Excav. Spread Footings
90	Fab. & Tie Steel Piers
50	Form Spread Footings
60	Reinforce Spread Footings
100	Fab. & Tie Steel Contin. Footings
70	Place Spread Footings
170	North Side Excav. Contin Ftg's
180	North Side Form Contin. Ftg's
110	Fab & Tie Steel Fdn. Walls
270	East Side Excav. Contin Ftg's

Orig Dur
2001 SEP OCT NOV DEC 2002 JAN FEB MAR APR MAY JUN J

TL-12 SAMP - Timescaled Logic Diagrams
Title: Time Scaled Logic Diagram
<< >> Run
Content | Date | Format | Tailoring | Pen | Size | Sort | Selection
Level 1 must meet All / Any of the following criteria:
Selection criteria:

Select if	Is	Low Value	High Value

Transfer... OK Cancel Help

Budget Codes Constr Cost Custom Dates Log Pred Res Succ WBS Help
ID 10 Clear Site Previous Next <<Less
OD 2 Pct 0.0 Cal 1 ES 01OCT01 EF 02OCT01 TF: 0
RD 2 Type Task LS 01OCT01 LF 02OCT01 FF: 0
ADC
Supr WBS
Classic Schedule Layout All Activities

Screen Image 8-47
Selection Window

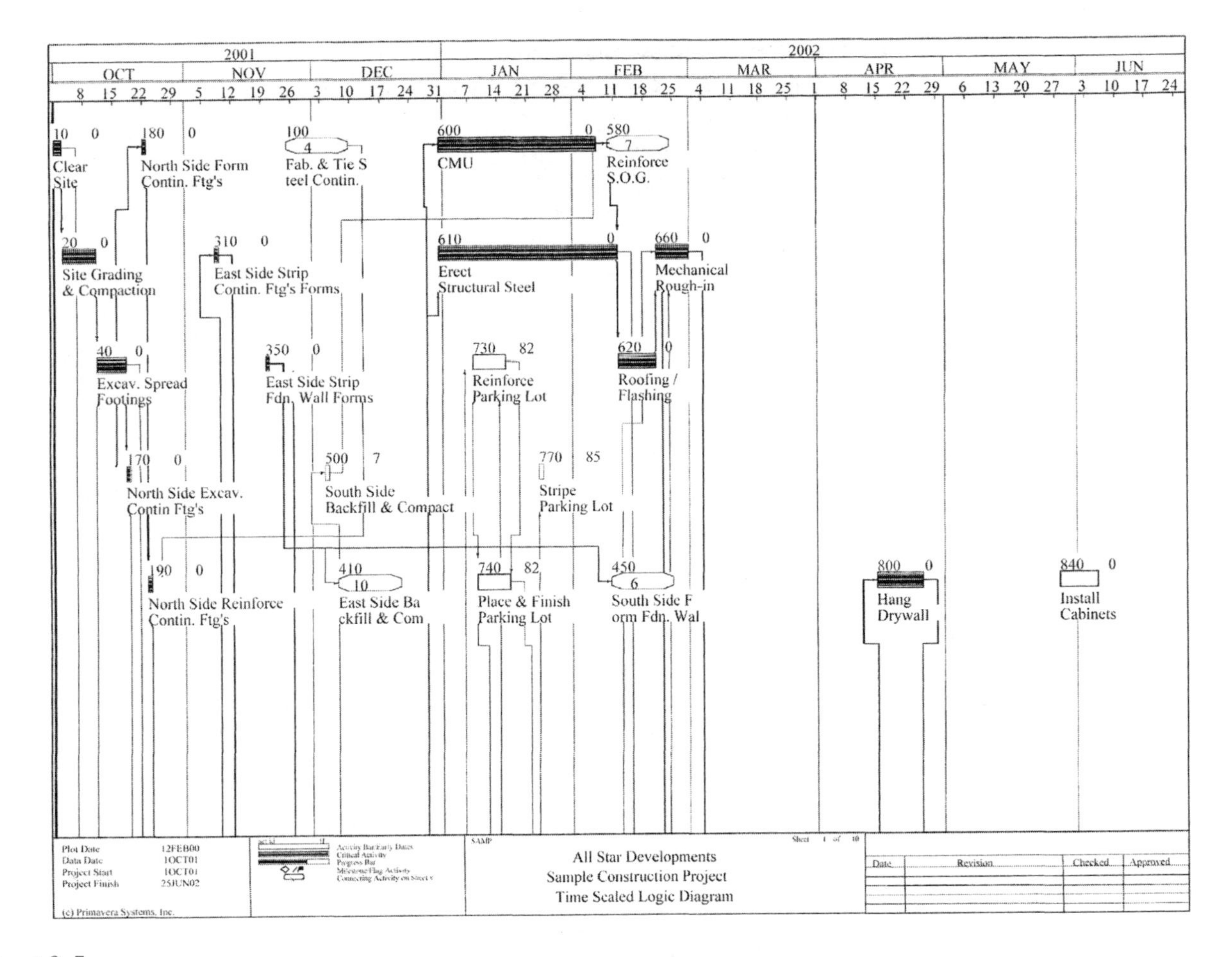

Output 8-5
Time Scaled Logic Diagram

The sort table in Screen Image 8-46 is used to organize the information vertically on the page. In that screen image the only sort is on Early Start, which will place the bars in the sequence that the tasks will be performed. Screen Image 8-47 offers selection choices. The input required and operation is identical to the selection window that was introduced and discussed in regard to tabular reports. Output 8-5 is a page out of the Timescaled logic diagram report.

Conclusion

The three tabular reports and two graphic reports detailed above are merely examples of the types of reports that could be produced. These reports would certainly be helpful in managing a project; however, they are by no means the only ones that should be used or would be required. By using various report windows, the scheduler can develop all the reports necessary to meet the needs of the project.

Suggested Exercises

1. Using the project that has been input and manipulated in the previous chapters, develop an early start report, a critical activity list report, and a responsibility report.

2. Pick a date roughly midway through the project and develop a 2-week look-ahead report. This report should list all of the activities that could start within the 2-week window.
3. If a color pen plotter or inkjet plotter is available, plot out a bar chart schedule as detailed in this chapter. Sort this graphic schedule by early start and total float.
4. If a plotter is not available, produce a useful bar chart schedule and route it to the line printer. The type size, bar size, and row spacing will need to be adjusted to ensure that the quality and readability are acceptable.
5. Produce a timescaled logic diagram using whatever output device is available.

Summarizing the Schedule

KEY TERMS

Hammock Activity A specialized activity that is used to summarize a chain of construction activities.

Milestone An activity that represents a significant point in time.

Flag An activity that signals the start or completion of an activity or group of activities.

Summary Schedules

The construction network schedule is an integral tool in effective project management. However, this type of schedule with its many activities and relationships imposes practical limitations due to its physical size and complexity. A network of this type may be easily understood by the individuals who created and update it. However, to many others the detail and complexity add unwanted information. To meet the needs of these users the construction network can be summarized into a simple bar chart schedule. The summary schedule can be configured to display a single activity that represents a number of activities. This type of schedule provides the user critical information about the project and points out where difficulties may lie.

The schedule can be summarized by adding summary activity codes or summary activities called hammocks. Regardless of the method selected, the summary activities will automatically reflect any changes that occur in the underlying construction activities. Both summarization methods have their advantages and disadvantages. Since hammocks are activities that are physically added to the schedule, they can be manipulated the same as any other activity. The activity code method will produce the same summary schedule but will not be able to accommodate some of the features that are associated with an activity.

Activity Codes Summary Schedule

The first step in producing an activity code summarization is to develop a matrix that relates all of the individual activities to a specific summary activity. Figure 9-1 is an example of a matrix. In that diagram every underlying construction activity is tied to a summary activity code. After the matrix has been developed, these activity codes need to be added to the Primavera Project Planner schedule. This activity code is created in the exact same fashion as were those done in Chapter 6. The process is begun by opening the *Activity Codes* dictionary window. This is done by clicking on the *Data* command followed by the *Activity Codes* option (Screen Image 9-1). From the *Activity Codes* window the new activity code which will be used for summarization and all of its associated valid entries can be entered. Screen Image 9-2 shows the *Activity Codes* window with the new activity code *SUMM*, which is the descriptor for the field, with its appropriate description. The valid entries, or summary activities, are entered in the *Values* box. These entries correspond to the items in the matrix in Figure 9-1.

After the summary activity codes have been added to the activity codes dictionary, it is necessary to add the appropriate activity code to the corresponding activity. The matrix in

1000 Sitework
- 10 Clear Site
- 20 Site Grading & Compaction

1010 Spread Footings
- 30 Fab. & Tie Steel Spread Ftg's
- 40 Excav. Spread Footings
- 50 Form Spread Footings
- 60 Reinforce Spread Footings
- 70 Place Spread Footings
- 80 Strip Forms - Spread Footings

1020 Piers
- 90 Fab. & Tie Steel Piers
- 100 Fab. & Tie Steel Contin. Footings
- 110 Fab. & Tie Steel Foundation Walls
- 120 Reinforce Piers
- 130 Form Piers
- 140 Place Piers
- 150 Strip Forms - Piers
- 160 Backfill Spread Footings

1030 North Contin Ftg's & Walls
- 170 North Side Excav. Contin Ftg's
- 180 North Side Form Contin. Ftg's
- 190 North Side Reinforce Contin. Ftg's
- 200 North Side Place Contin. Ftg's
- 210 North Side Strip Contin. Ftg's Forms
- 220 North Side Reinforce Fdn. Walls
- 230 North Side Form Fdn. Walls
- 240 North Side Place Fdn. Walls
- 250 North Side Strip Fdn. Wall Forms
- 260 North Side Backfill & Compact

1040 East Contin Ftg's & Walls
- 270 East Side Excav. Contin Ftg's
- 280 East Side Form Contin. Ftg's
- 290 East Side Reinforce Contin. Ftg's
- 300 East Side Place Contin. Ftg's
- 310 East Side Strip Contin. Ftg's Forms
- 320 East Side Reinforce Fdn. Walls
- 330 East Side Form Fdn. Walls
- 340 East Side Place Fdn. Walls
- 350 East Side Strip Fdn. Wall Forms
- 410 East Side Backfill & Compact

1050 South Contin Ftg's & Walls
- 360 South Side Excav. Contin Ftg's
- 370 South Side Form Contin. Ftg's
- 380 South Side Reinforce Contin. Ftg's
- 420 South Side Place Contin. Ftg's
- 430 South Side Strip Contin. Ftg's Forms
- 440 South Side Reinforce Fdn. Walls
- 450 South Side Form Fdn. Walls
- 460 South Side Place Fdn. Walls
- 490 South Side Strip Fdn Wall Forms
- 500 South Side Backfill & Compact

1060 West Contin Ftg's & Walls
- 390 West Side Excav. Contin Ftg's
- 470 West Side Form Contin. Ftg's
- 480 West Side Reinforce Contin. Ftg's
- 510 West Side Place Contin. Ftg's
- 520 West Side Strip Contin. Ftg's Forms
- 530 West Side Reinforce Fdn. Walls
- 540 West Side Form Fdn. Walls
- 550 West Side Place Fdn. Walls
- 560 West Side Strip Fdn. Wall Forms
- 570 West Side Backfill & Compact

1070 Plumbing
- 400 Underground Plumbing
- 700 Plumbing Top-out
- 860 Plumbing Trim

1080 Slab on Grade
- 580 Reinforce S.O.G.
- 590 Place & Finish S.O.G.

1090 Exterior Walls & Structure
- 600 CMU
- 610 Erect Structural Steel
- 620 Roofing / Flashing

1100 Exterior Finishes
- 630 Exterior Glass & Glazing
- 640 Overhead Doors

1110 Parking Lot
- 650 Grade Parking Lot
- 720 Form Parking Lot
- 730 Reinforce Parking Lot
- 740 Place & Finish Parking Lot
- 750 Strip Forms - Parking Lot
- 760 Place Curbs
- 770 Stripe Parking Lot

1120 Mechanical
- 660 Mechanical Rough-in
- 710 Electrical Rough-in
- 830 Mechanical Trim

1130 Electrical
- 670 Overhead Electrical Rough-in
- 850 Electrical Trim

1140 Interior Finishes
- 680 Ceiling Grid
- 690 Frame Interior Walls
- 800 Hang Drywall
- 810 Float, Tape & Paint
- 820 Hang & Trim Interior Doors
- 840 Install Cabinets
- 870 Ceiling Tiles
- 880 Carpet
- 890 Final Cleaning

1150 Landscape & Grading
- 780 Final Grading
- 790 Landscape

Figure 9-1
Summary Activity Matrix

Screen Image 9-1
Opening the Activity Codes Window

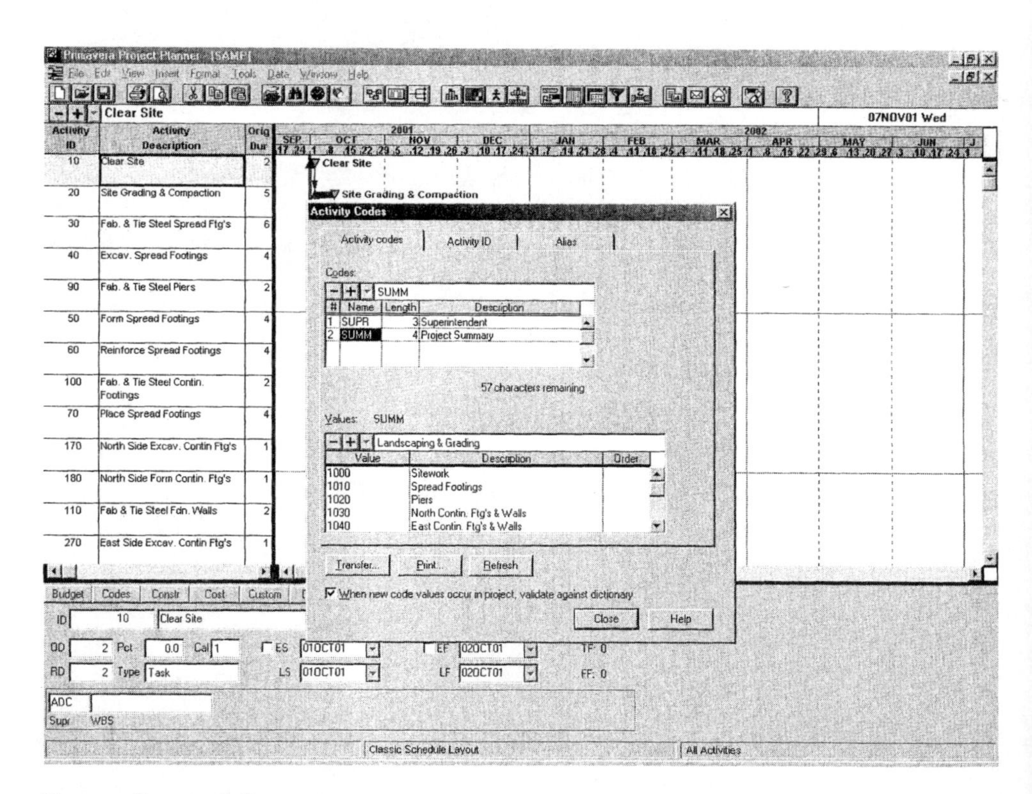

Screen Image 9-2
The Activity Codes Window

Figure 9-1 is a guide for this operation. In Screen Image 9-3 the pull-down menu is activated for Activity 10. Since Activity 10 corresponds to Activity Code 1000, sitework, that item is clicked to create the appropriate correlation. This operation needs to be repeated for every construction activity.

The next several steps deal with modifying the main input screen to summarize and display the summary activity bars. First, open the *Organize* window by selecting the *Format* command followed by the *Organize* option as shown in Screen Image 9-4 or by clicking the organize icon. From the *Organize* window enter the appropriate activity code in the *Group by* field. Depending upon the use of the output and the output device, it may be necessary to change the font and font color. The other item within the *Organize* window is the sort order. Listing the activities in early start order is usually preferable as it shows the sequence in which the work will be performed. Screen Image 9-5 shows the modified *Organize* window. When this is completed, click the *Organize Now* button to execute the new sorting and grouping. The project schedule, with the activities grouped by activity code, is shown in Screen Image 9-6.

The next step is to combine the grouped bars into a single bar. This is done by selecting the *Format* command followed by the *Summarize All* option as shown in Screen Image 9-7.

From this window, the activity code can be selected that will be used to summarize the schedule. The selected summary level must match the grouping of the activity codes. In Screen Image 9-8 the *Project Summary* is selected. By selecting this entry within both the *Summarize All* and *Organize* window, a bar chart summary schedule will be created when the *Summarize* button is clicked. Screen Image 9-9 shows the summarized main input screen. The indentation of the bar thickness shows periods of inactivity. The plumbing bar on Screen Image 9-9 shows how that activity has three distinct phases separated by periods of inactivity. The inactivity can be the result of gaps in the construction process, weekends, or holidays. The logic for necking of the bars can be manipulated. To modify this aspect of the schedule, select the *Format* command followed by the *Summary Bar* option as shown

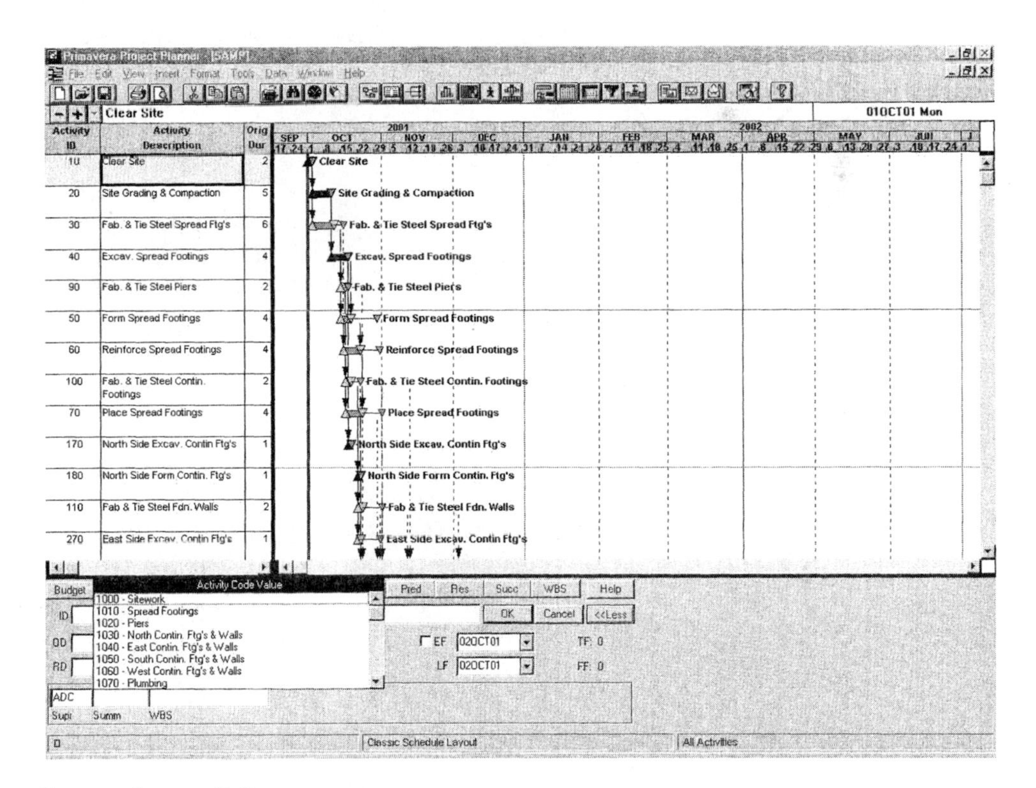

Screen Image 9-3
Adding Activity Codes to the Activities

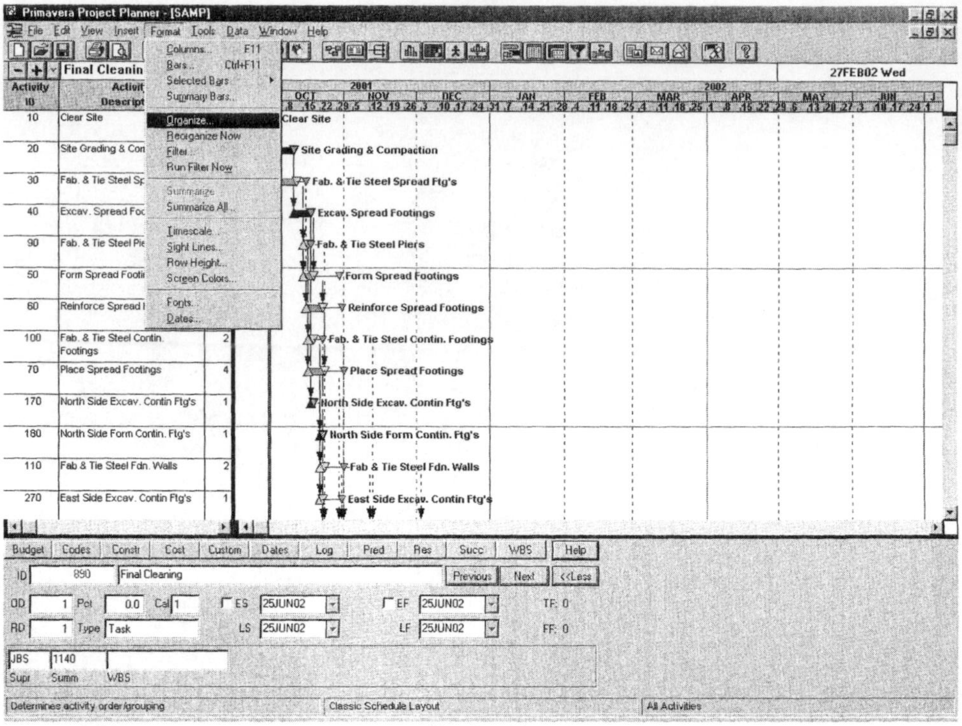

Primavera Project Planner - [SAMP]

File Edit View Insert Format Tools Data Window Help

Final Cleanin

Columns... F11
Bars... Ctrl+F11
Selected Bars
Summary Bars...
Organize...
Reorganize Now
Filter...
Run Filter Now
Summarize
Summarize All...
Timescale...
Sight Lines...
Row Height...
Screen Colors...
Fonts...
Dates...

27FEB02 Wed

Activity ID	Activity Description
10	Clear Site
20	Site Grading & Con
30	Fab. & Tie Steel Sp
40	Excav. Spread Foc
90	Fab. & Tie Steel Pie
50	Form Spread Footi
60	Reinforce Spread I
100	Fab. & Tie Steel Contin. Footings
70	Place Spread Footings
170	North Side Excav. Contin Ftg's
180	North Side Form Contin. Ftg's
110	Fab & Tie Steel Fdn. Walls
270	East Side Excav. Contin Ftg's

Clear Site
Site Grading & Compaction
Fab. & Tie Steel Spread Ftg's
Excav. Spread Footings
Fab. & Tie Steel Piers
Form Spread Footings
Reinforce Spread Footings
Fab. & Tie Steel Contin. Footings
Place Spread Footings
North Side Excav. Contin Ftg's
North Side Form Contin. Ftg's
Fab & Tie Steel Fdn. Walls
East Side Excav. Contin Ftg's

Budget Codes Constr Cost Custom Dates Log Pred Res Succ WBS Help

ID 890 Final Cleaning Previous Next <<Less

OD 1 Pct 0.0 Cal 1 ES 25JUN02 EF 25JUN02 TF: 0
RD 1 Type Task LS 25JUN02 LF 25JUN02 FF: 0

JBS 1140
Supr Summ WBS

Determines activity order/grouping | Classic Schedule Layout | All Activities

Screen Image 9-4
Opening the Organize Window

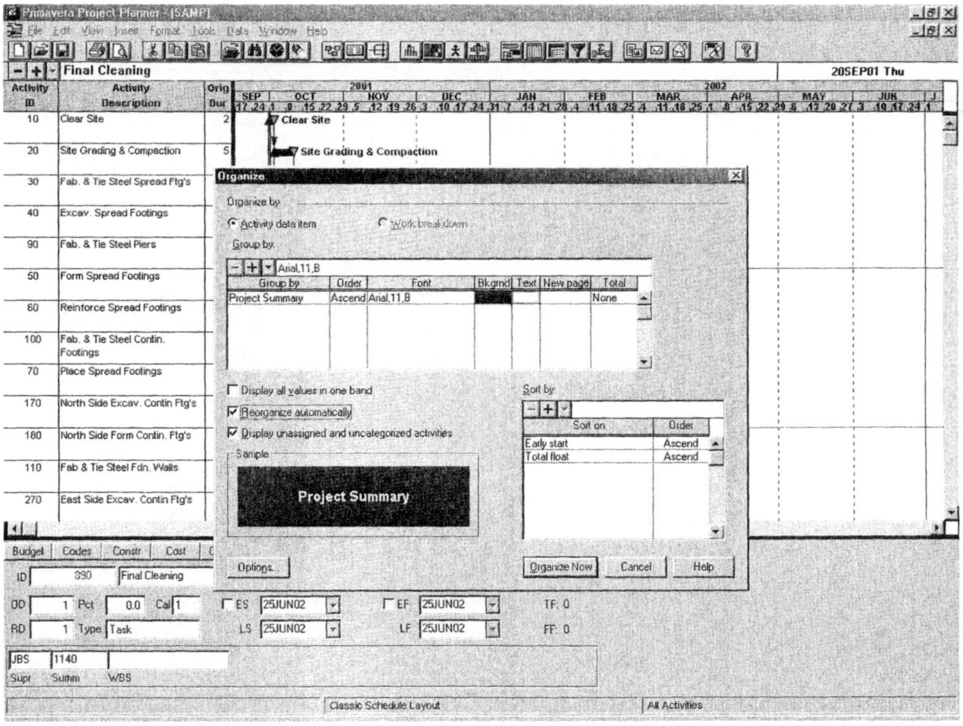

Primavera Project Planner - [SAMP]

File Edit View Insert Format Tools Data Window Help

Final Cleaning

20SEP01 Thu

Activity ID	Activity Description	Orig Dur
10	Clear Site	2
20	Site Grading & Compaction	5
30	Fab. & Tie Steel Spread Ftg's	
40	Excav. Spread Footings	
90	Fab. & Tie Steel Piers	
50	Form Spread Footings	
60	Reinforce Spread Footings	
100	Fab. & Tie Steel Contin. Footings	
70	Place Spread Footings	
170	North Side Excav. Contin Ftg's	
180	North Side Form Contin. Ftg's	
110	Fab & Tie Steel Fdn. Walls	
270	East Side Excav. Contin Ftg's	

Clear Site
Site Grading & Compaction

Organize

Organize by
Activity data item
Work breakdown
Group by
Arial,11,B

Group by	Order	Font	Bkgrnd	Text	New page	Total
Project Summary	Ascend	Arial,11,B				None

Display all values in one band
Reorganize automatically
Display unassigned and uncategorized activities
Sample
Project Summary

Sort by

Sort on	Order
Early start	Ascend
Total float	Ascend

Options... Organize Now Cancel Help

Budget Codes Constr Cost

ID 890 Final Cleaning

OD 1 Pct 0.0 Cal 1 ES 25JUN02 EF 25JUN02 TF: 0
RD 1 Type Task LS 25JUN02 LF 25JUN02 FF: 0

JBS 1140
Supr Summ WBS

Classic Schedule Layout | All Activities

Screen Image 9-5
Modified Organize Window

Screen Image 9-6
Activities Grouped by Activity Code

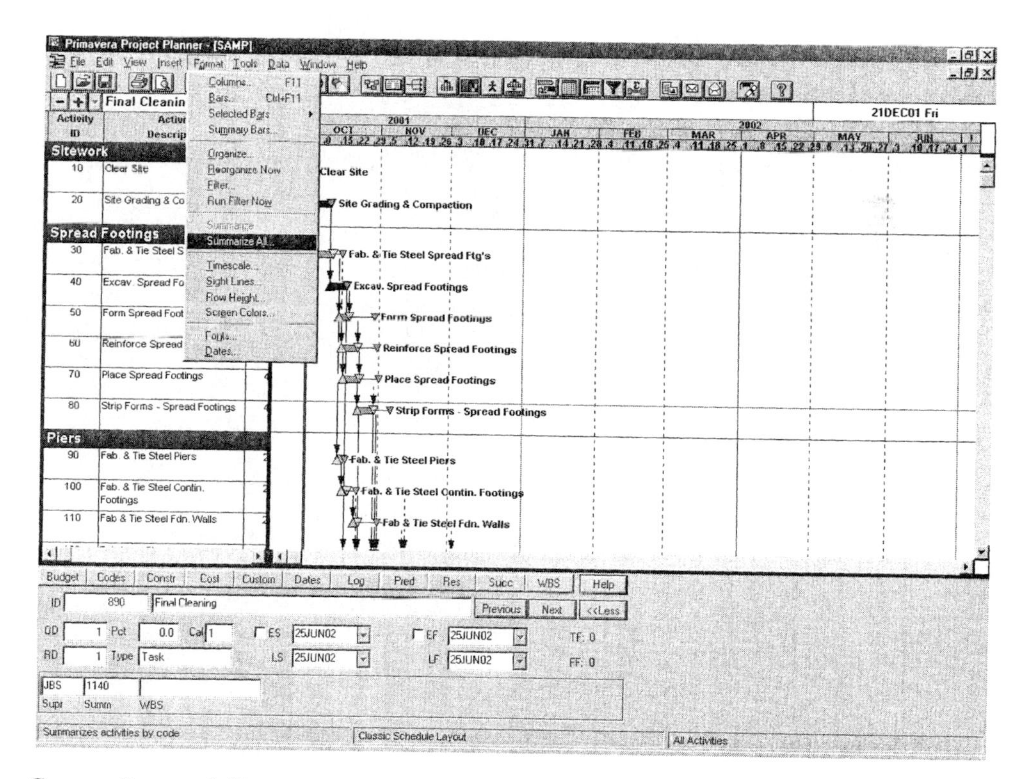

Screen Image 9-7
Opening the Summarize All Window

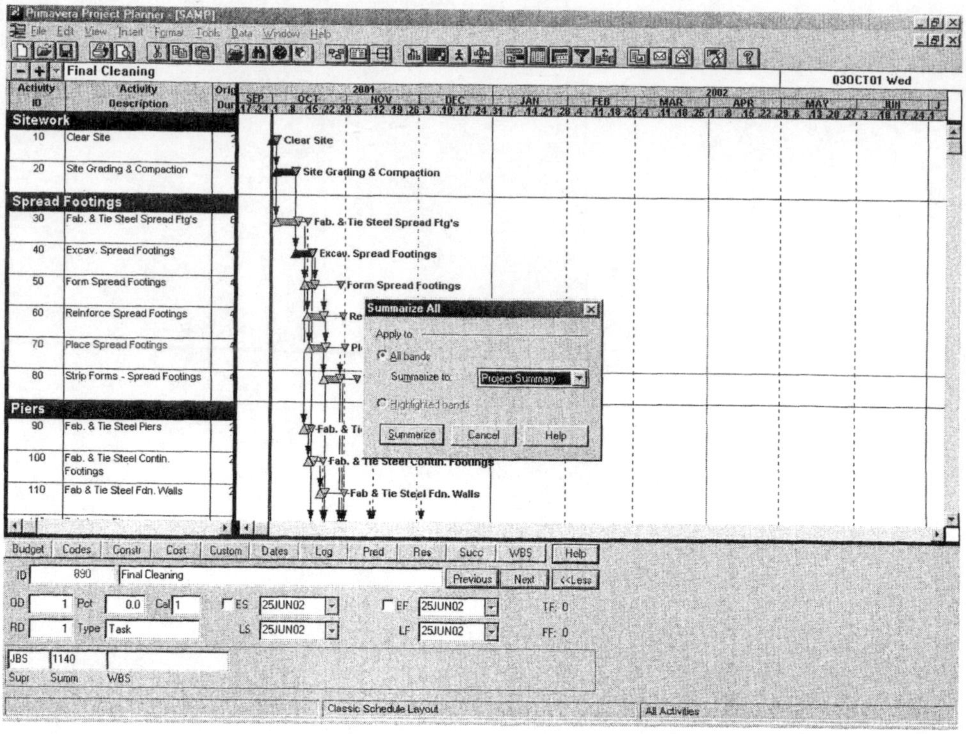

Primavera Project Planner - [SAMP]
File Edit View Insert Format Tools Data Window Help
Final Cleaning
03OCT01 Wed

Activity ID	Activity Description	Orig Dur
Sitework		
10	Clear Site	
20	Site Grading & Compaction	
Spread Footings		
30	Fab. & Tie Steel Spread Ftg's	
40	Excav. Spread Footings	
50	Form Spread Footings	
60	Reinforce Spread Footings	
70	Place Spread Footings	
80	Strip Forms - Spread Footings	
Piers		
90	Fab. & Tie Steel Piers	
100	Fab. & Tie Steel Contin. Footings	
110	Fab & Tie Steel Fdn. Walls	

2001: SEP OCT NOV DEC — 2002: JAN FEB MAR APR MAY JUN

Clear Site
Site Grading & Compaction
Fab. & Tie Steel Spread Ftg's
Excav. Spread Footings
Form Spread Footings
Fab. & Tie Steel Contin. Footings
Fab & Tie Steel Fdn. Walls

Summarize All
Apply to
All bands
Summarize to: Project Summary
Highlighted bands
Summarize Cancel Help

Budget Codes Constr Cost Custom Dates Log Pred Res Succ WBS Help
ID 890 Final Cleaning Previous Next <<Less
OD 1 Pct 0.0 Cal 1 ES 25JUN02 EF 25JUN02 TF: 0
RD 1 Type Task LS 25JUN02 LF 25JUN02 FF: 0
JBS 1140
Supr Summ WBS
Classic Schedule Layout All Activities

Screen Image 9-8
Summarize All Window

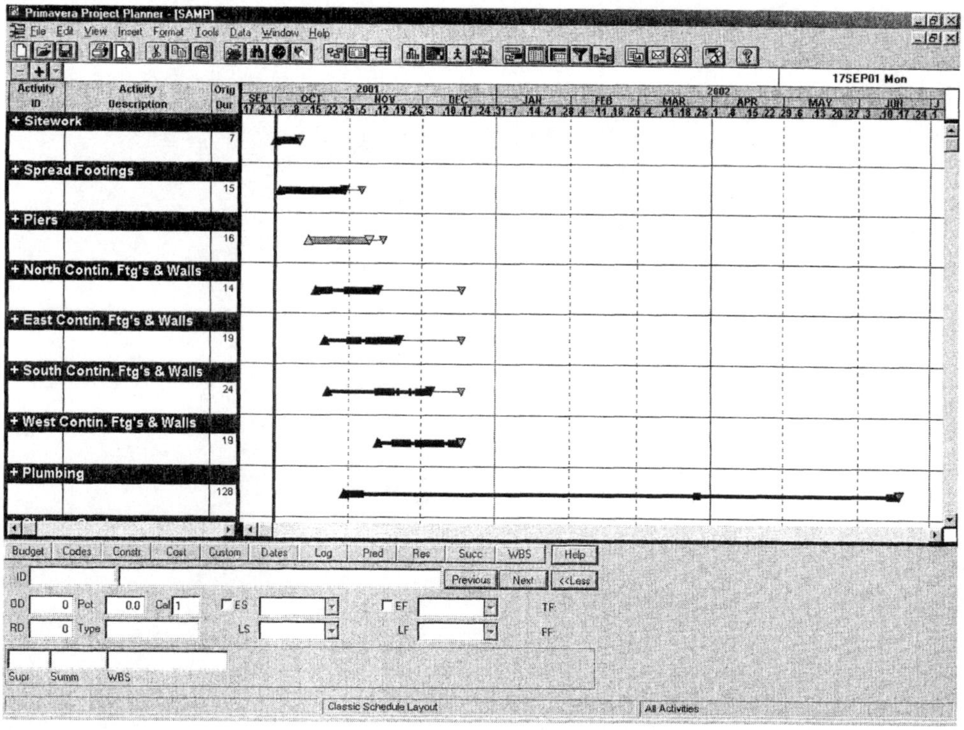

Primavera Project Planner - [SAMP]
File Edit View Insert Format Tools Data Window Help
17SEP01 Mon

Activity ID	Activity Description	Orig Dur
+ Sitework		
		7
+ Spread Footings		
		15
+ Piers		
		16
+ North Contin. Ftg's & Walls		
		14
+ East Contin. Ftg's & Walls		
		19
+ South Contin. Ftg's & Walls		
		24
+ West Contin. Ftg's & Walls		
		19
+ Plumbing		
		128

Budget Codes Constr Cost Custom Dates Log Pred Res Succ WBS Help
ID Previous Next <<Less
OD 0 Pct 0.0 Cal 1 ES EF TF
RD 0 Type LS LF FF
Supr Summ WBS
Classic Schedule Layout All Activities

Screen Image 9-9
Summarized Schedule

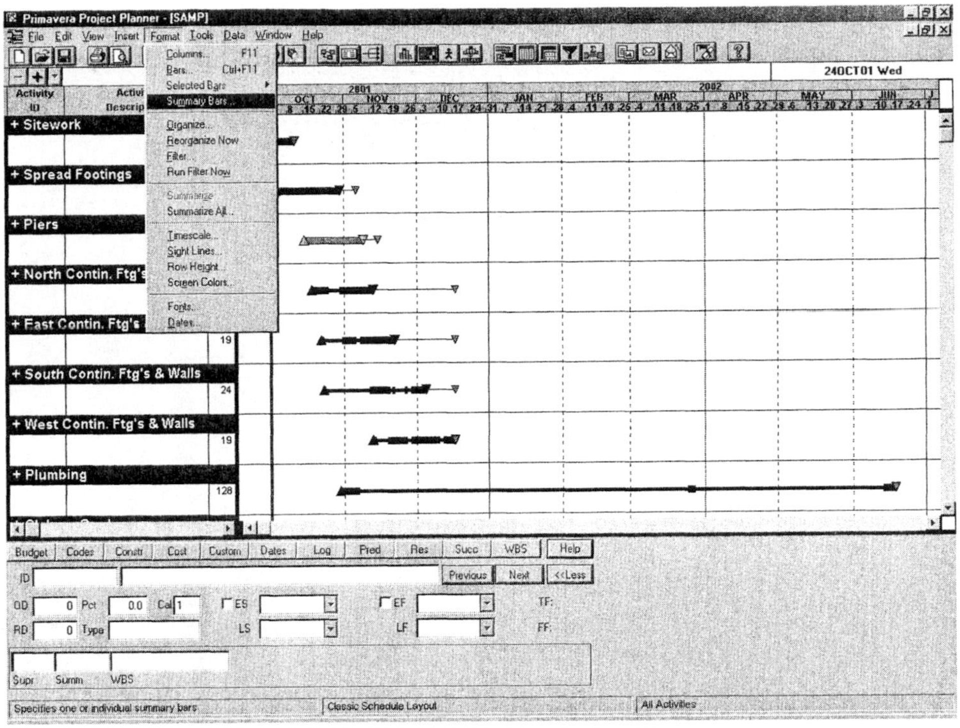

Screen Image 9-10
Opening the Summary Bar Window

in Screen Image 9-10. The *Summary Bar* window in Screen Image 9-11 has necking set for all periods of inactivity. By clicking on the two suboptions, the *Holidays* and *Weekends* can be added or removed from what is necked. In Screen Image 9-12 the necking option has been completely removed. When the *OK* button is clicked, the *Summary Bar* window closes and the modified summary schedule appears. Screen Image 9-13 shows the summary schedule with no periods of inactivity shown. In the instance of plumbing one cannot tell when phases of the work will begin or end. This diminishes the value of the schedule when there are activities with long periods of inactivity. The summary schedule can be printed using the methodology discussed in Chapter 8.

Hammock Activities

Hammock activities are special activities that are added to the schedule for the specific purpose of summarizing a chain of activities. Figure 9-2 shows three hammock activities that would be added to the construction schedule. Activity 1000, sitework, is the summary activity for the chain that starts with the beginning of clear site (Activity 10) and ends with the completion of site grading and compaction (Activity 20). The second hammock, Activity 1010, spread footings, begins with the start of Activity 30 (fabricate & tie steel spread footings) and extends through the completion of Activity 80 (strip forms–spread footings). The third hammock activity, Activity 1020, piers, commences with the start of Activity 90 (fabricate & tie steel piers) and extends through the completion of Activity 160 (backfill spread footings).

In Figure 9-2, all of the hammock activities begin with the start of one of the construction activities and end with the completion of one of the construction activities. This is the most common application; however, a hammock can also begin with the completion of an activity and extend through the completion of another activity. Likewise, a hammock

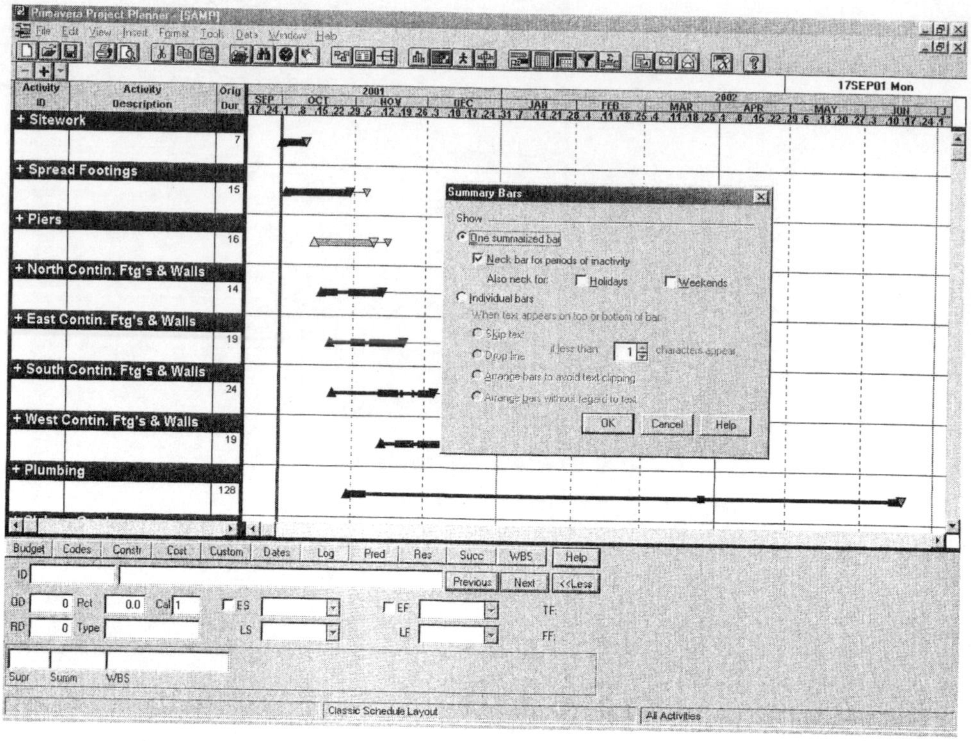

Primavera Project Planner - [SAMP]
File Edit View Insert Format Tools Data Window Help
17SEP01 Mon
Activity ID | Activity Description | Orig Dur
2001 SEP OCT NOV DEC 2002 JAN FEB MAR APR MAY JUN
+ Sitework 7
+ Spread Footings 15
+ Piers 16
+ North Contin. Ftg's & Walls 14
+ East Contin. Ftg's & Walls 19
+ South Contin. Ftg's & Walls 24
+ West Contin. Ftg's & Walls 19
+ Plumbing 128

Summary Bars
Show
One summarized bar
Neck bar for periods of inactivity
Also neck for: Holidays Weekends
Individual bars
When text appears on top or bottom of bar
Skip text
Drop line if less than 1 characters appear
Arrange bars to avoid text clipping
Arrange bars without regard to text
OK Cancel Help

Budget Codes Constr Cost Custom Dates Log Pred Res Succ WBS Help
ID Previous Next <<Less
OD 0 Pct 0.0 Cal 1 ES EF TF
RD 0 Type LS LF FF
Supr Summ WBS
Classic Schedule Layout All Activities

Screen Image 9-11
Summary Bar Window

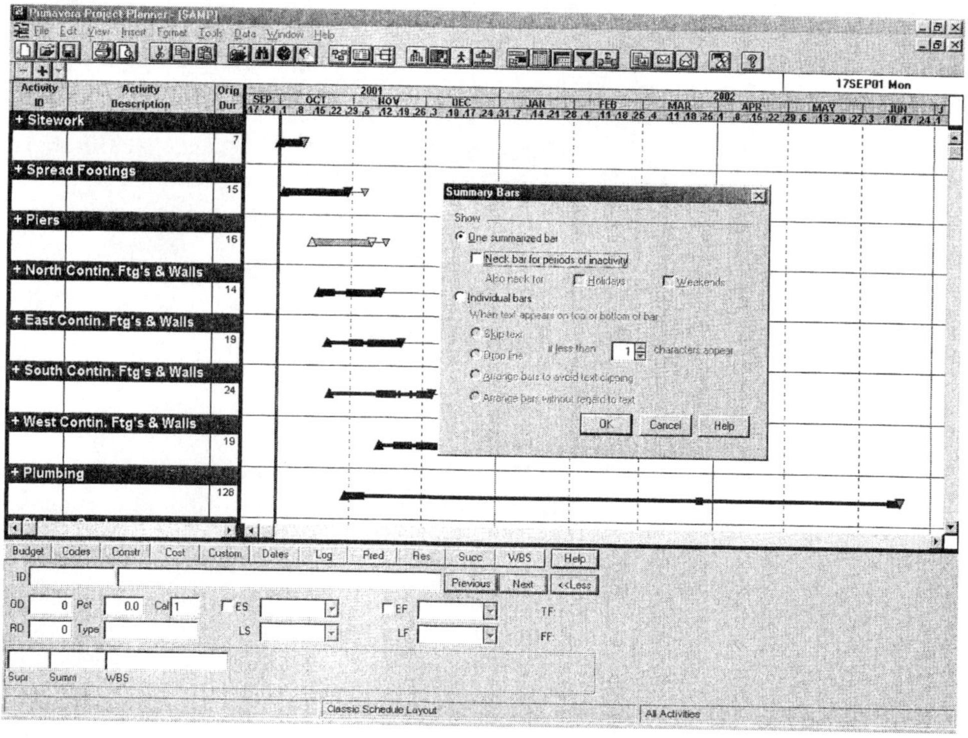

Primavera Project Planner - [SAMP]
File Edit View Insert Format Tools Data Window Help
17SEP01 Mon
Activity ID | Activity Description | Orig Dur
2001 SEP OCT NOV DEC 2002 JAN FEB MAR APR MAY JUN
+ Sitework 7
+ Spread Footings 15
+ Piers 16
+ North Contin. Ftg's & Walls 14
+ East Contin. Ftg's & Walls 19
+ South Contin. Ftg's & Walls 24
+ West Contin. Ftg's & Walls 19
+ Plumbing 128

Summary Bars
Show
One summarized bar
Neck bar for periods of inactivity
Also neck for: Holidays Weekends
Individual bars
When text appears on top or bottom of bar
Skip text
Drop line if less than 1 characters appear
Arrange bars to avoid text clipping
Arrange bars without regard to text
OK Cancel Help

Budget Codes Constr Cost Custom Dates Log Pred Res Succ WBS Help
ID Previous Next <<Less
OD 0 Pct 0.0 Cal 1 ES EF TF
RD 0 Type LS LF FF
Supr Summ WBS
Classic Schedule Layout All Activities

Screen Image 9-12
Summary Bar Window without Necking

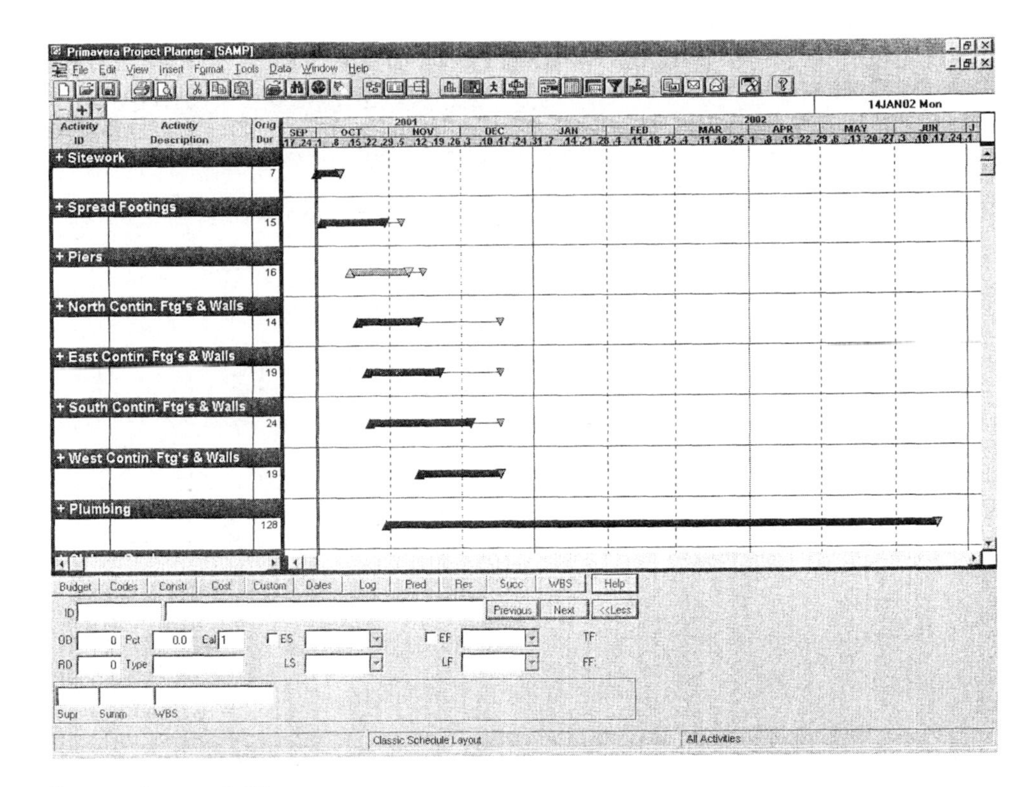

Screen Image 9-13
Summary Schedule without Necking

could begin with the start of an activity and extend through the start of one of the later activities. Whichever relationship is used, the hammock activity description must reflect the activities being summarized. In addition, it is a good practice to hammock only contiguous activities. The duration of the hammock will not take into consideration periods of inactivity.

Hammock activities are similar to the other activities on the schedule in that they have early dates, late dates, and float. However, these items are determined by using a different process than a forward and backward pass. These dates and floats are determined after the forward and backward pass has been completed. The early and late dates come from the early and late dates of the underlying activities. If the hammock begins with the completion of an activity, the early finish of that activity would be the early start of the hammock activity and the late finish date of the construction activity would become the late start of the hammock activity as shown in Figure 9-3. If the hammock activity begins with the start of one of the construction activities, then the early start and late start of that activity would become the early start and late start of the hammock activity. The completion of the hammock activity is handled the same way. If the hammock activity ends with the beginning of one of the construction activities, then the early start and late start of the construction activity would become the early finish and late finish of the hammock activity. In Figure 9-2, Activity 1010 gets its early start of 2 and its late start of 3 from the early start of Activity 30. At the opposite end of that chain, the early finish of 17 and late finish of 24 corresponds with the finish of Activity 80. Once the early dates have been found, the duration of the hammock activity can be determined. For the purpose of consistency, the duration of a hammock activity is the difference between the early dates. Formula 9-1 shows how the hammock activity duration would be determined.

$$\text{Hammock Duration} = \text{Early Finish} - \text{Early Start}$$

Formula 9-1

Once the duration has been found, the floats can be calculated using the process detailed in Chapter 4.

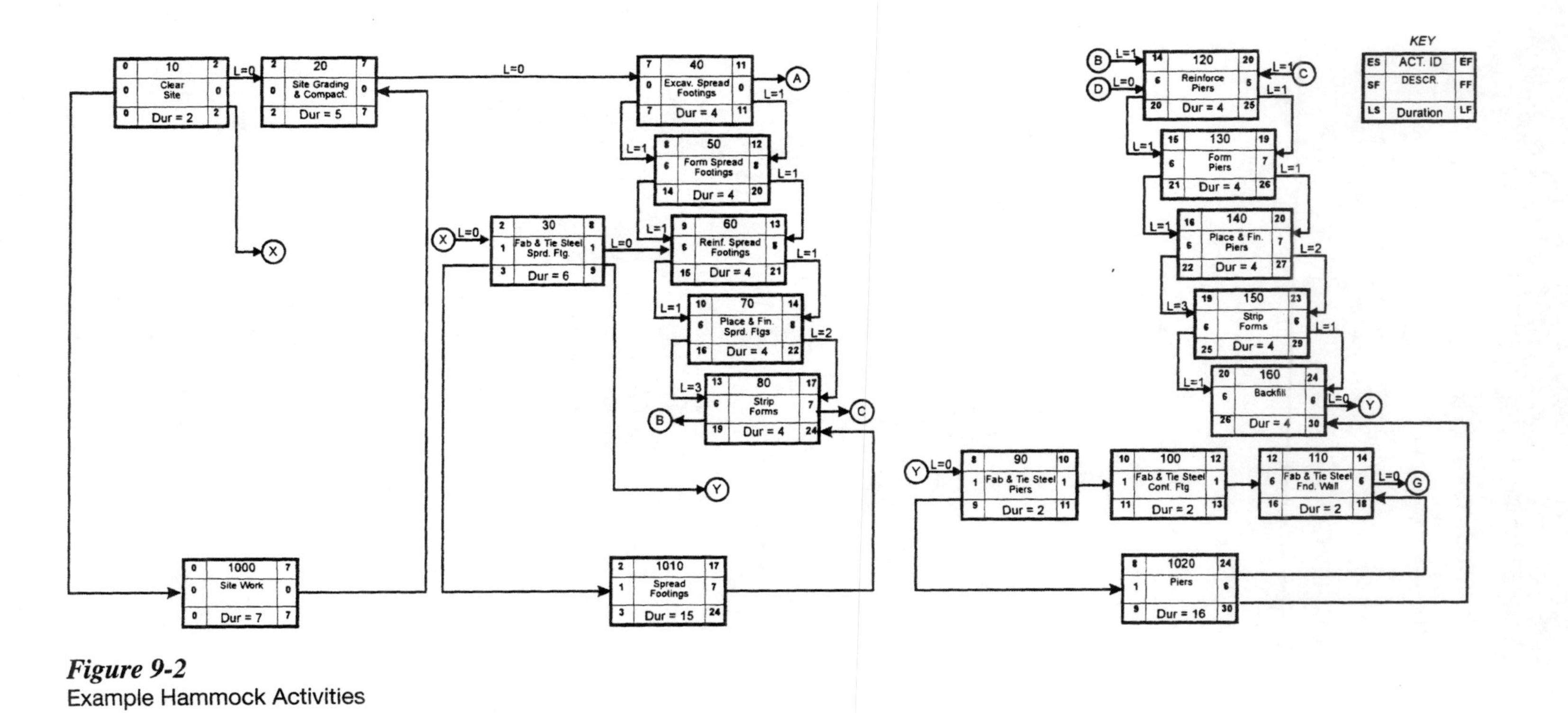

Figure 9-2
Example Hammock Activities

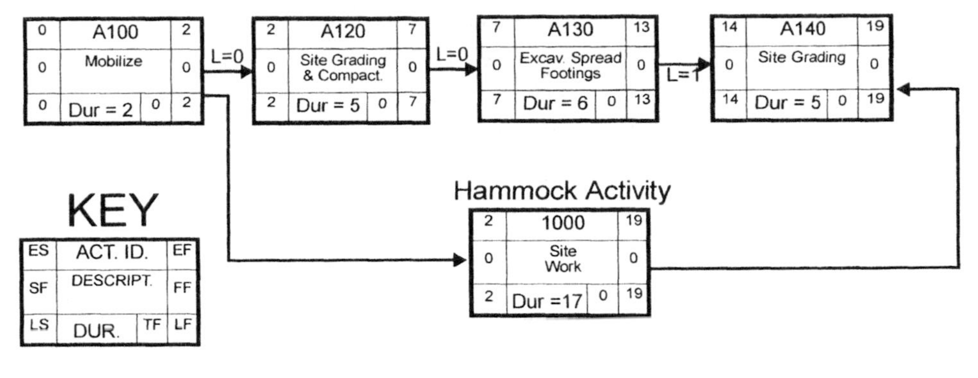

Figure 9-3
Hammock Activities

Hammock Activities in Primavera Project Planner Schedules

Hammock activities are added to the Primavera Project Planner schedule in much the same way as were task activities. To facilitate data management, a consistent activity numbering system needs to be developed. For example, the construction activities could be numbered from 10 through 990 and the hammock activities from 1000 to 1990. This practice will make the generation of reports less cumbersome. When the selection window is used from the report generator, the construction or summary activities can easily be included by using the within range operator to select the desired set of activities. From the main input screen the activity ID and description need to be entered for the hammock activity.

To add a hammock activity, click the plus button directly below the row of icons and enter the desired activity ID, press the *Enter* key, and then click on the description field on the activity form. If the automatic numbering option is being used, it should be turned off prior to adding the hammock activities. If this is not done Primavera Project Planner will automatically assign the next activity ID rather than the number required by the numbering scheme. Once this is done, tab over to the *Type* field and, using the down arrow, select *Hammock* as the activity type.

Once a hammock activity has been added, it is necessary to add the relationships that will designate its start and finish. To begin this operation, click on the *Succ* and *Pred* buttons to open the predecessor and successor windows. From these windows the relationships can be drawn. In Screen Image 9-14 hammock activity 1000 is active, Activity 10 is specified as the predecessor with a start-to-start relationship with no lag, while Activity 20 is specified as a successor with a finish-to-finish relationship and no lag. After these relationships have been added, any changes in Activity 10 or 20 will automatically be reflected in the hammock activity. In this simple example the hammocked activities are contiguous. Using contiguous activities is the preferable method to deal with summary activities. However, that is not always possible. When the activities are not contiguous it is often difficult to determine which activity will be the first activity in the chain and which will be the last. In addition, these activities may change position as actual work progresses on the project. In these situations all of the starts of the individual activities in the chain can be listed as a predecessor with the start-to-start relationship. Conversely, all of the finishes of the activities can be listed as successors with finish-to-finish relationships. If this is done Primavera Project Planner will look at all of the early start dates and select the earliest early start date as the early start date of the hammock and will evaluate all of the early finish dates and select the latest as the early finish date of the hammock. Screen Image 9-15 shows how

Primavera Project Planner - [SAMP]
File Edit View Insert Format Tools Data Window Help
Sitework
17SEP01 Mon

Activity ID	Activity Description	Orig Dur
850	Electrical Trim	4
860	Plumbing Trim	2
870	Ceiling Tiles	3
880	Carpet	1
890	Final Cleaning	1
1000	Sitework	1

Sitework
Electrical

Predecessors
Activity: 1000 - Sitework
SS - Start to start

Activity ID	Rel	Lag	TF	Description
10	SS	0	0	Clear Site

Successors
Activity: 1000 - Sitework
FF - Finish to finish

Activity ID	Rel	Lag	TF	Description
20	FF	0	0	Site Grading & Compaction

Budget Codes Constr Cost Custom Dates Log Pred Res Succ WBS Help
ID 1000 Sitework OK Cancel <<Less
OD 1 Pct 0.0 Cal 1 ES 01OCT01 EF 01OCT01 TF: 0
RD 1 Type Hammock LS 01OCT01 LF 01OCT01 FF: 0
Supr Summ WBS
Classic Schedule Layout All Activities

Screen Image 9-14
Adding a Simple Hammock Activity

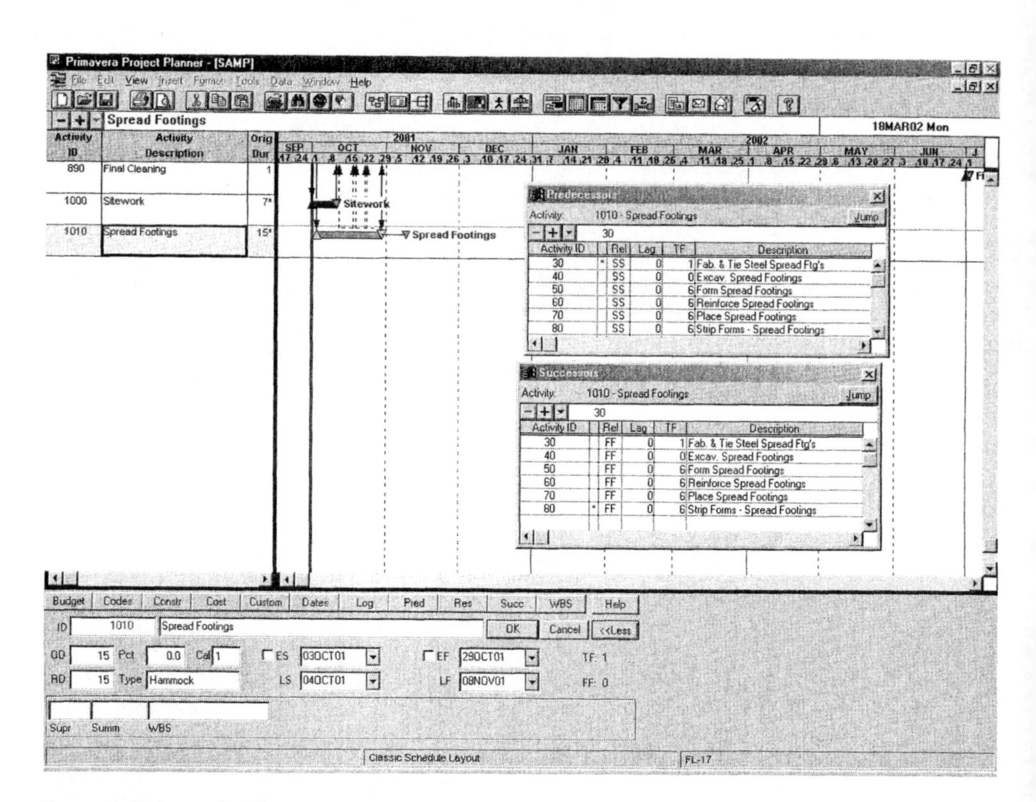

Screen Image 9-15
Adding a Complex Hammock Activity

all of the individual activities are included in hammock Activity 1010. The starts are listed as predecessors, while the finishes are listed as successors.

When summarizing the project schedule, multiple hammock summary schedules may be required. Since all of the activities are maintained in a single database, it may be necessary to designate which hammock goes with which summary schedule. One way to accomplish this is to develop another set of activity codes. For example, a project manager may need to have the schedule summarized by CSI (Construction Specification Institute) division. For this summary, all of the concrete activities would be summarized into one hammock activity, all of the masonry activities would be combined into another hammock activity, and so on through all of the sixteen divisions. In addition to this summary, the contract requires the contractor to provide a schedule that corresponds to the contractor's draw schedule. This draw schedule, or schedule of values, defines the sales price for the specific items of construction or segments of the project. The total of all of the items on the schedule of values must equal the contract amount. Since all the hammock activities are stored as part of the construction logic network, it is necessary to designate which hammock activities go to which summary schedules. This could easily be done by adding a new activity code field or blocking a new group of activity ID's. Output 9-1 is an example of tabular output that would be generated from selecting only the hammock activities. These activities are denoted on the report by asterisks next to the original duration and remaining duration columns. This allows the user to quickly identify which activities are construction activities and which are hammock activities. This is particularly helpful when both types of activities are included on the same report.

All Star Developments — PRIMAVERA PROJECT PLANNER — Sample Construction Project

REPORT DATE 18MAR00 RUN NO. 29 16:51 — START DATE 1OCT01 FIN DATE 1JUL02

Hammock Activities — DATA DATE 1OCT01 PAGE NO. 1

ACTIVITY ID	ORIG DUR	REM DUR	CAL	%	CODE	ACTIVITY DESCRIPTION	EARLY START	EARLY FINISH	LATE START	LATE FINISH	TOTAL FLOAT
1000	7*	7*	1	0		Sitework	1OCT01	10OCT01	1OCT01	10OCT01	0
1010	15*	15*	1	0		Spread Footings	3OCT01	29OCT01	4OCT01	8NOV01	1
1020	16*	16*	1	0		Piers	15OCT01	8NOV01	16OCT01	20NOV01	1
1030	16*	16*	1	0		North Contin. Ftg's & Walls	18OCT01	14NOV01	18OCT01	20DEC01	0
1040	22*	22*	1	0		East Contin. Ftg's & Walls	22OCT01	28NOV01	31OCT01	20DEC01	6
1050	27*	27*	1	0		South Contin. Ftg's & Walls	23OCT01	10DEC01	12NOV01	20DEC01	11
1060	22*	22*	1	0		West Contin. Ftg's & Walls	13NOV01	20DEC01	21NOV01	20DEC01	5
1070	131*	131*	1	0		Plumbing	30OCT01	20JUN02	13DEC01	20JUN02	25
1080	4*	4*	1	0		Slab on Grade	24DEC01	31DEC01	24DEC01	31DEC01	0
1090	30*	30*	1	0		Exterior Walls & Structure	7JAN02	26FEB02	7JAN02	26FEB02	0
1100	3*	3*	1	0		Exterior Finishes	18FEB02	20FEB02	21FEB02	26FEB02	3
1110	17*	17*	1	0		Parking Lot	2JAN02	30JAN02	28MAY02	1JUL02	82
1120	56*	56*	1	0		Mechanical	27FEB02	5JUN02	27FEB02	18JUN02	0
1130	63*	63*	1	0		Electrical	27FEB02	18JUN02	27FEB02	18JUN02	0
1140	62*	62*	1	0		Interior Finishes	13MAR02	1JUL02	13MAR02	1JUL02	0
1150	4*	4*	1	0		Landscape & Grading	30JAN02	5FEB02	25JUN02	1JUL02	82

Output 9-1
Tabular Hammock Schedule Report

Using Filters to Create Bar Chart Summary Schedules

The primary purpose for hammocking activities is to develop bar chart schedules. This abbreviated type of schedule can easily be understood by people that may not be versed in PDM scheduling. In a bar chart summary schedule, the hammock activities are shown as bars without any of their relationships. These schedules can be produced using the graphic schedule option by selecting the hammock activities from the selection tab. The other possibility is to create a bar chart schedule from the main input screen.

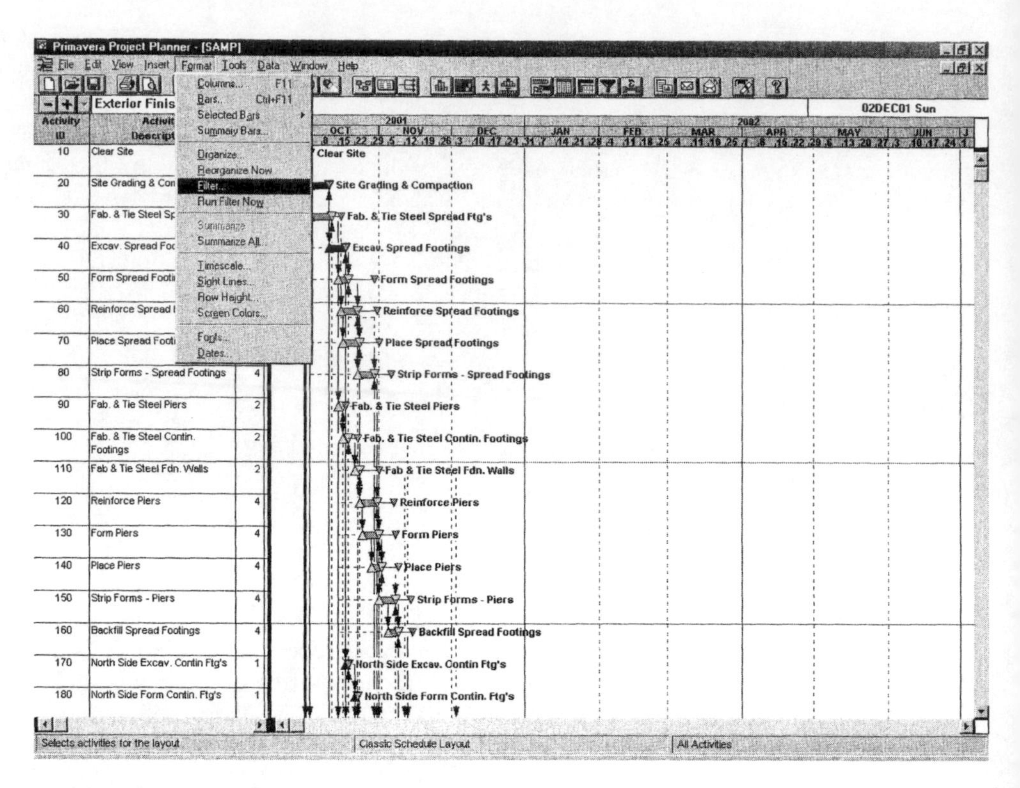

Screen Image 9-16
Opening the Filter Window

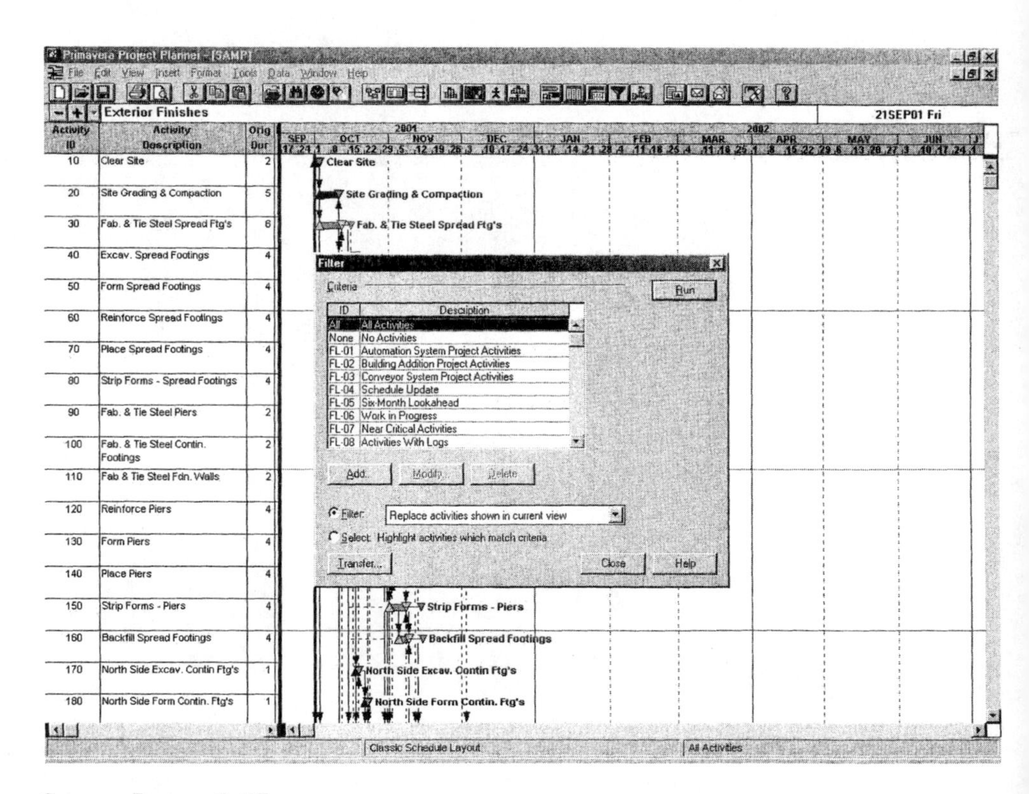

Screen Image 9-17
Filter Window

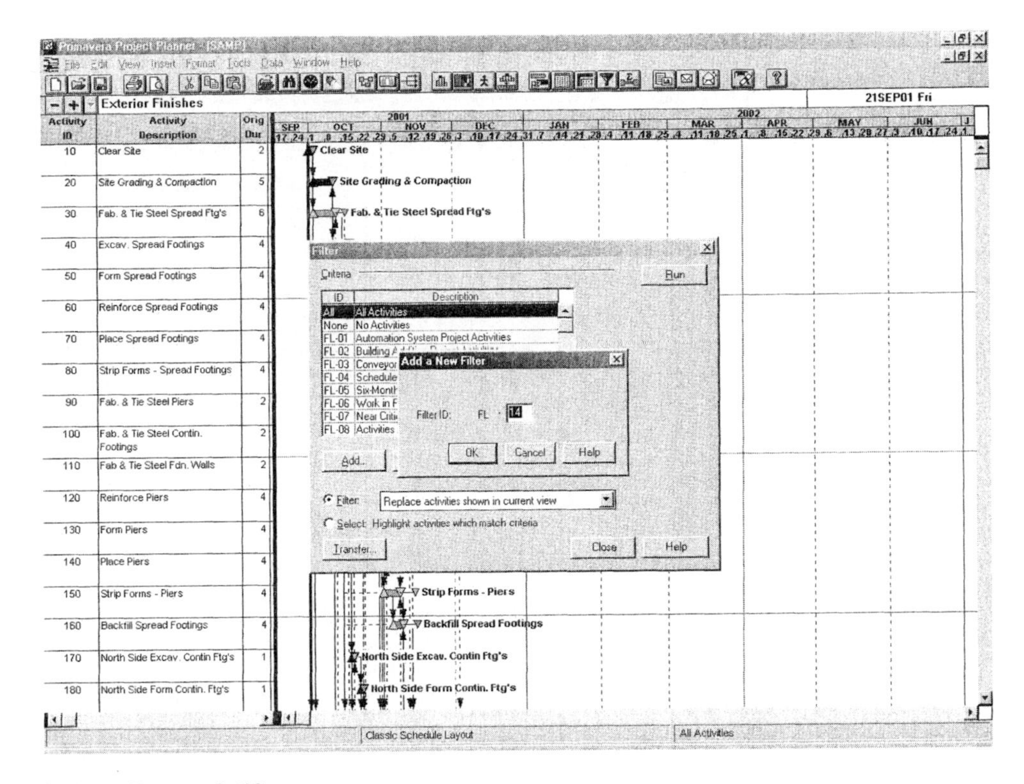

Screen Image 9-18
Add a New Filter Window

Depending on how the sort option is set up on the *Organize* window, the hammock activities may be moving around on the schedule. If the sort is set to early start, the hammock activities will be distributed throughout the schedule. Primavera Project Planner has a filter feature that allows for activities to be temporarily filtered on or off the screen.

The *Filter* window is opened by selecting the *Format* command followed by the *Filter* option as shown in Screen Image 9-16. The *Filter* window as shown in Screen Image 9-17 has a list of the previously developed filter specifications.

To create a new set of filter specifications click the *Add* button. This will prompt the user to accept the automatically assigned filter specification number. In Screen Image 9-18 the prompt is to accept the fourteenth set of filter specifications. If the OK button is clicked, the *Filter Specification* window will open. This window is identical in operation and appearance to the selection option within the report specifications. The specification title and selection criteria need to be entered. In Screen Image 9-19 the activity ID's are selected that fall between 1000 and 1990, which would select all of the hammock activities. When the OK button is clicked, the main input screen returns and only the hammock activities are shown as in Screen Image 9-20. This screen image can be routed to a printer to produce a printout of the bar chart schedule.

To restore the detailed activities, open the *Filter* window and create a new set of filter specifications. The difference here will be that the *Filter Specification* will be left blank. This will include all of the activities on the screen. In Screen Image 9-21 the new filter is named *Show All Activities* and the *Filter Specifications* window has no selection criteria.

Milestones

Milestones are special activities that have no duration. They represent a significant point or event in the life of the project. Milestones are usually shown on the bar chart schedule as a diamond or triangle. Since milestones are used to mark significant points in the project,

Screen Image 9-19
Filter Specification Window

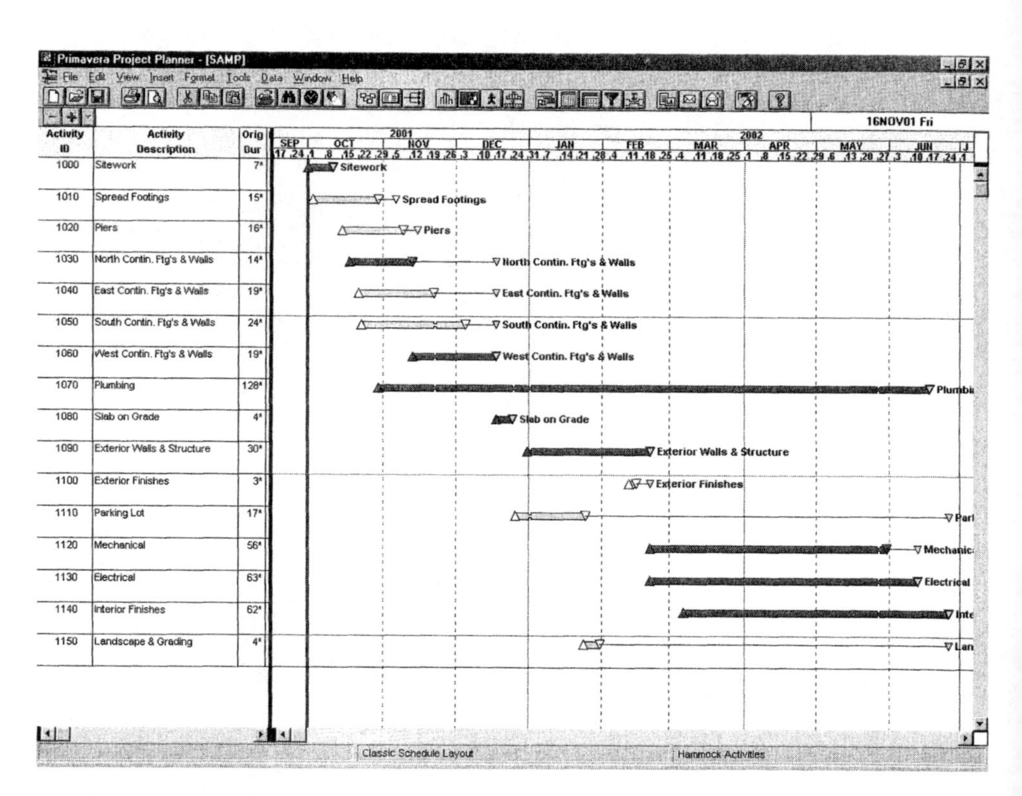

Screen Image 9-20
Hammock Activity Bar Chart Schedule

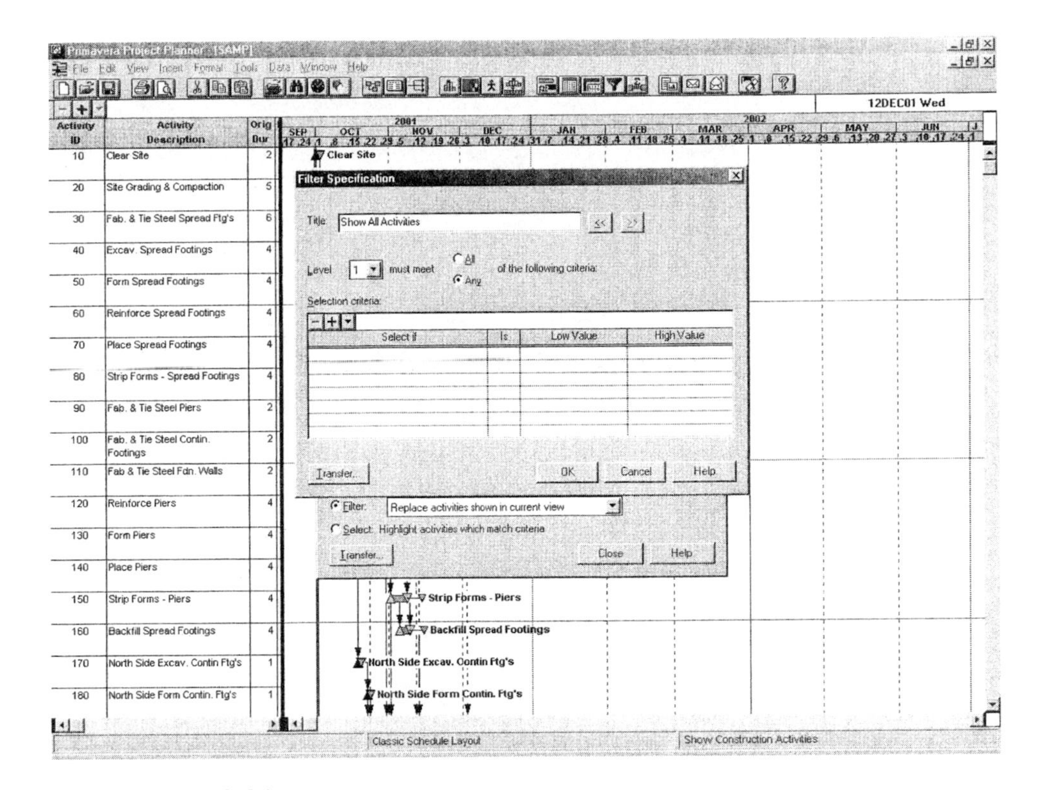

Screen Image 9-21
Restoring All Activities

their dates may be arbitrarily set, rather than determined by a chain of predecessors. An example of a milestone could be the promised delivery date for the fabricated reinforcing steel. A milestone such as this may not have a predecessor, but would have a fixed start date based upon the fabricator's promised delivery date. However, a milestone such as that would have a successor, the placement of the reinforcing steel. This linking of milestones to actual construction activities allows the project management team to evaluate the implications of missed delivery dates on the remainder of the project. Through this proactive approach, contingency plans can be developed that will minimize the impact of the missed delivery date.

In Figure 9-4, the promised delivery date is at the end of Day 1 of the project, which will be entered as the early start for the milestone activity. Since the duration of the milestone is 0, the early finish would be the same as the early start. Since the early finish of the milestone plus the lag is less than the early start of the successor activity, it has no impact on the activity's early start. If this had not been true, the early start of the successor activity would have changed and a new forward and backward pass would have been required. When milestones are added to the schedule a new forward and backward pass needs to be performed to ensure that their impact is reflected in the schedule. The promised delivery dates from vendor may move the critical path. If the start of the milestone is constrained by a promise date, then it will be an exception to the rule that all activities without a predecessor could be the first activity on the project.

Adding Milestones to the Primavera Project Planner Schedule

When using Primavera Project Planner, there are start milestones and finish milestones. Even though milestones are special activities, they still behave in the same fashion as would any activity. If a milestone does not have a predecessor, it will assume that it could be the

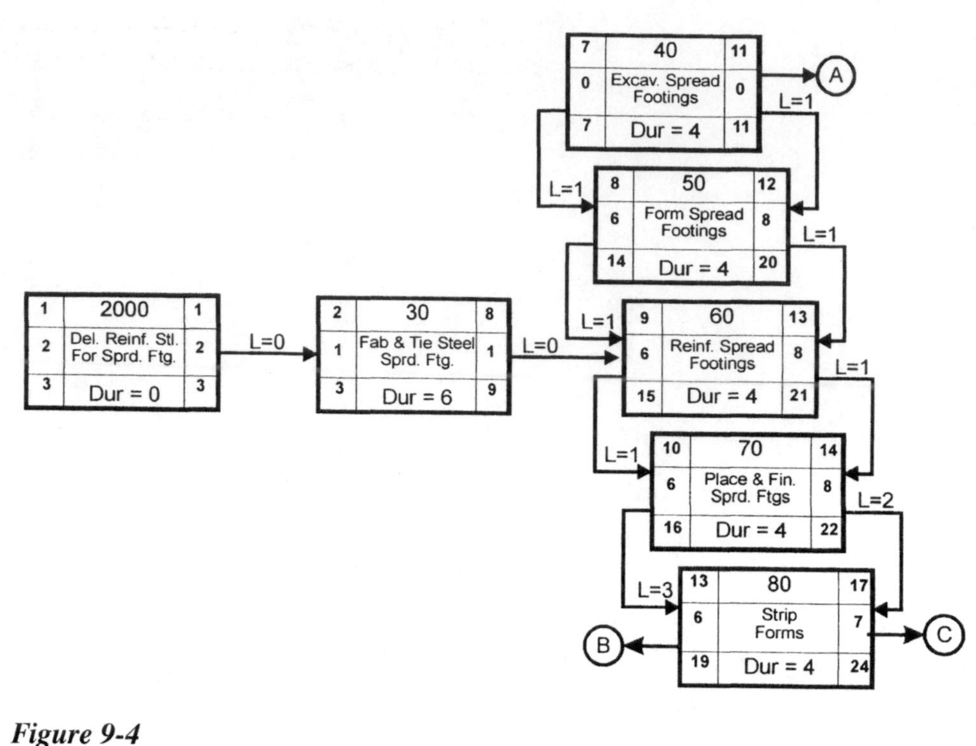

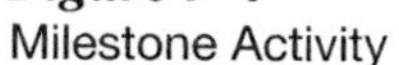

Figure 9-4
Milestone Activity

first activity in the schedule and have an early start date of the first day of the project. The same holds true if the milestone does not have a successor. In this situation, the late finish will be assigned the last day of the project. To overcome this difficulty, it is necessary to constrain either the early start or finish of the milestone. If it is necessary to constrain the finish dates of the milestone, then a finish milestone should be used.

To add a milestone to the schedule, several steps need to be followed. Starting with the main input screen, click on the plus button directly under the row of icons to add a new activity. For this example, 2000 will be entered in the *Activity ID* field. This is followed by inputting the desired description in the adjacent field. Since this is a milestone, remaining duration and percent complete are left at their default values. The next step is to designate this activity as a *Start milestone*. This is done by clicking on the *Type* field and using the down arrow to activate the list of acceptable activity types and selecting *Start milestone*. When this input is completed, the activity form should resemble the one shown in Screen Image 9-22.

The next step is to input the successor for this milestone. In this example, the successor is Activity 30 with a finish-to-start relationship and no lag. This can be added to this activity by clicking the *Succ* button from the activity form and manually inputting this information, as shown in Screen Image 9-23. The final step in adding the milestone is to constrain the start of the milestone. In this example, the start day was constrained to the end of Day 1 as specified by the fabricator. This workday format translates to October 2 on the project schedule. Therefore, the early start of this milestone activity needs to be fixed as October 2. This information is input by opening the *Constraints* window. This window is opened by clicking the *Constr* button on the activity form. From the constraints window click the *Early constraint* box and the *Start* option, which will activate the date input box. The constrained start date can be entered via the keyboard or by clicking the down arrow and selecting the date from the project calendar. In Screen Image 9-24 the start date of milestone 2000 has been constrained to 02 October 01. Once the constraint date has been entered, this window can be closed and the schedule recalculated. Once the schedule has been recalculated, the milestone will have early and late dates as well as a float time. Screen Image 9-25 shows the placement of the milestone on the main input screen schedule.

Screen Image 9-22
Adding a Milestone

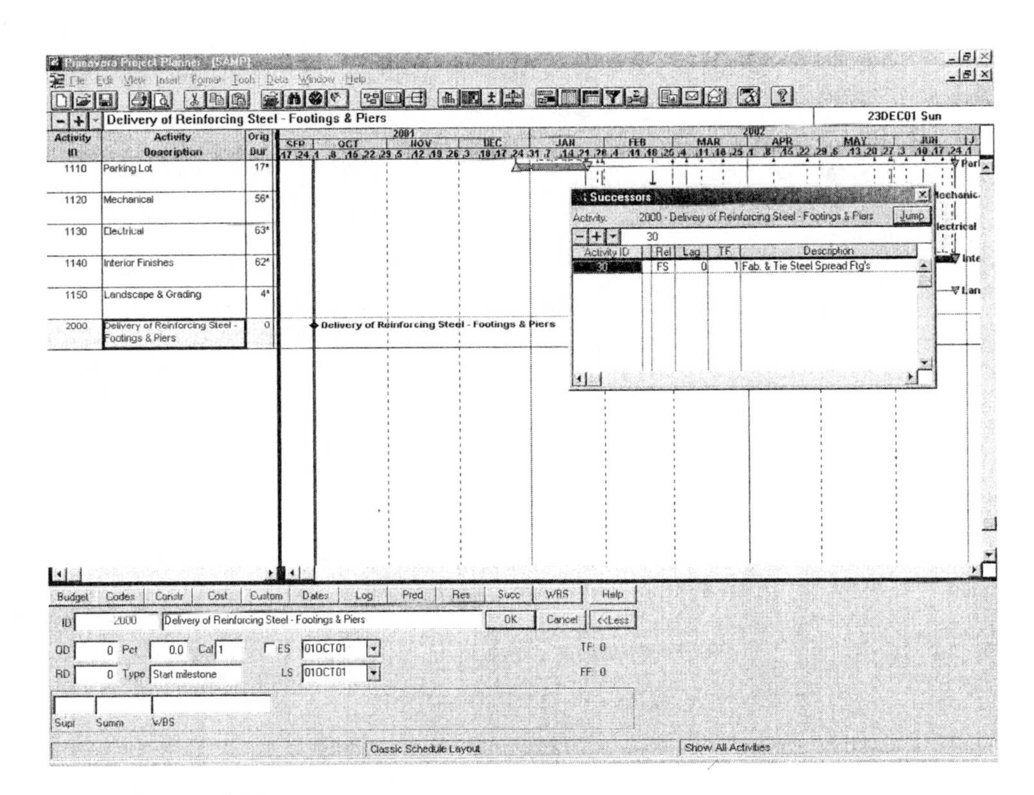

Screen Image 9-23
Adding a Successor to a Milestone

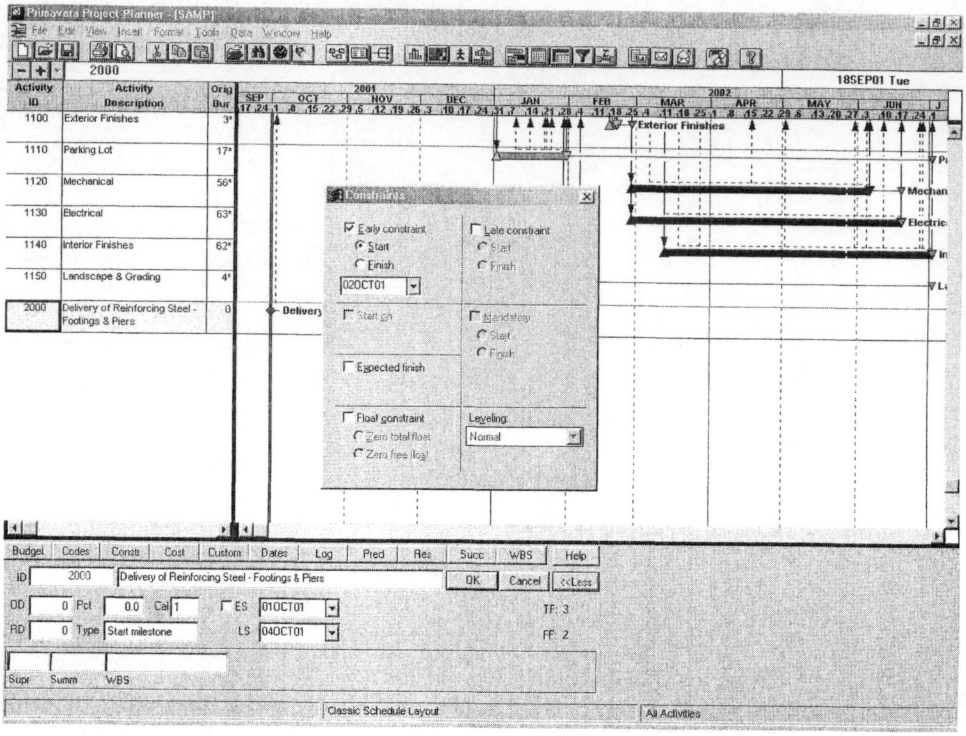

Primavera Project Planner - [SAMP]

File Edit View Insert Format Tools Data Window Help

2000

18SEP01 Tue

Activity ID	Activity Description	Orig Dur
1100	Exterior Finishes	3*
1110	Parking Lot	17*
1120	Mechanical	56*
1130	Electrical	63*
1140	Interior Finishes	62*
1150	Landscape & Grading	4*
2000	Delivery of Reinforcing Steel - Footings & Piers	0

2001 SEP OCT NOV DEC 2002 JAN FEB MAR APR MAY JUN

Exterior Finishes

Delivery

Mechan

Electric

Constraints

Early constraint

Start

Finish

02OCT01

Start on

Expected finish

Float constraint

Zero total float

Zero free float

Late constraint

Start

Finish

Mandatory

Start

Finish

Leveling:

Normal

Budget Codes Constr Cost Custom Dates Log Pred Res Succ WBS Help

ID 2000 Delivery of Reinforcing Steel - Footings & Piers OK Cancel <<Less

OD 0 Pct 0.0 Cal 1 ES 01OCT01 TF: 3

RD 0 Type Start milestone LS 04OCT01 FF: 2

Supr Summ WBS

Classic Schedule Layout All Activities

Screen Image 9-24
Adding a Constraint

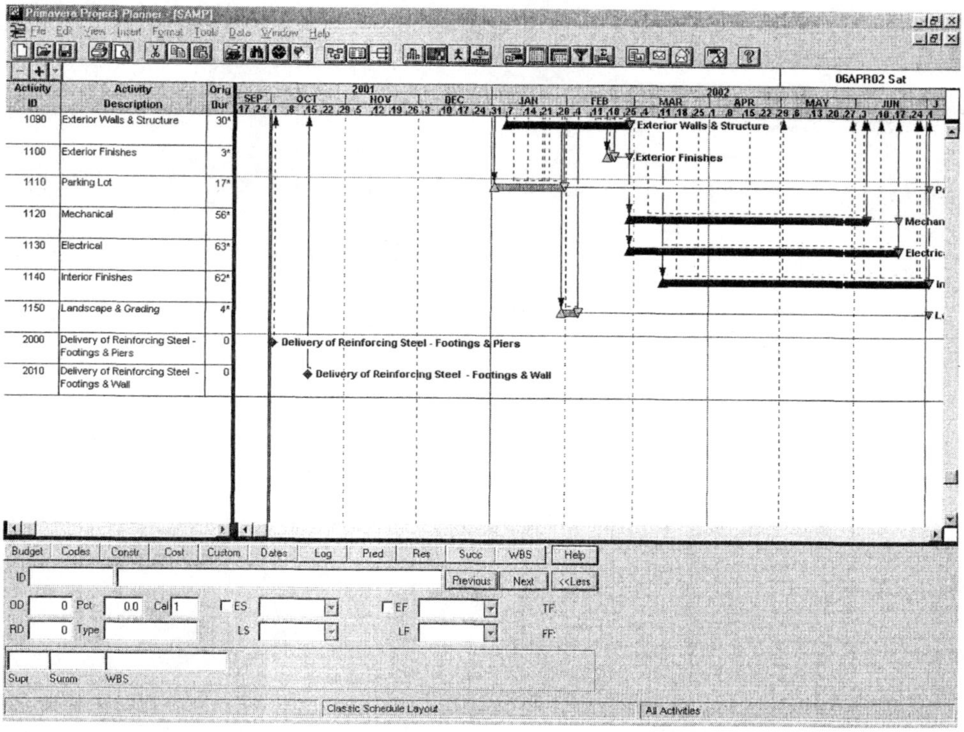

Primavera Project Planner - [SAMP]

File Edit View Insert Format Tools Data Window Help

06APR02 Sat

Activity ID	Activity Description	Orig Dur
1090	Exterior Walls & Structure	30*
1100	Exterior Finishes	3*
1110	Parking Lot	17*
1120	Mechanical	56*
1130	Electrical	63*
1140	Interior Finishes	62*
1150	Landscape & Grading	4*
2000	Delivery of Reinforcing Steel - Footings & Piers	0
2010	Delivery of Reinforcing Steel - Footings & Wall	0

2001 SEP OCT NOV DEC 2002 JAN FEB MAR APR MAY JUN

Exterior Walls & Structure

Exterior Finishes

Mechan

Electric

Delivery of Reinforcing Steel - Footings & Piers

Delivery of Reinforcing Steel - Footings & Wall

Budget Codes Constr Cost Custom Dates Log Pred Res Succ WBS Help

ID Previous Next <<Less

OD 0 Pct 0.0 Cal 1 ES EF TF:

RD 0 Type LS LF FF:

Supr Summ WBS

Classic Schedule Layout All Activities

Screen Image 9-25
Milestone Added and Recalculated

Milestones can also be used to track submittals. More and more general contractors are relying on subcontractors to complete larger portions of the project. While this scheme has many advantages, one of the disadvantages is the need for increased diligence in handling required submittals and approvals. In addition to the most obvious submittal, the request for payment, there are countless submittals required for colors, carpet, brick, and the like. When the actual work is being performed by the subcontractors and they have a supply and install contract, they also have responsibility for preparing the required submittal. However, the general contractor is always the conduit to the owner or architect. Therefore, the general contractor must contact the subcontractors about the required submittals, and often they must contact a supplier to get the needed information or samples. Because of the quantity of required submittals and all of the people handling them, the potential for one to get lost, misplaced, or held up is fairly high. Therefore, the contractor needs a methodology to track and expedite the required submittals in a timely manner. By adding milestones to the construction schedule to denote these events, the contractor has a timely reminder of what submittals need to be prepared and when. By adding these items to the schedule the impact of a missed or delayed submittal or approval can be measured. To add this type of reminder to the schedule, the chain of events and its timing need to be detailed. For example purposes the following times have been allocated by contract or as practical times to facilitate the delivery of the bathroom cabinets.

Description	Duration (days)
Rev. 0 shop drawings from owner	14
Rev. 1 shop drawings from shop	14
Rev. 1 shop drawings from owner	14
Fabrication of cabinets	30

Figure 9-5 is a graphic presentation of these activities. Since milestones have no duration, the lags are used to show the intervening time between the events.

In Screen Image 9-26 the cabinet procurement process milestones have been added and milestone 2100 is being related to milestone 2110 with a finish-to-start relationship with a 14-day lag. After all of the relationships and lags have been added, the schedule needs to be recalculated. The first milestone in the chain will have an early start of the beginning of the project and (since it was not constrained) a late start of the latest possible date to submit the drawings. If the start of the first milestone in the chain is constrained to the calculated late start and the schedule is recalculated, all of the milestones in the chain will be placed on the critical path. A far better practice is to constrain the milestone some number of days before the late start date. This gives the chain float and acts as a reminder of what needs to be done. In Screen Image 9-27 the first activity in the cabinet procurement chain is being set to 15 October 01. Screen Image 9-28 is the cabinet procurement chain after the initial activity was constrained and the schedule recalculated. Output 9-2 is a list of the milestones that were added to the schedule. This report was generated by selecting only those activities that fall within the appropriate range of activity ID's for milestones. Screen Image 9-29 is the selection screen for the report specifications.

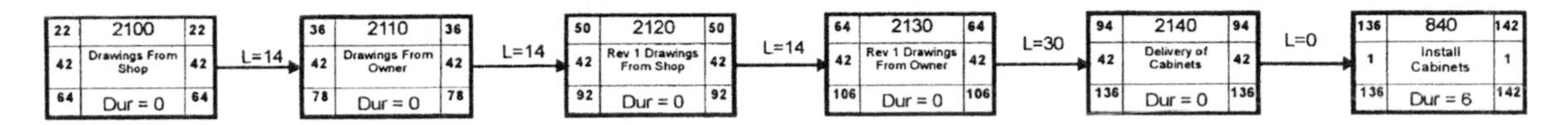

Figure 9-5
Cabinet Procurement Process

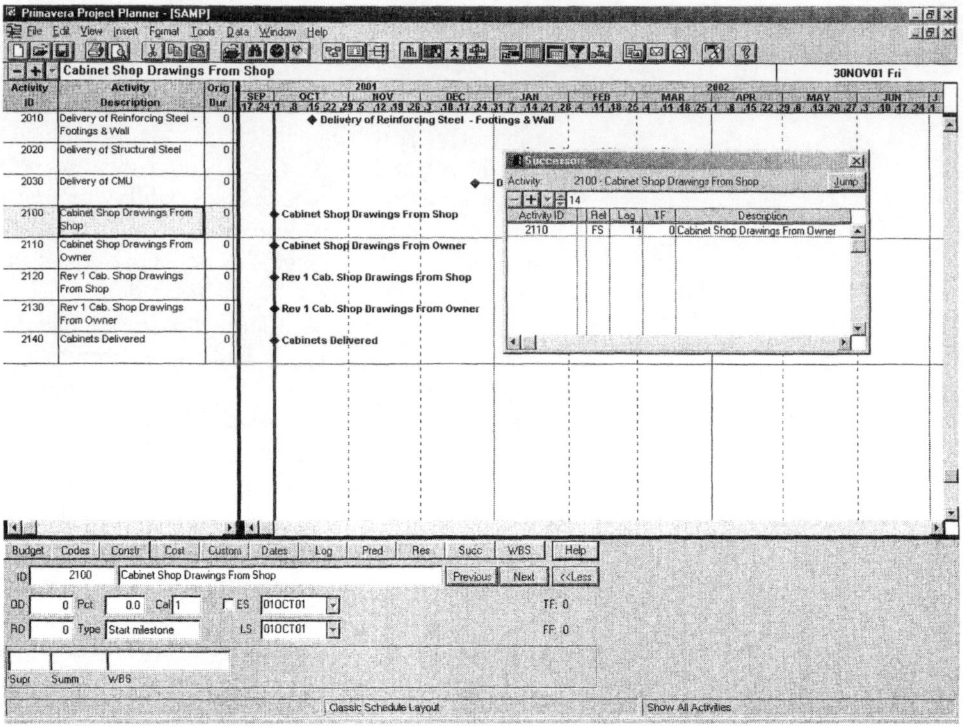

Primavera Project Planner - [SAMP]

File Edit View Insert Format Tools Data Window Help

Cabinet Shop Drawings From Shop 30NOV01 Fri

Activity ID	Activity Description	Orig Dur
2010	Delivery of Reinforcing Steel - Footings & Wall	0
2020	Delivery of Structural Steel	0
2030	Delivery of CMU	0
2100	Cabinet Shop Drawings From Shop	0
2110	Cabinet Shop Drawings From Owner	0
2120	Rev 1 Cab. Shop Drawings From Shop	0
2130	Rev 1 Cab. Shop Drawings From Owner	0
2140	Cabinets Delivered	0

Successors

Activity: 2100 - Cabinet Shop Drawings From Shop Jump

Activity ID	Rel	Lag	TF	Description
2110	FS	14	0	Cabinet Shop Drawings From Owner

Budget Codes Constr Cost Custom Dates Log Pred Res Succ WBS Help

ID 2100 Cabinet Shop Drawings From Shop Previous Next <<Less

OD 0 Pct 0.0 Cal 1 ES 01OCT01 TF: 0

RD 0 Type Start milestone LS 01OCT01 FF: 0

Supr Summ WBS

Classic Schedule Layout Show All Activities

Screen Image 9-26
Milestone Chain Added and Schedule Recalculated

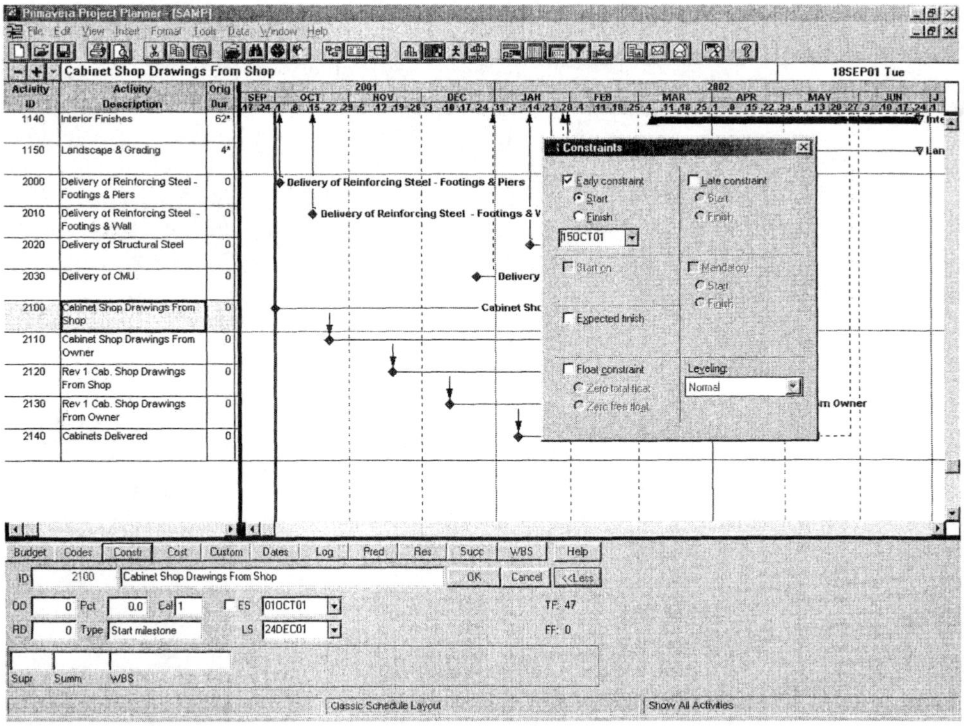

Primavera Project Planner - [SAMP]

File Edit View Insert Format Tools Data Window Help

Cabinet Shop Drawings From Shop 18SEP01 Tue

Activity ID	Activity Description	Orig Dur
1140	Interior Finishes	62*
1150	Landscape & Grading	4*
2000	Delivery of Reinforcing Steel - Footings & Piers	0
2010	Delivery of Reinforcing Steel - Footings & Wall	0
2020	Delivery of Structural Steel	0
2030	Delivery of CMU	0
2100	Cabinet Shop Drawings From Shop	0
2110	Cabinet Shop Drawings From Owner	0
2120	Rev 1 Cab. Shop Drawings From Shop	0
2130	Rev 1 Cab. Shop Drawings From Owner	0
2140	Cabinets Delivered	0

Constraints

Early constraint: Start / Finish — 15OCT01

Late constraint: Start / Finish

Start on

Mandatory: Start / Finish

Expected finish

Float constraint: Zero total float / Zero free float

Leveling: Normal

Budget Codes Constr Cost Custom Dates Log Pred Res Succ WBS Help

ID 2100 Cabinet Shop Drawings From Shop OK Cancel <<Less

OD 0 Pct 0.0 Cal 1 ES 01OCT01 TF: 47

RD 0 Type Start milestone LS 24DEC01 FF: 0

Supr Summ WBS

Classic Schedule Layout Show All Activities

Screen Image 9-27
Setting Constraints

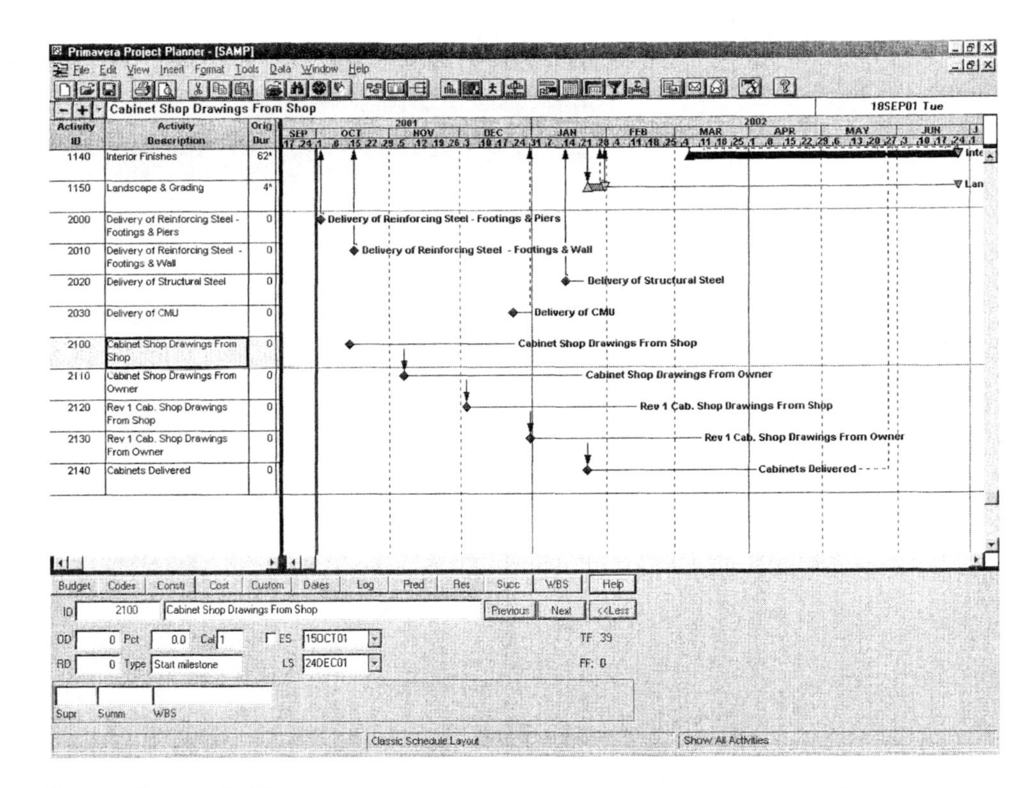

Screen Image 9-28
Milestone Chain Added and Schedule Recalculated

All Star Developments — PRIMAVERA PROJECT PLANNER — Sample Construction Project

REPORT DATE 18MAR00 RUN NO. 33 19:36 — START DATE 1OCT01 FIN DATE 1JUL02

Milestone Report — DATA DATE 1OCT01 PAGE NO. 1

ACTIVITY ID	ORIG DUR	REM DUR	CAL	%	CODE	ACTIVITY DESCRIPTION	EARLY START	EARLY FINISH	LATE START	LATE FINISH	TOTAL FLOAT
2000	0	0	1	0		Delivery of Reinforcing Steel - Footings & Piers	2OCT01*		4OCT01		2
2010	0	0	1	0		Delivery of Reinforcing Steel - Footings & Wall	17OCT01*		18OCT01		1
2020	0	0	1	0		Delivery of Structural Steel	15JAN02*		28JAN02		7
2030	0	0	1	0		Delivery of CMU	24DEC01*		7JAN02		6
2100	0	0	1	0		Cabinet Shop Drawings From Shop	15OCT01*		31DEC01		42
2110	0	0	1	0		Cabinet Shop Drawings From Owner	7NOV01		24JAN02		42
2120	0	0	1	0		Rev 1 Cab. Shop Drawings From Shop	4DEC01		19FEB02		42
2130	0	0	1	0		Rev 1 Cab. Shop Drawings From Owner	31DEC01		14MAR02		42
2140	0	0	1	0		Cabinets Delivered	24JAN02		9APR02		42

Output 9-2
Milestone Activities

Flags

Flags are used to mark points in time when a specific phase of the project will be completed. For example, a flag could be used to show when the foundation of the building will be completed or when the building will be dried in. On the sample schedule introduced in Chapter 4, the foundation will be completed when Activity 590, place and finish slab on grade, is completed. By adding a flag to the schedule, the user can quickly spot this event.

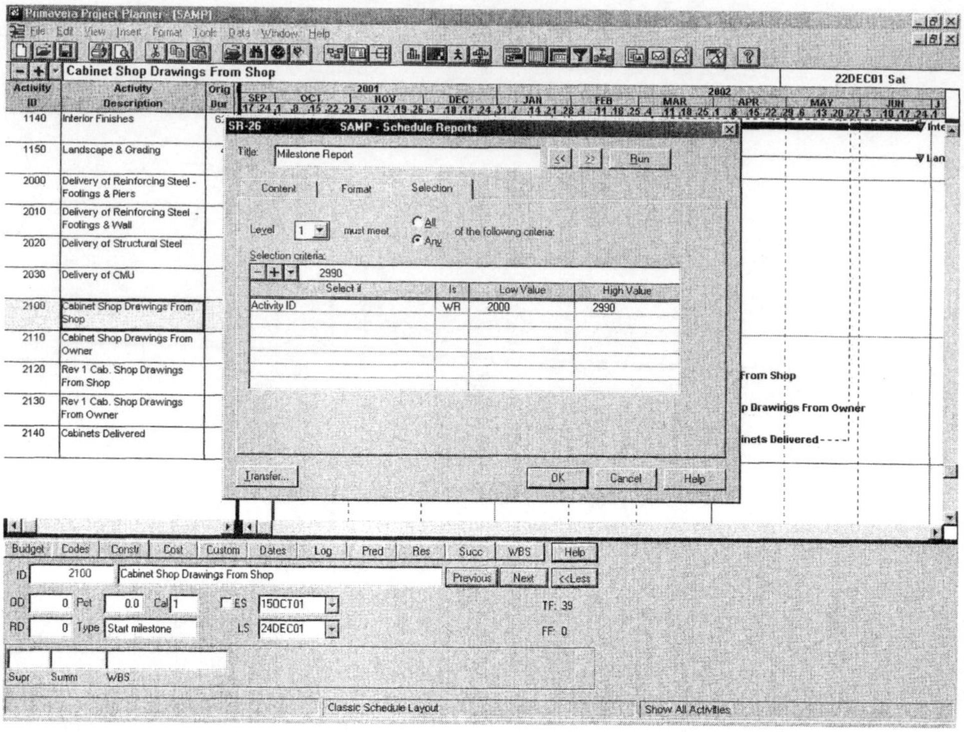

Primavera Project Planner - [SAMP]
File Edit View Insert Format Tools Data Window Help
Cabinet Shop Drawings From Shop 22DEC01 Sat

Activity ID	Activity Description	Orig Dur
1140	Interior Finishes	6
1150	Landscape & Grading	
2000	Delivery of Reinforcing Steel - Footings & Piers	
2010	Delivery of Reinforcing Steel - Footings & Wall	
2020	Delivery of Structural Steel	
2030	Delivery of CMU	
2100	Cabinet Shop Drawings From Shop	
2110	Cabinet Shop Drawings From Owner	
2120	Rev 1 Cab. Shop Drawings From Shop	
2130	Rev 1 Cab. Shop Drawings From Owner	
2140	Cabinets Delivered	

SR-26 SAMP - Schedule Reports
Title: Milestone Report << >> Run
Content | Format | Selection
Level 1 must meet All / Any of the following criteria:
Selection criteria:
2990

Select if	Is	Low Value	High Value
Activity ID	WR	2000	2990

Transfer... OK Cancel Help

Budget Codes Constr Cost Custom Dates Log Pred Res Succ WBS Help
ID 2100 Cabinet Shop Drawings From Shop Previous Next <<Less
OD 0 Pct 0.0 Cal 1 ES 15OCT01 TF: 39
RD 0 Type Start milestone LS 24DEC01 FF: 0
Supr Summ WBS
Classic Schedule Layout Show All Activities

Screen Image 9-29
Selecting Report Specifications

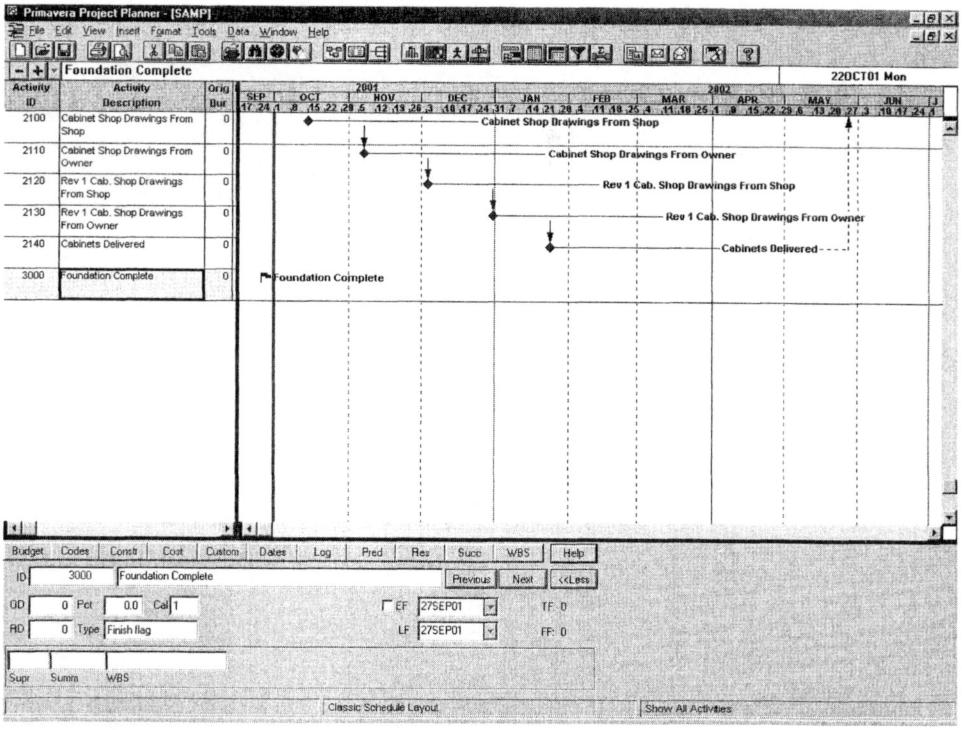

Primavera Project Planner - [SAMP]
File Edit View Insert Format Tools Data Window Help
Foundation Complete 22OCT01 Mon

Activity ID	Activity Description	Orig Dur
2100	Cabinet Shop Drawings From Shop	0
2110	Cabinet Shop Drawings From Owner	0
2120	Rev 1 Cab. Shop Drawings From Shop	0
2130	Rev 1 Cab. Shop Drawings From Owner	0
2140	Cabinets Delivered	0
3000	Foundation Complete	0

Cabinet Shop Drawings From Shop
Cabinet Shop Drawings From Owner
Rev 1 Cab. Shop Drawings From Shop
Rev 1 Cab. Shop Drawings From Owner
Cabinets Delivered
Foundation Complete

Budget Codes Constr Cost Custom Dates Log Pred Res Succ WBS Help
ID 3000 Foundation Complete Previous Next <<Less
OD 0 Pct 0.0 Cal 1 EF 27SEP01 TF: 0
RD 0 Type Finish flag LF 27SEP01 FF: 0
Supr Summ WBS
Classic Schedule Layout Show All Activities

Screen Image 9-30
Adding a Flag Activity

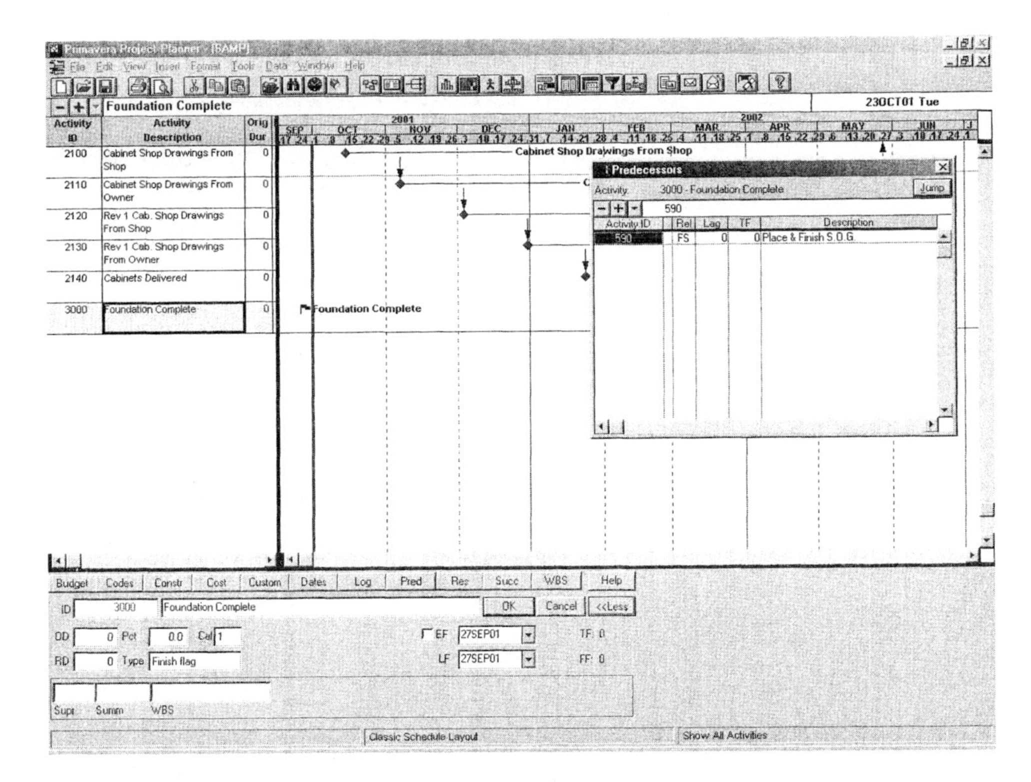

Screen Image 9-31
Predecessor Added for Flag Placement

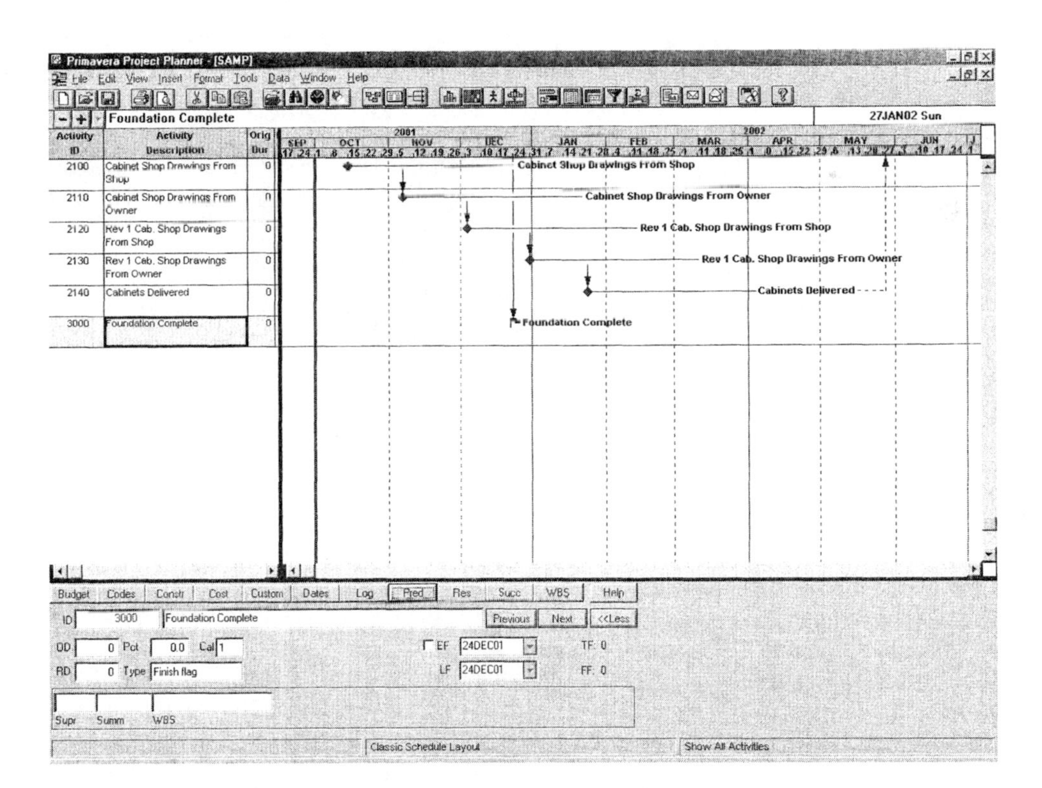

Screen Image 9-32
Flag Added and Schedule Recalculated

To add a flag to the Primavera Project Planner schedule, start with the main input screen and click the plus button below the row of icons to add a new activity. The new activity for this example will be Activity 3000, and it creates a new block of activity ID's. If the flag were to show the *beginning* of a phase of the project, a starting flag would be used. Screen Image 9-30 is an example of the required input on the activity form to initiate the finish flag activity.

Once the flag activity has been added, it is necessary to show how this flag relates to the project network schedule. Since the flag in this example shows when the foundation is completed, Activity 3000 is a successor to Activity 590, the last foundation activity.

However, instead of going back to that activity and specifying Activity 3000 as its successor, the predecessor window can be opened and Activity 590 can be listed as a predecessor to Activity 3000. To do this, click the *Pred* button from the activity form to open the *Predecessors* window, as shown in Screen Image 9-31. From that window, specify Activity 590 as a predecessor with a finish-to-start relationship.

After the predecessors have been added, the window needs to be closed, the *OK* button clicked on the activity form, and the schedule recalculated. This will place the flag in its proper location on the schedule. Screen Image 9-32 shows what the schedule would look like after the flag has been added.

Conclusion

Through the use of hammock activities or activity codes, the construction schedule can be summarized into a more manageable number of activities. While there are no specific rules governing what detailed activities would constitute a specific hammock activity, clearly some items need to be considered. First, the activities that are included in the hammock should be contiguous. This ensures that they do indeed represent a distinct portion of the project. If the activities are not contiguous, the summary bars will extend through portions of the project when no activity relating to that hammock is occurring. Second, the descriptions used for the hammock activities need to clearly relate what is being included within the hammock. If these guidelines are followed, a useful summary schedule can be developed and maintained with virtually no effort.

Flags and milestone activities are special activities that have no duration. These activities are included in the schedule to enhance its usefulness by highlighting portions completed or events during the project.

Suggested Exercises

1. Using the simulation project, add 15 hammock activities that will serve as summary activities. These activities should represent the entire project and should summarize only contiguous activities.
2. Add three chains of milestones that will represent how a particular item will be submitted to the owner for approval, procured, fabricated, delivered, and installed. At minimum, the chain of activities should include an activity denoting submission of shop drawings, receipt of approved shop drawings, and delivery of the item.
3. Add flags to the schedule to denote when the foundation is completed, when the superstructure is completed, and when the building is dried in.
4. Create a graphic schedule showing hammock activities, milestones, and flags.
5. Develop an activity code summary item matrix. Identify 15 summary activity codes and include all of the underlying construction activities into one of these activity codes.
6. Generate a summary bar chart schedule using activity codes.

Updating the Schedule

KEY TERMS

Actual Progress The completed percentage of the project based on actual field observations and quantification.

Target Schedules

The construction schedule is and should be a dynamic document. As construction progresses, actual start dates replace the planned dates and the implications of these dates on the future of the project need to be analyzed. As actual dates are entered in Primavera Project Planner, the planned dates are lost. For example, when an early start date is replaced with an actual start date, that early start date is lost. This approach ensures that the data presented within the schedule is the most current. In the most practical sense, once an element of the schedule is no longer valid, it is of no use in the future plans of the project. However, that original schedule does serve as a good baseline schedule and as a tool to improve future project planning. Primavera Project Planner supports this concept through the use of target schedules. A target schedule is a copy of a schedule that can be superimposed onto the current schedule in order to show any deviations.

Establishing a Target Schedule within Primavera Project Planner

The first step in establishing a target schedule is to select the *Tools* command, followed by the *Project Utilities/Targets* option, as shown in Screen Image 10-1. These commands will open the *Targets* window, as shown in Screen Image 10-2. The only required entry on this window is a new file name for the target schedule. Just as with the current schedule, the file name is limited to four alphanumeric characters. In Screen Image 10-3 the *Target 1* has been set to *SAM1*. Within the *Target* window the *Update* button will globally update the target schedules to that of the current schedule. When the *OK* button is clicked the user will be prompted to verify the desired action. Screen Image 10-4 is the prompt to create target schedule *SAM1*. If the action is appropriate, click the *YES* button, otherwise click the *NO* button, which will return control back to the main input screen. When the specified action on the *Target* window is complete, there will be no obvious impact on the main input screen. However, when the action is completed there could be as many as two target schedules and one current schedule. At that point in time all of these schedules would be identical.

To superimpose the target bars on the schedule, a number of modifications need to be made. This is done by selecting the *Format* command from the main command line followed by the *Bars* option, as shown in Screen Image 10-5. This will open the *Bars* window as shown in Screen Image 10-6. On that sample screen there are four bar descriptions. Within that window the *Position* column specifies the order of bar placement within a specific activity row on the screen. In that screen the *Early Bar* and *Float Bar* share position 1. When this occurs, the items are placed on top of each other. In Screen Image 10-5, Activity 70 shows this very well. The early bar is a thick bar in position 1 and the float bar extends out from that bar also in position 1. The other two bars designated on Screen Image 10-6 are not shown since the *Visible* box is not checked.

The *Key* option specifies which bar will be used for summarization. Only one bar may be given this designation. The *Progress* option, if checked, will fill the completed portion

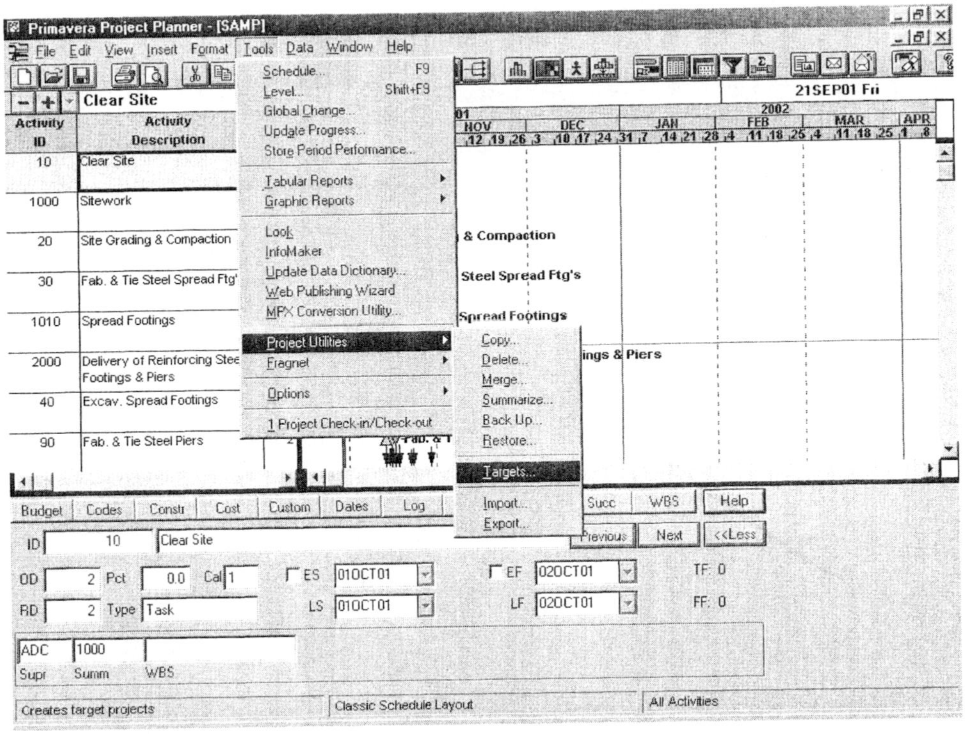

Primavera Project Planner - [SAMP]
File Edit View Insert Format Tools Data Window Help
Clear Site
21SEP01 Fri

Activity ID	Activity Description
10	Clear Site
1000	Sitework
20	Site Grading & Compaction
30	Fab. & Tie Steel Spread Ftg'
1010	Spread Footings
2000	Delivery of Reinforcing Stee Footings & Piers
40	Excav. Spread Footings
90	Fab. & Tie Steel Piers

Schedule... F9
Level... Shift+F9
Global Change...
Update Progress...
Store Period Performance...
Tabular Reports
Graphic Reports
Look
InfoMaker
Update Data Dictionary...
Web Publishing Wizard
MPX Conversion Utility...
Project Utilities
Fragnet
Options
1 Project Check-in/Check-out

Copy...
Delete...
Merge...
Summarize...
Back Up...
Restore...
Targets...
Import...
Export...

2001 2002 NOV DEC JAN FEB MAR APR
& Compaction
Steel Spread Ftg's
Spread Footings
ings & Piers

Budget Codes Constr Cost Custom Dates Log Succ WBS Help
ID 10 Clear Site Previous Next <<Less
OD 2 Pct 0.0 Cal 1 ES 01OCT01 EF 02OCT01 TF: 0
RD 2 Type Task LS 01OCT01 LF 02OCT01 FF: 0
ADC 1000
Supr Summ WBS
Creates target projects | Classic Schedule Layout | All Activities

Screen Image 10-1
Opening the Targets Window

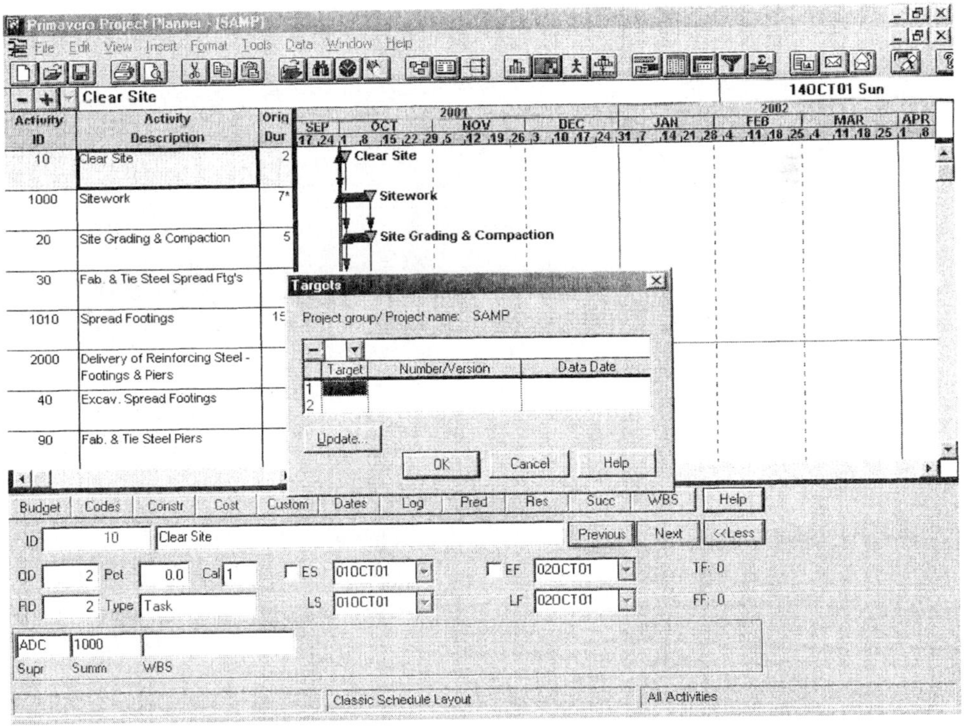

Primavera Project Planner - [SAMP]
File Edit View Insert Format Tools Data Window Help
Clear Site
14OCT01 Sun

Activity ID	Activity Description	Orig Dur
10	Clear Site	2
1000	Sitework	7*
20	Site Grading & Compaction	5
30	Fab. & Tie Steel Spread Ftg's	
1010	Spread Footings	15
2000	Delivery of Reinforcing Steel - Footings & Piers	
40	Excav. Spread Footings	
90	Fab. & Tie Steel Piers	

2001 2002 SEP OCT NOV DEC JAN FEB MAR APR
Clear Site
Sitework
Site Grading & Compaction

Targets
Project group/ Project name: SAMP

	Target	Number/Version	Data Date
1			
2			

Update...
OK Cancel Help

Budget Codes Constr Cost Custom Dates Log Pred Res Succ WBS Help
ID 10 Clear Site Previous Next <<Less
OD 2 Pct 0.0 Cal 1 ES 01OCT01 EF 02OCT01 TF: 0
RD 2 Type Task LS 01OCT01 LF 02OCT01 FF: 0
ADC 1000
Supr Summ WBS
Classic Schedule Layout | All Activities

Screen Image 10-2
Initial Targets Window

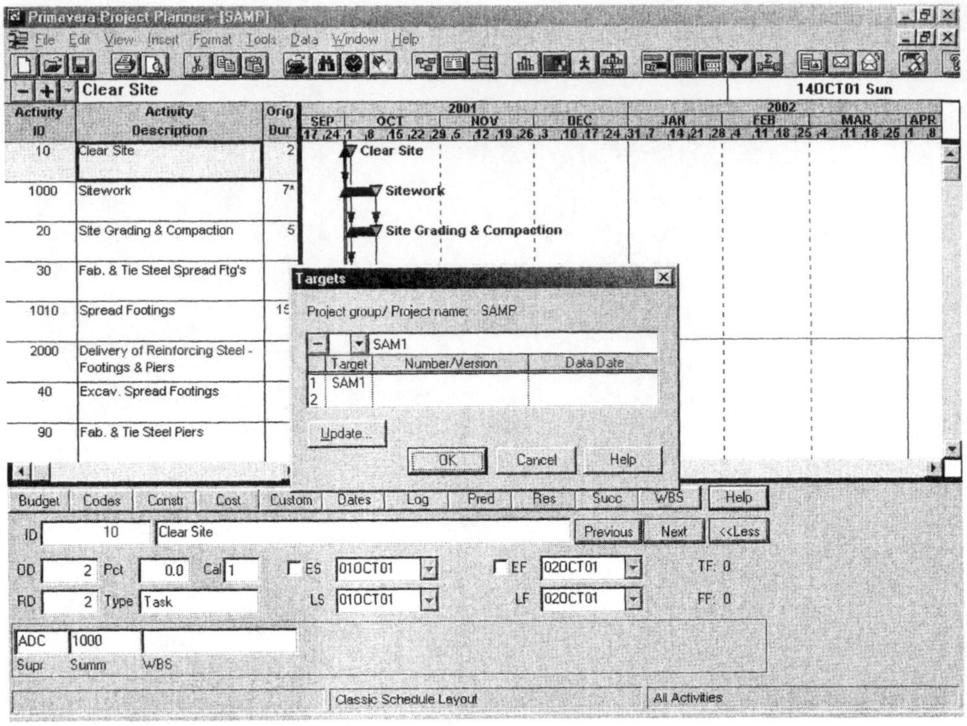

Primavera Project Planner - [SAMP]
File Edit View Insert Format Tools Data Window Help
Clear Site 14OCT01 Sun

Activity ID	Activity Description	Orig Dur
10	Clear Site	2
1000	Sitework	7*
20	Site Grading & Compaction	5
30	Fab. & Tie Steel Spread Ftg's	
1010	Spread Footings	15
2000	Delivery of Reinforcing Steel - Footings & Piers	
40	Excav. Spread Footings	
90	Fab. & Tie Steel Piers	

2001 SEP OCT NOV DEC 2002 JAN FEB MAR APR
Clear Site
Sitework
Site Grading & Compaction

Targets
Project group/ Project name: SAMP
SAM1

	Target	Number/Version	Data Date
1	SAM1		
2			

Update...
OK Cancel Help

Budget Codes Constr Cost Custom Dates Log Pred Res Succ WBS Help
ID 10 Clear Site Previous Next <<Less
OD 2 Pct 0.0 Cal 1 ES 01OCT01 EF 02OCT01 TF: 0
RD 2 Type Task LS 01OCT01 LF 02OCT01 FF: 0
ADC 1000
Supr Summ WBS
Classic Schedule Layout All Activities

Screen Image 10-3
Creating Target SAM1

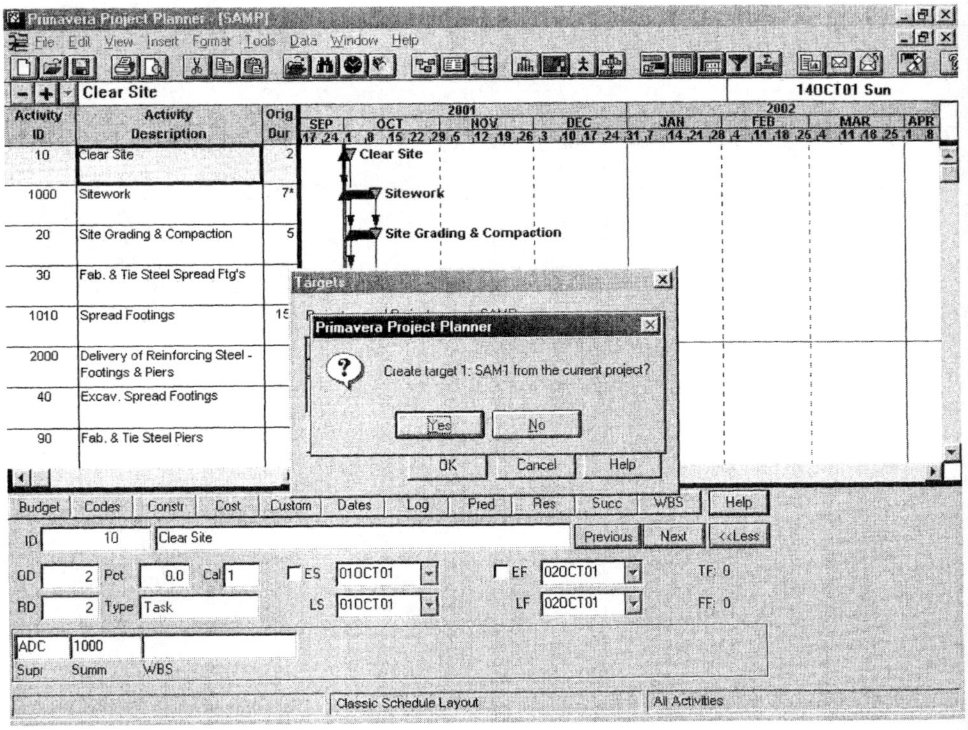

Primavera Project Planner - [SAMP]
File Edit View Insert Format Tools Data Window Help
Clear Site 14OCT01 Sun

Activity ID	Activity Description	Orig Dur
10	Clear Site	2
1000	Sitework	7*
20	Site Grading & Compaction	5
30	Fab. & Tie Steel Spread Ftg's	
1010	Spread Footings	15
2000	Delivery of Reinforcing Steel - Footings & Piers	
40	Excav. Spread Footings	
90	Fab. & Tie Steel Piers	

2001 SEP OCT NOV DEC 2002 JAN FEB MAR APR
Clear Site
Sitework
Site Grading & Compaction

Targets
Primavera Project Planner
Create target 1: SAM1 from the current project?
Yes No
OK Cancel Help

Budget Codes Constr Cost Custom Dates Log Pred Res Succ WBS Help
ID 10 Clear Site Previous Next <<Less
OD 2 Pct 0.0 Cal 1 ES 01OCT01 EF 02OCT01 TF: 0
RD 2 Type Task LS 01OCT01 LF 02OCT01 FF: 0
ADC 1000
Supr Summ WBS
Classic Schedule Layout All Activities

Screen Image 10-4
Confirming Target Creation

Screen Image 10-5
Opening the Bars Window

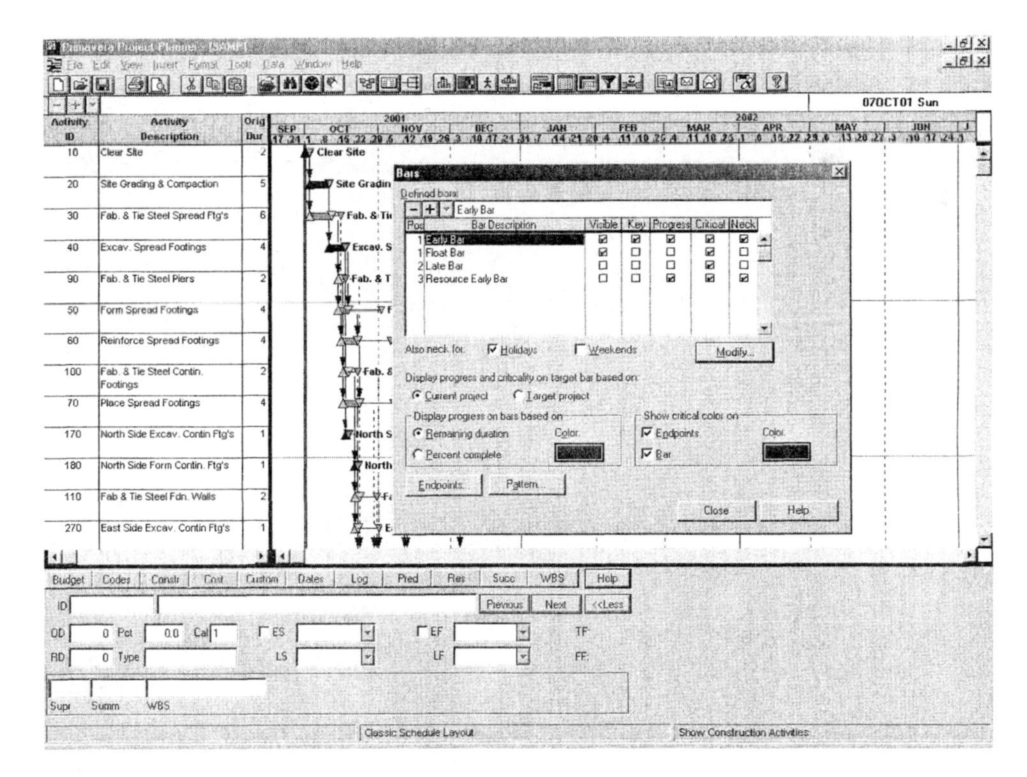

Screen Image 10-6
Initial Bars Window

of the bar with the color specified within the *Display progress on bars based on* box. This box is found in the lower left-hand corner of the *Bars* window. The *Critical* option will fill the uncompleted bars that are on the critical path with the color specified within the *Show critical color on* box, which is found in the lower right hand corner of the *Bars* window. The *Neck* option allows for the bar thickness to be reduced to designate *Holidays* or *Weekends*. Directly below the *Defined bars* table is where the user specifies which nonwork periods will be necked.

The progress shown on the bars can be based on either the *Remaining duration* or *Percent complete*. If an activity has an original duration of 10 days and it is 50% complete and the *Percent complete* option is selected, then half of the bar will be filled with the designated color. There will be a total disregard for the actual number of days required to reach the 50 percent point. With the *Percent complete* option it would appear that it took 5 days to reach that point even if it took more or less. Conversely if the *Remaining duration* option is used, the amount of the bar filled is the original duration minus the remaining duration, which would be the actual number of days that were used. Suppose that it took 6 days to reach the 50 percent mark; then 6 days would be shown as complete and 4 remaining. Both of these methods have their shortcomings. With the percent complete method there is the assumption that the rate of work is constant across the activity. The *Remaining duration* process can make the activity look further ahead of schedule than it actually is. To overcome these obstacles, the solution is to continually evaluate the remaining duration for all of the incomplete activities. If the most realistic remaining duration is entered on the schedule, then the *Remaining duration* shading method will present the most likely schedule status. The continual evaluation of the remaining duration should be part of the schedule updating process. The project planner and manager need to be constantly aware of the planned and actual rate of work.

In order to display the target schedule, a new bar will be added. This is done by clicking the plus button, which will change the window to one similar to that shown in Screen Image 10-7. The first item that must be addressed is to give the new bar the correct

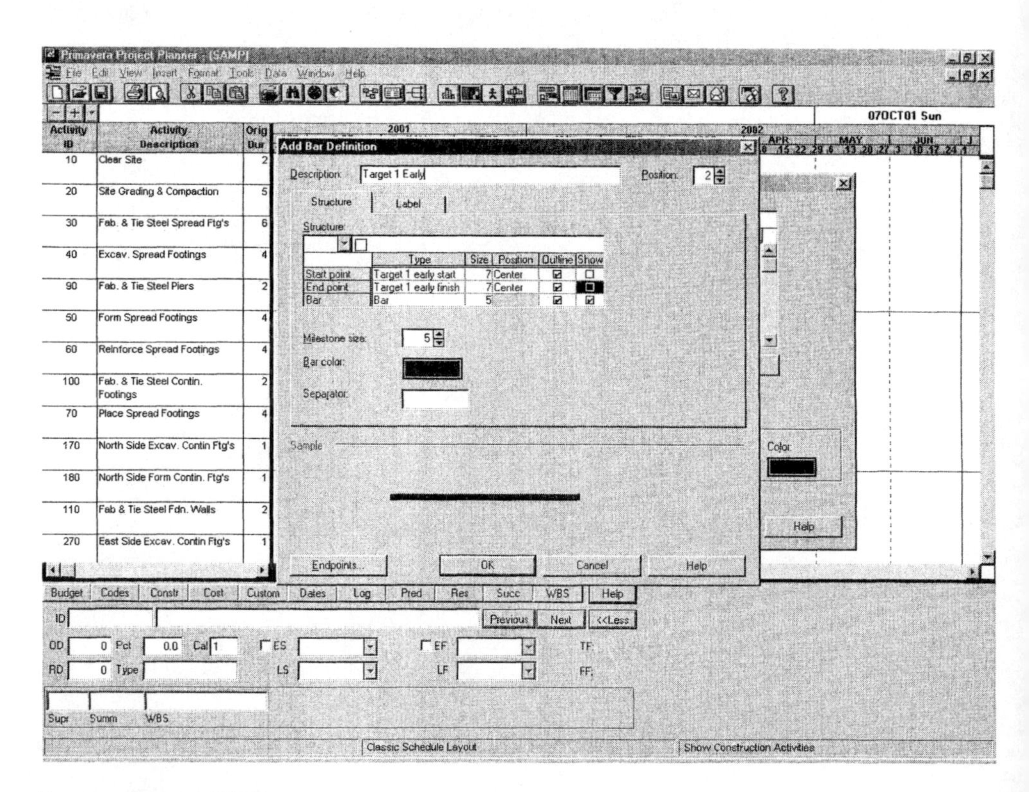

Screen Image 10-7
Add Bar Definition Window

name. Typically, when this window opens the bar name is *Early Bar*, which is not always appropriate. In Screen Image 10-7 the name has been changed to *Target 1 Early*. The next operation is to specify what defines the starting point and ending of the bar. Through the use of the pull-down menu, all of the possible selections can be seen and selected. In Screen Image 10-7, the *Start point* has been defined as *Target 1 early start*. This point will be the early start for all activities in project SAM1, the Target 1 schedule. The *End point* has been specified as the *Target 1early finish*. The starting and ending points on the current schedule are shown as triangles, while the target schedules get squares. To turn this feature off, simply remove the check from the *Show* column. This will make the target bar a simple bar.

The only other entry of interest is the bar thickness. If a thicker bar is desired increase the size, and if a smaller one is desired select a smaller number in the *Size* column. When the bar is properly configured, click the *OK* button to close this window, which will return to the previous window. The only remaining item to check now is the position. In Screen Image 10-8, a float bar for *Target 1* has been configured. The main differences between this and the previous bar are the starting and ending points, which deal with the late dates, and the thickness of the bar. The thickness has been reduced enough so that it will be a line. In Screen Image 10-9, both the *Target 1 Early* and the *Target 1 Float* have been specified as visible and in position 2. Highlighting any of the bars in the *Bar description* column and clicking on the *Modify* button will allow the user to modify any of the elements of the bar. When all of the bars have been properly configured, click the *Close* button to return to the main input screen. Screen Image 10-10 has the target bars added. Since nothing has been updated, both of the schedules are identical.

The target schedule that has been created can be easily modified. Opening up project SAM1 does this as if it were any other project. However, there should be only a limited number of reasons for modifying a target schedule. For example, if a change order were issued that substantially changes the scope of the project, this would invalidate the original schedule and would justify modifications. Another reason may be that a recovery plan

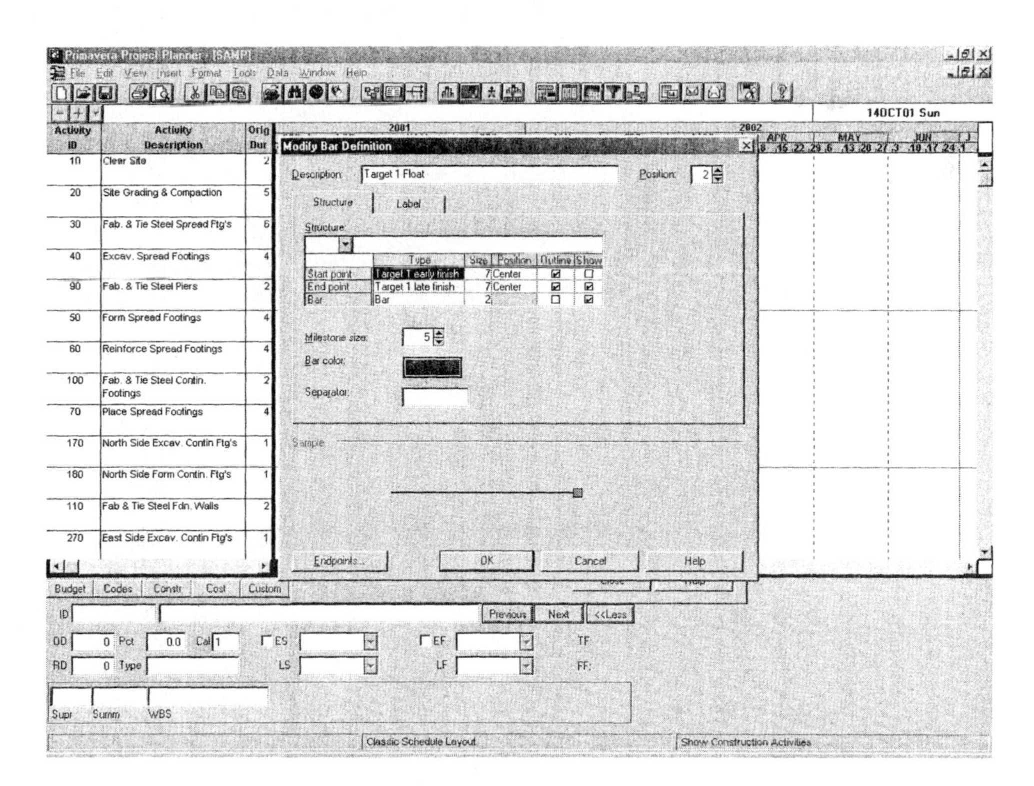

Screen Image 10-8
Float Bar Specifications

Screen Image 10-9
Bars Window with Target Bars

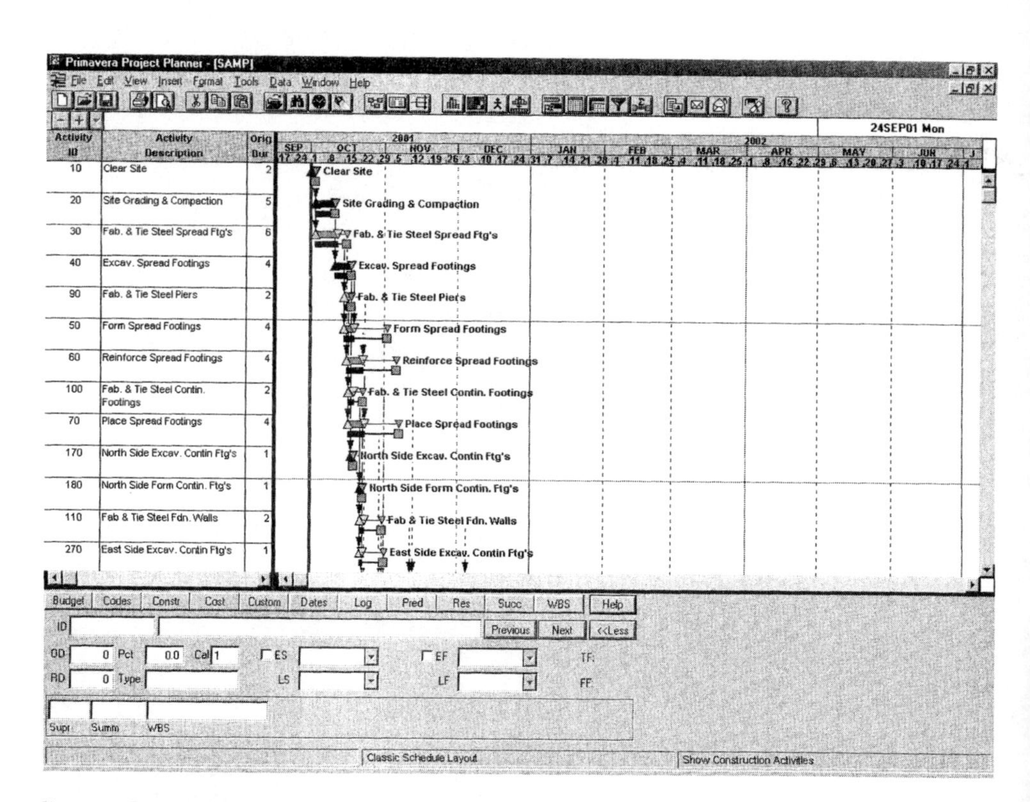

Screen Image 10-10
Bar Chart Schedule with Target Bars

to a deviation has been developed. This recovery plan would now become the target schedule. Primavera Project Planner will allow for there to be two target schedules. One of the targets may be the original schedule that will not be modified during the project. The second target will be that plan with modification based on revisions, new information, or deviations. This scheme will allow the contractor not only to compare the project's planned progress to its actual progress; in fixed-sum contracting the quality and accuracy of the original plan is the basis for the indirect costs and will ultimately impact the project profitability. This scheme makes it easy for the contractor to evaluate the quality of that plan.

Developing an Update Report

From a practical perspective the foremen and superintendents in the field are the most qualified to assess actual progress. Since they are the people closest to the work, they are the ones most able to provide actual start, finish, and progress information. Through the use of an update report, these people can easily gather and record information in a format that facilitates updating the schedule. Screen Image 10-11 is the first screen of the tabular report options. A new feature that has been added to this screen is the *Update line*. This addition adds blank lines to the report so that the people in the field can enter actual dates and progress information. In Screen Image 10-12, the report has been configured so that every superintendent has a sheet that contains the activities under their direct control. Finally, Screen Image 10-13 limits the activities to only the construction activities. In Chapter 9 the activities were blocked into a series of numbers. This selection criterion removes all of the flags, milestones, and hammock activities that have been added to the schedule. Output 10-1 is an example of the update report.

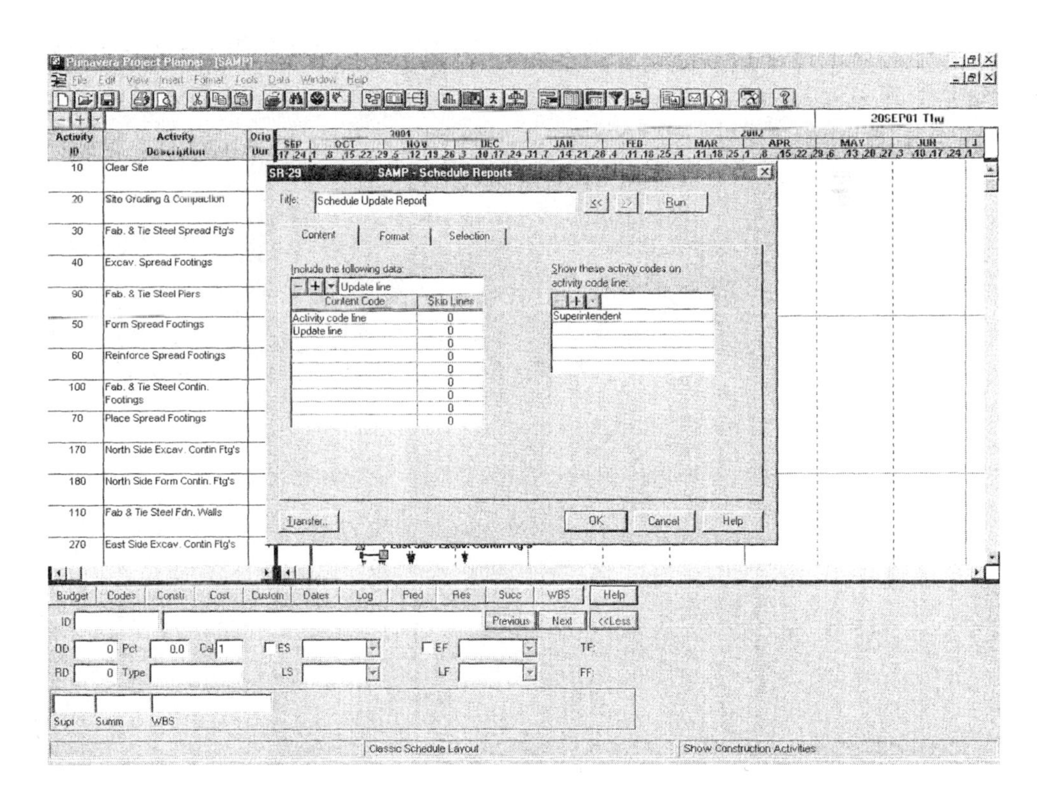

Screen Image 10-11
Content Window for Update Report

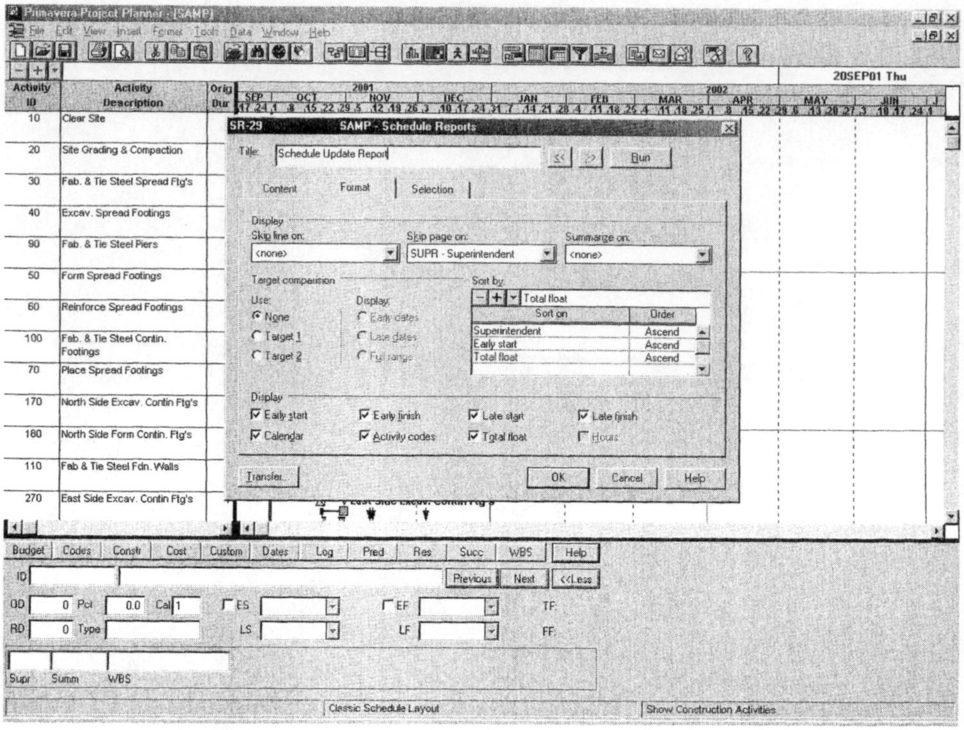

Primavera Project Planner - [SAMP]
File Edit View Insert Format Tools Data Window Help
20SEP01 Thu

Activity ID	Activity Description
10	Clear Site
20	Site Grading & Compaction
30	Fab. & Tie Steel Spread Ftg's
40	Excav. Spread Footings
90	Fab. & Tie Steel Piers
50	Form Spread Footings
60	Reinforce Spread Footings
100	Fab. & Tie Steel Contin. Footings
70	Place Spread Footings
170	North Side Excav. Contin Ftg's
180	North Side Form Contin. Ftg's
110	Fab & Tie Steel Fdn. Walls
270	East Side Excav. Contin Ftg's

SR-29 SAMP - Schedule Reports
Title: Schedule Update Report
Run
Content | Format | Selection
Display
Skip line on: <none>
Skip page on: SUPR - Superintendent
Summarize on: <none>
Target comparison
Use: None / Target 1 / Target 2
Display: Early dates / Late dates / Full range
Sort by: Total float

Sort on	Order
Superintendent	Ascend
Early start	Ascend
Total float	Ascend

Display
Early start, Early finish, Late start, Late finish, Calendar, Activity codes, Total float, Hours
Transfer... OK Cancel Help

Budget Codes Constr Cost Custom Dates Log Pred Res Succ WBS Help
ID Previous Next <<Less
OD 0 Pct 0.0 Cal 1 ES EF TF
RD 0 Type LS LF FF
Supr Summ WBS
Classic Schedule Layout | Show Construction Activities

Screen Image 10-12
Format Window for Update Report

Primavera Project Planner - [SAMP]
File Edit View Insert Format Tools Data Window Help
20SEP01 Thu

SR-29 SAMP - Schedule Reports
Title: Schedule Update Report
Run
Content | Format | Selection
Level 1 must meet All / Any of the following criteria:
Selection criteria: 990

Select if	Is	Low Value	High Value
Activity ID	LT	990	

Transfer... OK Cancel Help

Budget Codes Constr Cost Custom Dates Log Pred Res Succ WBS Help
ID Previous Next <<Less
OD 0 Pct 0.0 Cal 1 ES EF TF
RD 0 Type LS LF FF
Supr Summ WBS
Classic Schedule Layout | Show Construction Activities

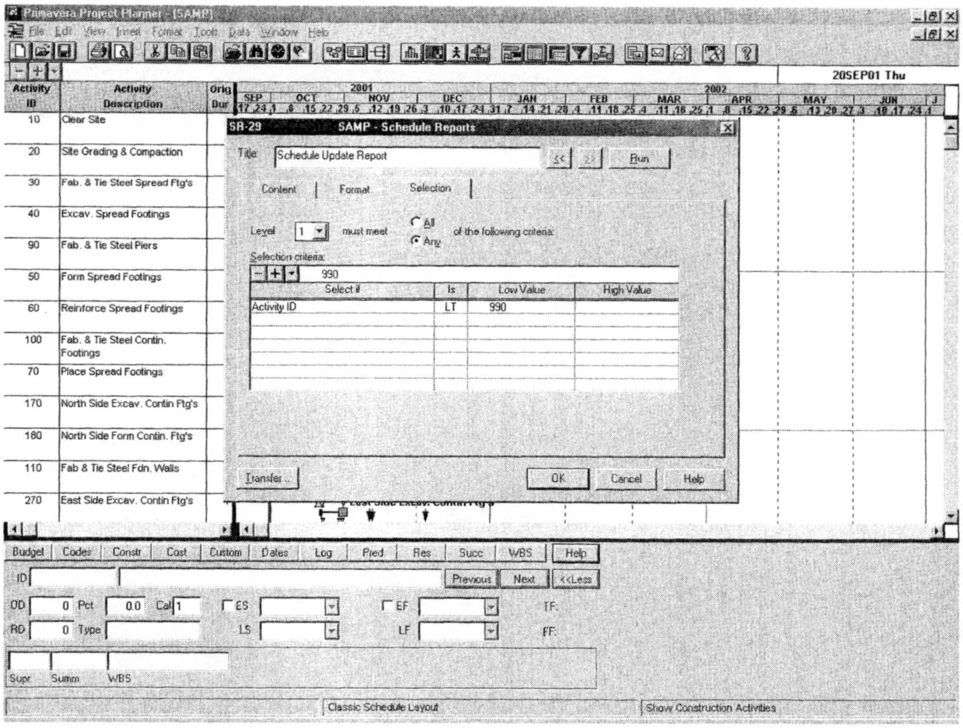

Screen Image 10-13
Selection Window for Update Report

All Star Developments PRIMAVERA PROJECT PLANNER Sample Construction Project

REPORT DATE 24MAR00 RUN NO. 34 START DATE 1OCT01 FIN DATE 1JUL02
11:40
Schedule Update Report DATA DATE 1OCT01 PAGE NO. 1

Albert Concrete

ACTIVITY ID	ORIG DUR	REM DUR	CAL	%	CODE	ACTIVITY DESCRIPTION	EARLY START	EARLY FINISH	LATE START	LATE FINISH	TOTAL FLOAT
10	2	2	1	0	ADC	Clear Site	1OCT01	2OCT01	1OCT01	2OCT01	0
20	5	5	1	0	ADC	Site Grading & Compaction	3OCT01	10OCT01	3OCT01	10OCT01	0
30	6	6	1	0	ADC	Fab. & Tie Steel Spread Ftg's	3OCT01	11OCT01	4OCT01	15OCT01	1
40	4	4	1	0	ADC	Excav. Spread Footings	11OCT01	17OCT01	11OCT01	17OCT01	0
90	2	2	1	0	ADC	Fab. & Tie Steel Piers	15OCT01	16OCT01	16OCT01	17OCT01	1
50	4	4	1	0	ADC	Form Spread Footings	15OCT01	18OCT01	24OCT01	1NOV01	6
60	4	4	1	0	ADC	Reinforce Spread Footings	16OCT01	22OCT01	25OCT01	5NOV01	6
100	2	2	1	0	ADC	Fab. & Tie Steel Contin. Footings	17OCT01	18OCT01	18OCT01	22OCT01	1
70	4	4	1	0	ADC	Place Spread Footings	17OCT01	23OCT01	29OCT01	6NOV01	6
170	1	1	1	0	ADC	North Side Excav. Contin Ftg's	18OCT01	18OCT01	18OCT01	18OCT01	0
180	1	1	1	0	ADC	North Side Form Contin. Ftg's	22OCT01	22OCT01	22OCT01	22OCT01	0
110	2	2	1	0	ADC	Fab & Tie Steel Fdn. Walls	22OCT01	23OCT01	29OCT01	30OCT01	4
270	1	1	1	0	ADC	East Side Excav. Contin Ftg's	22OCT01	22OCT01	31OCT01	31OCT01	6
190	1	1	1	0	ADC	North Side Reinforce Contin. Ftg's	23OCT01	23OCT01	23OCT01	23OCT01	0
80	4	4	1	0	ADC	Strip Forms - Spread Footings	23OCT01	29OCT01	1NOV01	8NOV01	6
360	1	1	1	0	ADC	South Side Excav. Contin Ftg's	23OCT01	23OCT01	12NOV01	12NOV01	11
200	1	1	1	0	ADC	North Side Place Contin. Ftg's	24OCT01	24OCT01	24OCT01	24OCT01	0
120	4	4	1	0	ADC	Reinforce Piers	24OCT01	30OCT01	5NOV01	12NOV01	6
130	4	4	1	0	ADC	Form Piers	25OCT01	31OCT01	6NOV01	13NOV01	6
140	4	4	1	0	ADC	Place Piers	29OCT01	1NOV01	7NOV01	14NOV01	6
210	1	1	1	0	ADC	North Side Strip Contin. Ftg's Forms	30OCT01	30OCT01	30OCT01	30OCT01	0
220	1	1	1	0	ADC	North Side Reinforce Fdn. Walls	31OCT01	31OCT01	31OCT01	5NOV01	0
280	1	1	1	0	ADC	East Side Form Contin. Ftg's	31OCT01	31OCT01	1NOV01	1NOV01	1
230	3	3	1	0	ADC	North Side Form Fdn. Walls	1NOV01	6NOV01	1NOV01	6NOV01	0
290	1	1	1	0	ADC	East Side Reinforce Contin. Ftg's	1NOV01	1NOV01	5NOV01	5NOV01	1
150	4	4	1	0	ADC	Strip Forms - Piers	1NOV01	7NOV01	13NOV01	19NOV01	6
300	1	1	1	0	ADC	East Side Place Contin. Ftg's	5NOV01	5NOV01	6NOV01	6NOV01	1
160	4	4	1	0	ADC	Backfill Spread Footings	5NOV01	8NOV01	14NOV01	20NOV01	6
240	1	1	1	0	ADC	North Side Place Fdn. Walls	7NOV01	7NOV01	7NOV01	7NOV01	0
310	1	1	1	0	ADC	East Side Strip Contin. Ftg's Forms	8NOV01	8NOV01	12NOV01	12NOV01	1
320	1	1	1	0	ADC	East Side Reinforce Fdn. Walls	12NOV01	12NOV01	13NOV01	15NOV01	1
370	1	1	1	0	ADC	South Side Form Contin. Ftg's	12NOV01	12NOV01	13NOV01	13NOV01	1
250	1	1	1	0	ADC	North Side Strip Fdn. Wall Forms	13NOV01	13NOV01	13NOV01	13NOV01	0
380	1	1	1	0	ADC	South Side Reinforce Contin. Ftg's	13NOV01	13NOV01	14NOV01	14NOV01	1

Output 10-1
Update Report

All Star Developments PRIMAVERA PROJECT PLANNER Sample Construction Project

REPORT DATE 24MAR00 RUN NO. 34 START DATE 1OCT01 FIN DATE 1JUL02
11:40
Schedule Update Report DATA DATE 1OCT01 PAGE NO. 2

Albert Concrete

ACTIVITY ID	ORIG DUR	REM DUR	CAL	%	CODE	ACTIVITY DESCRIPTION	EARLY START	EARLY FINISH	LATE START	LATE FINISH	TOTAL FLOAT
390	1	1	1	0	ADC	West Side Excav. Contin Ftg's	13NOV01	13NOV01	21NOV01	21NOV01	5
330	3	3	1	0	ADC	East Side Form Fdn. Walls	14NOV01	19NOV01	14NOV01	19NOV01	0
420	1	1	1	0	ADC	South Side Place Contin. Ftg's	14NOV01	14NOV01	15NOV01	15NOV01	1
260	1	1	1	0	ADC	North Side Backfill & Compact	14NOV01	14NOV01	20DEC01	20DEC01	20
340	1	1	1	0	ADC	East Side Place Fdn. Walls	20NOV01	20NOV01	20NOV01	20NOV01	0
430	1	1	1	0	ADC	South Side Strip Contin. Ftg's Forms	20NOV01	20NOV01	21NOV01	21NOV01	1
470	1	1	1	0	ADC	West Side Form Contin. Ftg's	21NOV01	21NOV01	26NOV01	26NOV01	1
440	1	1	1	0	ADC	South Side Reinforce Fdn. Walls	21NOV01	21NOV01	27NOV01	27NOV01	2
480	1	1	1	0	ADC	West Side Reinforce Contin. Ftg's	26NOV01	26NOV01	27NOV01	27NOV01	1
350	1	1	1	0	ADC	East Side Strip Fdn. Wall Forms	27NOV01	27NOV01	27NOV01	27NOV01	0
510	2	2	1	0	ADC	West Side Place Contin. Ftg's	27NOV01	28NOV01	28NOV01	29NOV01	1
450	1	1	1	0	ADC	South Side Form Fdn. Walls	28NOV01	28NOV01	28NOV01	28NOV01	0
410	1	1	1	0	ADC	East Side Backfill & Compact	28NOV01	28NOV01	20DEC01	20DEC01	13
460	1	1	1	0	ADC	South Side Place Fdn. Walls	29NOV01	29NOV01	29NOV01	29NOV01	0
520	1	1	1	0	ADC	West Side Strip Contin. Ftg's Forms	4DEC01	4DEC01	5DEC01	5DEC01	1
490	2	2	1	0	ADC	South Side Strip Fdn Wall Forms	5DEC01	6DEC01	5DEC01	6DEC01	0
530	1	1	1	0	ADC	West Side Reinforce Fdn. Walls	5DEC01	5DEC01	6DEC01	11DEC01	1
540	3	3	1	0	ADC	West Side Form Fdn. Walls	10DEC01	12DEC01	10DEC01	12DEC01	0
500	1	1	1	0	ADC	South Side Backfill & Compact	10DEC01	10DEC01	20DEC01	20DEC01	7
550	1	1	1	0	ADC	West Side Place Fdn. Walls	13DEC01	13DEC01	13DEC01	13DEC01	0
560	1	1	1	0	ADC	West Side Strip Fdn. Wall Forms	19DEC01	19DEC01	19DEC01	19DEC01	0
570	1	1	1	0	ADC	West Side Backfill & Compact	20DEC01	20DEC01	20DEC01	20DEC01	0
580	3	3	1	0	ADC	Reinforce S.O.G.	24DEC01	27DEC01	24DEC01	27DEC01	0
590	1	1	1	0	ADC	Place & Finish S.O.G.	31DEC01	31DEC01	31DEC01	31DEC01	0
650	5	5	1	0	ADC	Grade Parking Lot	2JAN02	9JAN02	28MAY02	4JUN02	82
720	3	3	1	0	ADC	Form Parking Lot	10JAN02	15JAN02	5JUN02	12JUN02	82
730	5	5	1	0	ADC	Reinforce Parking Lot	14JAN02	21JAN02	6JUN02	13JUN02	82
740	1	1	1	0	ADC	Place & Finish Parking Lot	15JAN02	22JAN02	13JUN02	17JUN02	85
750	1	1	1	0	ADC	Strip Forms - Parking Lot	21JAN02	24JAN02	19JUN02	19JUN02	85
760	2	2	1	0	ADC	Place Curbs	28JAN02	29JAN02	20JUN02	24JUN02	82

Output 10-1
(cont.)

All Star Developments PRIMAVERA PROJECT PLANNER Sample Construction Project

REPORT DATE 24MAR00 RUN NO. 34 START DATE 1OCT01 FIN DATE 1JUL02
11:40
Schedule Update Report DATA DATE 1OCT01 PAGE NO. 3

Joe Bob Steel

ACTIVITY ID	ORIG DUR	REM DUR	CAL	%	CODE	ACTIVITY DESCRIPTION	EARLY START	EARLY FINISH	LATE START	LATE FINISH	TOTAL FLOAT
610	12	12	1	0	JBS	Erect Structural Steel	15JAN02	14FEB02	28JAN02	14FEB02	7
680	3	3	1	0	JBS	Ceiling Grid	13MAR02	18MAR02	13MAR02	18MAR02	0
690	6	6	1	0	JBS	Frame Interior Walls	19MAR02	27MAR02	19MAR02	27MAR02	0
800	8	8	1	0	JBS	Hang Drywall	18APR02	1MAY02	18APR02	1MAY02	0
810	16	16	1	0	JBS	Float, Tape & Paint	2MAY02	30MAY02	2MAY02	30MAY02	0
840	6	6	1	0	JBS	Install Cabinets	3JUN02	11JUN02	3JUN02	11JUN02	0
820	2	2	1	0	JBS	Hang & Trim Interior Doors	3JUN02	4JUN02	19JUN02	20JUN02	10
870	3	3	1	0	JBS	Ceiling Tiles	24JUN02	26JUN02	24JUN02	26JUN02	0
880	1	1	1	0	JBS	Carpet	27JUN02	27JUN02	27JUN02	27JUN02	0
890	1	1	1	0	JBS	Final Cleaning	1JUL02	1JUL02	1JUL02	1JUL02	0

All Star Developments PRIMAVERA PROJECT PLANNER Sample Construction Project

REPORT DATE 24MAR00 RUN NO. 34 START DATE 1OCT01 FIN DATE 1JUL02
11:40
Schedule Update Report DATA DATE 1OCT01 PAGE NO. 4

John Doe

ACTIVITY ID	ORIG DUR	REM DUR	CAL	%	CODE	ACTIVITY DESCRIPTION	EARLY START	EARLY FINISH	LATE START	LATE FINISH	TOTAL FLOAT
400	5	5	1	0	JED	Underground Plumbing	30OCT01	6NOV01	13DEC01	20DEC01	25
780	2	2	1	0	JED	Final Grading	30JAN02	31JAN02	25JUN02	26JUN02	82
770	1	1	1	0	JED	Stripe Parking Lot	30JAN02	30JAN02	1JUL02	1JUL02	85
790	2	2	1	0	JED	Landscape	4FEB02	5FEB02	27JUN02	1JUL02	82
660	5	5	1	0	JED	Mechanical Rough-in	27FEB02	6MAR02	27FEB02	12MAR02	0
670	8	8	1	0	JED	Overhead Electrical Rough-in	27FEB02	12MAR02	27FEB02	12MAR02	0
710	12	12	1	0	JED	Electrical Rough-in	28MAR02	17APR02	28MAR02	17APR02	0
700	3	3	1	0	JED	Plumbing Top-out	28MAR02	2APR02	19JUN02	20JUN02	45
830	3	3	1	0	JED	Mechanical Trim	3JUN02	5JUN02	13JUN02	18JUN02	7
850	4	4	1	0	JED	Electrical Trim	12JUN02	18JUN02	12JUN02	18JUN02	0
860	2	2	1	0	JED	Plumbing Trim	19JUN02	20JUN02	19JUN02	20JUN02	0

All Star Developments PRIMAVERA PROJECT PLANNER Sample Construction Project

REPORT DATE 24MAR00 RUN NO. 34 START DATE 1OCT01 FIN DATE 1JUL02
11:40
Schedule Update Report DATA DATE 1OCT01 PAGE NO. 5

WIlliam Roe

ACTIVITY ID	ORIG DUR	REM DUR	CAL	%	CODE	ACTIVITY DESCRIPTION	EARLY START	EARLY FINISH	LATE START	LATE FINISH	TOTAL FLOAT
600	22	22	1	0	WER	CMU	7JAN02	12FEB02	7JAN02	12FEB02	0
620	6	6	1	0	WER	Roofing / Flashing	18FEB02	26FEB02	18FEB02	26FEB02	0
640	3	3	1	0	WER	Overhead Doors	18FEB02	20FEB02	21FEB02	26FEB02	3
630	1	1	1	0	WER	Exterior Glass & Glazing	18FEB02	18FEB02	26FEB02	26FEB02	5

Output 10-1
(cont.)

Updating the Schedule in Primavera Project Planner

The actual activity information is entered on the activity form. Immediately adjacent to the *ES* and *EF* fields are blank boxes. If these boxes are clicked on, a check will be entered in the box and the *ES* field will change to *AS*, to designate that that field now contains the actual start. The *EF* will turn to *AF* to designate the actual finish date. Another item that must be entered is the percent compete in the *Pct* field. The percent complete needs to be based on actual field progress. Using any measurement other than actual progress will result in questionable results. In Screen Image 10-14, Activity 10 has been given an actual start and finish date and has been statused at 100 percent complete. In Screen Image 10-15, Activity 20 has an actual start date and has been statused at 50 percent complete. Finally, Screen Image 10-16 shows the actual status for Activity 30. When the activity is incomplete, Primavera Project Planner will establish the remaining duration if one is not manually entered. The remaining duration is found by multiplying the percent remaining times the original duration.

Once the schedule has been updated with the actual start, finish, and percent complete information, it is necessary to recalculate the schedule. This will modify the early and late dates on all of the activities that have not begun or ended, to reflect the impact of the started and completed activities. The schedule can be calculated by clicking on the schedule icon, pressing the *F9* key, or selecting the *Tools/Schedule* command from the command line. Any of these actions will open the Schedule window as shown in Screen Image 10-17. The *Data date* up until this point has always been the starting date of the project. That assumption changes once the schedule has been updated. After the schedule has been updated, this field needs to have the date that the actual field progress was measured. This date will move the

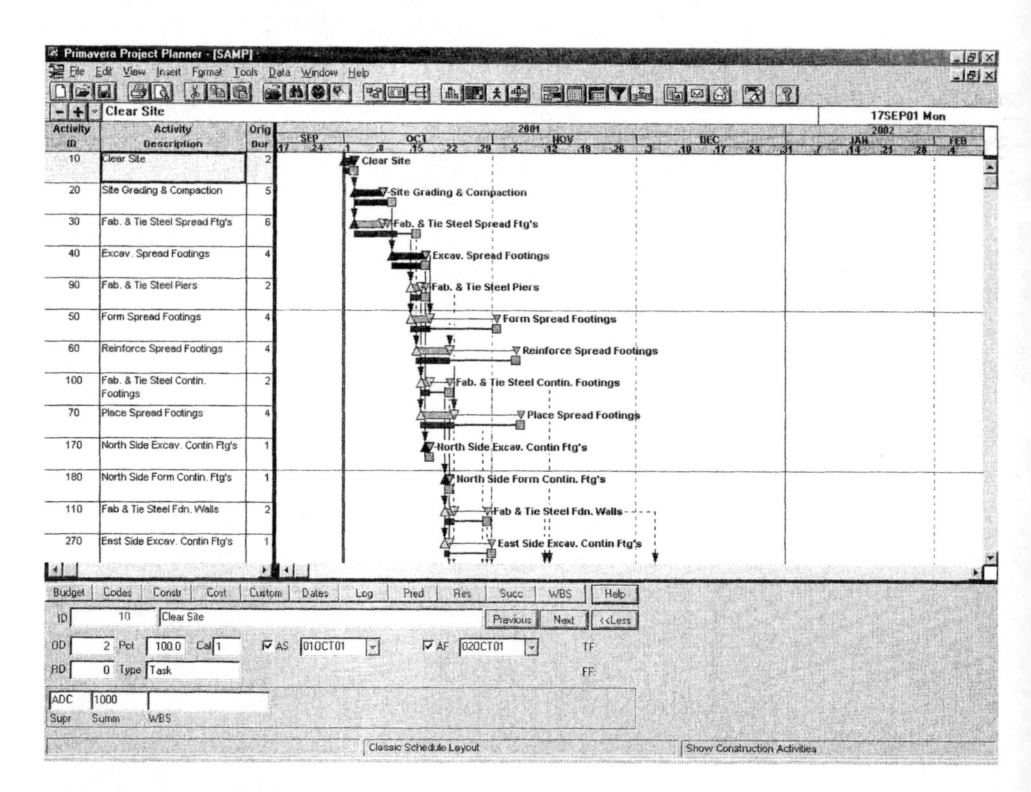

Screen Image 10-14
Activity 10 Status

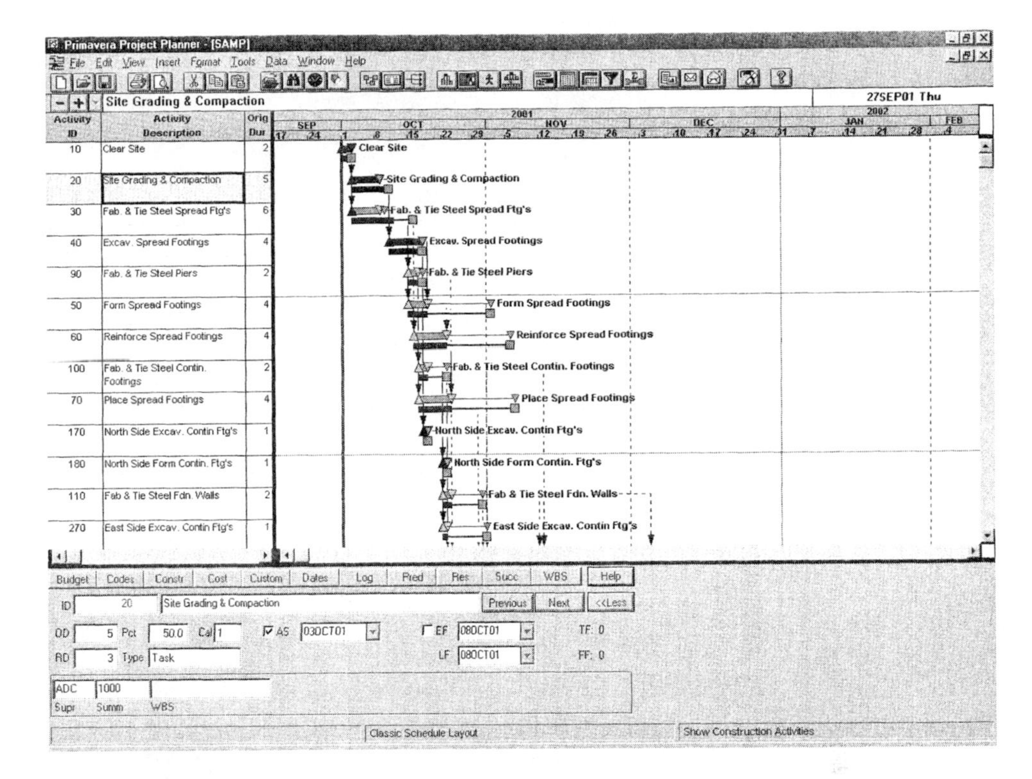

Screen Image 10-15
Activity 20 Status

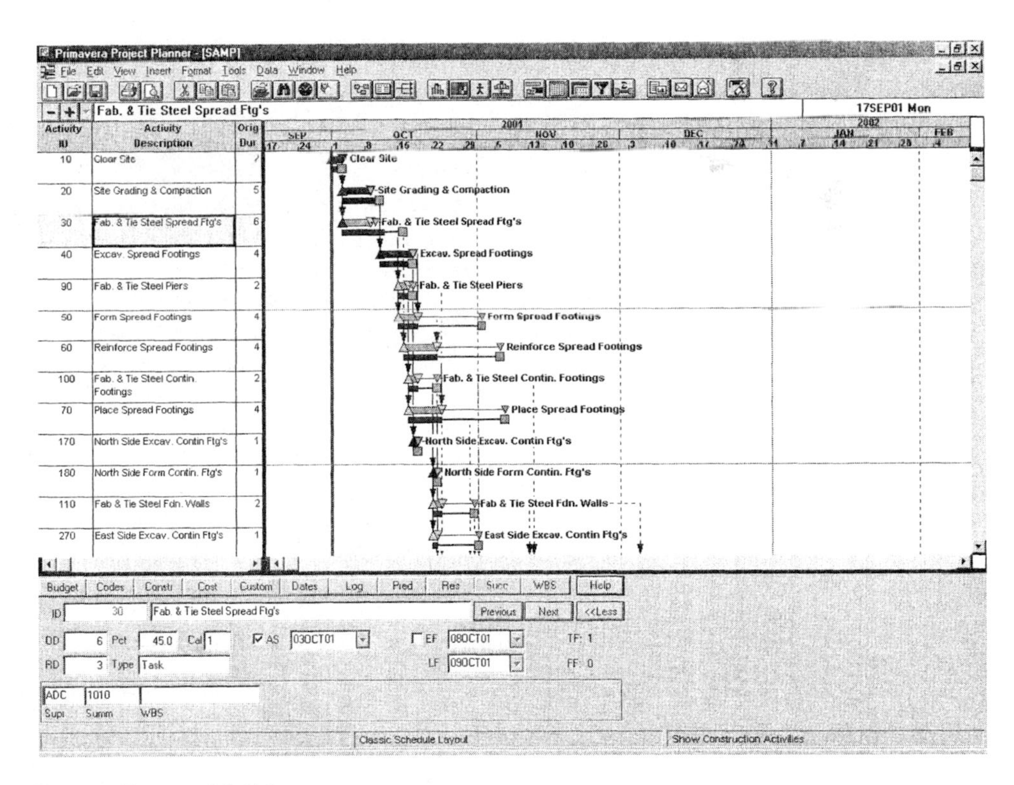

Screen Image 10-16
Activity 30 Status

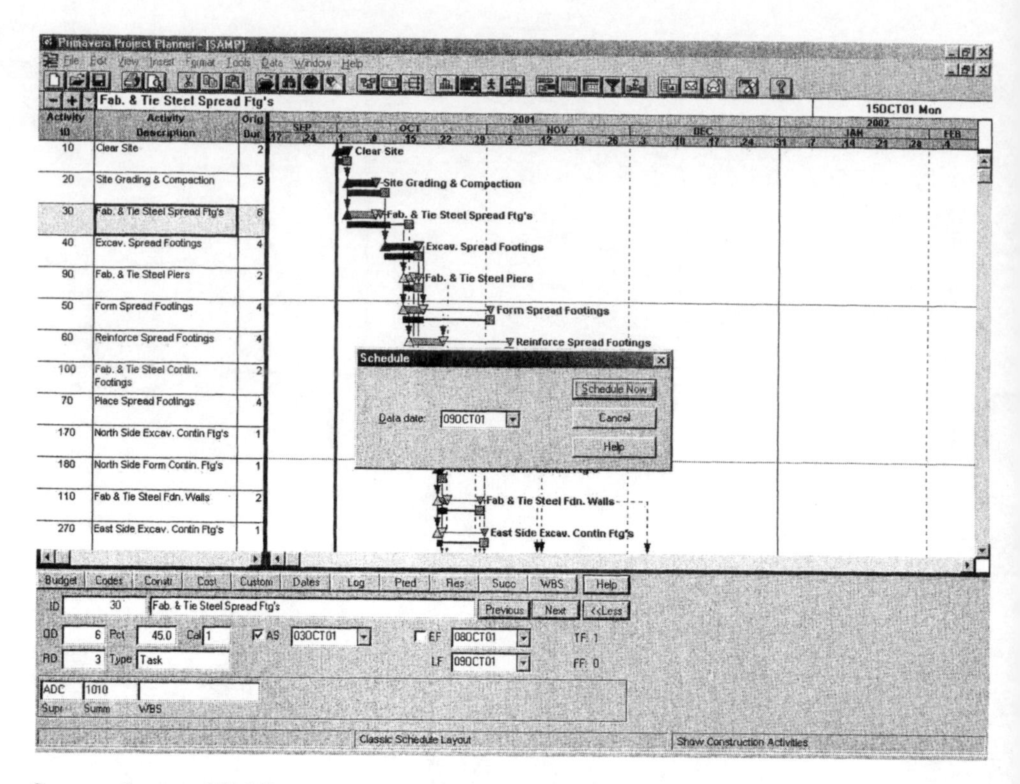

Screen Image 10-17
Recalculating the Schedule

status bar from the first day of the project to the appropriate date. In Screen Image 10-18 the *Data date* is being changed to 9 October of 2001. After the data date has been entered, the *Schedule Now* button can be clicked to recalculate the schedule.

After the project has been statused, all incomplete and unstarted activities are to the right of the status line. This is done to show that the earliest that an activity can begin is at the status bar. Even if an activity is not yet started and behind schedule, the early start date will move to the data date. It would be illogical to have early start dates that are in the past. The same holds true for activities that are ahead of schedule; the completed portion of the activity will be to the left of the status bar and the uncompleted portion will be to the right. Once again, the current schedule needs to show the most accurate information. This feature clearly shows the advantages of using target schedules. By showing the target and current schedule, it can be shown how the activities on the current schedule relate to the original plan. In Screen Image 10-19 this relationship is shown. For example, Activity 40 on that screen image is behind schedule in relation to the original plan. This is shown by the start of the target bar being to the left of the start of the current bar. In this schedule, it is apparent that things have not been going as planned and some investigation as to the reason needs to be performed. Some items that could contribute to the delays could be low productivity, late material deliveries, labor shortage, weather, or poor supervision. From an investigation into the cause of the delays a remediation plan can be developed. If no changes are made, the project will finish behind schedule. By identifying deviations as early as possible, the potential for getting back on schedule is maximized. This also points out the need to measure progress on a regular basis. The time interval between measuring progress is a function of the size and duration of the project. The shorter the duration of the project, the shorter the amount of time between measuring progress should be. Conversely, the longer the project duration, the longer the interval between measuring progress can be. Typically, the increment between measuring progress is one to 2 weeks. It is also important to make sure that progress has been accurately and constantly measured. The holding back of progress measurement for a "rainy day" needs to be discouraged. This process, if done

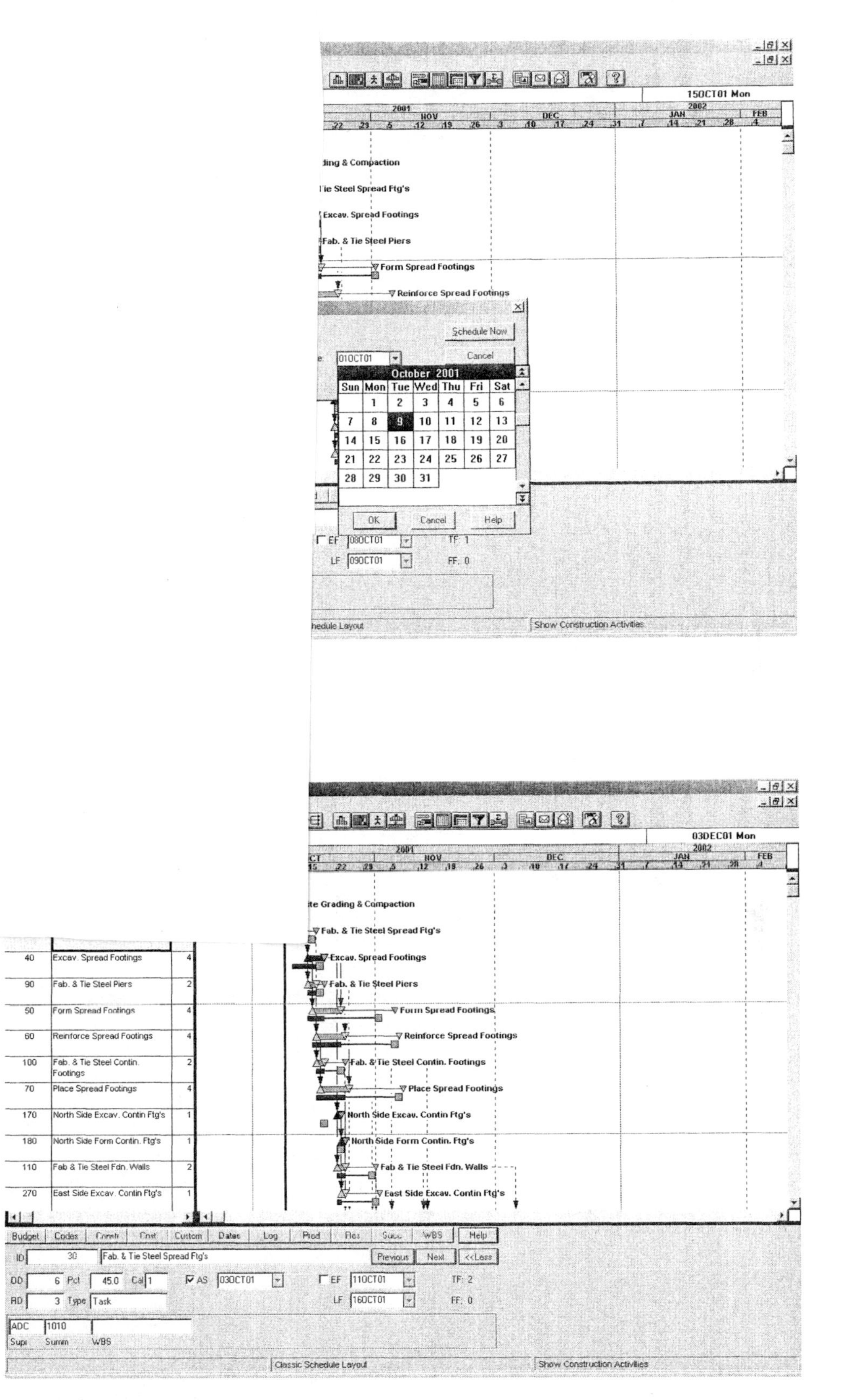

Screen Image 10-19
Updated Project Schedule

by a number of persons on the project will impact the percent complete and show the project behind schedule when it may be on or ahead of schedule. The project manager may implement some recovery procedures that are expensive when nothing at all may have been required.

Suggested Exercises

1. Using the construction schedule built up in earlier chapters, create a target schedule. This schedule will serve as the baseline schedule.
2. Take the first 20 activities and status them. Make some of them complete and some of them partially completed. A good way to practice this exercise is to status the even-numbered activities as starting 2 days late and the odd-numbered activities as starting 2 days early.

Resource Loading

KEY TERMS

Resource A construction element or asset that is needed directly or indirectly in the construction process.

Peak Craft Personnel The maximum number of persons who will be working on the project.

Equivalent Craft Personnel The mathematically determined number of persons working during an increment of time.

Resources

Resources are assets used in the construction process. Examples of construction resources are materials, craft labor, and equipment. The greatest challenge in construction management is to bring together all of the required resources in the correct quantity at the best time. The early and late dates derived from the forward and backward pass had an underlying assumption that the contractor can have any construction resource at any time and in any quantity. That assumption is impractical and cost prohibitive. The early and late dates give the construction manager a range of possible dates in which to perform the work. The objective is for the task to begin and end so that the construction resources are used in the most effective and efficient manner.

Contractors who have their craft personnel waiting for materials are clearly wasting money. This nonproductive time adds cost to the project and adversely impacts the morale of the workforce. If this happens on a regular basis, it will become an expectation, and no one will pay any attention to the scheduled sequence of work. This attitude will permeate the field craft persons, subcontractors, and vendors, which will further delay the project. At the opposite extreme, having too much material and not enough of the proper craft personnel can have the same results. If the materials arrive at a pace faster than they can be installed, the materials must be stored at the project site. Then the stored materials must be moved from storage when needed for installation. This storing and moving adds unnecessary cost to the project and increases the likelihood of damage and theft. It is up to the job site management team to strike the right balance so that the arrival of the resources is orchestrated to maximize productivity and profits.

The construction network diagram, in addition to detailing how the project will be assembled, can be used to determine when and in what quantity resources are needed. If all of the construction project resources are coordinated and brought together at the right time, the likelihood for completing the project on time and within budget is maximized. The process of attaching resources to specific construction activities is referred to as resource loading. Prior to loading any resources, decisions need to be made concerning the level of detail required to effectively plan and monitor the use of the resources. Developing a resource plan to a level of detail that cannot be compared with actual data will not serve any purpose once the construction process commences. In the simplest terms, if the financial system can only track costs to a certain level, then planning resources to a greater level of detail only adds work and will not enhance decision making. For example, the project may be small enough and simple enough that only the total number of craft persons working on the project needs to be tracked and monitored. On the other hand, the project may be large enough and complicated enough that the individual crafts requirements need to be known. In that scenario the number of masons, carpenters, pipe fitters, laborers, boilermakers, and the like would need to be known on a daily or weekly basis. By knowing this information, the hiring process can begin early enough to ensure that it will be organized and that properly trained craft persons can be recruited for the project. Finally, by analyzing the resource requirements, the construction process may be changed so that the resources can be used more efficiently. The desire is to bring the needed craft personnel onto the project, have them execute the work, and then have them move on to the next project. If the construction project has a person coming on the project, being furloughed, and then brought back on the project, productivity will suffer.

Resource Loading

The first step in loading resources is to review and organize the project estimate. The estimate needs to be organized in such a fashion that the quantity, unit cost, or total cost can easily be found. Figure 11-1 is the estimate for Activity 600, CMU, from the sample project introduced in Chapter 3. Using that sample coupled with the information about crew M1and M2 the following can be derived:

Activity 600 Crew M1
Placing Block 04-221-15

Count	Craft	Wage Rate	$/Crew Hour
3	Mason	$14.75	$44.25
2	Mason Helper	$9.75	$19.50
5	TOTAL		$63.75
	Crew Average Wage Rate	$12.75	

Activity 600 Crew M2
Rubbing Block 04-710-11

Count	Craft	Wage Rate	$/Crew Hour
2	Mason Helper	$9.75	$19.50
2	TOTAL		$19.50
	Crew Average Wage Rate	$9.75	

Description	Workhours
Activity 600–Place Block	
Masons	631
Mason Helpers	421
Activity 600–Rubbing Block	
Mason Helpers	145

ACTIVITY 600 - CMU

Project: Office Warehouse
Location: Anywhere
Items Activity 600

Estimate: 100
Sheet: 6 of 75
By / Check: LHF/ADC

Cost Code	Description	Q.T.O.	Waste Factor	Purch. Qty.	Unit	Crew	Prod Rate	Crew Average Wage Rate	Work Hours	Unit Cost					Total Cost				Total
										Labor	Material	Equipment	Labor	Sub	Labor	Material	Equipment	Sub	
04-221-15	Concrete Block	8154	0.05	8562	Each	M1	0.129	12.75	1052	$1.64	$1.05	$0.14			$13,413.00	$8,990.10	$1,100.79	$0.00	$23,503.89
04-221-35	Mortar	16	0.3	21	CY	W/ Block					$62.50	$65.00				$1,312.50	$1,040.00		$2,352.50
04-221-45	Reinforcing	6	0.05	6	MLF	W/ Block					$99.75					$598.50			$598.50
04-710-11	Rub Block	145	0.2	174	SQS	M2	1	9.75	145	$9.75	$2.00				$1,413.75	$348.00			$1,761.75
	Total														$14,826.75	$11,249.10	$2,140.79	$0.00	$28,216.64

Figure 11-1
Activity 600 Estimate

With crew M1 three-fifths of the workhours are allocated to masons, since there are three masons on the crew. The remaining two-fifths of the workhours are designated for mason helpers. These workhours can now be spread over the duration of the activity to develop a profile for the number and utilization of masons and helpers. In Figure 11-2, the workhours have been spread over the duration of the activity using a straight-line distribution. Using that distribution, the mason workhours per day are found by taking the 631 mason workhours and dividing it by the 22-day duration. That yields roughly 28.7 mason workhours per day. In this example the equivalent masons are found by dividing the number of mason workhours in a day by the length of the standard workday. Using a 10-hour workday there are roughly three masons required per day. The same calculation can be performed with regard to any craft to determine how many persons of that craft would be required on a specific day.

The craft requirements can be graphically displayed and tracked. Figure 11-3 is a graphic presentation of the required mason workhours. This graph represents the planned incremental and cumulative mason workhours. After construction commences the actual information could be added to the graph. If the cumulative actual workhours line is above the planned, that would mean that the workhours are being consumed faster than planned. That is not a bad scenario as long as progress is being earned faster than planned. If the

	Day Count											
Description	1	2	3	4	5	6	7	8	9	10	11	12
Placing Block 04-221-15												
Crew M1												
Masons	28.7	28.7	28.7	28.7	28.7	28.7	28.7	28.7	28.7	28.7	28.7	28.7
Mason Helpers	19.1	19.1	19.1	19.1	19.1	19.1	19.1	19.1	19.1	19.1	19.1	19.1
Rubbing Block 04-71011												
Crew M2												
Mason Helpers	6.6	6.6	6.6	6.6	6.6	6.6	6.6	6.6	6.6	6.6	6.6	6.6
Masons												
Work Hours	28.7	28.7	28.7	28.7	28.7	28.7	28.7	28.7	28.7	28.7	28.7	28.7
Cummulative Work Hours	28.7	57.4	86.0	114.7	143.4	172.1	200.8	229.5	258.1	286.8	315.5	344.2
Equivalent	3.0	3.0	3.0	3.0	3.0	3.0	3.0	3.0	3.0	3.0	3.0	3.0
Mason Helpers												
Work Hours	25.7	25.7	25.7	25.7	25.7	25.7	25.7	25.7	25.7	25.7	25.7	25.7
Cummulative Work Hours	25.7	51.5	77.2	102.9	128.6	154.4	180.1	205.8	231.5	257.3	283.0	308.7
Equivalent	3	3	3	3	3	3	3	3	3	3	3	3

	Day Count									
Description	13	14	15	16	17	18	19	20	21	22
Placing Block 04-221-15										
Crew M1										
Masons	28.7	28.7	28.7	28.7	28.7	28.7	28.7	28.7	28.7	28.7
Mason Helpers	19.1	19.1	19.1	19.1	19.1	19.1	19.1	19.1	19.1	19.1
Rubbing Block 04-710-11										
Crew M2										
Mason Helpers	6.6	6.6	6.6	6.6	6.6	6.6	6.6	6.6	6.6	6.6
Masons										
Work Hours	28.7	28.7	28.7	28.7	28.7	28.7	28.7	28.7	28.7	28.7
Cummulative Work Hours	372.9	401.5	430.2	458.9	487.6	516.3	545.0	573.6	602.3	631.0
Equivalent	3.0	3.0	3.0	3.0	3.0	3.0	3.0	3.0	3.0	3.0
Mason Helpers										
Work Hours	25.7	25.7	25.7	25.7	25.7	25.7	25.7	25.7	25.7	25.7
Cummulative Work Hours	334.5	360.2	385.9	411.6	437.4	463.1	488.8	514.5	540.3	566.0
Equivalent	3	3	3	3	3	3	3	3	3	3

Figure 11-2
Workhour Distribution for Masons and Helpers

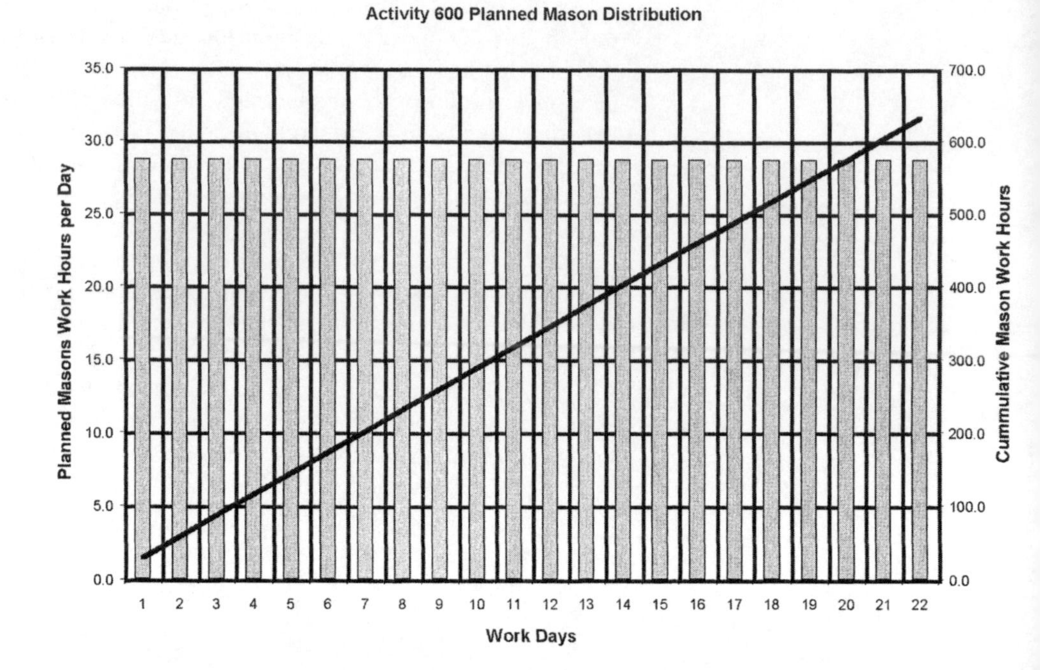

Figure 11-3
Activity 600 Straight-line Mason Distribution

workhours are being consumed faster than planned and progress is going as planned, that would mean that the productivity rate is slower than estimated and there is a clear potential for a cost overrun on that activity.

The utilization of workhours can be planned using distributions other than straight-line. In Figure 11-4 a bell distribution is shown. In that figure the shape of the cumulative curve is referred to as an S shaped curve. The previous examples are models of how resources

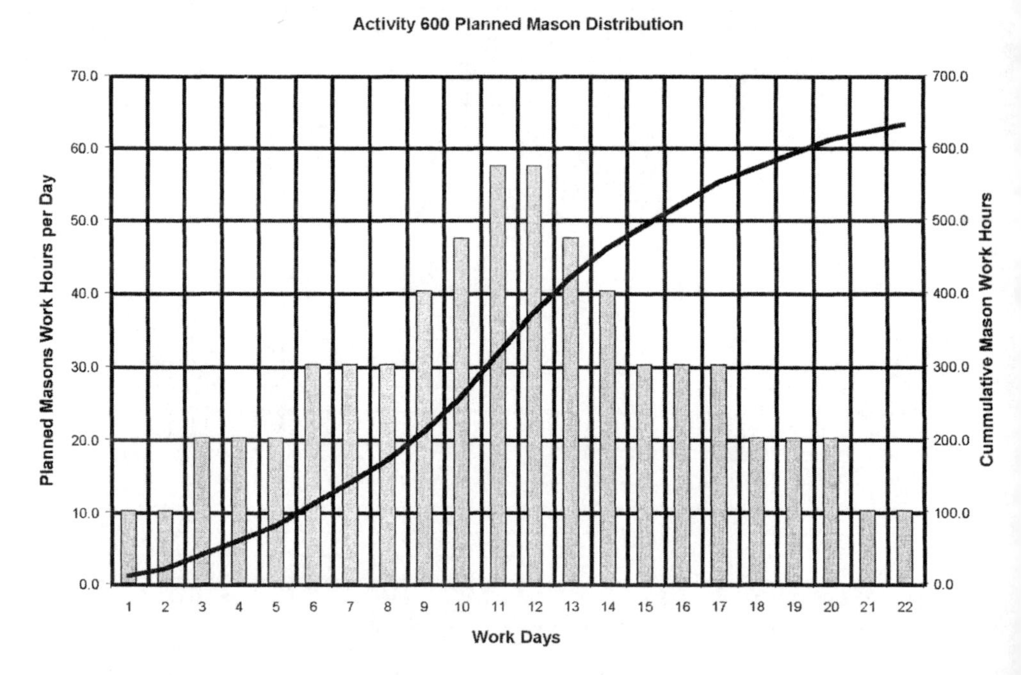

Figure 11-4
Activity 600 Bell-Shaped Mason Distribution

could be distributed. This process can be performed on every activity on the schedule and then the like crafts could be combined and a planned distribution for all of the crafts could be developed. The shape of these distributions could become the basis for when to start an activity in relation to its early start and late start. By moving the start dates around, the cumulative distribution can be changed. The distributions for all crafts will vary throughout the project. For example, the concrete-finishers would be early in the project while the landscapers would be at the end of the project. Regardless of when the craft people are needed on the project, it is important for their cumulative curve to be smooth. If there are inclines and then flat places followed by inclines, that means that people are being brought onto the project followed by inactivity for that craft and then that craft is being brought back onto the project. This practice is very poor and leads to low productivity and poor worker morale. With a smooth curve the particular craft is being brought onto the project, is performing a segment of the work, and then is being sent to the next project. A sequence such as this is efficient and keeps craft people busy. As one can see, the possibility of performing a resource distribution, accumulation, and presentation manually would be virtually impossible. This is one of the reasons that computerized scheduling software is so popular.

Defining Resources

Primavera Project Planner manages resource information very easily. The first step in the process requires determining which resources are to be loaded and what level of detail is required to effectively plan and monitor the project. One important item to remember is that the actual information needs to be gathered at the same level of detail as the planned. If the actual data cannot be gathered by craft, planning by craft is of little value. In that situation, once construction started the actual resource utilization and its planned usage could not be compared. Without this, the ability to identify any deviations is lost. If the only actual information being gathered is total workhours, then only one labor resource should be created, and that distribution would be field labor rather than a specific craft.

For example purposes seven labor resources will be defined and manipulated. These resources will represent the crafts that will be directly hired by the general contractor. If a subcontractor is performing the work, resource information is virtually impossible to get. These people have no vested interest in providing workhour information, and if they did it most likely would be understated.

The first step in loading resources into a Primavera Project Planner schedule is to define the resources that are to be tracked. This is done via the *Resources* window. Selecting the *Data* command followed by the *Resources* option, as shown in Screen Image 11-1, opens this window. The *Resources* window shown in Screen Image 11-2 has three distinct input areas: *Resources, Limits* and *Prices*. The *Resources* table is where the individual resources are defined and their behavior is specified. To add a resource to this box, click the plus button in the top left-hand corner of the input box. The *Resource* column is for defining a unique resource descriptor for each resource. This descriptor is limited to no more than eight alphanumeric characters. The *Units* field is where the units for that resource are defined. If the resource being added is labor, then workhours would be an appropriate unit. The input for the driving column is a check mark. If a resource is defined as driving, its availability and quantity dictate the activity duration rather than the specified duration. For example, if a resource was defined as driving, and the resource availability is 20 workhours per day and there were 200 workhours remaining, the remaining duration would automatically be 10 days regardless of the number that was entered when the activity was input. The *Base* column is where the resource is tied to a specific calendar. The *Description* column is where the descriptor is defined.

The *Limits* input box is where the availability of a specific resource is defined. This box allows the user to specify the available quantity of a specific resource for a given time period. This will allow the user to determine the impact of resource shortages as well as a mechanism

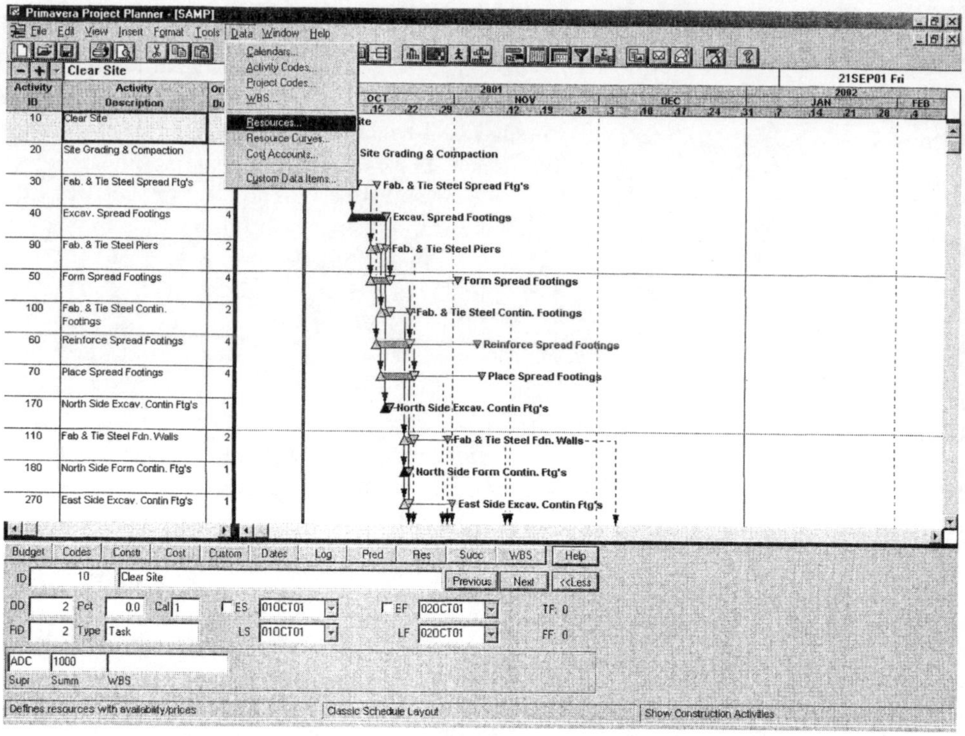

Primavera Project Planner - [SAMP]

File Edit View Insert Format Tools Data Window Help

Data menu: Calendars... | Activity Codes... | Project Codes... | WBS... | Resources... | Resource Curves... | Cost Accounts... | Custom Data Items...

Clear Site — 21SEP01 Fri

Activity ID	Activity Description	Orig Dur
10	Clear Site	
20	Site Grading & Compaction	
30	Fab. & Tie Steel Spread Ftg's	
40	Excav. Spread Footings	4
90	Fab. & Tie Steel Piers	2
50	Form Spread Footings	4
100	Fab. & Tie Steel Contin. Footings	2
60	Reinforce Spread Footings	4
70	Place Spread Footings	4
170	North Side Excav. Contin Ftg's	1
110	Fab & Tie Steel Fdn. Walls	2
180	North Side Form Contin. Ftg's	1
270	East Side Excav. Contin Ftg's	1

Budget | Codes | Constr | Cost | Custom | Dates | Log | Pred | Res | Succ | WBS | Help

ID 10 Clear Site — Previous | Next | <<Less

OD 2 Pct 0.0 Cal 1 ES 01OCT01 EF 02OCT01 TF: 0

RD 2 Type Task LS 01OCT01 LF 02OCT01 FF: 0

ADC 1000 — Supr Summ WBS

Defines resources with availability/prices | Classic Schedule Layout | Show Construction Activities

Screen Image 11-1
Opening Resources Window

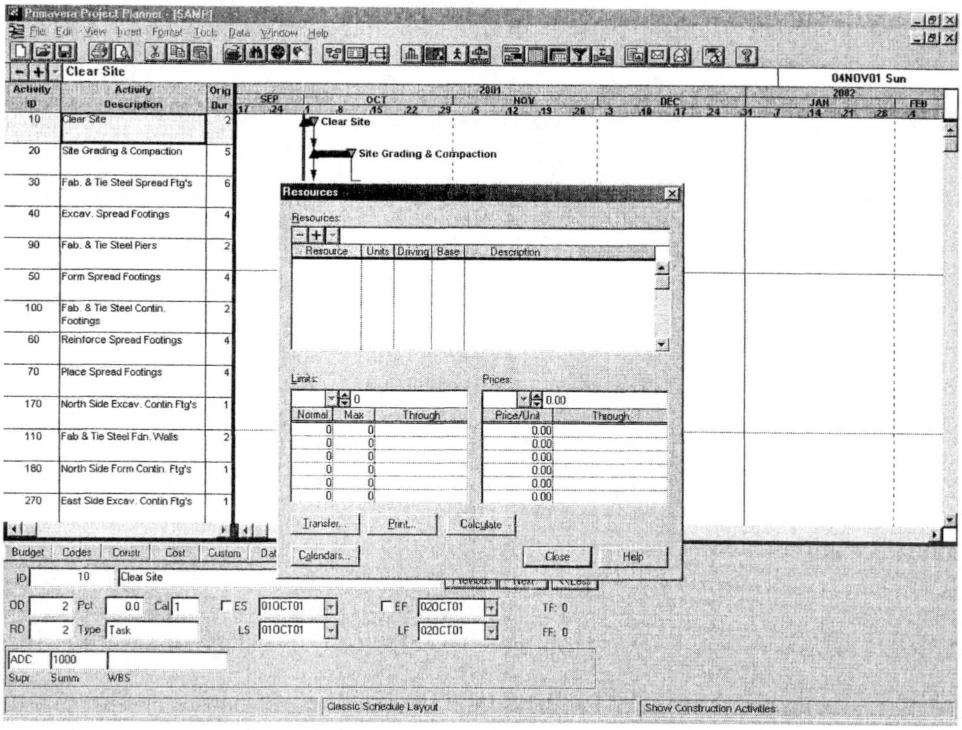

Primavera Project Planner - [SAMP]

File Edit View Insert Format Tools Data Window Help

Clear Site — 04NOV01 Sun

Activity ID	Activity Description	Orig Dur
10	Clear Site	2
20	Site Grading & Compaction	5
30	Fab. & Tie Steel Spread Ftg's	6
40	Excav. Spread Footings	4
90	Fab. & Tie Steel Piers	2
50	Form Spread Footings	4
100	Fab. & Tie Steel Contin. Footings	2
60	Reinforce Spread Footings	4
70	Place Spread Footings	4
170	North Side Excav. Contin Ftg's	1
110	Fab & Tie Steel Fdn. Walls	2
180	North Side Form Contin. Ftg's	1
270	East Side Excav. Contin Ftg's	1

Resources

Resources: Resource | Units | Driving | Base | Description

Limits: 0 — Normal | Max | Through

Prices: 0.00 — Price/Unit | Through

Transfer... | Print... | Calculate | Calendars... | Close | Help

ID 10 Clear Site

OD 2 Pct 0.0 Cal 1 ES 01OCT01 EF 02OCT01 TF: 0

RD 2 Type Task LS 01OCT01 LF 02OCT01 FF: 0

ADC 1000 — Supr Summ WBS

Classic Schedule Layout | Show Construction Activities

Screen Image 11-2
Blank Resources Window

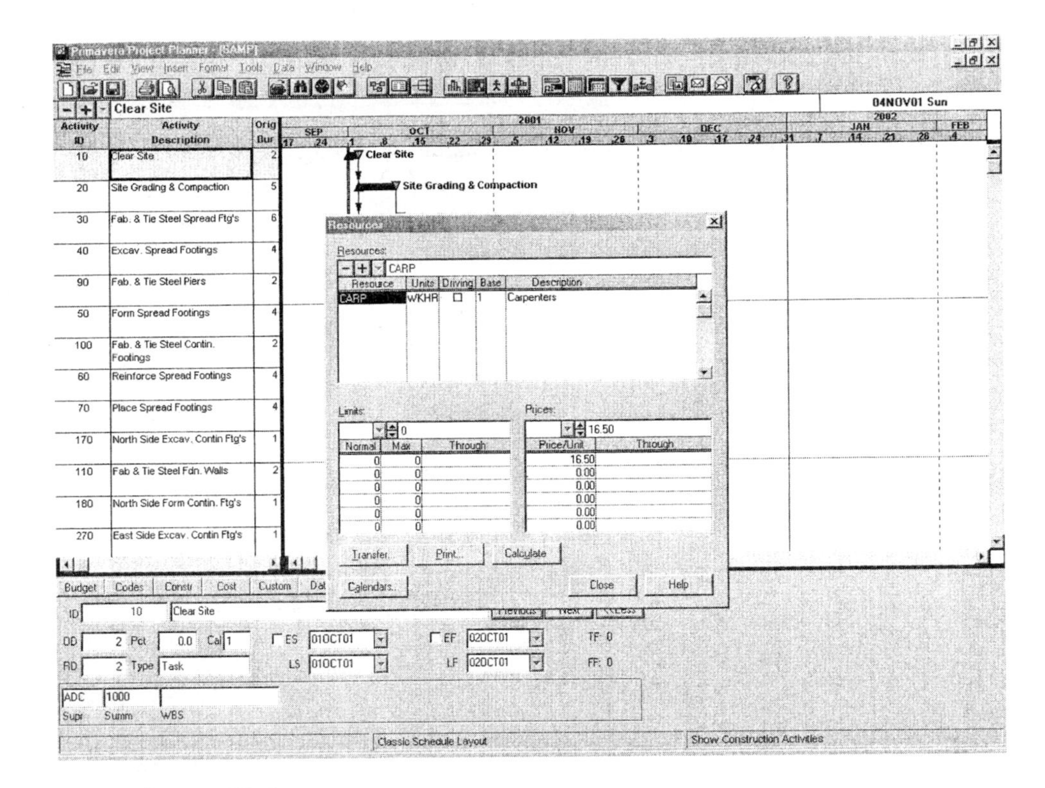

Screen Image 11-3
Defining Carpenter Resource

to smooth out the resource curves. The *Normal* and *Max* limit is where the desired and maximum availability of a resource is specified. It is important that the units be consistent with the units specified on the *Resource* table. For example, if three carpenters is the *Normal* limit and workhours are the unit and the standard work-day is 10 hours, then the *Normal* limit would be set to 30. The *Through* column allows the user to specify time periods for which the limits will be accurate. The final input box addresses the unit price for the individual resources. By inputting the unit pricing information, Primavera Project Planner will automatically calculate the total cost for each of the resources. This will save time when inputting the costs for the activities. For example, if it is anticipated that carpenters are to get a 25 cent per hour pay raise on March 1 to $16.75 per hour, that information would be loaded into this box. The first line would contain *Price/Unit* of $16.50 through 1 March. The second line would contain $16.75 in the *Price/Unit* column and the project completion date as the through date.

Screen Image 11-3 demonstrates how the carpenter resource would be added to the *Resources* window. In that screen image, CARP is the active resource with a wage rate of $16.50 per carpenter workhour throughout the entire project. Since, there are no *Limits* specified, that means there is an infinite supply of available carpenters. Screen Image 11-4 shows the resource dictionary with seven different labor resources or crafts defined. Resources other than labor can be defined. However, their units must be consistent. For example, a backhoe could be specified as a resource and tied to every activity that requires the use of a backhoe. This would help the contractor determine how many backhoes would be required and when. Materials could also be tracked as a resource, but once again, unit consistency would be essential. Generic materials could not be a resource since the units would be a combination of a variety of items. However, concrete could be a resource and cubic yards would be the units. If resources are added without units, the unit would then become dollars or cost.

The *Transfer* button allows for the resources to be copied from another project. The *Calculate* button recalculates the cost if the unit price is changed. For example, if the unit cost for carpenters is changed to $16.75, the *Calculate* button would search for every activity

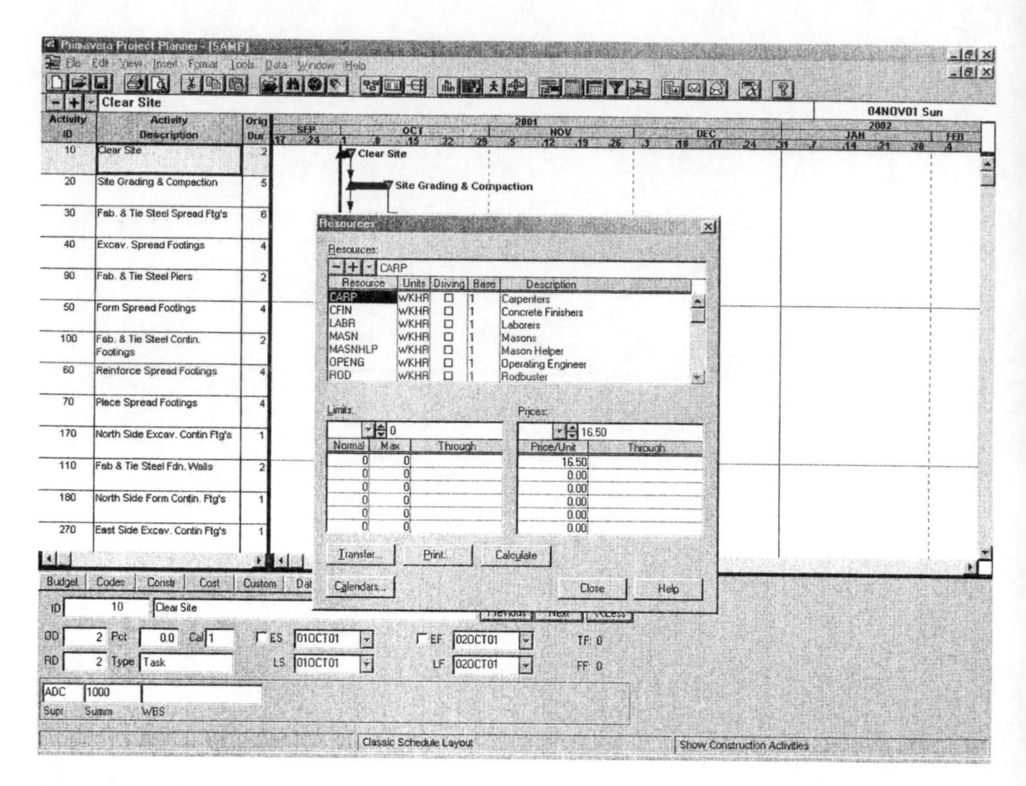

Screen Image 11-4
Multiple Resources Defined

with a carpenter resource and multiply the number of required resources by the new unit cost and update the cost.

The *Calendars* button allows the user to change the calendar that is attached to the individual resources. The *Print* button can be used to print out a list of all of the resources along with their cost and availability. Output 11-1 is a sample printout of the resource dictionary. Once all of the resources have been added, clicking the *Close* button closes the *Resources* window, which will return the program to the main input screen. When this is done the user will be prompted concerning updating the impact of the changes on the *Resources* window. Screen Image 11-5 is an example of that prompt.

Defining Cost Accounts

The contractor's chart of accounts is composed of the cost codes that are used to identify elements of construction. The chart of accounts should have been used in the preparation of the estimate and should be used to capture actual cost during construction. By being consistent between the estimate and the actual construction cost, a mechanism exists to compare the planned cost with the actual cost. This consistency should be applied from project to project. This allows for the gathering of historical data and the comparison of historical costs to current costs.

The chart of accounts needs to be established prior to preparing the estimate; it should simply be an enhancement of the chart of accounts that has been used on all of the previous projects. Typically some digit in the cost code is used to designate if the costs are for labor, material, equipment, or subcontractors. Using the following cost code as an example, a trailing digit is used to designate whether it is a labor, materials, equipment, or subcontract cost element. In the scheme shown below a trailing 1 is for labor, a trailing 2 is for materials, a trailing 3 is for equipment, and a trailing 4 is for subcontract costs.

```
---------------------------------------------------------------------------------
                          PRIMAVERA PROJECT PLANNER

Date 29MAR00              -----SUMMARY OF RESOURCES-----                   Page   1

SAMP - Sample Construction Project
---------------------------------------------------------------------------------
CARP       Carpenters                                  WKHR  Base calendar = 1

                            Limits
             Normal          Max           Through
           ----------     ----------     ----------

                            Price
                      Per Unit       Through
                     ----------     ----------
                         16.50

CFIN       Concrete Finishers                          WKHR  Base calendar = 1

                            Limits
             Normal          Max           Through
           ----------     ----------     ----------

                            Price
                      Per Unit       Through
                     ----------     ----------
                         12.75

LABR       Laborers                                    WKHR  Base calendar = 1

                            Limits
             Normal          Max           Through
           ----------     ----------     ----------

                            Price
                      Per Unit       Through
                     ----------     ----------
                         10.75

MASN       Masons                                      WKHR  Base calendar = 1

                            Limits
             Normal          Max           Through
           ----------     ----------     ----------

                            Price
                      Per Unit       Through
                     ----------     ----------
                         14.75

MASNHLP    Mason Helper                                WKHR  Base calendar = 1

                            Limits
             Normal          Max           Through
           ----------     ----------     ----------

                            Price
                      Per Unit       Through
                     ----------     ----------
                          9.75

OPENG      Operating Engineer                          WKHR  Base calendar = 1

                            Limits
             Normal          Max           Through
           ----------     ----------     ----------

                            Price
                      Per Unit       Through
                     ----------     ----------
                         15.00

ROD        Rodbuster                                   WKHR  Base calendar = 1

                            Limits
             Normal          Max           Through
           ----------     ----------     ----------

                            Price
                      Per Unit       Through
                     ----------     ----------
                         13.25
```

Output 11-1
Printout of Resource Dictionary

02.200.60 Backfilling walls
- 02.200.60.1 Labor backfilling walls
- 02.200.60.2 Materials backfilling walls
- 02.200.60.3 Equipment backfilling walls
- 02.200.60.4 Subcontract backfilling walls

The chart of accounts is loaded into Primavera Project Planner through the *Cost Accounts* window. This window is opened from the main input screen by selecting the *Data* command and the *Cost Accounts* option as shown in Screen Image 11-6. The *Cost Accounts* window has two distinct input areas, *Categories* and *Titles*, as shown in Screen Image 11-7. The *Categories* box is used to define the trailing digit that is used to designate the cost as labor, material, equipment, or subcontract. Following the precedent established in the

Screen Image 11-5
Prompt upon Closing Resources Window

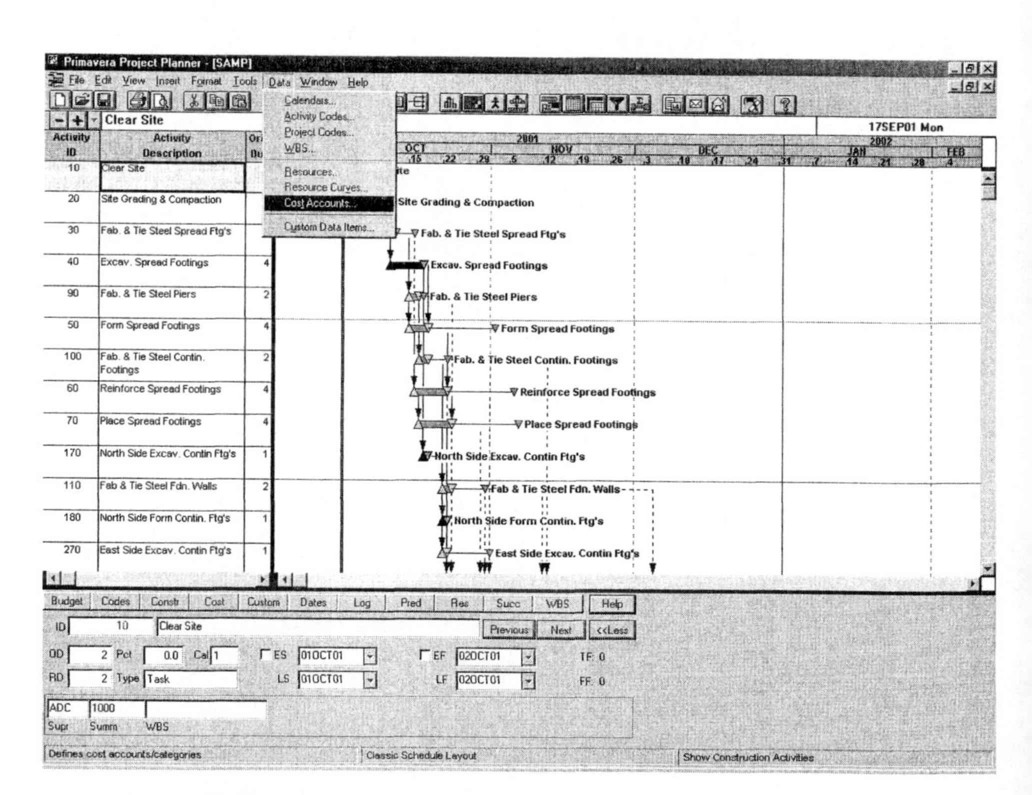

Screen Image 11-6
Opening the Cost Accounts Window

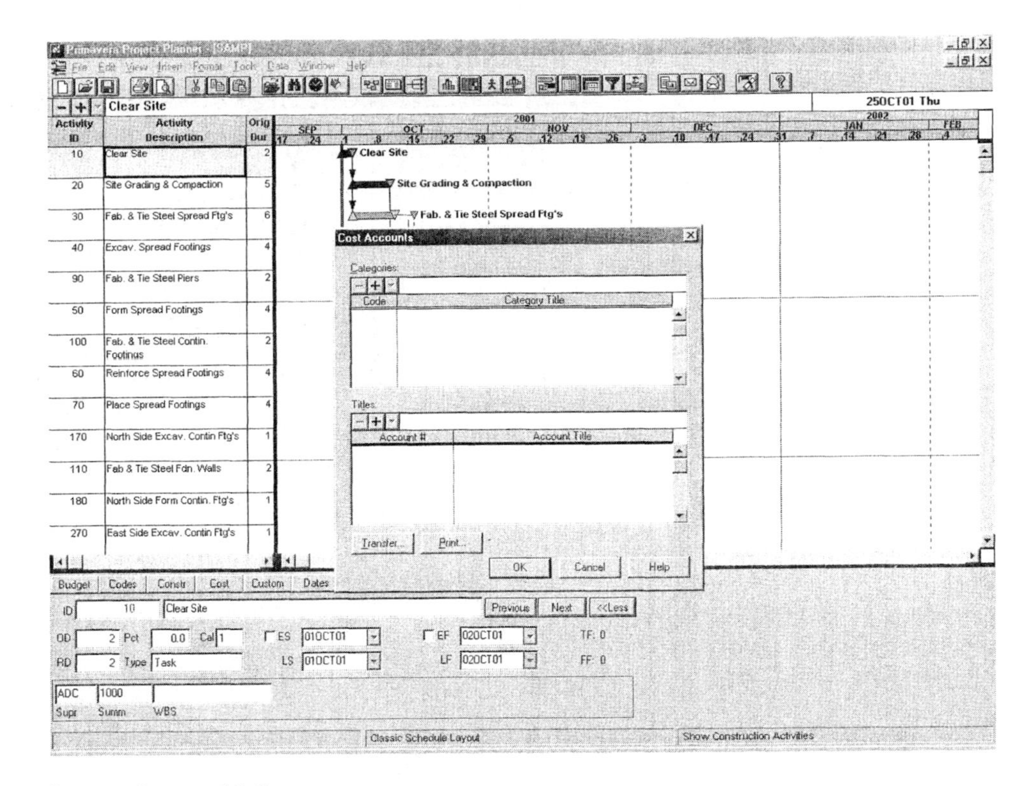

Screen Image 11-7
Blank Cost Accounts Window

previous example, 1 will be for labor, 2 will be for material, 3 will be for equipment, and 4 for subcontract costs. Screen Image 11-8 details the cost categories entry.

The actual chart of accounts is loaded in the *Titles* box. The *Account #* field is where the maximum 11-character cost code is entered. Its associated description would be entered in the adjacent *Account Title* field. Screen Image 11-9 shows the *Cost Accounts* window with the cost categories and associated cost codes. Clicking the *Print* button will initiate the printing process for the chart of accounts. Output 11-2 is a sample printout of the chart of accounts. The top of the report contains the cost categories, while the remainder of the report lists all of the cost codes and their associated descriptions. Since the chart of accounts should be basically the same from project to project, they can and should be transferred rather than input whenever a new project is set up.

To transfer the chart of accounts, click on the *Transfer* button on the *Cost Accounts* window. This will open the *Transfer* window, as shown in Screen Image 11-10. From this window select the project to copy the chart of accounts from. If the from project is not in the default folder, then use the *Folders* box to find the correct folder.

Loading Resources on the Individual Activities

All of the previous Primavera Project Planner work so far has been focused on preparing to load resources. Now the attention turns to actually loading the resource information about the specific activities. For the current time the only resources to be loaded will be craft labor workhours.

The resources are loaded on the individual activities by first clicking on that activity on the main input screen and pressing the *Res* button on the activity form to open the *Resources*

Primavera Project Planner - [SAMP]
File Edit View Insert Format Tools Data Window Help
Clear Site
25OCT01 Thu
Activity ID | Activity Description | Orig Dur
10 Clear Site 2
20 Site Grading & Compaction 5
30 Fab. & Tie Steel Spread Ftg's 6
40 Excav. Spread Footings 4
90 Fab. & Tie Steel Piers 2
50 Form Spread Footings 4
100 Fab. & Tie Steel Contin. Footings 2
60 Reinforce Spread Footings 4
70 Place Spread Footings 4
170 North Side Excav. Contin Ftg's 1
110 Fab & Tie Steel Fdn. Walls 2
180 North Side Form Contin. Ftg's 1
270 East Side Excav. Contin Ftg's 1
Clear Site
Site Grading & Compaction
Fab. & Tie Steel Spread Ftg's
Cost Accounts
Categories:
1
Code | Category Title
1 Labor
2 Material
3 Equipmen
4 Subs
Titles:
Account # | Account Title
Transfer... Print...
OK Cancel Help
Budget Codes Constr Cost Custom Dates
ID 10 Clear Site
Previous Next <<Less
OD 2 Pct 0.0 Cal 1 ES 01OCT01 EF 02OCT01 TF: 0
RD 2 Type Task LS 01OCT01 LF 02OCT01 FF: 0
ADC 1000
Supr Summ WBS
Classic Schedule Layout
Show Construction Activities

Screen Image 11-8
Updated Cost Accounts Window

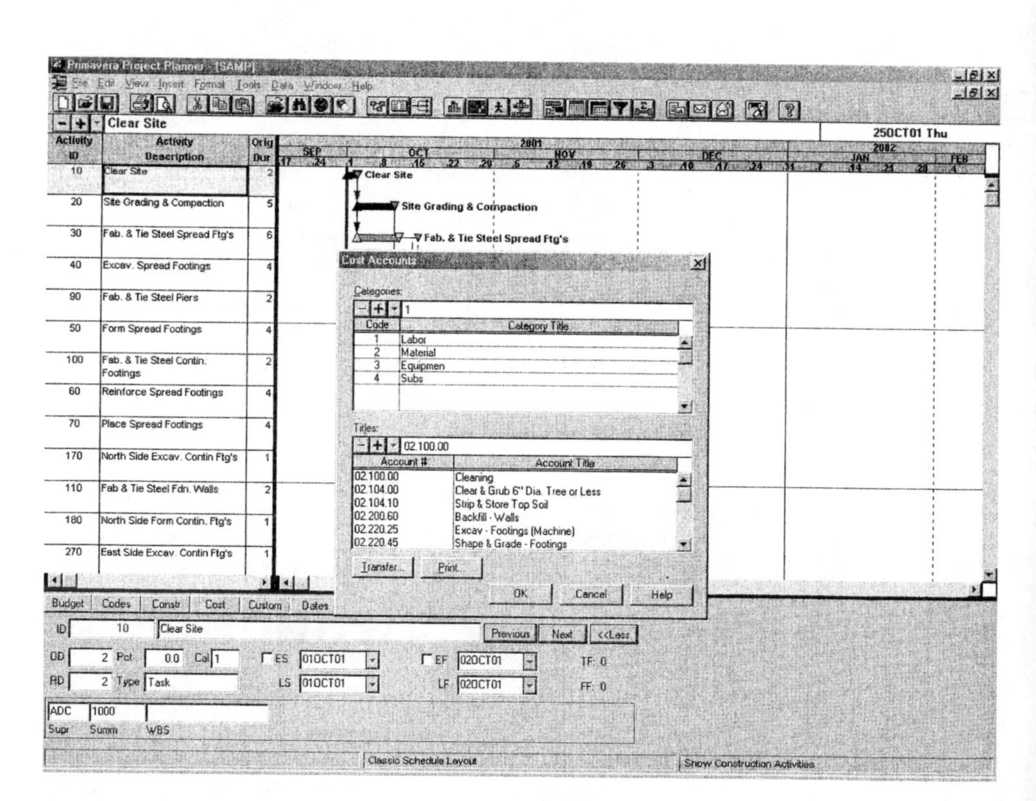

Screen Image 11-9
Multiple Cost Accounts Defined

PRIMAVERA PROJECT PLANNER

Date 18DEC00 -----SUMMARY OF COST ACCOUNTS----- Page 1

SAMP - Sample Construction Project

Cost Categories:

Code	Category Title	Code	Category Title
1	Labor	2	Material
3	Equipmen	4	Subs

Cost Account Titles:

Cost Account #	Account Title
02.100.00	Cleaning
02.104.00	Clear & Grub 6" Dia. Tree or Less
02.104.10	Strip & Store Top Soil
02.200.60	Backfill - Walls
02.220.25	Excav - Footings (Machine)
02.220.45	Shape & Grade - Footings
02.220.50	Backfill - Spread Footings
02.220.55	Fine Grading - Site
02.262.00	Fill & Compaction 6" Lifts
02.504.00	Landscape
02.580.00	Parking Lot Stripe
03.111.00	Form - Footings
03.114.00	Form - Foundation Walls
03.126.00	Forms - Piers
03.130.00	Form - Curbs
03.159.00	Strip Forms
03.206.00	Reinforce - footings
03.206.10	Reinforce Walls
03.207.00	Snap Ties
03.216.00	Reinforce - Slab on Grade
03.226.00	Reinforce - Piers
03.306.00	Concrete - Footings
03.306.10	Concrete - Piers
03.307.00	Concrete Walls
03.310.10	Concrete - Slab on Grade
03.310.20	Trowel Finish & Curing Compound
03.310.30	Concrete - Curing Compound
03.313.00	Concrete - Curbs
04.158.00	Reinforce Concrete Block
04.221.15	8" Concrete Block
04.221.35	Mortar
04.221.45	Horizontal & Vertical Reinforcement
04.710.11	Clean & Rub Concrete Block
05.110.00	Structural Steel
06.532.00	Shop Constructed Cabinets
07.410.00	Roofing & Flashing
08.210.01	Wood Interior Doors
08.410.00	Store Front Glass
08.690.00	Overhead Doors
08.710.00	Interior Door Hardware
09.251.00	Metal Stud Framing
09.253.00	Gyp. Wall Board
09.254.00	Float, Tape & Paint
09.510.10	Ceiling Grid
09.510.50	Ceiling Tiles
09.852.00	Carpet
15.100.00	Plumbing
15.200.00	Mechanical
16.100.00	Electrical

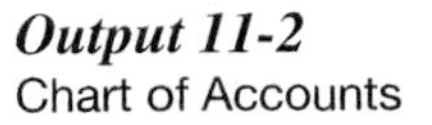

Output 11-2
Chart of Accounts

window shown in Screen Image 11-11. To add a resource, click the plus button or click in the blank space adjacent to the field descriptions. Either of these methods will create a new and unique resource column. While there are a number of fields for each resource only a few of these fields require input. As with any of the other windows, clicking on most of the fields will activate pull-down menus that give all of the possible entries into that field. The first item that must be entered is the resource name. By clicking on the *Resource* field and using the pull-down menu, a list of all of the possible resource entries can be found as shown in Screen Image 11-12. From that list the required resource can be selected by clicking on that resource. The next entry is for the *Cost Acct/Category*. This field requires two entries, one of which the pull-down menus can be used for, and the other the user must know. On Screen Image 11-13, the *Cost Acct/Category* field has been clicked and the

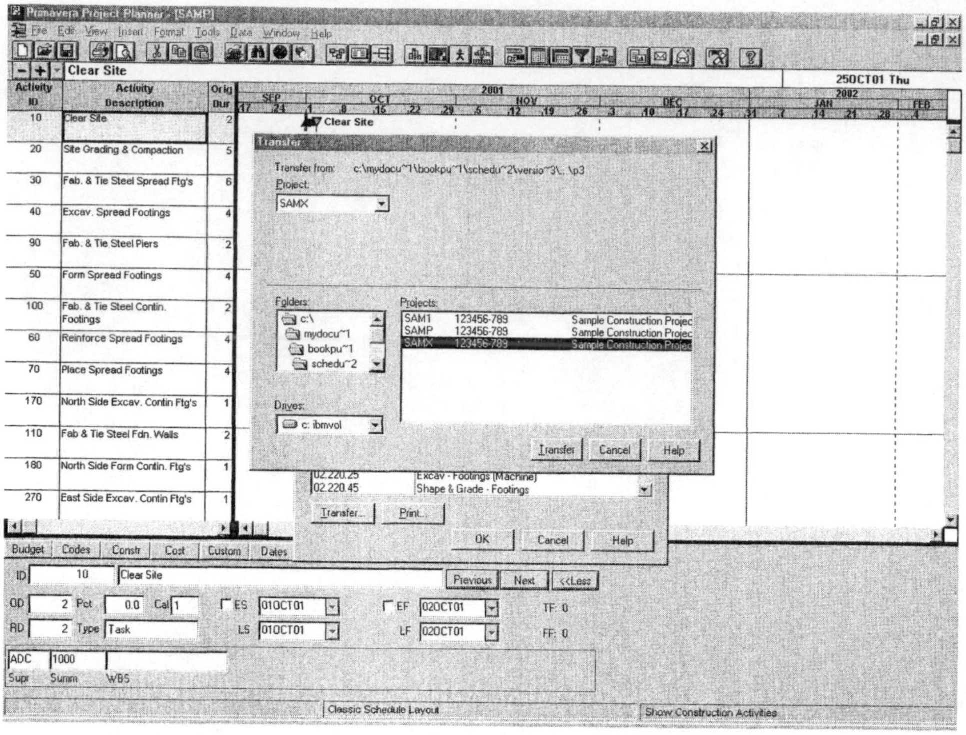

Primavera Project Planner - [SAMP]
File Edit View Insert Format Tools Data Window Help
Clear Site
25OCT01 Thu

Activity ID	Activity Description	Orig Dur
10	Clear Site	2
20	Site Grading & Compaction	5
30	Fab. & Tie Steel Spread Ftg's	6
40	Excav. Spread Footings	4
90	Fab. & Tie Steel Piers	2
50	Form Spread Footings	4
100	Fab. & Tie Steel Contin. Footings	2
60	Reinforce Spread Footings	4
70	Place Spread Footings	4
170	North Side Excav. Contin Ftg's	1
110	Fab & Tie Steel Fdn. Walls	2
180	North Side Form Contin. Ftg's	1
270	East Side Excav. Contin Ftg's	1

Transfer
Transfer from: c:\mydocu~1\bookpu~1\schedu~2\versio~3\...\p3
Project: SAMX
Folders: c:\ mydocu~1 bookpu~1 schedu~2
Projects:
SAM1 123456-789 Sample Construction Projec
SAMP 123456-789 Sample Construction Projec
SAMX 123456-789 Sample Construction Projec
Drives: c: ibmvol
Transfer Cancel Help

02.220.25 Excav - Footings (Machine)
02.220.45 Shape & Grade - Footings
Transfer... Print...
OK Cancel Help

Budget Codes Constr Cost Custom Dates
ID 10 Clear Site Previous Next <<Less
OD 2 Pct 0.0 Cal 1 ES 01OCT01 EF 02OCT01 TF: 0
RD 2 Type Task LS 01OCT01 LF 02OCT01 FF: 0
ADC 1000
Supr Summ WBS
Classic Schedule Layout Show Construction Activities

Screen Image 11-10
Transferring the Chart of Accounts

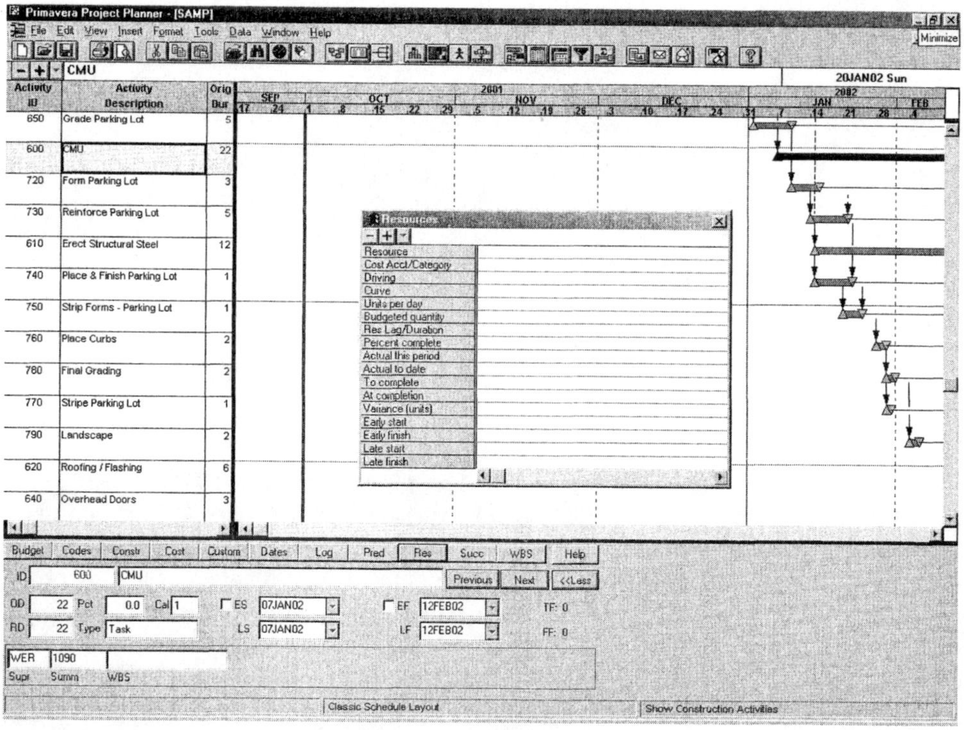

Primavera Project Planner - [SAMP]
File Edit View Insert Format Tools Data Window Help
CMU
20JAN02 Sun

Activity ID	Activity Description	Orig Dur
650	Grade Parking Lot	5
600	CMU	22
720	Form Parking Lot	3
730	Reinforce Parking Lot	5
610	Erect Structural Steel	12
740	Place & Finish Parking Lot	1
750	Strip Forms - Parking Lot	1
760	Place Curbs	2
780	Final Grading	2
770	Stripe Parking Lot	1
790	Landscape	2
620	Roofing / Flashing	6
640	Overhead Doors	3

Resources
Resource
Cost Acct/Category
Driving
Curve
Units per day
Budgeted quantity
Res Lag/Duration
Percent complete
Actual this period
Actual to date
To complete
At completion
Variance (units)
Early start
Early finish
Late start
Late finish

Budget Codes Constr Cost Custom Dates Log Pred Res Succ WBS Help
ID 600 CMU Previous Next <<Less
OD 22 Pct 0.0 Cal 1 ES 07JAN02 EF 12FEB02 TF: 0
RD 22 Type Task LS 07JAN02 LF 12FEB02 FF: 0
WER 1090
Supr Summ WBS
Classic Schedule Layout Show Construction Activities

Screen Image 11-11
Resources Window

Primavera Project Planner - [SAMP]
File Edit View Insert Format Tools Data Window Help
CMU
01OCT01 Mon

Activity ID	Activity Description	Orig Dur
650	Grade Parking Lot	5
600	CMU	22
720	Form Parking Lot	3
730	Reinforce Parking Lot	5
610	Erect Structural Steel	12
740	Place & Finish Parking Lot	1
750	Strip Forms - Parking Lot	1
760	Place Curbs	2
780	Final Grading	2
770	Stripe Parking Lot	1
790	Landscape	2
620	Roofing / Flashing	6
640	Overhead Doors	3

2001 DEC 3 10 17 24 31 2002 JAN 7 14 21 28 FEB 4 11 18 25 MAR 4 11 18 25 APR 1 8 15 22
CMU
Flashing
Overhead Doors

Resources
Resource
CARP - Carpenters
CFIN - Concrete Finishers
LABR - Laborers
MASN - Masons
MASNHLP - Mason Helper
OPENG - Operating Engineer
ROD - Rodbuster
Actual this period 0
Actual to date 0
To complete 0
At completion 0
Variance (units) 0
Early start 07JAN02
Early finish 12FEB02
Late start 07JAN02
Late finish 12FEB02

Budget Codes Constr Cost Custom Dates Log Pred Res Succ WBS Help
ID 600 CMU OK Cancel <<Less
OD 22 Pct 0.0 Cal 1 ES 07JAN02 EF 12FEB02 TF: 0
RD 22 Type Task LS 07JAN02 LF 12FEB02 FF: 0
WER 1090
Supr Summ WBS
Classic Schedule Layout
Show Construction Activities

Screen Image 11-12
Resource Pull-Down Menu

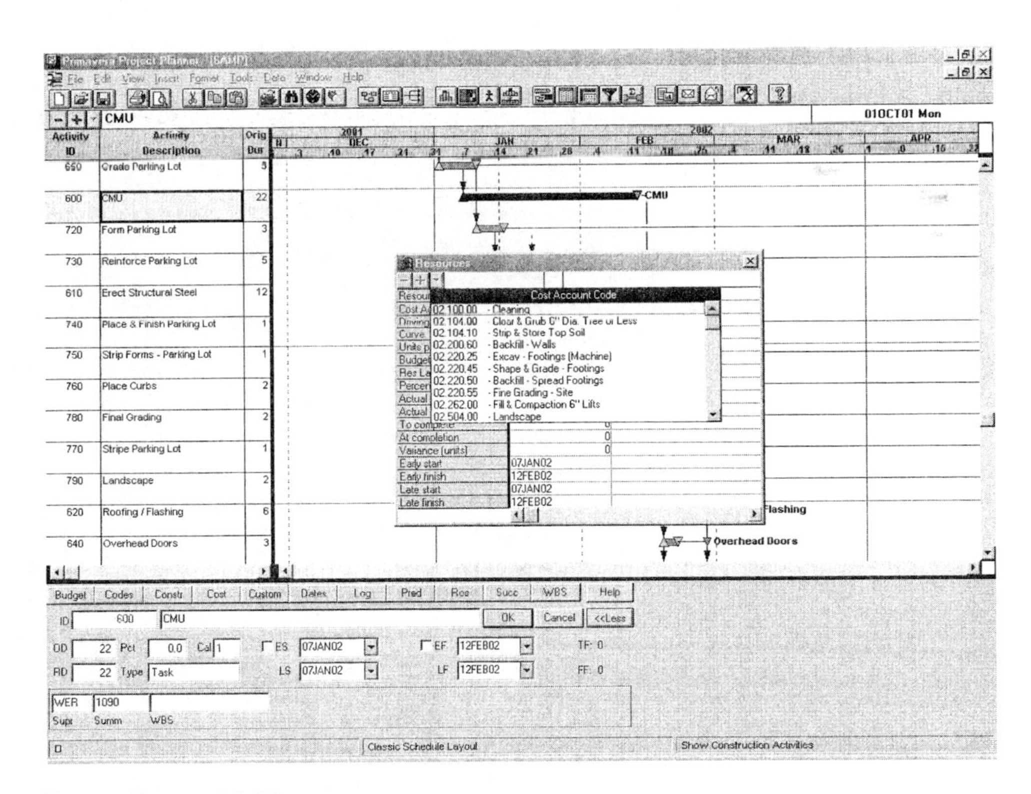

Screen Image 11-13
Cost Accounts Pull-Down Menu

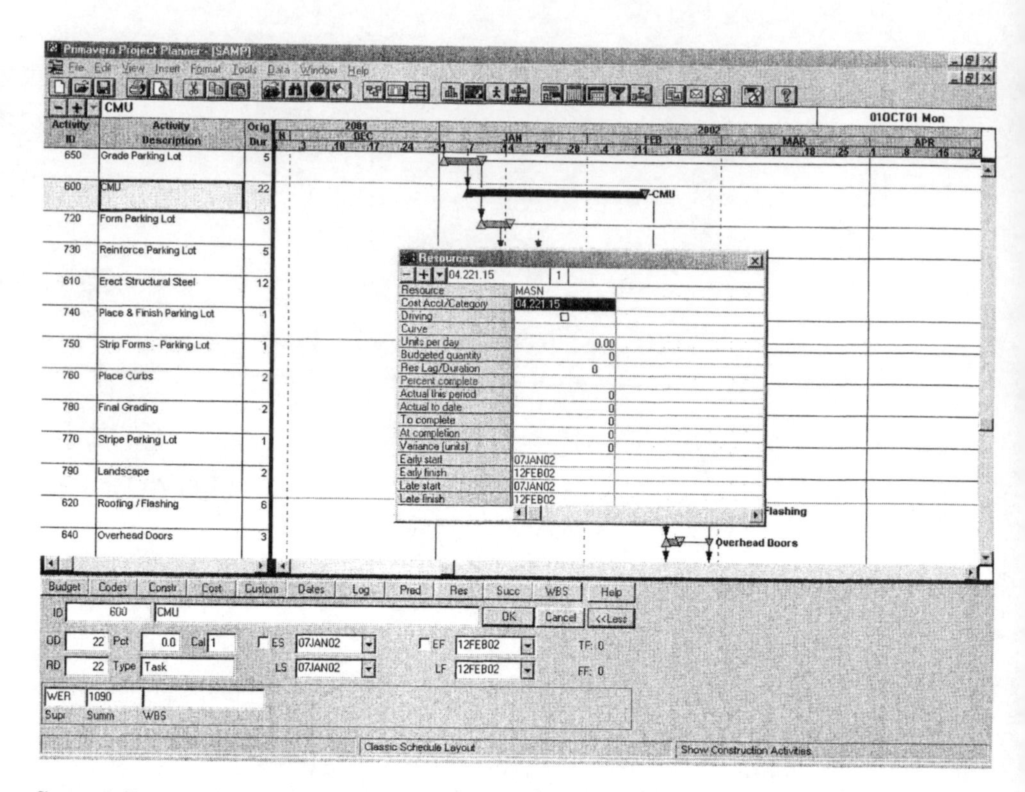

Screen Image 11-14
Resource Window with Cost Category Entered

pull-down menu activated. From that list the correct cost code can be selected. However, if one looks closely at that screen image, there are two input areas for this one field. There is the relatively long field to the left and the rather short field in the middle. The small field in the middle is for the cost category. This is one of the few entries that does not have an available pull-down menu. Instead, the user must remember the cost categories. The correct cost category needs to be entered in that field. These categories were defined with the cost accounts for this example, and they are 1 for labor, 2 for materials, 3 for equipment, and 4 for subcontracts. Screen Image 11-14 shows the MASN resource with cost code 04.221.15.1 for Activity 600. The trailing digit of the cost code is 1, which designates this as a labor item.

The third entry on the *Resource* window is whether the resource is driving. If a resource is designated as driving, the duration of the activity is set by the availability of the resource rather than a manually input duration. When the resources were defined this same question was asked. This input field allows the user to override the input from the resource dictionary.

This entry is followed by designating a resource curve, which addresses how the specific resource will be distributed across the duration of the activity. Primavera Project Planner comes with nine predefined resource distribution curves. These curves can be found and modified by opening the *Display of Defined Curves* window. Selecting the *Data* command followed by the *Resource Curves* option as shown in Screen Image 11-15 opens this window. Once that window is opened, the predefined curves can be displayed by clicking the arrow buttons in the top right-hand corner of the window. Screen Images 11-16, 11-17, and 11-18 show the nine predefined distribution curves. If no curve is entered on the resource window, a linear distribution is assumed.

The resource quantity can be entered in a number of different ways in the *Resources* window. The first method is to enter the quantity from the estimate in the *Budgeted quantity* field. Using cost code 04.221.15.1 and the estimate shown in Figure 11-1 as an example, there are 1052 workhours budgeted for placing the concrete block. Crew M1 consists of

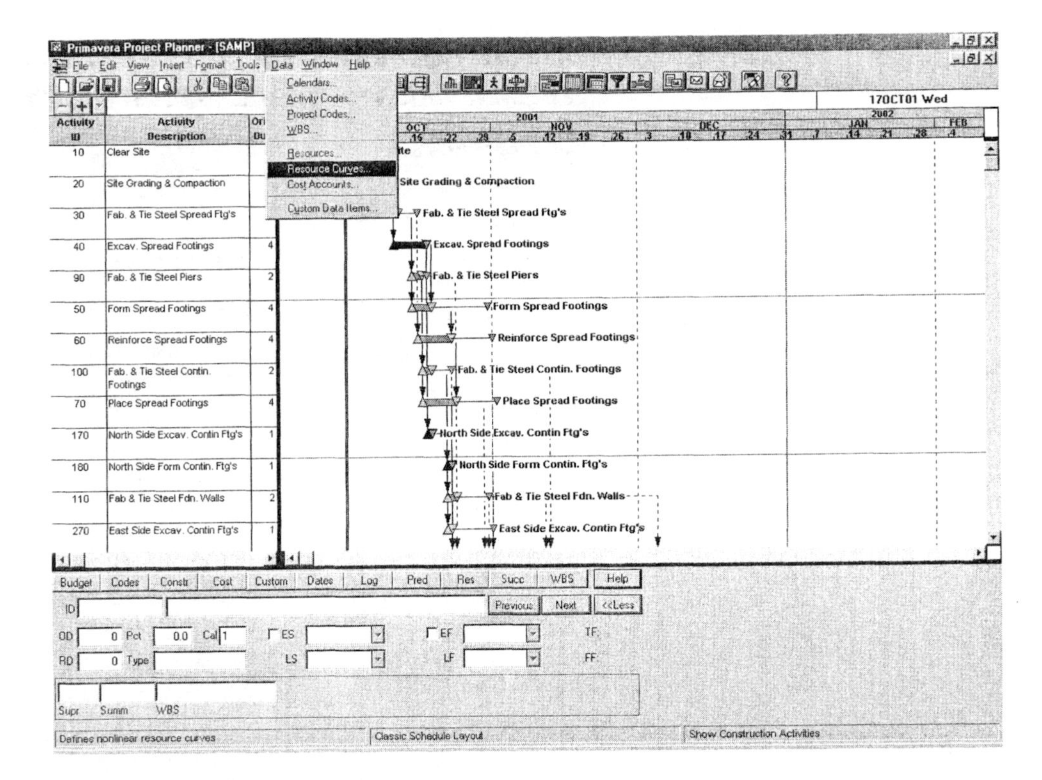

Screen Image 11-15
Opening Resource Distribution Curves

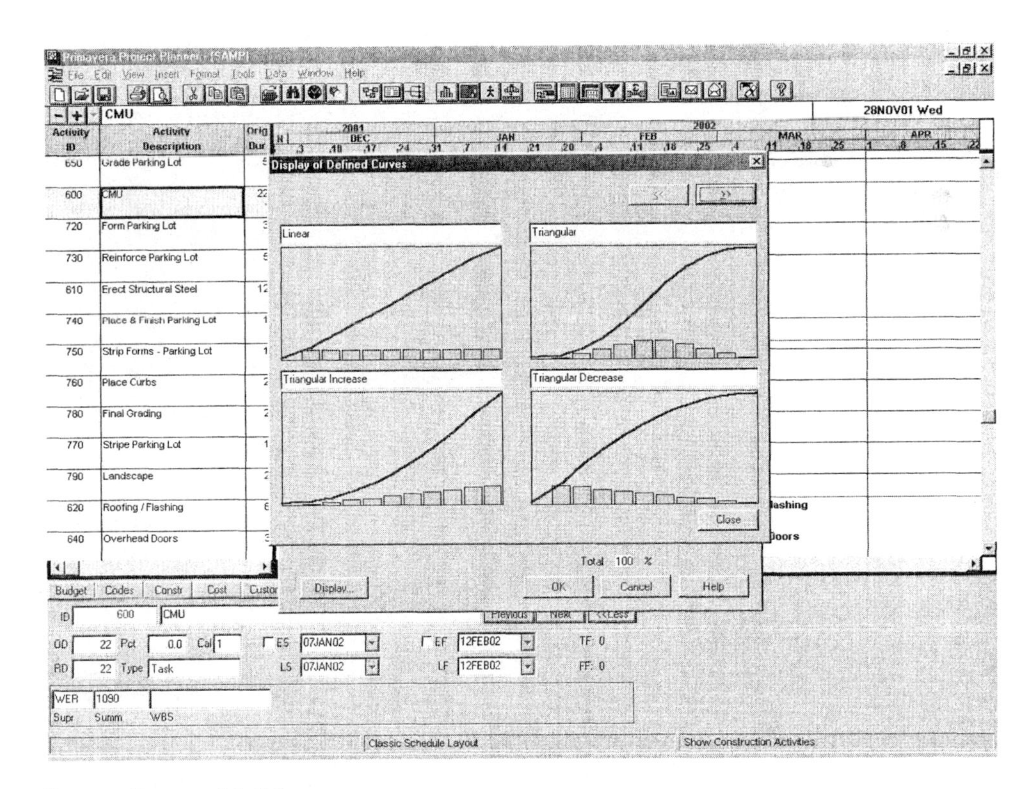

Screen Image 11-16
Resource Distribution Curves

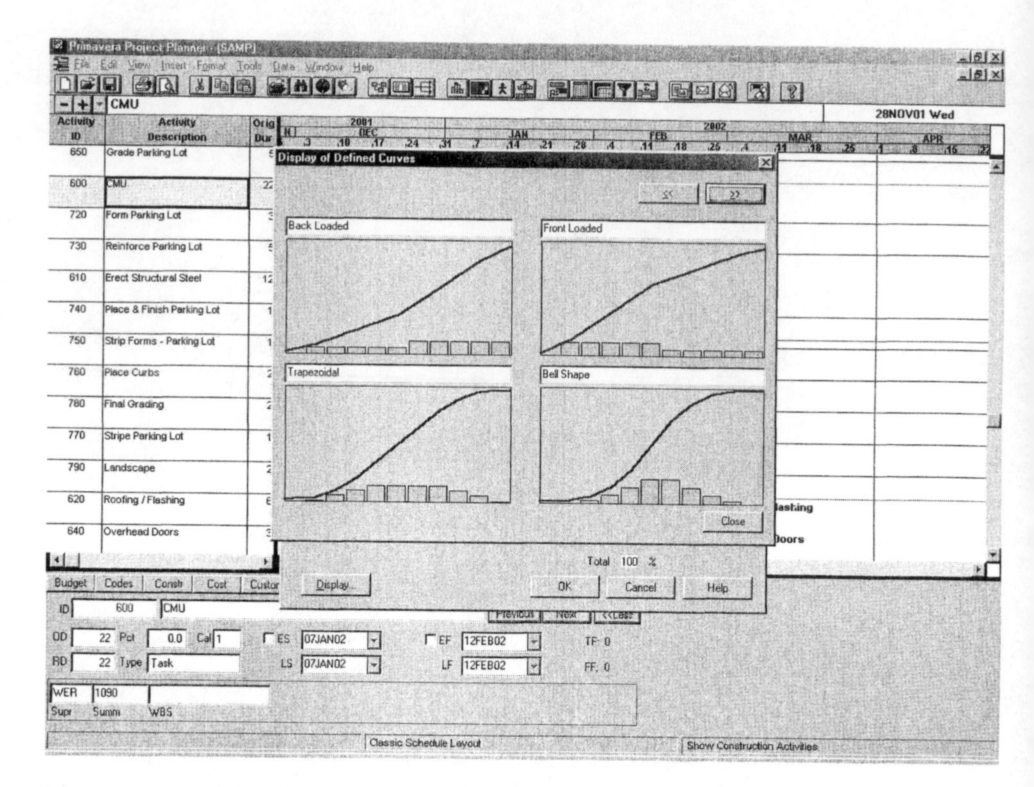

Screen Image 11-17
Resource Distribution Curves

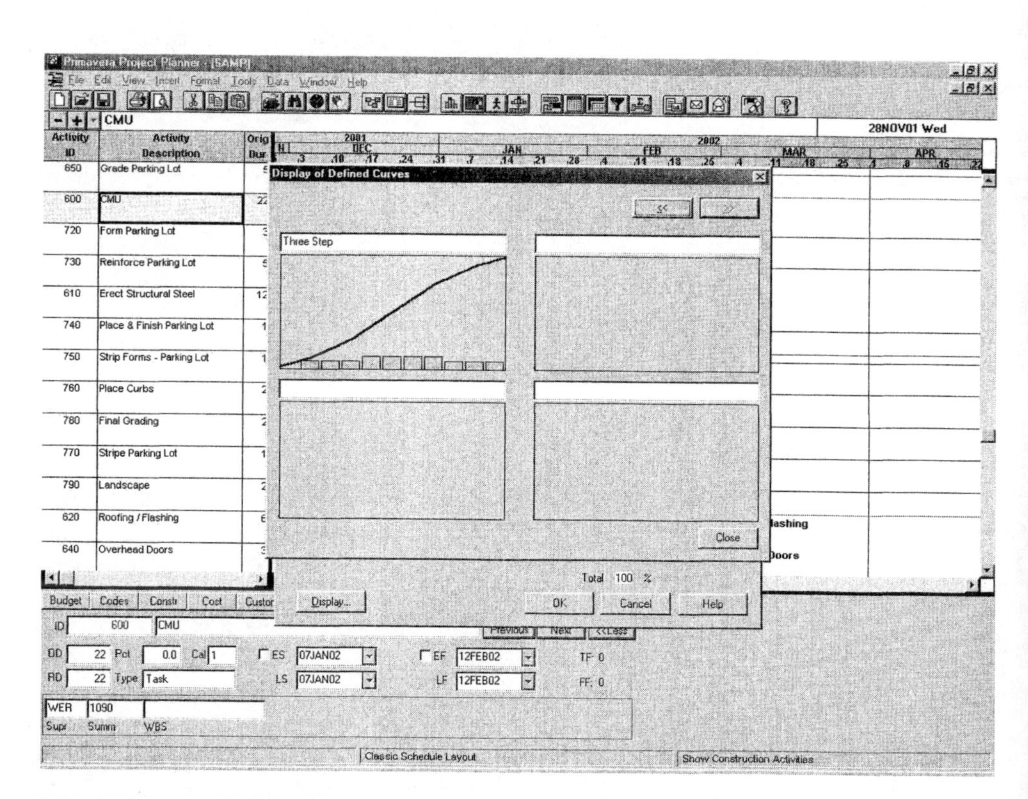

Screen Image 11-18
Resource Distribution Curves

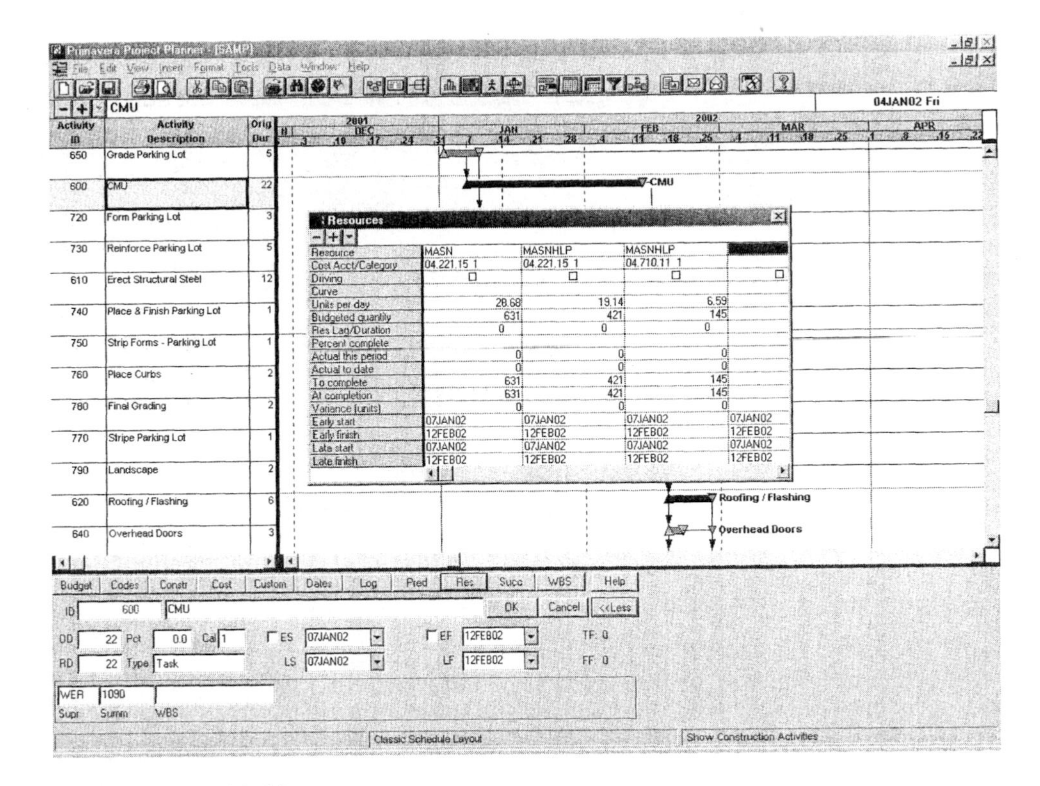

Screen Image 11-19
Multiple Cost Account and Resource

three masons and two mason helpers, therefore three-fifths of the workhours (631) are allocated to masons and two-fifths (421) of the workhours to mason helpers. The third resource entry associated with this activity is the labor associated with rubbing the concrete block. This cost account is 04.710.11.1 and the associated labor resource is mason helpers. Screen Image 11-19 shows how these three cost accounts and their associated resources would be entered. The estimated workhours for each of the cost accounts/resources are typically entered in the *Budgeted quantity* field. If this input method is used, Primavera Project Planner will perform a number of different calculations. The first of which is to divide the *Budgeted quantity* by the original duration (*OD*) to come up with the *Units per day*. The user could have entered the *Units per day* and Primavera Project Planner would have multiplied that quantity by the original duration (*OD*) and entered that amount as the *Budgeted quantity*. Another automatic occurrence is the *At completion* field, which is set equal to the *Budgeted quantity*. From there the *At completion* quantity is subtracted from the *Actual to date* quantity and the resulting amount was entered in the *To complete* field. Currently the *At completion* and *To complete* are the same since the *Actual to date* is zero. The final calculation that is automatically performed is to subtract the *At completion* from the *Budgeted quantity* to develop the *Variance*. In Screen Image 11-19 the variance is zero since those two quantities are the same.

Once all of the resources for a particular activity have been entered, the *Resource* window can be closed by double clicking on the Primavera Project Planner logo in the top left-hand corner of the window or the *X* in the right-hand corner. If additional activities require resource input, simply click on that activity on the main input screen while the resource window is active.

While the automatic calculations are simple, as work commences these calculations may not be appropriate and might need to be changed. The rules for calculations in the *Resources* windows is controlled by the actions specified in the *Automatic Cost/Resource Calculation Rules* window. Selecting the *Tools* command followed by the *Options* and *Autocost Rules* as shown in Screen Image 11-20 opens this window.

Primavera Project Planner - [SAMP]
File Edit View Insert Format Tools Data Window Help
CMU
04OCT01 Thu

Activity ID	Activity Description	Orig Dur
490	South Side Strip Fdn Wall Forms	
530	West Side Reinforce Fdn. Wal	
540	West Side Form Fdn. Walls	
500	South Side Backfill & Compact	
550	West Side Place Fdn. Walls	
560	West Side Strip Fdn. Wall Forms	
570	West Side Backfill & Compact	
580	Reinforce S.O.G.	
590	Place & Finish S.O.G.	
650	Grade Parking Lot	
600	CMU	
720	Form Parking Lot	
730	Reinforce Parking Lot	5

RL-09 SAMP - Resource Loading Reports
Title: Activity 600 Mason Resource Profile
Run
Resource Selection | Format | Selection | Timescale
Generate report in P3 | Font... | Arial, Regular, 8
Export report in CSV format to the file | C:\P3WIN\P3OUT\P3.csv | Browse
Organize
Organize by: SUPR - Superintendent
Group/subtotal by: <none>
Summary report
Options
Truncate activity description
Repeat activity title on all horizontal sheets
Display row totals | Left | Right
Page width: 132 | Column width: 5
Divide by: 1 | Decimals: 0
Sort by: Sort on | Order
Transfer... | OK | Cancel | Help

Forms, dn. Walls, Walls, ckfill & Compact, . Walls, p Fdn. Wall Forms, ckfill & Compact, rce S.O.G., ce & Finish S.O.G.

Budget | Codes | Constr | Cost | Custom | Dates | Log | Pred | Res | Succ | WBS | Help
ID 600 CMU | Previous | Next | <<Less
OD 22 Pct 0.0 Cal 1 | ES 07JAN02 | EF 12FEB02 | TF: 0
RD 22 Type Task | LS 07JAN02 | LF 12FEB02 | FF: 0
WER 1090
Supr Summ WBS
Classic Schedule Layout | Show Construction Activities

Screen Image 11-20
Opening Autocost Rules

Primavera Project Planner - [SAMP]
File Edit View Insert Format Tools Data Window Help
CMU
15DEC01 Sat

Activity ID	Activity Description	Orig Dur
600	CMU	22
720	Form Parking Lot	3
730	Reinforce Parking Lot	5
610	Erect Structural Steel	12
740	Place & Finish Parking Lot	1
750	Strip Forms - Parking Lot	1
760	Place Curbs	2
780	Final Grading	2
770	Stripe Parking Lot	1
790	Landscape	2
620	Roofing / Flashing	6
640	Overhead Doors	3
630	Exterior Glass & Glazing	1

Automatic Cost/Resource Calculation Rules for SAMP
Identifier: | Date:
Link remaining duration and schedule percent complete
Freeze resource units per time period
Add actual to ETC | Subtract actual from EAC
Allow negative ETC
When quantities change, use current unit prices to recompute costs:
Budget | Actual to date | Estimate to complete
Use the update percent complete against budget to estimate:
Actual quantity to date | Actual cost to date
Link budget and EAC for non-progressed activities
Calculate variance as: Budget - EAC | EAC - budget
Perform these calculations during each schedule computation
Apply these rules when moving from one Cell / Resource to another
Transfer... | Default... | OK | Cancel | Help

CMU, tural Steel, Roofing / Flashing, Overhead Doors, Exterior Glass & Glazing

Budget | Codes | Constr | Cost | Custom | Dates | Log | Pred | Res | Succ | WBS | Help
ID 600 CMU | Previous | Next | <<Less
OD 22 Pct 0.0 Cal 1 | ES 07JAN02 | EF 12FEB02 | TF: 0
RD 22 Type Task | LS 07JAN02 | LF 12FEB02 | FF: 0
WER 1090
Supr Summ WBS
Classic Schedule Layout | Show Construction Activities

Screen Image 11-21
Autocost Rules Window

The *Automatic Cost/Resource Calculation Rules* window as shown in Screen Image 11-21 has a number of items that need to be investigated. If the *Link remaining duration and schedule percent complete* option is checked, there will be a linear linkage between the percent complete and remaining duration. For example, if this option is checked and there is a 10-day activity that is statused at 50 percent complete, the remaining duration would automatically be calculated as 5 days and entered in the *RD* field. Conversely, if that same activity was given a 3-day remaining duration, the activity would be progressed at 70 percent complete. When this item is checked there are two underlying assumptions that are questionable. First there is the assumption that a linear relationship exists between the remaining duration and percent complete. With that assumption it is assumed that all of the workhours expended on the activity will generate an equal amount of progress. The other assumption is that the productivity rate from the estimate is true and accurate. Both of these assumptions are questionable and run contrary to good project controls practices. The field construction staff personnel should critically evaluate the number of days to complete every activity on a regular basis. This critical evaluation should yield the best estimate of the remaining duration. As construction progresses and actual productivity information becomes available, that information is a better source of information than the original estimate. The estimate productivity information is based on a standard or average condition, while that gathered as the project progresses takes into consideration the chemistry of the specific project. Similarly, the percent complete should be based on what percentage of the work on that activity is actually in-place. If there is a disparity between the percent complete and the remaining duration, an evaluation should be performed to find the reason. By tying the percent complete to the remaining duration, the ability to spot deviations in the field productivity is hidden.

The *Freeze resource units per time period* fixes the *Units per day* entry on the *Resources* window. If this is left unchecked, there is a continual calculation of dividing the budget quantity by the original duration. With this field unchecked the units per day measurement is dynamic and changes as information is revised.

The *Add actual to ETC* and *Subtract actual from EAC* are tools for forecasting. The first of the options is preferred and follows good project controls practices. If that option is selected, the *At completion* amount on the *Resources* window is found by adding the *Actual to date* to the *To complete*. This option requires the project team to constantly monitor the remaining work and determine its cost. Then the cost of the remaining work can be added to the dollars that have already been expended to determine the most likely final cost. The latter option requires the project team to develop the estimated final cost and subtract that amount from the actual to date. This is a much more complicated process and in many circumstances lacks the diligence of surveying the remaining work and using judgment to determine its cost.

The *Allow negative ETC* option gives the user the option as to how to address items that are over budget. The only way to get a negative *Estimated to complete* is for the project budget to be less than the *Expended to date*. In a literal sense if this is checked, it would mean that actual costs would decline as the project progressed. Once a cost becomes actual, how can it reduce? If this is left unchecked, when any items are over budget, the *At completion* amount is automatically set equal to the *Actual to date*. This tells the project team that the lowest possible cost is that which is already expended. While it is not recommended that this option be checked, the only advantage to using this option is that items that are over budget will stand out, pointing out items that need further investigation.

The *When quantities change, use current unit price to recompute costs* choice deals with the multiplication of the resource quantity by the unit price specified within the *Resources* window. If all of these items are checked, the costs for these items will automatically be recalculated. The problem with selecting these features is the underlying assumption that the estimated unit cost is the actual unit cost. In reality there will be deviations between the planned unit cost and the actual unit cost. The best approach is to check all of the options during the input of the resources, then remove the check from the *Actual to date* and the

Estimate to complete; this will require the user to input the actual units and cost rather than just the units. One of the project controls benchmarks is to compare actual to planned unit cost. If the actual resource quantity and cost are separate, this evaluation can be easily performed.

The *Use the update percent complete against budget to estimate* options allow the user to have the program enter actual resource costs and quantities based on actual percent complete. If either of these potions is selected, the actual cost is estimated based on the percent complete. For example, if the resources for a particular activity are 100 workhours and $1,000.00 and both quantity and cost options are selected, and the activity is statused at 10 percent complete, the actual to date units would be set to 10 workhours and the cost to $100.00. Selecting these options circumvents good project controls practices. If these are selected, then there is the underlying assumption that the unit progress and unit cost in the estimate are what is actually being experienced in the field. That is a poor assumption and denies the existence of the actual information. With that option the actual costs are an estimate rather than an actual cost. When one sees the word actual, there is the assumption that it is a true actual cost and not an estimated actual cost.

The *Link budget and EAC for non-progressed activities* option allows for faster input during the planning stages. If this item is checked, when the budget is changed the estimate at complete amount is automatically changed to match that amount. This allows for less input. However, once that activity has commenced, this linkage is lost. The two items below this entry deal with how the variance is calculated and whether a positive variance is good or bad. If the *Budget - EAC* is selected, a positive variance is good and a negative means that the current forecast is for the item to be over budget. On the other hand, if the *EAC - Budget* is selected, a positive variance means that the forecast cost is over budget and a negative variance means the forecast is under budget. The latter of these two options is the simplest to understand.

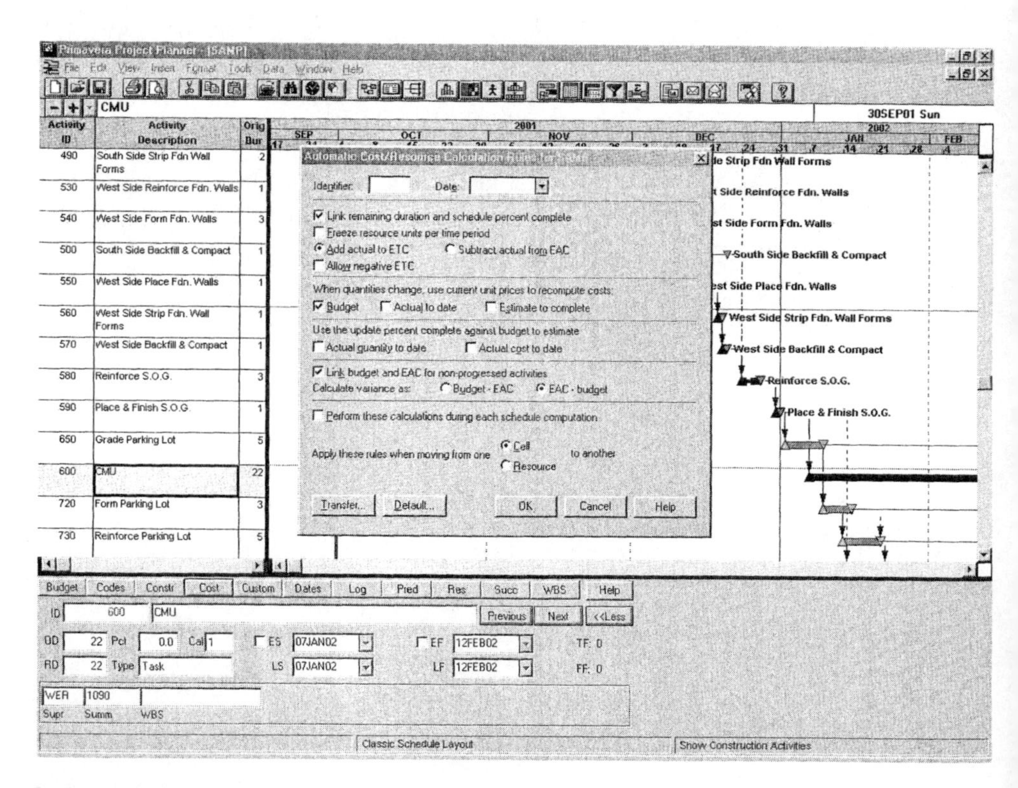

Screen Image 11-22
Recommended Autocost Rules Window

If the *Perform these calculations during each schedule computation* option is checked, the resultant calculations are transferred to hammocks. If there are no hammocks used on the schedule, there is no reason to check this option. The final option on the *Automatic Cost/Resource Calculation Rules* window deals with when to apply these rules. The default setting is *Cell*, and with that option anytime the user moves from one cell to another on the *Resource* window these rules are automatically applied. The other option is to apply these rules when moving from one resource to another. The only reason to select the latter option is if you are using an older slow computer and the continual recalculation slows down the input process.

Screen Image 11-22 shows the recommended settings for the autocost rules once the project starts. These recommendations may seem to make the operation more cumbersome, but they are geared toward enhancing the project controls process.

Producing Resource Loading Reports

The underlying reason for loading all of the resources is to be able to determine what resources are required, in what quantity, and when. This information can be found by generating tabular resource loading reports. The tabular resource loading reports are generated by selecting the *Reports* command followed by the *Tabular, Resource*, and *Loading* options as shown in Screen Image 11-23. The resource loading tabular reports are generated in a fashion similar to the tabular schedule reports. Just as with the tabular schedule reports, the *Resource Loading Reports* window contains the list of all of the previously developed report specifications as seen in Screen Image 11-24. Since these specifications are for different schedules and may not be applicable for the current project, the *Add* button is clicked and the *Add a New Report* window is activated, as shown in Screen Image 11-25. This

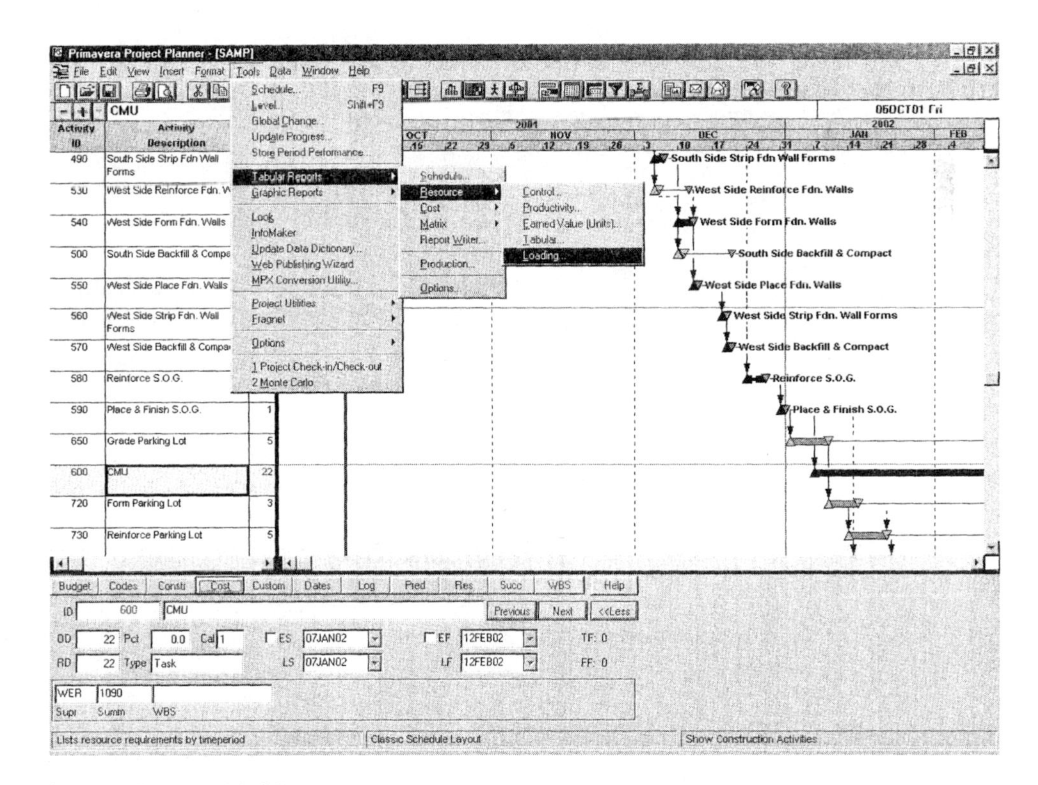

Screen Image 11-23
Resource Loading Reports

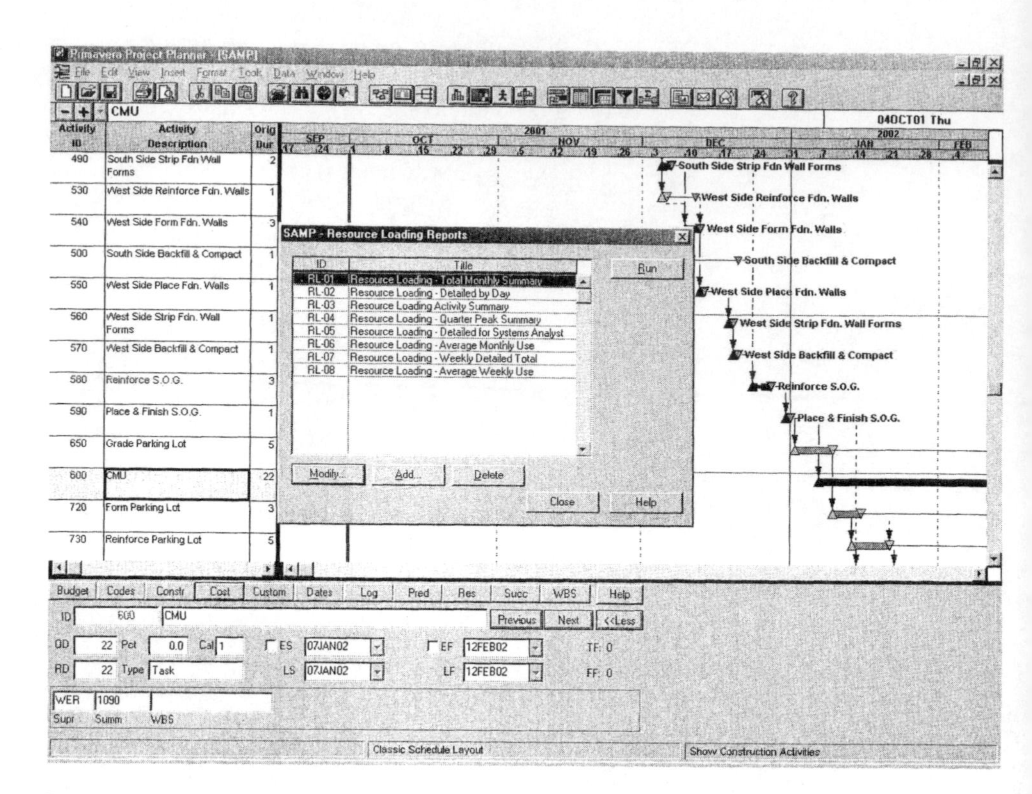

Screen Image 11-24
Resource Loading Reports List

report identifier is simply the next set of resource loading report specifications. In Screen Image 11-25, RL-09 designates that this set of specifications is the ninth set that has been developed. If this identifier is acceptable, click the *OK* button and the *Resource Loading Reports* window will be activated, as shown in Screen Image 11-26.

Just as with the tabular reports options, a series of tabs must be worked through in order to generate the desired report. For example purposes, only the masons in Activity 600 will be investigated. Just as with all of the other report specifications, the first thing to do is change the report title to reflect the use of the report specifications. The *Resource Selection* tab is used to define which resources are to be included in the tabular report. The *Profile if* column is where the selection criteria are specified. Listed below are the acceptable criteria:

EQ	Equal to
NE	Not equal to
GT	Greater than
LT	Less than
WR	Within range
NR	Not within range

The remaining four columns are used to define what resource item to include in the report. There are a number of columns since there are a variety of ways to define the resource item. Screen Image 11-27 illustrates how the mason and mason helper resource would be selected.

Once all of the desired resource selection criteria have been entered, the *Format* tab should be clicked. Screen Image 11-28 is an example of this window before any modifications have been made. This window allows the user to organize the tabular resource loading

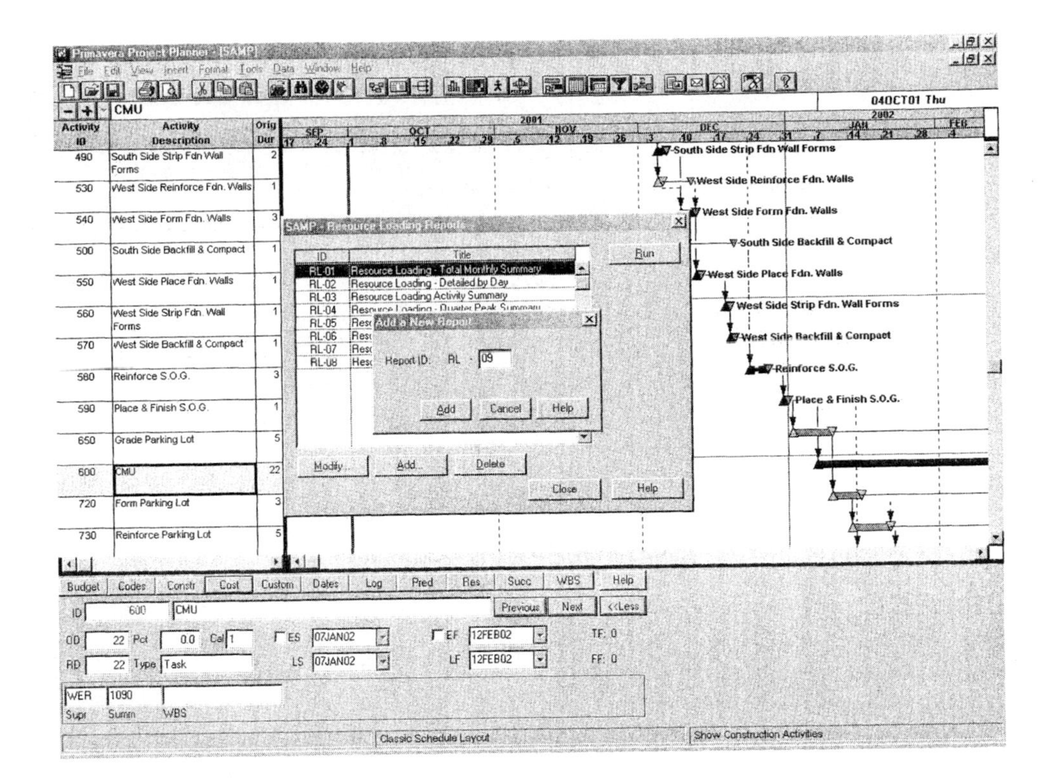

Screen Image 11-25
Add a New Report Window

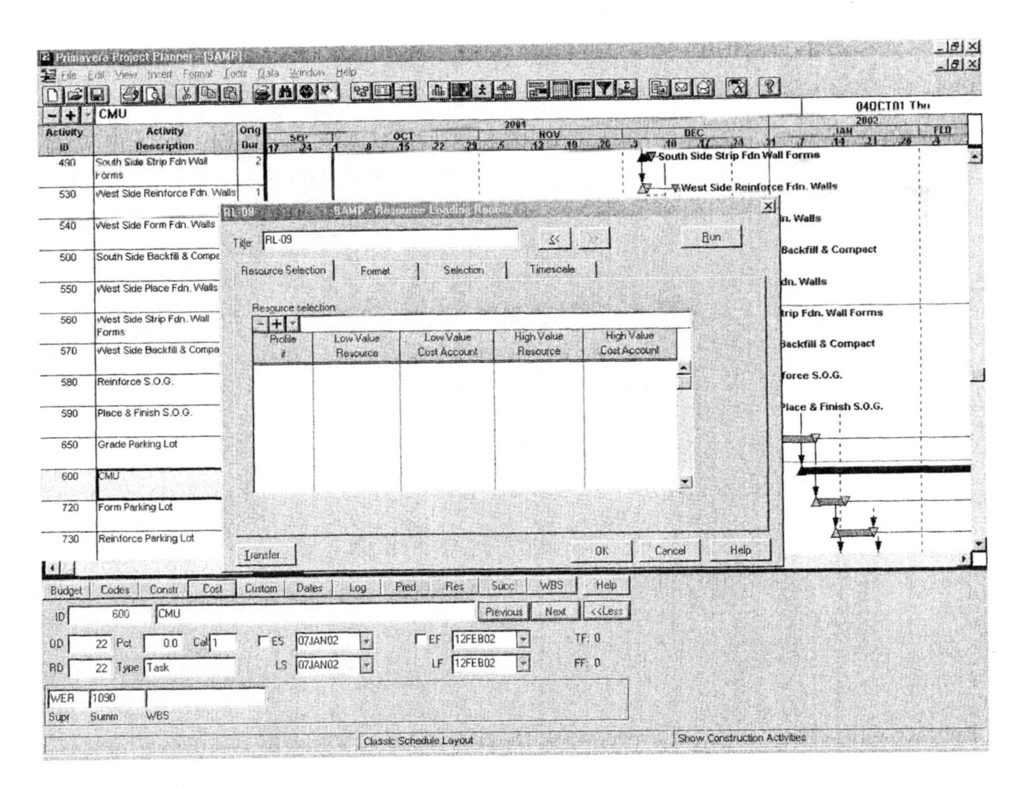

Screen Image 11-26
Resource Selection Window

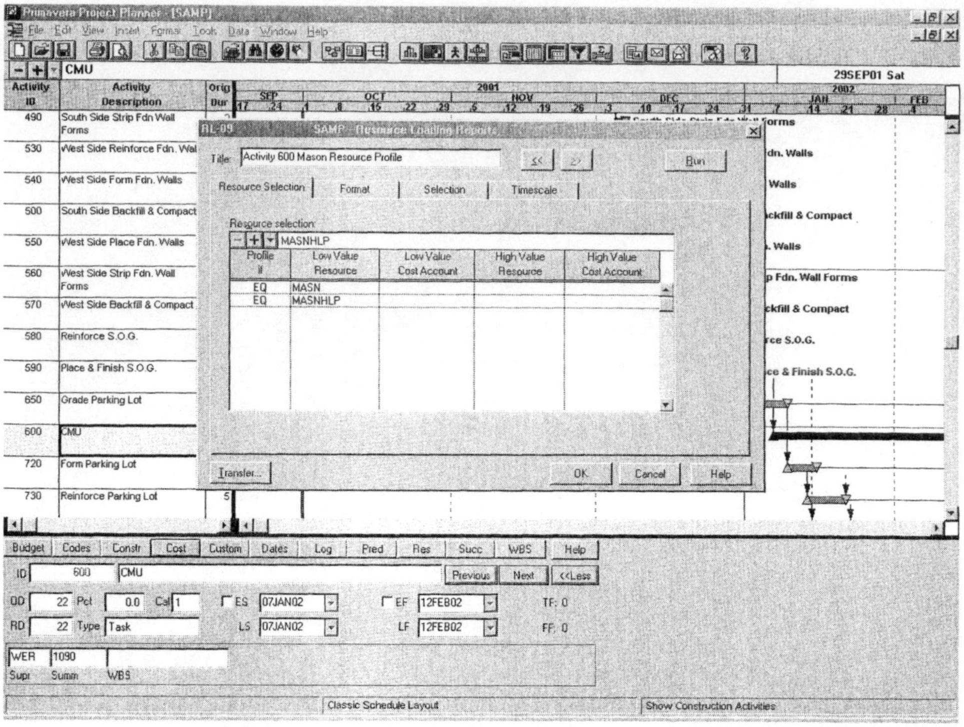

Primavera Project Planner - [SAMP]
File Edit View Insert Format Tools Data Window Help
CMU
29SEP01 Sat

Activity ID	Activity Description	Orig Dur
490	South Side Strip Fdn Wall Forms	
530	West Side Reinforce Fdn. Wal	
540	West Side Form Fdn. Walls	
500	South Side Backfill & Compact	
550	West Side Place Fdn. Walls	
560	West Side Strip Fdn. Wall Forms	
570	West Side Backfill & Compact	
580	Reinforce S.O.G.	
590	Place & Finish S.O.G.	
650	Grade Parking Lot	
600	CMU	
720	Form Parking Lot	
730	Reinforce Parking Lot	5

2001 SEP 17 24 OCT 1 8 15 22 29 NOV 5 12 19 26 DEC 3 10 17 24 31 2002 JAN 7 14 21 28 FEB 4

Forms
dn. Walls
Walls
ckfill & Compact
. Walls
p Fdn. Wall Forms
ckfill & Compact
ce S.O.G.
ce & Finish S.O.G.

RL-09 SAMP - Resource Loading Reports

Title: Activity 600 Mason Resource Profile << >> Run

Resource Selection | Format | Selection | Timescale

Resource selection:
MASNHLP

Profile #	Low Value Resource	Low Value Cost Account	High Value Resource	High Value Cost Account
EQ	MASN			
EQ	MASNHLP			

Transfer... OK Cancel Help

Budget | Codes | Constr | Cost | Custom | Dates | Log | Pred | Res | Succ | WBS | Help

ID 600 CMU Previous Next <<Less

OD 22 Pct 0.0 Cal 1 ES 07JAN02 EF 12FEB02 TF: 0

RD 22 Type Task LS 07JAN02 LF 12FEB02 FF: 0

WER 1090

Supr Summ WBS

Classic Schedule Layout | Show Construction Activities

Screen Image 11-27
Mason Resource Selected

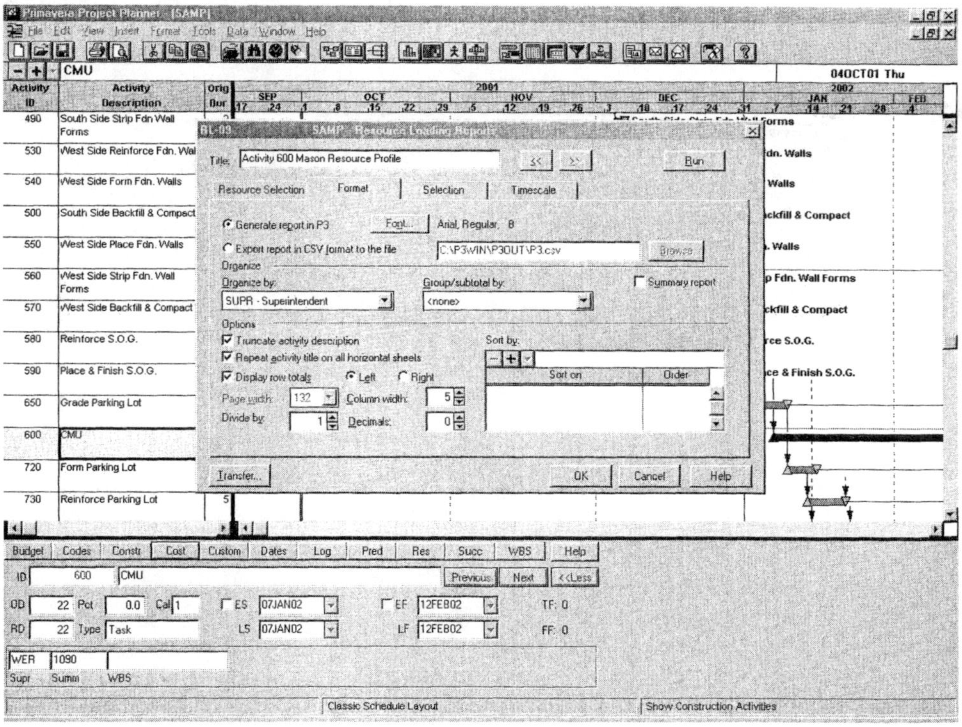

Primavera Project Planner - [SAMP]
File Edit View Insert Format Tools Data Window Help
CMU
04OCT01 Thu

Activity ID	Activity Description	Orig Dur
490	South Side Strip Fdn Wall Forms	
530	West Side Reinforce Fdn. Wal	
540	West Side Form Fdn. Walls	
500	South Side Backfill & Compact	
550	West Side Place Fdn. Walls	
560	West Side Strip Fdn. Wall Forms	
570	West Side Backfill & Compact	
580	Reinforce S.O.G.	
590	Place & Finish S.O.G.	
650	Grade Parking Lot	
600	CMU	
720	Form Parking Lot	
730	Reinforce Parking Lot	5

2001 SEP 17 24 OCT 1 8 15 22 29 NOV 5 12 19 26 DEC 3 10 17 24 31 2002 JAN 7 14 21 28 FEB 4

Forms
dn. Walls
Walls
ckfill & Compact
. Walls
p Fdn. Wall Forms
ckfill & Compact
ce S.O.G.
ce & Finish S.O.G.

RL-09 SAMP - Resource Loading Reports

Title: Activity 600 Mason Resource Profile << >> Run

Resource Selection | Format | Selection | Timescale

Generate report in P3 Font... Arial, Regular, 8

Export report in CSV format to the file C:\P3WIN\P3OUT\P3.csv Browse

Organize

Organize by: SUPR - Superintendent

Group/subtotal by: <none>

Summary report

Options

Truncate activity description

Repeat activity title on all horizontal sheets

Display row totals Left Right

Page width: 132 Column width: 5

Divide by: 1 Decimals: 0

Sort by:

Sort on	Order

Transfer... OK Cancel Help

Budget | Codes | Constr | Cost | Custom | Dates | Log | Pred | Res | Succ | WBS | Help

ID 600 CMU Previous Next <<Less

OD 22 Pct 0.0 Cal 1 ES 07JAN02 EF 12FEB02 TF: 0

RD 22 Type Task LS 07JAN02 LF 12FEB02 FF: 0

WER 1090

Supr Summ WBS

Classic Schedule Layout | Show Construction Activities

Screen Image 11-28
Resource Format Tab

report into a desired format. The first options deal with what type of report to generate. The *Generate report in P3* will produce a report just as would any of the previously discussed reports. The *CSV* format develops a file that can be opened using a spreadsheet program and will be discussed in a later chapter.

This feature is very handy since many of the spreadsheet programs have very enhanced graphic capabilities allowing the tabular data to be converted into graphic data with a minimal amount of effort. Another unique feature of this tab is the *Divide by* option found in the lower left-hand corner. This allows for the resource quantities to be divided by a user-defined constant. In a tabular resource loading report all of the dates are listed across the top of the page and their activities and corresponding resource quantities are listed down the page. This format results in rather long reports that require multiple pages of output. If the column width for each of the dates could be reduced, more dates would fit on a single page, which would ultimately reduce the size of the report. This could be accomplished by dividing the resource quantities by 10, 100, or 1000, which would reduce the required column width.

Another useful feature of the *Divide by* option would be to set the time and then divide by the hours in that standard time scale to get equivalent persons. For example, if the time scale is set to weeks and the *Divide by* is set to 40, then the resource loading report would reflect equivalent people rather than workhours. Screen Image 11-29 is an example of the updated Format tab. The *Organize by* option is where the user specifies the vertical organization of the report and sets the subtotals. In Screen Image 11-29 that option has been set to *RES* which will list all activities within a specific resource. The *Sort by* table specifies the order within the *Organize* categories.

The *Selection* tab for the tabular resource loading report is identical to the one that was used in the previous tabular and graphic schedule reports. This window is used to select items in or out of the report. In Screen Image 11-30, a selection criterion is developed to include only Activity 600.

The final window is the *Timescale* tab, as shown in Screen Image 11-31. This window allows the user to define the scale of the report as well as how much of the project will be included on the report. The default start and end dates are the beginning and end of the project. However, if only a portion of the project is being looked at, the dates can be changed to correspond with that desired time frame. The *Unit* chosen will be the smallest units that will be placed horizontally across the top of the page. If the *Divide by* option was used on the *Format* tab, it is important that it is compatible with this scale. For example, if the resources were divided by 40 to develop equivalent manpower, it is essential that the *Unit* be *week*. If the resources were divided by 10 for the length of the workday, then the timescale unit should be *day*.

Once all of the windows have been completed, the *Run* button can be clicked to execute the report specifications. The user will be prompted about where to route the output, just as with all of the other reports. Output 11-3 is an example of the resource loading report for masons. This report tells how many workhours are planned for mason on a daily basis for Activity 600. That report matches the one that was manually generated in Figure 11-2.

Just as a resource loading report was generated to show one craft, such a report could be produced to show the loading of all of the crafts on the project. This would be done by including all the labor resources. Screen Image 11-32 is an example of the *Resource Selection* window that would be required to include all construction crafts on a resource loading report. The *Format* tab is identical to the one in the previous example. The only difference in the *Selection* tab is that all activities are included so there are no criteria specified. Output 11-4 is an excerpt of the resource loading report that includes all of the resources. In Screen Image 11-33 the timescale *Unit* for this report required resources per week to be known, as well as what activities generated the need for the resources.

Screen Image 11-29
Modified Format Window

Screen Image 11-30
Activity Selection Window

Screen Image 11-31
Timescale Tab

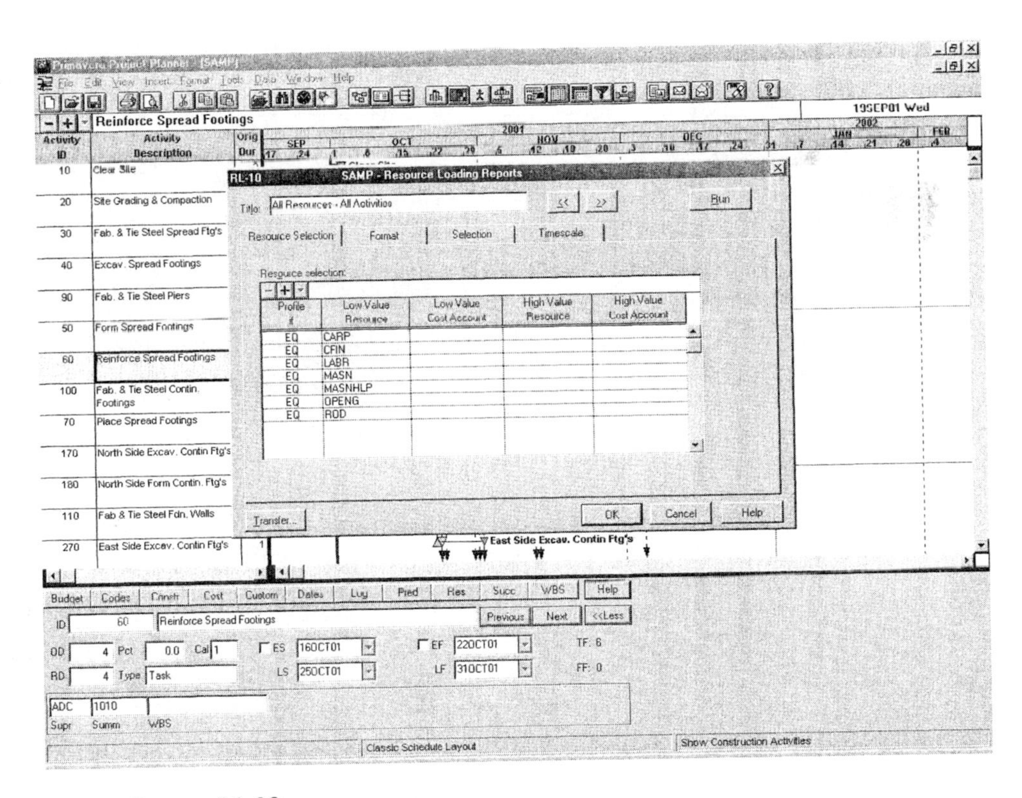

Screen Image 11-32
Resource Selection Window—All Crafts

All Star Developments | PRIMAVERA PROJECT PLANNER | Sample Construction Project

REPORT DATE 04APR00 RUN NO. 68 09:58 | RESOURCE LOADING REPORT | START DATE 01OCT01 FIN DATE 01JUL02

Activity 600 Mason Resource Profile | TOTAL USAGE FOR DAY | DATA DATE 01OCT01 PAGE NO. 1-1

ACT ID	DESC	TOTAL	1JAN 2002	2JAN 2002	3JAN 2002	7JAN 2002	8JAN 2002	9JAN 2002	10JAN 2002	14JAN 2002	15JAN 2002	16JAN 2002
MASN - Masons (WKHR)												
600	CMU	631				29	29	29	29	29	29	29
TOTAL	MASN	**631**				**29**	**29**	**29**	**29**	**29**	**29**	**29**
MASNHLP - Mason Helper (WKHR)												
600	CMU	566				26	26	26	26	26	26	26
TOTAL	**MASNHLP**	**566**				**26**	**26**	**26**	**26**	**26**	**26**	**26**
	REPORT TOTAL	**1197**				**54**	**54**	**54**	**54**	**54**	**54**	**54**

All Star Developments | PRIMAVERA PROJECT PLANNER | Sample Construction Project

REPORT DATE 04APR00 RUN NO. 68 09:58 | RESOURCE LOADING REPORT | START DATE 01OCT01 FIN DATE 01JUL02

Activity 600 Mason Resource Profile | TOTAL USAGE FOR DAY | DATA DATE 01OCT01 PAGE NO. 1-2

ACT ID	17JAN 2002	21JAN 2002	22JAN 2002	23JAN 2002	24JAN 2002	28JAN 2002	29JAN 2002	30JAN 2002	31JAN 2002	4FEB 2002	5FEB 2002	6FEB 2002	7FEB 2002
MASN - Masons (WKHR)													
600	29	29	29	29	29	29	29	29	29	29	29	29	29
TOTAL	**29**	**29**	**29**	**29**	**29**	**29**	**29**	**29**	**29**	**29**	**29**	**29**	**29**
MASNHLP - Mason Helper (WKHR)													
600	26	26	26	26	26	26	26	26	26	26	26	26	26
TOTAL	**26**	**26**	**26**	**26**	**26**	**26**	**26**	**26**	**26**	**26**	**26**	**26**	**26**
	54	**54**	**54**	**54**	**54**	**54**	**54**	**54**	**54**	**54**	**54**	**54**	**54**

All Star Developments | PRIMAVERA PROJECT PLANNER | Sample Construction Project

REPORT DATE 04APR00 RUN NO. 68 09:58 | RESOURCE LOADING REPORT | START DATE 01OCT01 FIN DATE 01JUL02

Activity 600 Mason Resource Profile | TOTAL USAGE FOR DAY | DATA DATE 01OCT01 PAGE NO. 1-3

ACT ID	11FEB 2002	12FEB 2002	13FEB 2002	14FEB 2002	18FEB 2002	19FEB 2002	20FEB 2002
MASN - Masons (WKHR)							
600	29	29					
TOTAL	**29**	**29**					
MASNHLP - Mason Helper (WKHR)							
600	26	26					
TOTAL	**26**	**26**					
	54	**54**					

Output 11-3
Mason and Mason Helper Loading Report

All Star Developments | PRIMAVERA PROJECT PLANNER | Sample Construction Project

REPORT DATE 04APR00 RUN NO. 69 | RESOURCE LOADING REPORT | START DATE 01OCT01 FIN DATE 01JUL02

11:17

DATA DATE 01OCT01 PAGE NO. 1- 1

All Resources - All Activities | TOTAL USAGE FOR WEEK

ACT ID	DESC	TOTAL	1OCT 2001	8OCT 2001	15OCT 2001	22OCT 2001	29OCT 2001	5NOV 2001	12NOV 2001	19NOV 2001	26NOV 2001
CARP - Carpenters (WKHR)											
50	Form Spread Footings	6			6						
180	North Side Form Cont	13				13					
130	Form Piers	105				26	79				
280	East Side Form Conti	8					8				
230	North Side Form Fdn.	100					33	67			
370	South Side Form Cont	10							10		
330	East Side Form Fdn.	60							40	20	
470	West Side Form Conti	9								9	
450	South Side Form Fdn.	40									40
540	West Side Form Fdn.	60									
720	Form Parking Lot	52									
690	Frame Interior Walls	26									
820	Hang & Trim Interior	512									
840	Install Cabinets	179									
TOTAL	CARP	**1180**			**6**	**39**	**120**	**67**	**50**	**29**	**40**
CFIN - Concrete Finishers (WKHR)											
70	Place Spread Footing	8			4	4					
200	North Side Place Con	15				15					
140	Place Piers	4					4				
300	East Side Place Cont	10						10			
240	North Side Place Fdn	2						2			
420	South Side Place Con	15							15		
340	East Side Place Fdn.	2								2	
510	West Side Place Cont	9									9
460	South Side Place Fdn	2									2
550	West Side Place Fdn.	2									
590	Place & Finish S.O.G	537									
740	Place & Finish Parki	470									
760	Place Curbs	378									
TOTAL	CFIN	**1454**			**4**	**19**	**4**	**12**	**15**	**2**	**11**
LABR - Laborers (WKHR)											
40	Excav. Spread Footin	51		13	38						
50	Form Spread Footings	6			6						
170	North Side Excav. Co	60			60						
270	East Side Excav. Con	45				45					
80	Strip Forms - Spread	10				8	3				
360	South Side Excav. Co	60				60					
210	North Side Strip Con	8					8				
150	Strip Forms - Piers	16					4	12			
230	North Side Form Fdn.	50					17	33			
160	Backfill Spread Foot	147						147			
310	East Side Strip Cont	5						5			
250	North Side Strip Fdn	40							40		
390	West Side Excav. Con	40							40		
260	North Side Backfill	20							20		
330	East Side Form Fdn.	30							20	10	
430	South Side Strip Con	9								9	
350	East Side Strip Fdn.	25									25
410	East Side Backfill &	100									100
450	South Side Form Fdn.	80									80
520	West Side Strip Cont	6									
490	South Side Strip Fdn	40									
500	South Side Backfill	20									
540	West Side Form Fdn.	30									
560	West Side Strip Fdn.	25									
570	West Side Backfill &	100									
750	Strip Forms - Parkin	20									
TOTAL	LABR	**1043**		**13**	**104**	**113**	**31**	**197**	**120**	**19**	**205**

Output 11-4

All Crafts Resource Loading Report

All Star Developments | PRIMAVERA PROJECT PLANNER | Sample Construction Project

REPORT DATE 04APR00 RUN NO. 69 11:17 | RESOURCE LOADING REPORT | START DATE 01OCT01 FIN DATE 01JUL02

All Resources - All Activities | TOTAL USAGE FOR WEEK | DATA DATE 01OCT01 PAGE NO. 2- 1

ACT ID	DESC	TOTAL	1OCT 2001	8OCT 2001	15OCT 2001	22OCT 2001	29OCT 2001	5NOV 2001	12NOV 2001	19NOV 2001	26NOV 2001
MASN - Masons (WKHR)											
600	CMU	631									
TOTAL	MASN	631									
MASNHLP - Mason Helper (WKHR)											
600	CMU	566									
TOTAL	MASNHLP	566									
OPENG - Operating Engineer (WKHR)											
40	Excav. Spread Footin	12		3	9						
170	North Side Excav. Co	13			13						
270	East Side Excav. Con	8				8					
360	South Side Excav. Co	13				13					
160	Backfill Spread Foot	10						10			
390	West Side Excav. Con	8							8		
260	North Side Backfill	150							150		
410	East Side Backfill &	12									12
500	South Side Backfill	150									
570	West Side Backfill &	20									
TOTAL	OPENG	396		3	22	21		10	158		12
ROD - Rodbuster (WKHR)											
30	Fab. & Tie Steel Spr	32	11	21							
90	Fab. & Tie Steel Pie	10			10						
60	Reinforce Spread Foo	10			8	3					
100	Fab. & Tie Steel Con	112			112						
110	Fab & Tie Steel Fdn.	24				24					
190	North Side Reinforce	20				20					
120	Reinforce Piers	8				4	4				
220	North Side Reinforce	40					40				
290	East Side Reinforce	20					20				
320	East Side Reinforce	40							40		
380	South Side Reinforce	20							20		
440	South Side Reinforce	40								40	
480	West Side Reinforce	20									20
530	West Side Reinforce	40									
580	Reinforce S.O.G.	170									
730	Reinforce Parking Lo	52									
TOTAL	ROD	658	11	21	130	51	64		60	40	20
	REPORT TOTAL	5928	11	37	266	242	219	286	403	90	288

Output 11-4
(cont.)

The length of the resource loading report can be reduced by running a summary report. With the summary report the detail of how the resources were loaded on each activity is lost. The summary report lists only the resource total for the timescale increment. The only difference in generating the summary report is on the *Format* tab.

If the *Summary report* option is checked, all of the detail is rolled up into one line on the report. Screen Image 11-34 shows the *Format tab* with that option checked. Output 11-5 is the resource summary schedule.

Just as the resources can be looked at by which activities use which resource and when, the report can be reorganized by activity. In that report all of the resources that are required by a specific activity can be seen. The major difference between this report and the previous ones is how the report is arranged. In Screen Image 11-35 the *Organize by* field is changed to *ACT* for activity. Output 11-6 is an excerpt of the report showing which resources are needed for each activity.

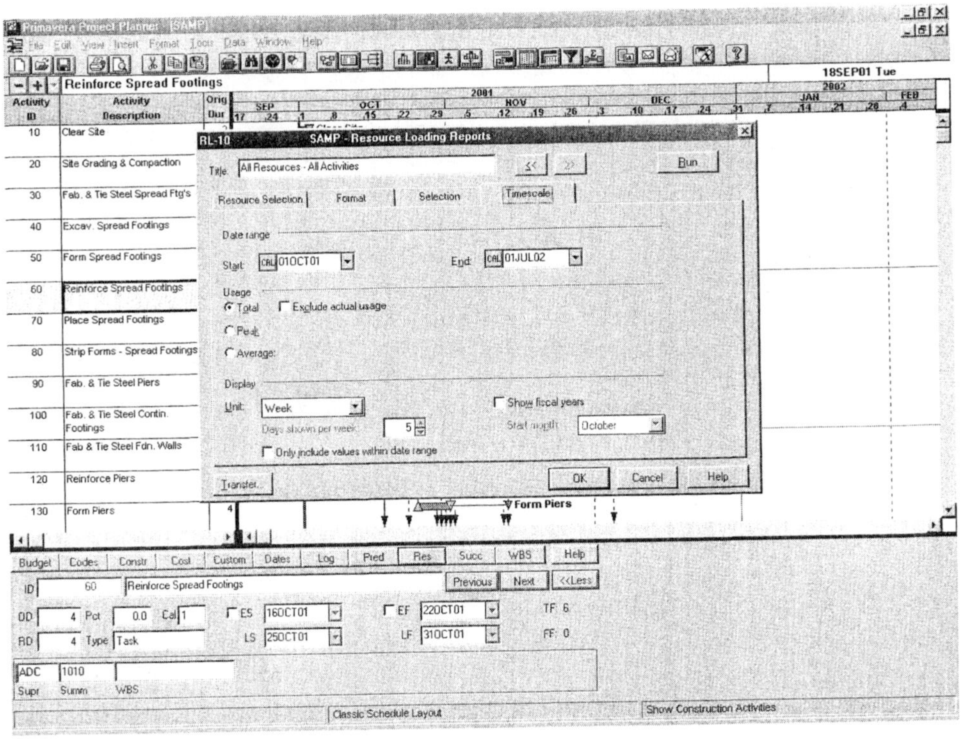

Primavera Project Planner - [SAMP]
File Edit View Insert Format Tools Data Window Help
18SEP01 Tue
Reinforce Spread Footings

Activity ID	Activity Description
10	Clear Site
20	Site Grading & Compaction
30	Fab. & Tie Steel Spread Ftg's
40	Excav. Spread Footings
50	Form Spread Footings
60	Reinforce Spread Footings
70	Place Spread Footings
80	Strip Forms - Spread Footings
90	Fab. & Tie Steel Piers
100	Fab. & Tie Steel Contin. Footings
110	Fab & Tie Steel Fdn. Walls
120	Reinforce Piers
130	Form Piers

RL-10 SAMP - Resource Loading Reports
Title: All Resources - All Activities
Run
Resource Selection | Format | Selection | Timescale
Date range
Start: 01OCT01 End: 01JUL02
Usage
Total Exclude actual usage
Peak
Average:
Display
Unit: Week Show fiscal years
Days shown per week: 5 Start month: October
Only include values within date range
Transfer... OK Cancel Help

Budget | Codes | Constr | Cost | Custom | Dates | Log | Pred | Res | Succ | WBS | Help
ID 60 Reinforce Spread Footings Previous Next <<Less
OD 4 Pct 0.0 Cal 1 ES 16OCT01 EF 22OCT01 TF 6
RD 4 Type Task LS 25OCT01 LF 31OCT01 FF 0
ADC 1010
Supr Summ WBS
Classic Schedule Layout Show Construction Activities

Screen Image 11-33
Resource Timescale Window—All Crafts

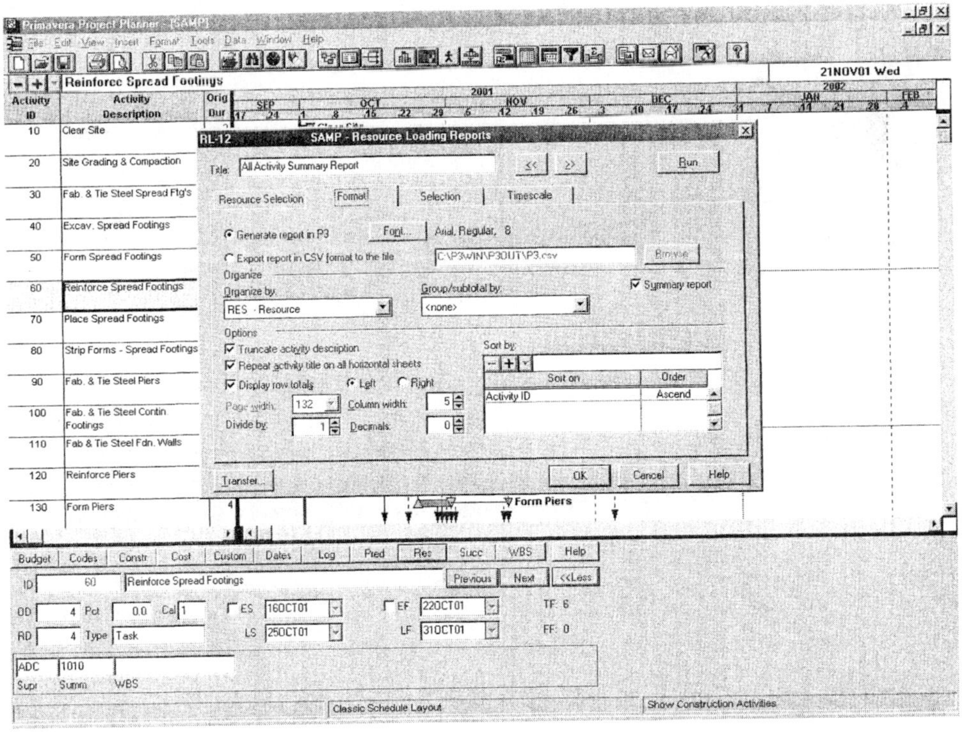

Primavera Project Planner - [SAMP]
File Edit View Insert Format Tools Data Window Help
21NOV01 Wed
Reinforce Spread Footings

Activity ID	Activity Description
10	Clear Site
20	Site Grading & Compaction
30	Fab. & Tie Steel Spread Ftg's
40	Excav. Spread Footings
50	Form Spread Footings
60	Reinforce Spread Footings
70	Place Spread Footings
80	Strip Forms - Spread Footings
90	Fab. & Tie Steel Piers
100	Fab. & Tie Steel Contin. Footings
110	Fab & Tie Steel Fdn. Walls
120	Reinforce Piers
130	Form Piers

RL-12 SAMP - Resource Loading Reports
Title: All Activity Summary Report
Run
Resource Selection | Format | Selection | Timescale
Generate report in P3 Font... Arial, Regular, 8
Export report in CSV format to the file C:\P3WIN\P3OUT\P3.csv Browse
Organize
Organize by: RES - Resource Group/subtotal by: <none> Summary report
Options
Truncate activity description
Repeat activity title on all horizontal sheets
Display row totals Left Right
Page width: 132 Column width: 5
Divide by: 1 Decimals: 0
Sort by:
Sort on Order
Activity ID Ascend
Transfer... OK Cancel Help

Budget | Codes | Constr | Cost | Custom | Dates | Log | Pred | Res | Succ | WBS | Help
ID 60 Reinforce Spread Footings Previous Next <<Less
OD 4 Pct 0.0 Cal 1 ES 16OCT01 EF 22OCT01 TF 6
RD 4 Type Task LS 25OCT01 LF 31OCT01 FF 0
ADC 1010
Supr Summ WBS
Classic Schedule Layout Show Construction Activities

Screen Image 11-34
Resource Summary Report Format

All Star Developments | PRIMAVERA PROJECT PLANNER | Sample Construction Project

REPORT DATE 04APR00 RUN NO. 70 11:27 | RESOURCE LOADING REPORT | START DATE 01OCT01 FIN DATE 01JUL02

All Activity Summary Report | TOTAL USAGE FOR WEEK | DATA DATE 01OCT01 PAGE NO. 1- 1

RESOURCE	RESOURCE DESCRIPTION	TOTAL	1OCT 2001	8OCT 2001	15OCT 2001	22OCT 2001	29OCT 2001	5NOV 2001	12NOV 2001	19NOV 2001
CARP	Carpenters	1180			6	39	120	67	50	29
CFIN	Concrete Finishers	1454			4	19	4	12	15	2
LABR	Laborers	1043		13	104	113	31	197	120	19
MASN	Masons	631								
MASNHLP	Mason Helper	566								
OPENG	Operating Engineer	396		3	22	21		10	158	
ROD	Rodbuster	658	11	21	130	51	64		60	40
	REPORT TOTAL	5928	11	37	266	242	219	286	403	90

All Star Developments | PRIMAVERA PROJECT PLANNER | Sample Construction Project

REPORT DATE 04APR00 RUN NO. 70 11:27 | RESOURCE LOADING REPORT | START DATE 01OCT01 FIN DATE 01JUL02

All Activity Summary Report | TOTAL USAGE FOR WEEK | DATA DATE 01OCT01 PAGE NO. 1- 2

RESOURCE	26NOV 2001	3DEC 2001	10DEC 2001	17DEC 2001	24DEC 2001	31DEC 2001	7JAN 2002	14JAN 2002	21JAN 2002	28JAN 2002	4FEB 2002	11FEB 2002
CARP	40		60				17	35				
CFIN	11		2			537		470		378		
LABR	205	46	50	125					20			
MASN							115	115	115	115	115	57
MASNHLP							103	103	103	103	103	51
OPENG	12		150	20								
ROD	20	40			170			42	10			
	288	86	262	145	170	537	235	764	248	596	218	109

All Star Developments | PRIMAVERA PROJECT PLANNER | Sample Construction Project

REPORT DATE 04APR00 RUN NO. 70 11:27 | RESOURCE LOADING REPORT | START DATE 01OCT01 FIN DATE 01JUL02

All Activity Summary Report | TOTAL USAGE FOR WEEK | DATA DATE 01OCT01 PAGE NO. 1- 3

RESOURCE	18FEB 2002	25FEB 2002	4MAR 2002	11MAR 2002	18MAR 2002	25MAR 2002	1APR 2002	8APR 2002	15APR 2002	22APR 2002	29APR 2002	6MAY 2002
CARP					13	13						
CFIN												
LABR												
MASN												
MASNHLP												
OPENG												
ROD												
					13	13						

All Star Developments | PRIMAVERA PROJECT PLANNER | Sample Construction Project

REPORT DATE 04APR00 RUN NO. 70 11:27 | RESOURCE LOADING REPORT | START DATE 01OCT01 FIN DATE 01JUL02

All Activity Summary Report | TOTAL USAGE FOR WEEK | DATA DATE 01OCT01 PAGE NO. 1- 4

RESOURCE	13MAY 2002	20MAY 2002	27MAY 2002	3JUN 2002	10JUN 2002	17JUN 2002	24JUN 2002	1JUL 2002
CARP				631	60			
CFIN								
LABR								
MASN								
MASNHLP								
OPENG								
ROD								
				631	60			

Output 11-5

Craft Loading Summary Report

All Star Developments — PRIMAVERA PROJECT PLANNER — Sample Construction Project

REPORT DATE 31MAR00 RUN NO. 61 18:37 — RESOURCE LOADING REPORT — START DATE 01OCT01 FIN DATE 01JUL02

DATA DATE 01OCT01 PAGE NO. 1- 1

Activity Resource Report — TOTAL USAGE FOR WEEK

RESOURCE	RESOURCE DESCRIPTION	TOTAL	1OCT 2001	8OCT 2001	15OCT 2001	22OCT 2001	29OCT 2001	5NOV 2001	12NOV 2001	19NOV 2001
30 - Fab. & Tie Steel Spread Ftg's										
ROD	Rodbuster	32	11	21						
TOTAL	**30**	**32**	**11**	**21**						
40 - Excav. Spread Footings										
LABR	Laborers	51		13	38					
OPENG	Operating Engineer	12		3	9					
TOTAL	**40**	**63**		**16**	**47**					
50 - Form Spread Footings										
CARP	Carpenters	6			6					
LABR	Laborers	6			6					
TOTAL	**50**	**12**			**12**					
60 - Reinforce Spread Footings										
ROD	Rodbuster	10			8	3				
TOTAL	**60**	**10**			**8**	**3**				
70 - Place Spread Footings										
CFIN	Concrete Finishers	8			4	4				
TOTAL	**70**	**8**			**4**	**4**				
80 - Strip Forms - Spread Footings										
LABR	Laborers	10				8	3			
TOTAL	**80**	**10**				**8**	**3**			
90 - Fab. & Tie Steel Piers										
ROD	Rodbuster	10			10					
TOTAL	**90**	**10**			**10**					
100 - Fab. & Tie Steel Contin. Footings										
ROD	Rodbuster	112			112					
TOTAL	**100**	**112**			**112**					
110 - Fab & Tie Steel Fdn. Walls										
ROD	Rodbuster	24				24				
TOTAL	**110**	**24**				**24**				
120 - Reinforce Piers										
ROD	Rodbuster	8				4	4			
TOTAL	**120**	**8**				**4**	**4**			
130 - Form Piers										
CARP	Carpenters	105				26	79			
TOTAL	**130**	**105**				**26**	**79**			
140 - Place Piers										
CFIN	Concrete Finishers	4					4			
TOTAL	**140**	**4**					**4**			

Output 11-6
Resources by Activity

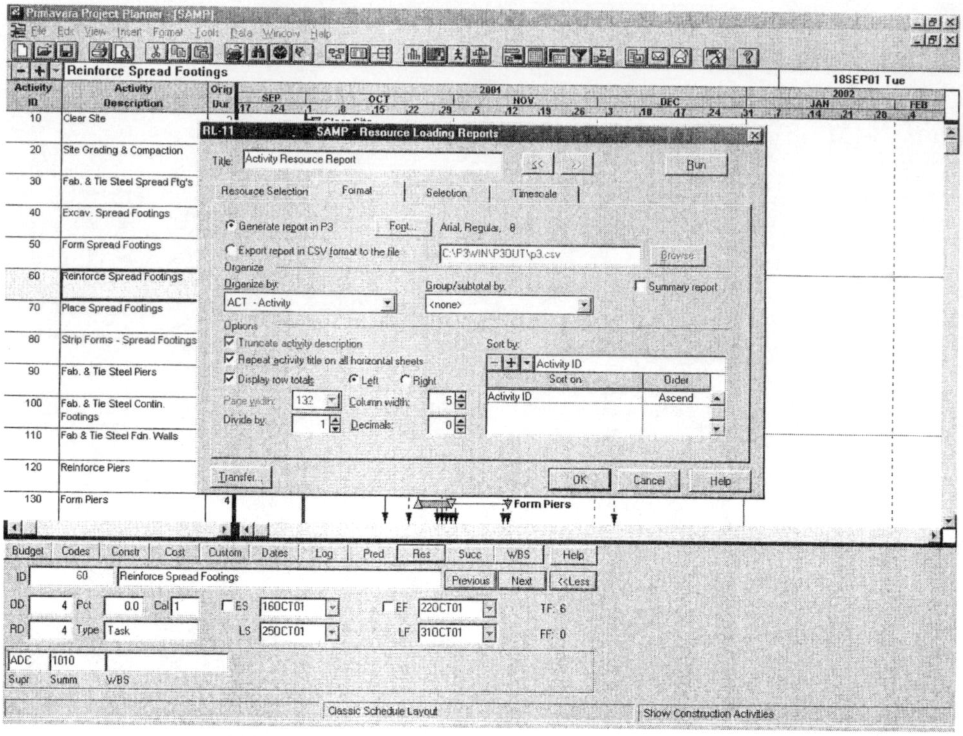

Screen Image 11-35
Choosing to Summarize by Activity

Graphic Resource Output

Primavera Project Planner graphic report options allow the user to automatically present the data from the cost loading report in a graphical format. The first step in the process is to select the *Tools* command followed by the *Graphic Reports* and *Resource and Cost* option as shown in Screen Image 11-36. Just as with all of the previous report specifications, the first window that is opened is one that contains a list of all of the previously developed specifications (Screen Image 11-37). Since project setup varies so much, the *Add* button is clicked to begin with a clean slate. The *Add* button precipitates the opening of the *Add a New Report* window as shown in Screen Image 11-38. That screen image shows the new report will be *RC-11*. The *RC* designation is for resource/cost. Once that input is accepted, Screen Image 11-39 should appear. This screen image contains eight tabs.

Several of these tabs are identical to the ones used when producing graphic schedule reports. The first tab, *Resource Selection*, is where the user designates which resources are to be included within the graphic report. This tab looks very similar to the one used when producing the tabular cost loading report and is used to identify which resources to include within the graphic output. The differences deal with how the information is combined and graphed. The first column, *Group #*, is used to designate if any of the resources will be grouped into a single graph. In Screen Image 11-39 each of the selected resources has its own Group # except for *MASN* and *MASNHLP*. These two resources will be combined into a single graph. The other difference between this tab and the one found in the tabular reports are the right two columns. The *Pattern* column is where the user specifies what fill pattern is to be used for filling in any required areas. The *Pen* column is where the place in the pen carousel is specified. When using a pen plotter, this specifies the color and line thickness.

The *Content* tab allows the user to specify the source of information. In Screen Image 11-40 the *Current* schedule with *Resource* data have been selected. Since

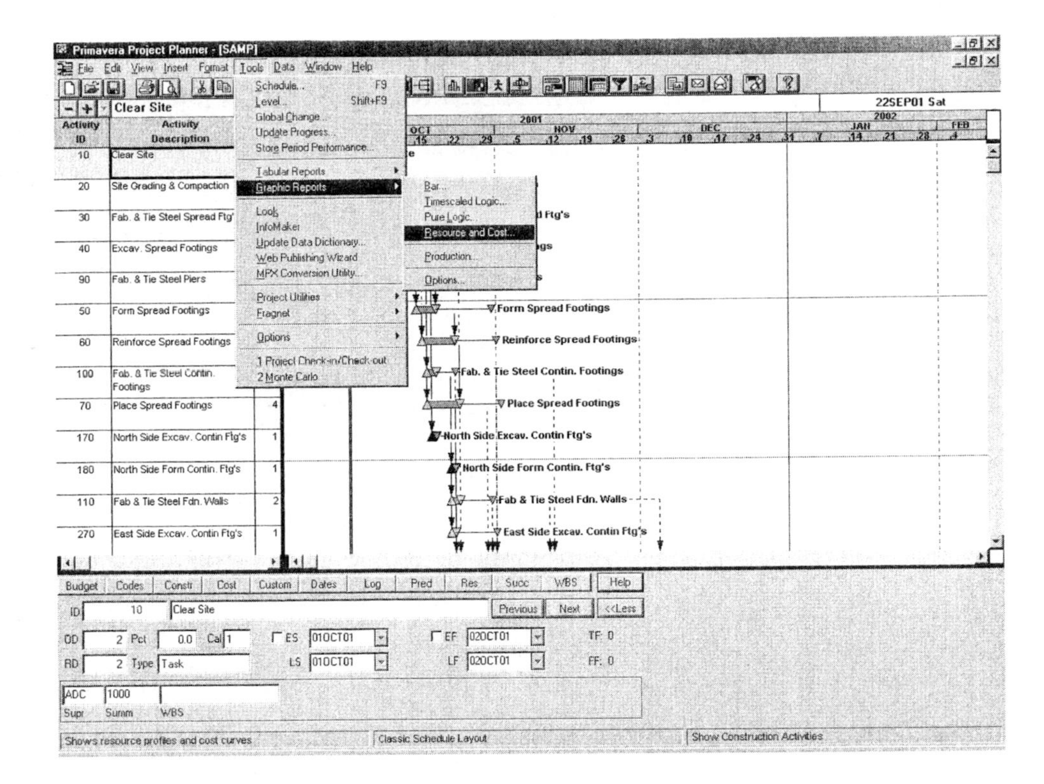

Screen Image 11-36
Opening Resource and Cost Window

Early Dates has been selected, the resources will be spread from the early start to early finish of the individual activities. The resource graphics can be produced either in the form of a histogram or a cumulative curve. In Screen Image 11-40 the *Both* option is selected so both of these curves will be plotted on a single graph. With that option the graph will have two Y-axes. One will be for the incremental quantities and the other for the cumulative quantities. The third tab is the *Date* tab; it is identical to the one used when generating graphic schedules. In Screen Image 11-41 the beginning and end dates for the time scale are specified.

The *Format* tab shown in Screen Image 11-42 is relatively self-explanatory. This tab allows the user to place vertical sight lines to improve the readability of the graph. In that screen image there are major sight lines set for every month and minor ones for every week. Screen Image 11-43 shows the *Title* tab. This tab allows the user to specify the title for the X and Y-axis. In addition, there is a box that allows the user to specify a name for the groups that were created with the *Resource Selection* tab. If no *Group* or *Title* is specified, the group title will be the resource name. The problem occurs when there are multiple resources included in the same group. If nothing is specified within the tab, the report will not have a title. Therefore, it is imperative for a title to be specified for grouped resources. In Screen Image 11-43, Group 4, which is the combination of two resources, is given the title of *Masons & Helpers*. This is the exact title that will appear at the top of the graph. The *Pen* (Screen Image 11-44) and *Size* (Screen Image 11-45) tabs allow the user to specify text size and color as well as the size, color, and fill pattern for the individual graphic items. The final tab is the *Selection* tab (Screen Image 11-46), which is identical to the selection criteria that have been used whenever generating output. When the specifications are complete, the *Run* button can be clicked. After a few moments of generating the image, the user will be prompted with the number of pages and given the option to continue or cancel the print job. Screen Image 11-47 is that prompt. Outputs 11-7, 11-8, and 11-9 are

Screen Image 11-37
Resource/Cost Graphics Window

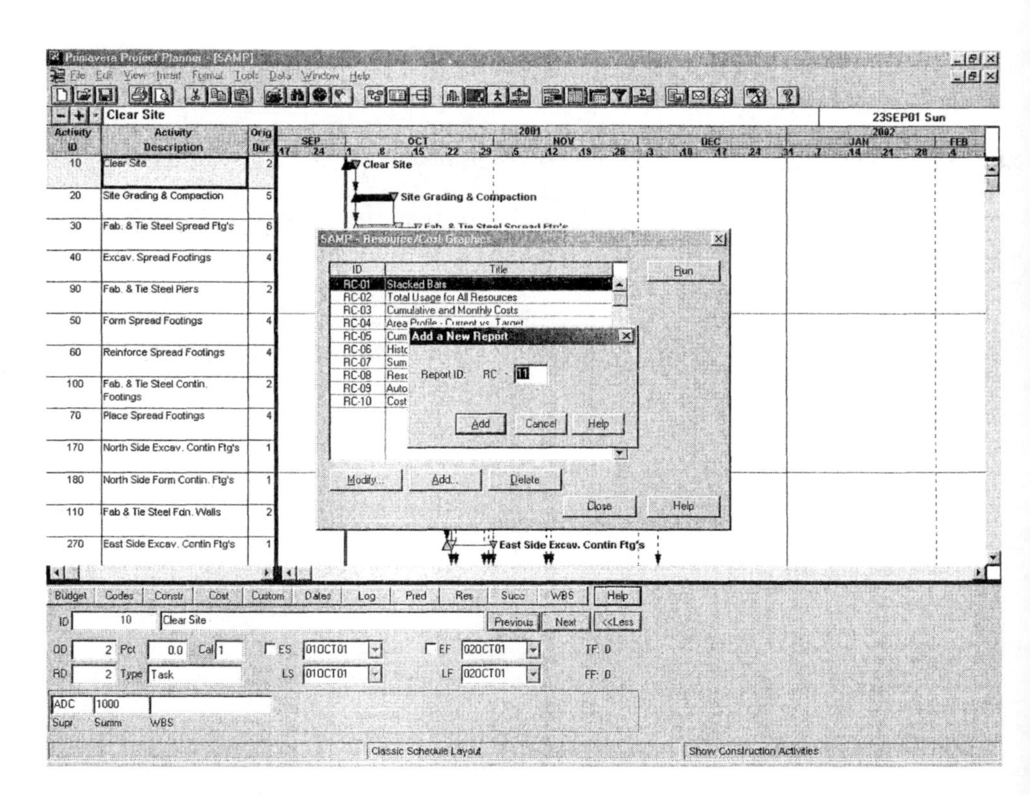

Screen Image 11-38
Add a New Report Window

Primavera Project Planner - [SAMP]
File Edit View Insert Format Tools Data Window Help
Clear Site
23SEP01 Sun

Activity ID	Activity Description
10	Clear Site
20	Site Grading & Compaction
30	Fab. & Tie Steel Spread Ftg's
40	Excav. Spread Footings
90	Fab. & Tie Steel Piers
50	Form Spread Footings
60	Reinforce Spread Footings
100	Fab. & Tie Steel Contin. Footings
70	Place Spread Footings
170	North Side Excav. Contin Ftg's
180	North Side Form Contin. Ftg's
110	Fab & Tie Steel Fdn. Walls
270	East Side Excav. Contin Ftg's

Orig 2001 SEP OCT NOV DEC 2002 JAN FEB 7 14 21 28 4

RC-11 SAMP - Resource/Cost Graphics
Title: Resource Curves << >> Run
Resource Selection | Content | Date | Format | Titles | Pen | Size | Selection
Resource selection:
ROD

Group #	Profile #	Low Value Resource	Low Value Cost Account	High Value Resource	High Value Cost Account	Pattern	Pen
1	EQ	CARP				Mask	0
2	EQ	CFIN				Mask	0
3	EQ	LABR				Mask	0
4	EQ	MASN				Mask	0
4	EQ	MASNHI P				Mask	0
5	EQ	OPENG				Mask	0
6	EQ	ROD				Mask	0

Transfer... OK Cancel Help

Budget | Codes | Constr | Cost | Custom | Dates | Log | Pred | Res | Succ | WBS | Help
ID 10 Clear Site Previous Next <<Less
OD 2 Pct 0.0 Cal 1 ES 01OCT01 EF 02OCT01 TF: 0
RD 2 Type Task LS 01OCT01 LF 02OCT01 FF: 0
ADC 1000
Supr Summ WBS
Classic Schedule Layout | Show Construction Activities

Screen Image 11-39
Resource Selection

Primavera Project Planner - [SAMP]
File Edit View Insert Format Tools Data Window Help
Clear Site
23SEP01 Sun

Activity ID	Activity Description
10	Clear Site
20	Site Grading & Compaction
30	Fab. & Tie Steel Spread Ftg's
40	Excav. Spread Footings
90	Fab. & Tie Steel Piers
50	Form Spread Footings
60	Reinforce Spread Footings
100	Fab. & Tie Steel Contin. Footings
70	Place Spread Footings
170	North Side Excav. Contin Ftg's
180	North Side Form Contin. Ftg's
110	Fab & Tie Steel Fdn. Walls
270	East Side Excav. Contin Ftg's

Orig 2001 SEP OCT NOV DEC 2002 JAN FEB 7 14 21 28 4

RC-11 SAMP - Resource/Cost Graphics
Title: Resource Curves << >> Run
Resource Selection | Content | Date | Format | Titles | Pen | Size | Selection
Schedule: ☑ Current ☐ Target 1 ☐ Target 2
Show data: ◉ Resource ○ Cost
Dates: ◉ Early ○ Late ○ Both
Divide by: 1
Curve
○ Cumulative ○ Histogram ◉ Both
Cumulative
☐ Display % on Y axis
Base % on:
Current early
☑ Display labels
Histogram
Format: ◉ Bar ○ Area
Use/Cost: ◉ Total ○ Peak ○ Average
Use calendar: Decimal places: 0
☐ Display data values
Transfer... OK Cancel Help

Budget | Codes | Constr | Cost | Custom | Dates | Log | Pred | Res | Succ | WBS | Help
ID 10 Clear Site Previous Next <<Less
OD 2 Pct 0.0 Cal 1 ES 01OCT01 EF 02OCT01 TF: 0
RD 2 Type Task LS 01OCT01 LF 02OCT01 FF: 0
ADC 1000
Supr Summ WBS
Classic Schedule Layout | Show Construction Activities

Screen Image 11-40
Content Tab

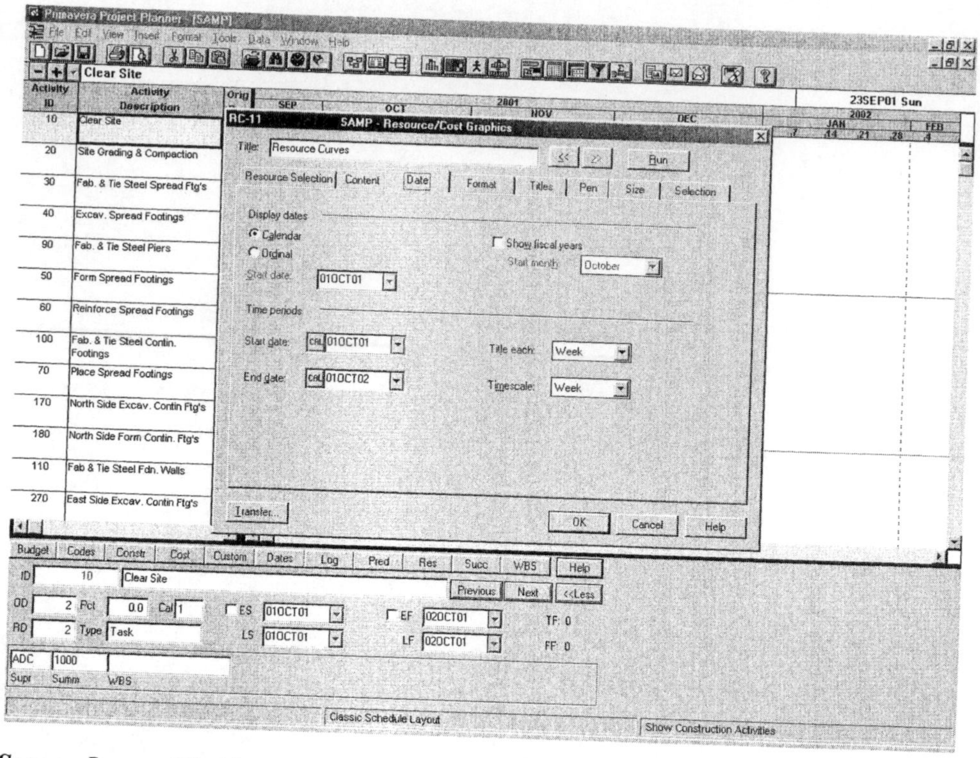

Primavera Project Planner - [SAMP]
File Edit View Insert Format Tools Data Window Help
Clear Site

Activity ID	Activity Description
10	Clear Site
20	Site Grading & Compaction
30	Fab. & Tie Steel Spread Ftg's
40	Excav. Spread Footings
90	Fab. & Tie Steel Piers
50	Form Spread Footings
60	Reinforce Spread Footings
100	Fab. & Tie Steel Contin. Footings
70	Place Spread Footings
170	North Side Excav. Contin Ftg's
180	North Side Form Contin. Ftg's
110	Fab & Tie Steel Fdn. Walls
270	East Side Excav. Contin Ftg's

Orig SEP OCT 2001 NOV DEC 23SEP01 Sun 2002 JAN FEB 7 14 21 28 4
RC-11 SAMP - Resource/Cost Graphics
Title: Resource Curves << >> Run
Resource Selection | Content | Date | Format | Titles | Pen | Size | Selection
Display dates
Calendar
Ordinal
Start date: 01OCT01
Show fiscal years
Start month: October
Time periods
Start date: CAL 01OCT01
End date: CAL 01OCT02
Title each: Week
Timescale: Week
Transfer... OK Cancel Help
Budget | Codes | Constr | Cost | Custom | Dates | Log | Pred | Res | Succ | WBS | Help
ID 10 Clear Site Previous Next <<Less
OD 2 Pct 0.0 Cal 1 ES 01OCT01 EF 02OCT01 TF: 0
RD 2 Type Task LS 01OCT01 LF 02OCT01 FF: 0
ADC 1000
Supr Summ WBS
Classic Schedule Layout Show Construction Activities

Screen Image 11-41
Date Tab

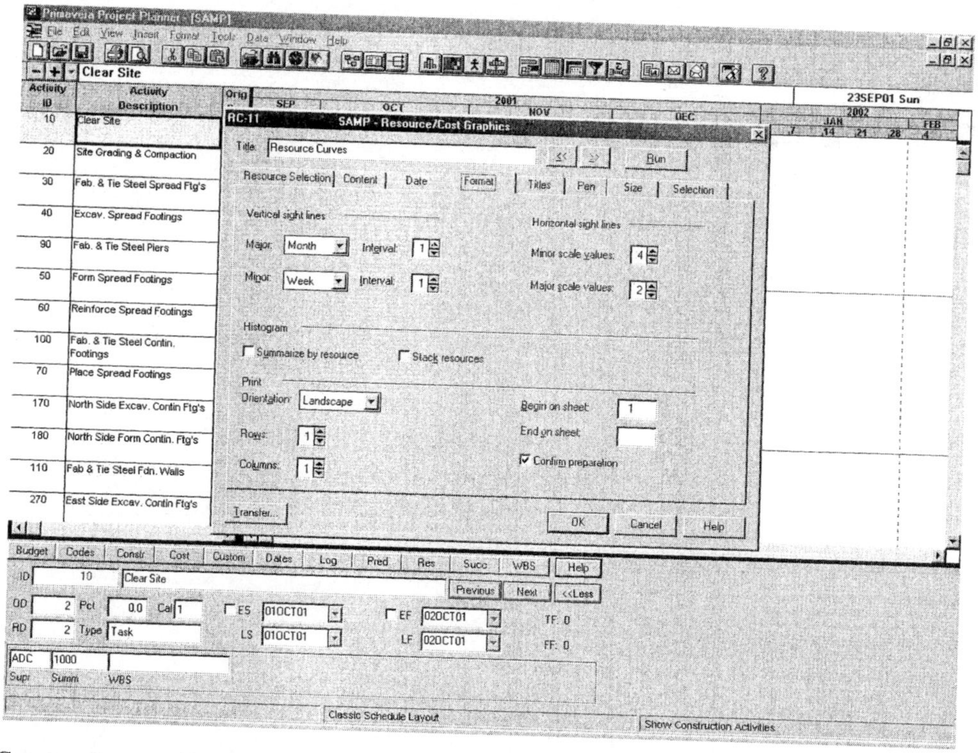

Primavera Project Planner - [SAMP]
File Edit View Insert Format Tools Data Window Help
Clear Site

Activity ID	Activity Description
10	Clear Site
20	Site Grading & Compaction
30	Fab. & Tie Steel Spread Ftg's
40	Excav. Spread Footings
90	Fab. & Tie Steel Piers
50	Form Spread Footings
60	Reinforce Spread Footings
100	Fab. & Tie Steel Contin. Footings
70	Place Spread Footings
170	North Side Excav. Contin Ftg's
180	North Side Form Contin. Ftg's
110	Fab & Tie Steel Fdn. Walls
270	East Side Excav. Contin Ftg's

Orig SEP OCT 2001 NOV DEC 23SEP01 Sun 2002 JAN FEB 7 14 21 28 4
RC-11 SAMP - Resource/Cost Graphics
Title: Resource Curves << >> Run
Resource Selection | Content | Date | Format | Titles | Pen | Size | Selection
Vertical sight lines
Major: Month Interval: 1
Minor: Week Interval: 1
Horizontal sight lines
Minor scale values: 4
Major scale values: 2
Histogram
Summarize by resource
Stack resources
Print
Orientation: Landscape
Rows: 1
Columns: 1
Begin on sheet: 1
End on sheet:
Confirm preparation
Transfer... OK Cancel Help
Budget | Codes | Constr | Cost | Custom | Dates | Log | Pred | Res | Succ | WBS | Help
ID 10 Clear Site Previous Next <<Less
OD 2 Pct 0.0 Cal 1 ES 01OCT01 EF 02OCT01 TF: 0
RD 2 Type Task LS 01OCT01 LF 02OCT01 FF: 0
ADC 1000
Supr Summ WBS
Classic Schedule Layout Show Construction Activities

Screen Image 11-42
Format Tab

Primavera Project Planner - [SAMP]
File Edit View Insert Format Tools Data Window Help
Clear Site
23SEP01 Sun

Activity ID	Activity Description
10	Clear Site
20	Site Grading & Compaction
30	Fab. & Tie Steel Spread Ftg's
40	Excav. Spread Footings
90	Fab. & Tie Steel Piers
50	Form Spread Footings
60	Reinforce Spread Footings
100	Fab. & Tie Steel Contin. Footings
70	Place Spread Footings
170	North Side Excav. Contin Ftg's
180	North Side Form Contin. Ftg's
110	Fab & Tie Steel Fdn. Walls
270	East Side Excav. Contin Ftg's

Orig 2001 SEP OCT NOV DEC 2002 JAN FEB 7 14 21 28 4

RC-11 SAMP - Resource/Cost Graphics

Title: Resource Curves << >> Run

Resource Selection | Content | Date | Format | Titles | Pen | Size | Selection

X axis title: Weeks Y axis title: Work Hours

Masons & Helpers

Group	Title	Max Y-Scale
4	Masons & Helpers	0

Title block
Placement: All ☑ Display project and filename ☑ Show legend
Line 1: All Star Developments
Line 2: Sample Construction Project
Rev title:

Transfer... OK Cancel Help

Budget | Codes | Constr | Cost | Custom | Dates | Log | Pred | Res | Succ | WBS | Help
ID 10 Clear Site Previous Next <<Less
OD 2 Pct 0.0 Cal 1 ES 01OCT01 EF 02OCT01 TF: 0
RD 2 Type Task LS 01OCT01 LF 02OCT01 FF: 0
ADC 1000
Supr Summ WBS
Classic Schedule Layout Show Construction Activities

Screen Image 11-43
Title Tab

Primavera Project Planner - [SAMP]
File Edit View Insert Format Tools Data Window Help
Clear Site
23SEP01 Sun

Activity ID	Activity Description
10	Clear Site
20	Site Grading & Compaction
30	Fab. & Tie Steel Spread Ftg's
40	Excav. Spread Footings
90	Fab. & Tie Steel Piers
50	Form Spread Footings
60	Reinforce Spread Footings
100	Fab. & Tie Steel Contin. Footings
70	Place Spread Footings
170	North Side Excav. Contin Ftg's
180	North Side Form Contin. Ftg's
110	Fab & Tie Steel Fdn. Walls
270	East Side Excav. Contin Ftg's

Orig 2001 SEP OCT NOV DEC 2002 JAN FEB 7 14 21 28 4

RC-11 SAMP - Resource/Cost Graphics

Title: Resource Curves << >> Run

Resource Selection | Content | Date | Format | Titles | Pen | Size | Selection

Elements:
2 Strikes

Item	Label	Pen	Line Weight	Line Type	Fill Pen	Fill Pattern
Current Early		1	2 Strikes	-------	3	XXXX
Current Late		1	2 Strikes	-------	2	Fill
Target 1 Early		1	2 Strikes	-------	1	Fill
Target 1 Late		1	2 Strikes	-------	1	Fill
Target 2 Early		1	2 Strikes	-------	1	Fill
Target 2 Late		1	2 Strikes	-------	1	Fill
Above Normal Limit		1			1	Fill
Above Maximum Limit		1			1	Fill
Normal Limit		1	1 Strike	-------		
Maximum Limit		1	1 Strike	-------		
X Axis Value		1				
Y Axis Value		1				

Transfer... OK Cancel Help

Budget | Codes | Constr | Cost | Custom | Dates | Log | Pred | Res | Succ | WBS | Help
ID 10 Clear Site Previous Next <<Less
OD 2 Pct 0.0 Cal 1 ES 01OCT01 EF 02OCT01 TF: 0
RD 2 Type Task LS 01OCT01 LF 02OCT01 FF: 0

Classic Schedule Layout Show Construction Activities

Screen Image 11-44
Pen Tab

Primavera Project Planner - [SAMP]

File Edit View Insert Format Tools Data Window Help

Clear Site

Activity ID	Activity Description
10	Clear Site
20	Site Grading & Compaction
30	Fab. & Tie Steel Spread Ftg's
40	Excav. Spread Footings
90	Fab. & Tie Steel Piers
50	Form Spread Footings
60	Reinforce Spread Footings
100	Fab. & Tie Steel Contin. Footings
70	Place Spread Footings
170	North Side Excav. Contin Ftg's
180	North Side Form Contin. Ftg's
110	Fab & Tie Steel Fdn. Walls
270	East Side Excav. Contin Ftg's

Orig 2001 SEP OCT NOV DEC 23SEP01 Sun 2002 JAN FEB 7 14 21 28 4

RC-11 SAMP - Resource/Cost Graphics

Title: Resource Curves << >> Run

Resource Selection | Content | Date | Format | Titles | Pen | Size | Selection

Specifications:

Item	Point Size
X Axis Value	9
Y Axis Value	9
X Axis Title	12
Y Axis Title	12
Data Value	9
Group Title	18
Title Block	9

Transfer... OK Cancel Help

Budget | Codes | Constr | Cost | Custom | Dates | Log | Pred | Res | Succ | WBS | Help

ID 10 Clear Site Previous Next <<Less

OD 2 Pct 0.0 Cal 1 ES 01OCT01 EF 02OCT01 TF: 0

RD 2 Type Task LS 01OCT01 LF 02OCT01 FF: 0

ADC 1000

Supr Summ WBS

Classic Schedule Layout Show Construction Activities

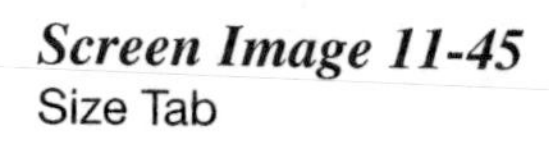

Screen Image 11-45
Size Tab

Primavera Project Planner - [SAMP]

File Edit View Insert Format Tools Data Window Help

Clear Site

Activity ID	Activity Description
10	Clear Site
20	Site Grading & Compaction
30	Fab. & Tie Steel Spread Ftg's
40	Excav. Spread Footings
90	Fab. & Tie Steel Piers
50	Form Spread Footings
60	Reinforce Spread Footings
100	Fab. & Tie Steel Contin. Footings
70	Place Spread Footings
170	North Side Excav. Contin Ftg's
180	North Side Form Contin. Ftg's
110	Fab & Tie Steel Fdn. Walls
270	East Side Excav. Contin Ftg's

Orig 2001 SEP OCT NOV DEC 23SEP01 Sun 2002 JAN FEB 7 14 21 28 4

RC-11 SAMP - Resource/Cost Graphics

Title: Resource Curves << >> Run

Resource Selection | Content | Date | Format | Titles | Pen | Size | Selection

Level 1 must meet All / Any of the following criteria

Selection criteria:

1000

Select if	Is	Low Value	High Value
Activity ID	LT	1000	

Transfer... OK Cancel Help

Budget | Codes | Constr | Cost | Custom | Dates | Log | Pred | Res | Succ | WBS | Help

ID 10 Clear Site Previous Next <<Less

OD 2 Pct 0.0 Cal 1 ES 01OCT01 EF 02OCT01 TF: 0

RD 2 Type Task LS 01OCT01 LF 02OCT01 FF: 0

ADC 1000

Supr Summ WBS

Classic Schedule Layout Show Construction Activities

Screen Image 11-46
Selection Tab

Screen Image 11-47
Output Prompt

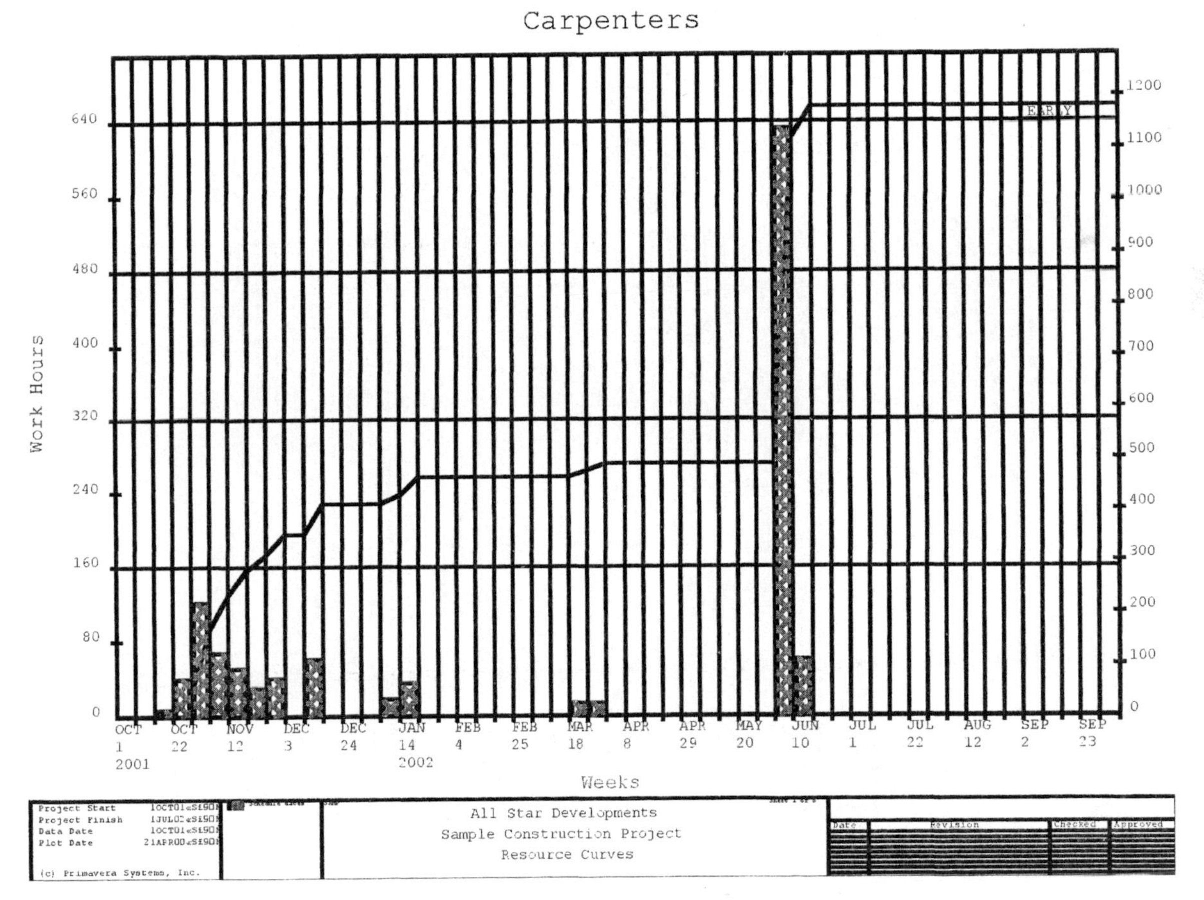

Output 11-7
Carpenters Resource Graphed

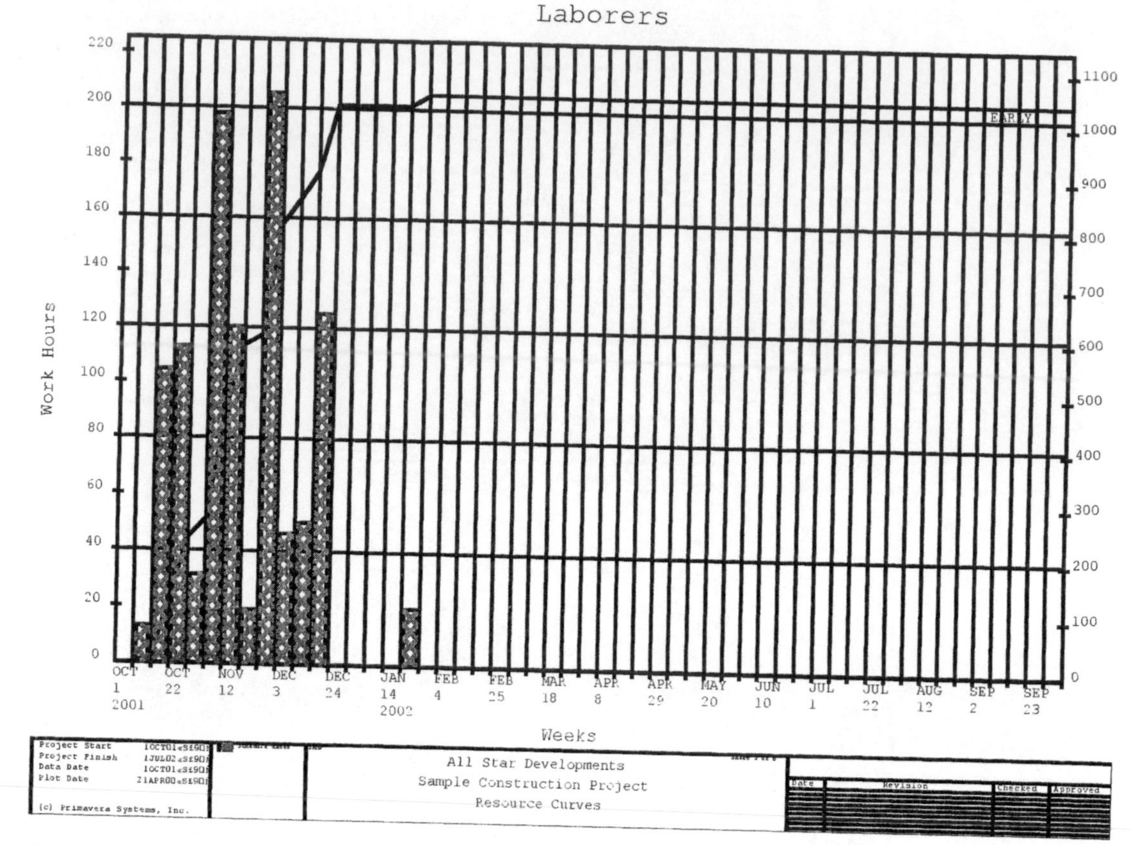

Output 11-8
Laborers Resource Graphed

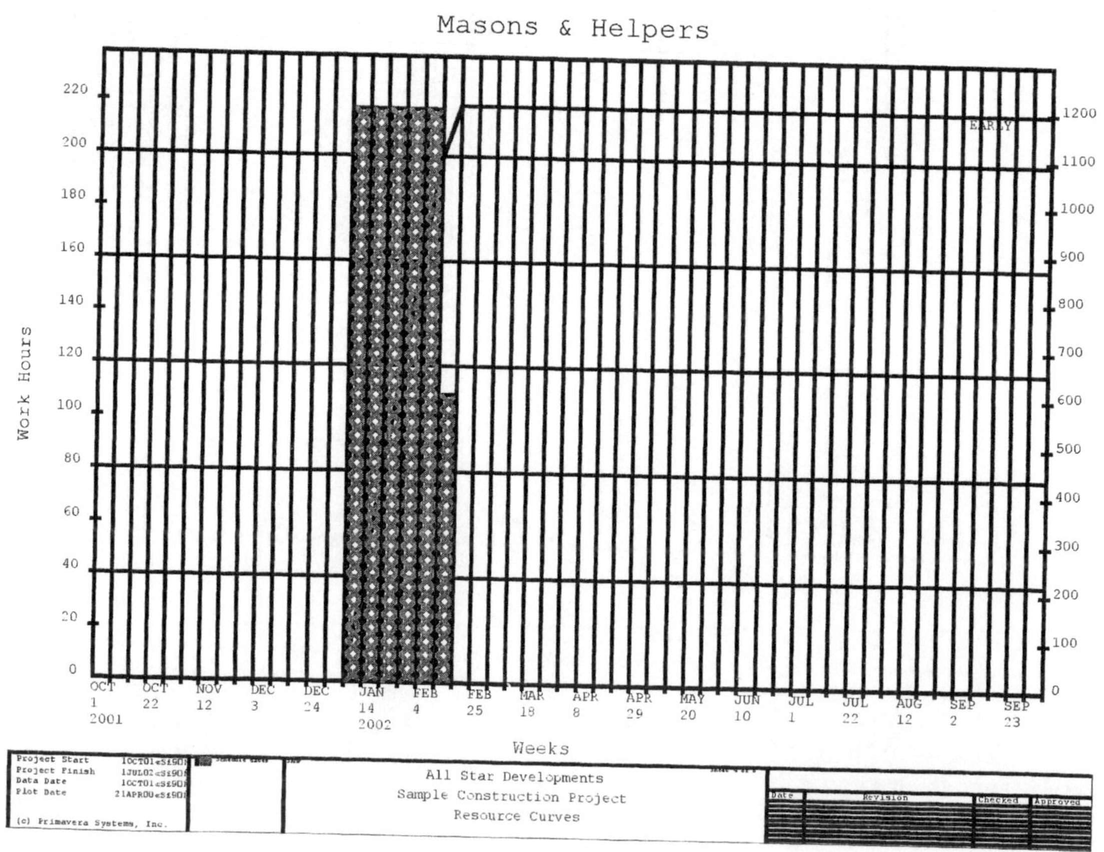

Output 11-9
Masons and Helpers Resources Graphed

excerpts of the graphs from these report specifications. Outputs 11-7 and 11-8 are for the carpenters and laborers. Output 11-8 is the combination of the mason and mason helpers resource.

A review of the first two graphs gives insight into how the start dates should be manipulated. Both the carpenters and laborers curves are not smooth. These curves have ups and downs, meaning that there are periods of inactivity. What this means is that some of the activity start dates need to be shifted in order to smooth out these curves. In the simplest of terms, some of the float will be used to smooth out the resource utilization. This will increase the efficiency of the construction process. Another possibility may be to evaluate the sequence of the activities to see if another sequence might allow for a more effective use of the resources.

Conclusion

Through the use of resource loading, the contractor can determine what resources are required in what quantity and when. The resource distribution is the primary tool that assists the contractor in determining the best time to start a specific activity. Activities on the critical path must begin and end on their required date. Activities with float can be shifted around to help maximize the efficiency of the construction process. Resources that are needed to complete the construction project—such as equipment or labor—need to be brought onto the project in an organized and sustainable fashion. As long as these resources are on the project, they need to be used and then removed when no longer required. Therefore, the shape of resource curves should be positively sloping until they reach a peak, then they should be negatively sloping until that resource is removed from the project. The worst scenario is a resource curve that is up and down throughout the project. This type of curve requires the same resource to be brought onto the job, removed, and brought back later. This process is inefficient, expensive, and causes low morale among the craft personnel. To overcome these difficulties, either the schedule logic needs to be changed or the start dates of those activities with float need to be moved from their early start date to a date that maximizes the effective use of the impacted resource. Through this iterative process, the start and finish dates can be fine-tuned to develop a list of the best start and finish dates.

Suggested Exercises

1. Within Primavera Project Planner, create a minimum of four resources (labor, material, equipment, and subcontracts) for the sample project from earlier exercises. If the estimate has labor information broken down into individual crafts, there should be a resource for each of the individual crafts.
2. Load a comprehensive set of cost accounts.
3. Generate a printout of the cost accounts.
4. Load the resources into the construction logic diagram.
5. Produce a tabular resource loading report for the labor workhours. If the resources were set up by craft, create a report for each of the crafts. Then plot the workhours over time to develop a logical plan for utilizing craft labor.

Cost Loading and Cash Flow

KEY TERMS

Progress Payments Incremental payments from the owner to the contractor for work on the project that is completed and in place.

Line of Credit A preestablished amount of money that a financial institution makes available to a contractor to cover immediate cash requirements.

Loading Cost Data on the Individual Activities

For many resources there is a direct correlation between the quantity of the resource and its cost. With labor, for example, if the carpenter workhours increase by 30 percent the associated cost should increase close to 30 percent. There could be minor deviations based upon pay differentials within the craft. However, if the unit cost is fixed, there would be a direct proportional impact. In that situation, if hours went up 30 percent, the costs would escalate an identical 30 percent since all workhours for that craft are paid the same and there is no pay differential within the craft. Often, however, because of units, control philosophy, or methods of execution certain items do not lend themselves well to this relationship. For example, a subcontracted activity would have a unit of each and a unit price that was equal to the contract amount. In that situation the concept of unit price is meaningless. Another example is with field materials; it may not be practical to establish a unit and unit price for every material. In those situations, only cost will be evaluated and monitored through the scheduling system. That does not diminish the importance of controlling material quantities; it is simply not in the scope of the scheduling system.

In order to add these cost-only items to the schedule, they need to be defined in the same fashion as were the resources. The difference is that no unit price or units will be loaded. The resources are defined via the *Resources* window, which is opened by selecting the *Data* command followed by the *Resources* option from the main command line as shown in Screen Image 12-1. In Screen Image 12-2, three new resources have been added, *DFE, DFM*, and *DFS*. These resources define direct field equipment, materials, and subcontracts. Since these are generic resources, there are no specific quantities and the units are dollars. Once all of the dollar-only resources have been added, click the *Close* button to return to the main input screen. Within Primavera Project Planner there is always a linkage between resources and cost. This can be demonstrated by opening the *Cost* and *Resources* window by using the *Cost* and *Res* buttons on the activity form. In Screen Image 12-3 both of these windows are opened for Activity 600. In that screen image, it can be seen that every resource has a corresponding entry on the *Cost* window. Conversely, if any new cost-only item is generated, a like entry is generated on the *Resources* window. If no unit costs are specified in the resource dictionary, then only the resource name, cost code, and cost category are transferred between the *Cost* and *Resources* window. Since unit cost masons and mason helpers were defined within the resources dictionary, their associated costs were automatically calculated and entered in the appropriate cell within the *Cost* window. The remaining cost columns were manually added using the same process that was used to add resources. The four material cost columns represent the cost for mortar, reinforcing, concrete block, and cleaning.

These material cost items have corresponding account code entries on the *Resources* window; however, there is no associated quantity. The calculation entries on the *Cost* window are handled by the exact same process as are the resources. The *At completion, Budgeted cost, Cost to complete*, and *Variance* are calculated as specified by the *Autocost* window,

Screen Image 12-1
Opening the Resources Window

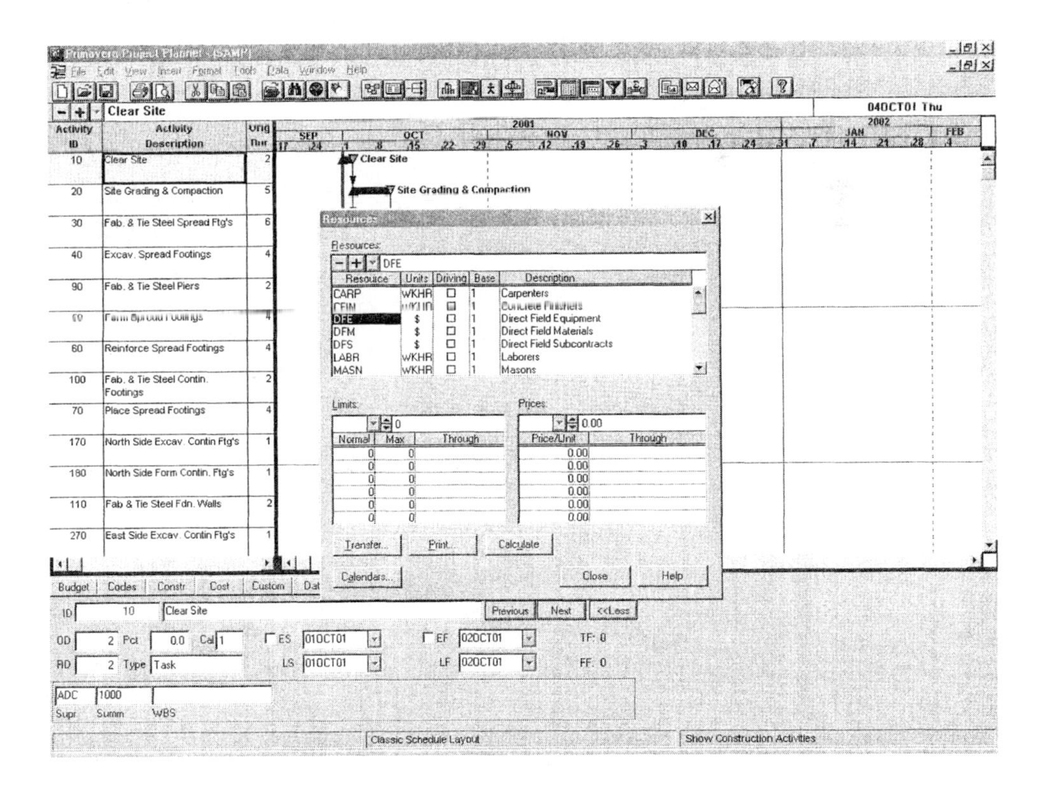

Screen Image 12-2
The Resources Window

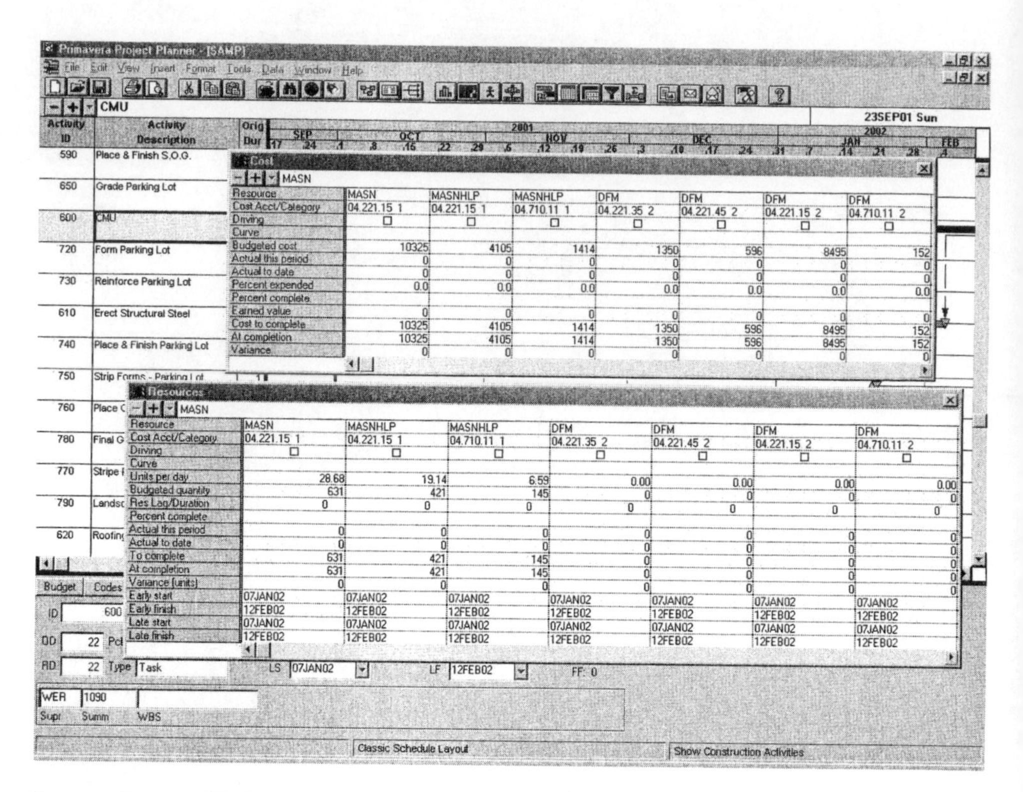

Screen Image 12-3
The Cost and Resources Windows

which was covered in Chapter 11. The cost loading operation needs to be completed for every construction activity on the schedule.

Cash Flow

Cash flow is perhaps one of the most important elements in the overall success of a general or specialty contractor. Since most construction organizations are thinly capitalized, it is essential for them to have a steady supply of cash. This supply of cash can come from completed work, loans, retained earnings, or capital contributions. Of these sources, completed work is the most desirable. The other items all have a real or opportunity cost associated with them. Borrowing funds from a financial institution adds interest expense to the cost of the project. Using retained earnings and capital contributions takes funds that could be used to a higher potential and utilizes them to cover short-term cash needs. Therefore, it is essential for the contractor to have a firm grasp on when and how much money will be received from the owners. Similarly, the contractor must have just as effective a grasp on how much money is owed and when those obligations must be paid.

To facilitate determining the cash requirements, a cash flow plan needs to be developed. This plan will forecast when cash will be received and when obligations need to be paid. By planning for these items, the general contractor can make arrangements in advance for a line of credit. Conversely, when there is a cash surplus, arrangements can be made to invest these funds. Conceptually, cash flow is a simple comparison of cash inflows and outflows. The items that complicate the conceptual model are the commercial terms between the contractor and subcontractors and suppliers and the terms between the contractor and the owner. The first step in the cash flow plan is to develop the inflows, which are a direct function of the schedule of values.

Schedule of Values

Because of the size, complexity, and duration of construction projects it is typical for contractors to receive partial payments during the construction process. These partial payments are based on some amount of the project being completed or materials stored on the project location. The payment arrangement in competitive bid projects is described in the general conditions of the contract. This section will address how often the contractor can submit a request for payment, what the retainage percentage is, how payments will be approved, and how many days the owner has to pay. In the bidding process, these items appear fixed; however, they are often open for negotiation.

Regardless of whether the project is being executed under competitive bid or negotiated lump sum terms, there is typically a schedule of values, which is a list of construction activities with a value attached. The owner will pay the contractor a specific amount of money when an item of work on the schedule of values is completed or partially completed. Items listed on the schedule of values may or may not have any correlation to the contractor's actual cost. There are two reasons for this discrepancy. First, the actual cost of that item of work must be increased proportionally to cover field indirect costs, office overhead, and profit. Second, the contractor has a vested interest in front-end loading the schedule of values. Front-end loading is applying a higher proportion of the project indirect costs and profit to those activities that will be completed the earliest. This provides many advantages to the contractor. First there is the retainage. If 10 percent of all progress payments is being retained by the owner to ensure the completion, the likelihood exists of a negative cash flow throughout the entire project. This would require the contractor to add capital throughout the duration of the project until the retainage is paid to the contractor. This added cash—typically in the form of bank loans—must be paid back with interest. These interest charges add cost to the project. If the need for borrowed capital can be delayed, some of the interest cost will be avoided. If the contractor has enough operating capital to cover the negative cash flow throughout the duration of the project, the contractor will be tying up critical assets without receiving just compensation. Just as the contractor is attempting to front-end load the schedule of values, the owner has just as strong of a vested interest in preventing this from happening. The owner who can avoid front-end loading will have less interest expense since the contractor will be financing a portion of the project. Another reason that owners will resist front-end loading is the fear that if the contractor receives too much of the profit up front, the motivation to complete the project will be diminished. This is why there usually are lively negotiations between the owner and contractor over the schedule of values.

Figure 12-1 is the estimated direct cost by activity for the sample project. In that figure the estimated direct costs are $611,631, field indirect costs are $157,000, home office cost allocation is $36,000, and the profit is $80,769, yielding a bid price, or cost to the owner, of $886,000. To the contractor, the entire bid amounts to potential cash inflows, while the direct cost, indirect costs and home office costs are cash outflows for the project. The difference between the planned cash inflows and planned cost is the anticipated profit. The initial challenge to the contractor is to subdivide the project into a list of payment activities and assign a value to each of these items. The sum of all of the payment activities must equal the bid amount. In the example cited above, the total of all of the payment activities would be $886,000.

There are any numbers of ways payment activities can be developed. However, it is best that they be large enough that they can be easily visualized and measured. Figure 12-2 is an example of how a schedule of values could be developed. That chart attempts to identify the contractor's cost for each of the payment activities and then to assign a sales price for each of these activities. The basis of this list is the summary matrix that is found in Figure 9-1. If the payment activities are listed in the order they will be executed, the contractor can easily inflate the sales price for the early activities and deflate the later activities. If this approach is used, the document must be limited to internal use by the contractor. Once the schedule of values has been prepared, submitted, and approved by the owner, it will become the basis

OFFICE WAREHOUSE PROJECT ACTIVITY ESTIMATE SUMMARY							
Activity	Description	Labor Work Hours	Dollars				Total Cost
			Labor	Material	Equipment	Subcontracts	
10	Clear Site					1,450	1,450
20	Site Grading & Compaction					957	957
30	Fab. & Tie Steel Spread Ftg's	32	424	377			801
40	Excav. Spread Footings	63	728		356		1,084
50	Form Spread Footings	12	164				164
60	Reinforce Spread Footings	10	133				133
70	Place Spread Footings	8	102	781			883
80	Strip Forms - Spread Footings	10	108				108
90	Fab. & Tie Steel Piers	10	133	2,107			2,240
100	Fab. & Tie Steel Contin. Footings	112	1,484	2,036			3,520
110	Fab & Tie Steel Fdn. Walls	24	318	424			742
120	Reinforce Piers	8	106				106
130	Form Piers	105	1,733	354			2,087
140	Place Piers	4	51	258			309
150	Strip Forms - Piers	16	172				172
160	Backfill Spread Footings	157	1,730	995	948		3,673
170	North Side Excav. Contin Ftg's	73	840		394		1,234
180	North Side Form Contin. Ftg's	13	215	345			560
190	North Side Reinforce Contin. Ftg's	20	265				265
200	North Side Place Contin. Ftg's	15	191	1,561			1,752
210	North Side Strip Contin. Ftg's Forms	8	86				86
220	North Side Reinforce Fdn. Walls	40	530				530
230	North Side Form Fdn. Walls	150	2,188	1,271			3,459
240	North Side Place Fdn. Walls	2	26	130			156
250	North Side Strip Fdn. Wall Forms	40	430				430
260	North Side Backfill & Compact	170	2,465	1,122	1,069		4,656
270	East Side Excav. Contin Ftg's	53	604		250		854
280	East Side Form Contin. Ftg's	8	132	218			350
290	East Side Reinforce Contin. Ftg's	20	265				265
300	East Side Place Contin. Ftg's	10	128	976			1,104
310	East Side Strip Contin. Ftg's Forms	5	54				54
320	East Side Reinforce Fdn. Walls	40	530	795			1,325
330	East Side Form Fdn. Walls	90	1,313	795			2,108
340	East Side Place Fdn. Walls	2	26	81			107
350	East Side Strip Fdn. Wall Forms	25	269				269
360	South Side Excav. Contin Ftg's	73	840		394		1,234
370	South Side Form Contin. Ftg's	10	165	221			386
380	South Side Reinforce Contin. Ftg's	20	265				265
390	West Side Excav. Contin Ftg's	48	550		250		800
400	Underground Plumbing					35,000	35,000
410	East Side Backfill & Compact	112	1,255	712	678		2,645
420	South Side Place Contin. Ftg's	15	191	1,561			1,752
430	South Side Strip Contin. Ftg's Forms	9	97				97
440	South Side Reinforce Fdn. Walls	40	530				530
450	South Side Form Fdn. Walls	120	1,520	1,071			2,591
460	South Side Place Fdn. Walls	2	26	130			156
470	West Side Form Contin. Ftg's	9	149	218			367
480	West Side Reinforce Contin. Ftg's	20	265				265
490	South Side Strip Fdn Wall Forms	40	430				430
500	South Side Backfill & Compact	170	2,465	1,122	1,069		4,656
510	West Side Place Contin. Ftg's	9	115	976			1,091
520	West Side Strip Contin. Ftg's Forms	6	65				65
530	West Side Reinforce Fdn. Walls	40	530				530
540	West Side Form Fdn. Walls	90	1,313	795			2,108
550	West Side Place Fdn. Walls	2	26	81			107
560	West Side Strip Fdn. Wall Forms	25	269				269
570	West Side Backfill & Compact	120	1,375	712	678		2,765
580	Reinforce S.O.G.	170	2,253	3,054			5,307
590	Place & Finish S.O.G.	537	6,847	11,148			17,995
600	CMU	1,197	15,844	10,593			26,437

Figure 12-1
Activity List (page 1)

OFFICE WAREHOUSE PROJECT ACTIVITY ESTIMATE SUMMARY							
		Labor Work Hours	Dollars				Total Cost
Activity	Description		Labor	Material	Equipment	Subcontracts	
610	Erect Structural Steel					58,200	58,200
620	Roofing / Flashing					43,981	43,981
630	Exterior Glass & Glazing					9,120	9,120
640	Overhead Doors					8,000	8,000
650	Grade Parking Lot					4,680	4,680
660	Mechanical Rough-in					25,000	25,000
670	Overhead Electrical Rough-in					28,000	28,000
680	Ceiling Grid					5,384	5,384
690	Frame Interior Walls	26	429	3,584			4,013
700	Plumbing Top-out					30,000	30,000
710	Electrical Rough-in					28,000	28,000
720	Form Parking Lot	52	858	214			1,072
730	Reinforce Parking Lot	52	689	1,149			1,838
740	Place & Finish Parking Lot	470	5,993	15,943			21,936
750	Strip Forms - Parking Lot	20	215				215
760	Place Curbs	378	4,820	1,010			5,830
770	Stripe Parking Lot					390	390
780	Final Grading					4,800	4,800
790	Landscape					9,141	9,141
800	Hang Drywall					28,137	28,137
810	Float, Tape & Paint					5,661	5,661
820	Hang & Trim Interior Doors	512	8,448	8,726			17,174
830	Mechanical Trim					28,000	28,000
840	Install Cabinets	179	2,954	13,645			16,599
850	Electrical Trim					27,000	27,000
860	Plumbing Trim					30,000	30,000
870	Ceiling Tiles					5,667	5,667
880	Carpet					16,752	16,752
890	Final Cleaning					1,200	1,200
	Sub Total	5,928	79,734	91,291	6,086	434,520	611,631
	Field Indirects						157,000
	Home Office Cost						36,600
	Total Cost						805,231
	Profit						80,769
	Bid Price						886,000

Figure 12-1
Activity List (page 2)

for all payments to the contractor unless change orders are issued. Change orders can add, delete, or change the value of any item on the schedule of values.

Progress Payments

The general contractor is typically paid by a process called progress payments. These payments, in their simplest form, allow the contractor to be compensated based upon the actual percentage of the payment milestone that is complete. For example, if the payment activity is 50 percent complete, the contractor is eligible to bill half of the scheduled value. In addition, contractors are typically eligible to receive compensation for materials that are stored on the project site. Any number of forms may be used for progress payments, but they typically will all contain the same information. Figure 12-3 is an example of an initial detail sheet for progress payments. This example lists all of the payment activities and their value. The total on the schedule of values must equal the contract amount.

At specified increments during the construction process, the contractor will submit an application for payment to the owner. To complete and submit this request, the contractor must determine the percent complete for each payment activity. Most of the forms require the contractor to certify that the percentages are accurate and all obligations of the contractor are current.

OFFICE WAREHOUSE PROJECT
ACTIVITY ESTIMATE SUMMARY

Line Item	Schedule of Values Description	Underlying Activity Id	Description	Underlying Direct Field Cost	Payment Activity DFC	% Of Direct Field Cost	Underlying Allocation of O, H & P 274,369	Estimated Actual Sales Price	Amount On Schedule Of Values
		10	Clear Site	1,450					
		20	Site Grading & Compaction	957					
1	Sitework				2,407	0.39%	1,080	3,487	$42,000.00
		30	Fab. & Tie Steel Spread Ftg's	801					
		40	Excav. Spread Footings	1,084					
		50	Form Spread Footings	164					
		60	Reinforce Spread Footings	133					
		70	Place Spread Footings	883					
		80	Strip Forms - Spread Footings	108					
2	Spread Footings				3,173	0.52%	1,423	4,596	$18,000.00
		90	Fab. & Tie Steel Piers	2,240					
		100	Fab. & Tie Steel Contin. Footings	3,520					
		110	Fab & Tie Steel Fdn. Walls	742					
		120	Reinforce Piers	106					
		130	Form Piers	2,087					
		140	Place Piers	309					
		150	Strip Forms - Piers	172					
		160	Backfill Spread Footings	3,673					
3	Piers				12,849	2.10%	5,764	18,613	$32,000.00
		170	North Side Excav. Contin Ftg's	1,234					
		180	North Side Form Contin. Ftg's	560					
		190	North Side Reinforce Contin. Ftg's	265					
		200	North Side Place Contin. Ftg's	1,752					
		210	North Side Strip Contin. Ftg's Forms	86					
		220	North Side Reinforce Fdn. Walls	530					
		230	North Side Form Fdn. Walls	3,459					
		240	North Side Place Fdn. Walls	156					
		250	North Side Strip Fdn. Wall Forms	430					
		260	North Side Backfill & Compact	4,656					
4	North Contin. Ftg's & Walls				13,128	2.15%	5,889	19,017	$30,000.00
		270	East Side Excav. Contin Ftg's	854					
		280	East Side Form Contin. Ftg's	350					
		290	East Side Reinforce Contin. Ftg's	265					
		300	East Side Place Contin. Ftg's	1,104					
		310	East Side Strip Contin. Ftg's Forms	54					
		320	East Side Reinforce Fdn. Walls	1,325					
		330	East Side Form Fdn. Walls	2,108					
		340	East Side Place Fdn. Walls	107					
		350	East Side Strip Fdn. Wall Forms	269					
		410	East Side Backfill & Compact	2,645					
5	East Contin. Ftg's & Walls				9,081	1.48%	4,074	13,155	$31,000.00
		360	South Side Excav. Contin Ftg's	1,234					
		370	South Side Form Contin. Ftg's	386					
		380	South Side Reinforce Contin. Ftg's	265					
		420	South Side Place Contin. Ftg's	1,752					
		430	South Side Strip Contin. Ftg's Forms	97					
		440	South Side Reinforce Fdn. Walls	530					
		450	South Side Form Fdn. Walls	2,591					
		460	South Side Place Fdn. Walls	156					
		490	South Side Strip Fdn Wall Forms	430					
		500	South Side Backfill & Compact	4,656					
6	South Contin. Ftg's & Walls				12,097	1.98%	5,427	17,524	$37,000.00

Figure 12-2
Schedule of Values Calculation (page 1)

OFFICE WAREHOUSE PROJECT
ACTIVITY ESTIMATE SUMMARY

Line Item	Schedule of Values Description	Underlying Activity Id	Description	Underlying Direct Field Cost	Payment Activity DFC	% Of Direct Field Cost	Underlying Allocation of O, H & P 274,369	Estimated Actual Sales Price	Amount On Schedule Of Values
		390	West Side Excav. Contin Ftg's	800					
		470	West Side Form Contin. Ftg's	367					
		480	West Side Reinforce Contin. Ftg's	265					
		510	West Side Place Contin. Ftg's	1,091					
		520	West Side Strip Contin. Ftg's Forms	65					
		530	West Side Reinforce Fdn. Walls	530					
		540	West Side Form Fdn. Walls	2,108					
		550	West Side Place Fdn. Walls	107					
		560	West Side Strip Fdn. Wall Forms	269					
		570	West Side Backfill & Compact	2,765					
7	West Contin. Ftg's & Walls				8,367	1.37%	3,753	12,120	$31,000.00
		400	Underground Plumbing	35,000					
		700	Plumbing Top-out	30,000					
		860	Plumbing Trim	30,000					
8	Plumbing				95,000	15.53%	42,616	137,616	$158,000.00
		580	Reinforce S.O.G.	5,307					
		590	Place & Finish S.O.G.	17,995					
9	Slab on grade				23,302	3.81%	10,453	33,755	$55,000.00
		600	CMU	26,437					
		610	Erect Structural Steel	58,200					
		620	Roofing / Flashing	43,981					
10	Exterior Walls & Structure				128,618	21.03%	57,696	186,314	$195,000.00
		630	Exterior Glass & Glazing	9,120					
		640	Overhead Doors	8,000					
11	Exterior Finishes				17,120	2.80%	7,680	24,800	$12,000.00
		650	Grade Parking Lot	4,680					
		720	Form Parking Lot	1,072					
		730	Reinforce Parking Lot	1,838					
		740	Place & Finish Parking Lot	21,936					
		750	Strip Forms - Parking Lot	215					
		760	Place Curbs	5,830					
		770	Stripe Parking Lot	390					
12	Parking Lot				35,961	5.88%	16,132	52,093	$25,000.00
		660	Mechanical Rough-in	25,000					
		710	Electrical Rough-in	28,000					
		830	Mechanical Trim	28,000					
13	Mechanical				81,000	13.24%	36,335	117,335	$70,000.00
		670	Overhead Electrical Rough-in	28,000					
		850	Electrical Trim	27,000					
14	Electrical				55,000	8.99%	24,672	79,672	$55,000.00
		680	Ceiling Grid	5,384					
		690	Frame Interior Walls	4,013					
		800	Hang Drywall	28,137					
		810	Float, Tape & Paint	5,661					
		820	Hang & Trim Interior Doors	17,174					
		840	Install Cabinets	16,599					
		870	Ceiling Tiles	5,667					
		880	Carpet	16,752					
		890	Final Cleaning	1,200					
15	Interior Finishes				100,587	16.45%	45,122	145,709	$85,000.00
		780	Final Grading	4,800					
		790	Landscape	9,141					
16	Landscape & Grading				13,941	2.28%	6,254	20,195	$10,000.00
	TOTAL				611,631	100.00%	823,107	886,000	$886,000.00

Figure 12-2
Schedule of Values Calculation (page 2)

CONTINUATION SHEET *AIA DOCUMENT G703* PAGE OF PAGES

AIA Document G702, APPLICATION AND CERTIFICATION FOR PAYMENT, containing
Contractor's signed Certification is attached.
In tabulation below, amounts are stated to the nearest dollar.
Use Column I on Contracts where variable retainage for line items may apply.

APPLICATION NUMBER:
APPLICATION DATE:
PERIOD FROM:
TO:
ARCHITECT'S PROJECT NO:

A	B	C	D	E	F	G		H	I
ITEM No.	DESCRIPTION OF WORK	SCHEDULED VALUE	WORK COMPLETED			TOTAL COMPLETED AND STORED TO DATE (D + E + F)	% G / C	BALANCE TO FINISH C - G	RETAINAGE
			Previous Applications	This Application Work in Place	Stored Materials (not in D or E)				
1	Sitework	$42,000.00				$0.00		$42,000.00	
2	Spread Footings	$18,000.00				$0.00		$18,000.00	
3	Piers	$32,000.00				$0.00		$32,000.00	
4	North Contin. Ftg's & Walls	$30,000.00				$0.00		$30,000.00	
5	East Contin. Ftg's & Walls	$31,000.00				$0.00		$31,000.00	
6	South Contin. Ftg's & Walls	$37,000.00				$0.00		$37,000.00	
7	West Contin. Ftg's & Walls	$31,000.00				$0.00		$31,000.00	
8	Plumbing	$158,000.00				$0.00		$158,000.00	
9	Slab on Grade	$55,000.00				$0.00		$55,000.00	
10	Exterior Walls & Structutre	$195,000.00				$0.00		$195,000.00	
11	Exterior Finishes	$12,000.00				$0.00		$12,000.00	
12	Parking Lot	$25,000.00				$0.00		$25,000.00	
13	Mechanical	$70,000.00				$0.00		$70,000.00	
14	Electrical	$55,000.00				$0.00		$55,000.00	
15	Interior Finishes	$85,000.00				$0.00		$85,000.00	
16	Landscape & Grading	$10,000.00				$0.00		$10,000.00	
		$886,000.00				$0.00		$886,000.00	

Figure 12-3
AIA Schedule of Values

It is important that the contractor address progress and payments in an honest and professional manner. While the owner has the ability to prevent or minimize front-end loading, owners typically lack the ability to evaluate actual percent compete. This is further complicated by the dynamics of the project. The contractor measures progress and prepares the payment request; however, by the time the owner gets the request and goes out to the site; those percents complete are no longer accurate. The contractor is charged with the responsibility to accurately assess the percent complete at the date the payment request is submitted. Since the contractor holds such an advantage in this situation, estimating future progress or inflating progress is clearly an unethical practice. Since most of these payment requests require a notarized certification, it is in essence a sworn affidavit. In that situation, overstating progress in not only unethical, it is fraudulent.

The request for payments form in Figure 12-3 requires the contractor to input the percent complete and to prorate an equivalent amount of that activity's sales price as earned value. The sum of these earned amounts plus the value of materials stored is the total amount that the contractor has earned. This earned amount less retainage and all previously billed amounts is the amount that the contractor is entitled to bill.

Figures 12-4 and 12-5 are examples of the first application for payment from the schedule of values diagrammed in Figure 12-3. In Figure 12-4, several of the payment activities

CONTINUATION SHEET *AIA DOCUMENT G703* PAGE 1 OF 1 PAGES

AIA Document G702, APPLICATION AND CERTIFICATION FOR PAYMENT, containing
Contractor's signed Certification is attached.
In tabulation below, amounts are stated to the nearest dollar.
Use Column I on Contracts where variable retainage for line items may apply.

APPLICATION NUMBER: 1
APPLICATION DATE: 30-Mar-97
PERIOD FROM: 25-Feb-97
TO: 30-Mar-97
ARCHITECT'S PROJECT NO: 121212

A	B	C	D	E	F	G		H	I
ITEM No.	DESCRIPTION OF WORK	SCHEDULED VALUE	WORK COMPLETED			TOTAL COMPLETED AND STORED TO DATE (D + E + F)	% G / C	BALANCE TO FINISH C - G	RETAINAGE
			Previous Applications	This Application: Work in Place	This Application: Stored Materials (not in D or E)				
1	Sitework	$42,000.00		$42,000.00		$42,000.00	100.00%	$0.00	$4,200.00
2	Spread Footings	$18,000.00		$12,000.00		$12,000.00	66.67%	$6,000.00	$1,200.00
3	Piers	$32,000.00		$16,000.00		$16,000.00	50.00%	$16,000.00	$1,600.00
4	North Contin. Ftg's & Walls	$30,000.00		$10,000.00		$10,000.00	33.33%	$20,000.00	$1,000.00
5	East Contin. Ftg's & Walls	$31,000.00				$0.00	0.00%	$31,000.00	$0.00
6	South Contin. Ftg's & Walls	$37,000.00				$0.00	0.00%	$37,000.00	$0.00
7	West Contin. Ftg's & Walls	$31,000.00				$0.00	0.00%	$31,000.00	$0.00
8	Plumbing	$158,000.00			$15,000.00	$15,000.00	9.49%	$143,000.00	$1,500.00
9	Slab on Grade	$55,000.00				$0.00	0.00%	$55,000.00	$0.00
10	Exterior Walls & Structutre	$195,000.00				$0.00	0.00%	$195,000.00	$0.00
11	Exterior Finishes	$12,000.00				$0.00	0.00%	$12,000.00	$0.00
12	Parking Lot	$25,000.00				$0.00	0.00%	$25,000.00	$0.00
13	Mechanical	$70,000.00			$10,000.00	$10,000.00	14.29%	$60,000.00	$1,000.00
14	Electrical	$55,000.00				$0.00	0.00%	$55,000.00	$0.00
15	Interior Finishes	$85,000.00				$0.00	0.00%	$85,000.00	$0.00
16	Landscape & Grading	$10,000.00				$0.00	0.00%	$10,000.00	$0.00
		$886,000.00	$0.00	$80,000.00	$25,000.00	$105,000.00	11.85%	$781,000.00	$10,500.00

AIA DOCUMENT G702 • CONTINUATION SHEET • APRIL 1978 EDITION • AIA® • © 1978
THE AMERICAN INSTITUTE OF ARCHITECTS, 1735 NEW YORK AVE., N.W., WASHINGTON, D.C. 2006

Figure 12-4
Payment Request

are progressed or have materials stored on site. In this example, the owner retains 10 percent of the requested payments. From Figure 12-4 the project—for payment purposes—is 11.85 percent complete. The contractor has earned $105,000 and must leave $10,500 on deposit with the owner for retainage. This would net the contractor a payment of $94,500 from the owner. This amount is found on the *Application and Certification for Payment* summary found in Figure 12-5. Prior to the owner paying this or any payment request, it typically must be approved by the architect. In addition, the owner is given a set number of days to pay this request.

Project Cash Inflows Plan

After the schedule of values has been determined, it is necessary to develop a cash inflow plan. Due to the uniqueness, size, and complexity of a construction project, this plan requires some sort of computerized spreadsheet program in addition to a scheduling software package. Typically, the terms established between the owner and contractor cannot be detailed in Primavera Project Planner. The basic elements of the cash flow can be generated with Primavera Project Planner, and the specifics of the project's commercial terms are detailed within a spreadsheet program. In addition, most spreadsheet programs have superior graphic capabilities.

The first step in developing the revenue plan is to add the payment activities to the construction schedule. In Primavera Project Planner, hammock activities work as the means to add these activities and relate them to underlying construction activities. The use of

APPLICATION AND CERTIFICATION FOR PAYMENT *AIA DOCUMENT G702* PAGE 1 OF 1 PAGES

TO (Owner): ALL STAR PROJECT: OFFICE WAREHOUSE PROJECT APPLICATION NO: 1 Distribution to: OWNER ARCHITECT CONTRACTOR

PERIOD FROM:

TO: 3/30/97

ATTENTION: I. B. DESIGNER CONTRACTOR FOR: ABC CONSTRUCTION

ARCHITECT'S PROJECT NO: 121212

CONTRACT DATE: 20-Feb-97

CONTRACTOR'S APPLICATION FOR PAYMENT

CHANGE ORDER SUMMARY		ADDITIONS	DEDUCTIONS
Change orders approved in previous months by Owner			
TOTAL			
Approved this Month			
Number	Date Approved		
TOTALS			
Net change by Change Orders			

Application is made for Payment, as shown below, in connection with the Contract Continuation Sheet, AIA Document G703 is attached.

The present status of the account for this Contract is as follows:

ORIGINAL CONTRACT SUM	$886,000.00
Net change by Change Orders	$0.00
CONTRACT SUM TO DATE	$886,000.00
TOTAL COMPLETED & STORED TO DATE (Column G on G703)	$105,000.00
Retainage 10 % or total in Column I on G703	$10,500.00
TOTAL EARNED LESS RETAINAGE	$94,500.00
LESS PREVIOUS CERTIFICATES FOR PAYMENT	$0.00
CURRENT PAYMENT DUE	$94,500.00

The undersigned Contractor to the best of his knowledge, information and belief the Work covered by this Application for Payment has been completed in accordance with the Contract Documents, that all amounts have been paid to him for Work for which previous Certificates for Payment were issued and payments received from the Owner, and that current payment shown herein is now due.

CONTRACTOR:

By: __________ Date: ______

State of: County of:

Subscribed and sworn to before me day of ,19

Notary Public

My Commission expires:

ARCHITECT'S CERTIFICATE FOR PAYMENT

In accordance with the Contract Documents, based on on-site observations and the data comprising the above application, the Architect certifies to the Owner that the Work has progressed to the point indicated; that to the best of his knowledge, information and belief, the quality of the Work is in accordance with the Contract Documents; and that the Contractor is entitled to payment of the AMOUNT CERTIFIED.

AMOUNT CERTIFIED $94,500.00

(Attach explanation if amount certified differs from the amount applied for.)

ARCHITECT:

By: __________ Date:

This Certificate is not negotiable. the AMOUNT CERTIFIED is payable only to the Contractor named herein. Issuance, payment and acceptance of payment are without prejudice to any rights of the Owner or Contractor under this Contract

AIA DOCUMENT G702 • CONTINUATION SHEET • APRIL 1978 EDITION • AIA® • © 1978
THE AMERICAN INSTITUTE OF ARCHITECTS, 1735 NEW YORK AVE., N.W., WASHINGTON, D.C. 2006

Figure 12-5
Application for Payment

hammock activities for an owner's schedule does not preclude using another set of hammock activities to develop a different summary schedule. For example, there may be a company-required summary schedule and a client-required summary schedule. Regardless of the format and use of the summary schedules, they must be tied to the underlying construction network. If multiple summary schedules are required, a block of activity IDs needs to be set aside for each of the summary schedules or an activity code designation needs to be defined and given values to represent each of the summary schedules. Either of these approaches allows for these items to be easily selected for output using the selection criteria.

When preparing client schedules it is a good practice for the payment activities on the schedule of values to be the same items on the construction schedule. This reduces confusion and allows the contractor to retain a greater level of confidentiality.

Once the owner's summary hammock schedule has been created as shown in Figure 12-6, the adding of the schedule of values cost information can begin. The first step is to set up a resource to accommodate these entries. This is done through the *Resources* window, which is opened by selecting the *Data* command followed by the *Resources* option. In Screen Image 12-4 the resource *DRAW* has been added without any limits or unit cost information. Closing this window will return the program to the main input screen. In Screen Image 12-5, the cost window has been opened by clicking on the *Cost* button on the activity form. In that screen image, hammock Activity 1000 has been highlighted, resource *DRAW* entered, and the amount from the schedule of values entered. This amount is entered only in the *Budgeted cost* cell. The remaining entries in the *Cost to Complete*

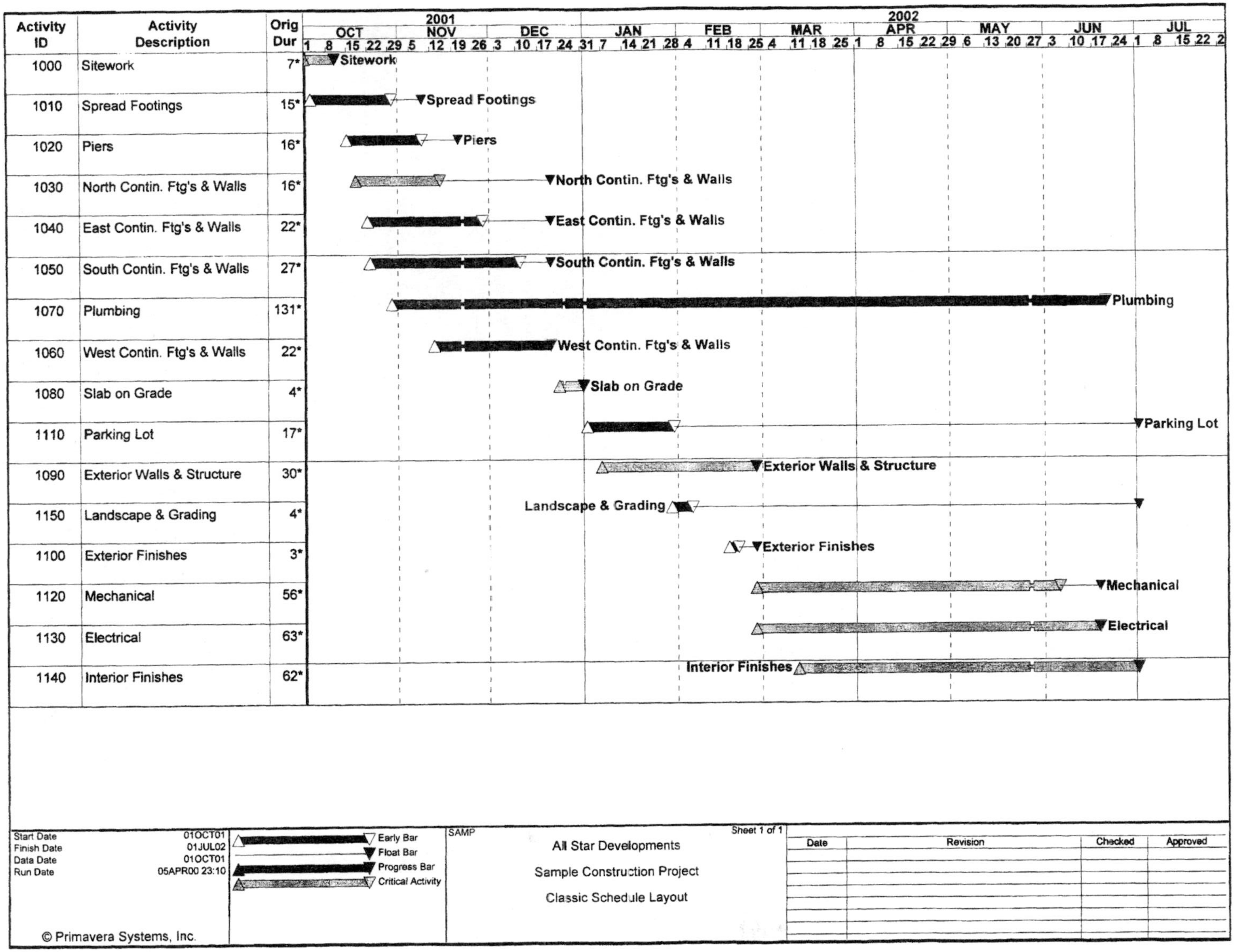
Activity ID
Activity Description
Orig Dur
2001
OCT
NOV
DEC
2002
JAN
FEB
MAR
APR
MAY
JUN
JUL
1 8 15 22 29 5 12 19 26 3 10 17 24 31 7 14 21 28 4 11 18 25 4 11 18 25 1 8 15 22 29 6 13 20 27 3 10 17 24 1 8 15 22 2
1000 Sitework 7*
1010 Spread Footings 15*
1020 Piers 16*
1030 North Contin. Ftg's & Walls 16*
1040 East Contin. Ftg's & Walls 22*
1050 South Contin. Ftg's & Walls 27*
1070 Plumbing 131*
1060 West Contin. Ftg's & Walls 22*
1080 Slab on Grade 4*
1110 Parking Lot 17*
1090 Exterior Walls & Structure 30*
1150 Landscape & Grading 4*
1100 Exterior Finishes 3*
1120 Mechanical 56*
1130 Electrical 63*
1140 Interior Finishes 62*
Sitework
Spread Footings
Piers
North Contin. Ftg's & Walls
East Contin. Ftg's & Walls
South Contin. Ftg's & Walls
Plumbing
West Contin. Ftg's & Walls
Slab on Grade
Parking Lot
Exterior Walls & Structure
Landscape & Grading
Exterior Finishes
Mechanical
Electrical
Interior Finishes
Start Date 01OCT01
Finish Date 01JUL02
Data Date 01OCT01
Run Date 05APR00 23:10
© Primavera Systems, Inc.
Early Bar
Float Bar
Progress Bar
Critical Activity
SAMP
Sheet 1 of 1
All Star Developments
Sample Construction Project
Classic Schedule Layout
Date
Revision
Checked
Approved

Figure 12-6
Client Summary Schedule

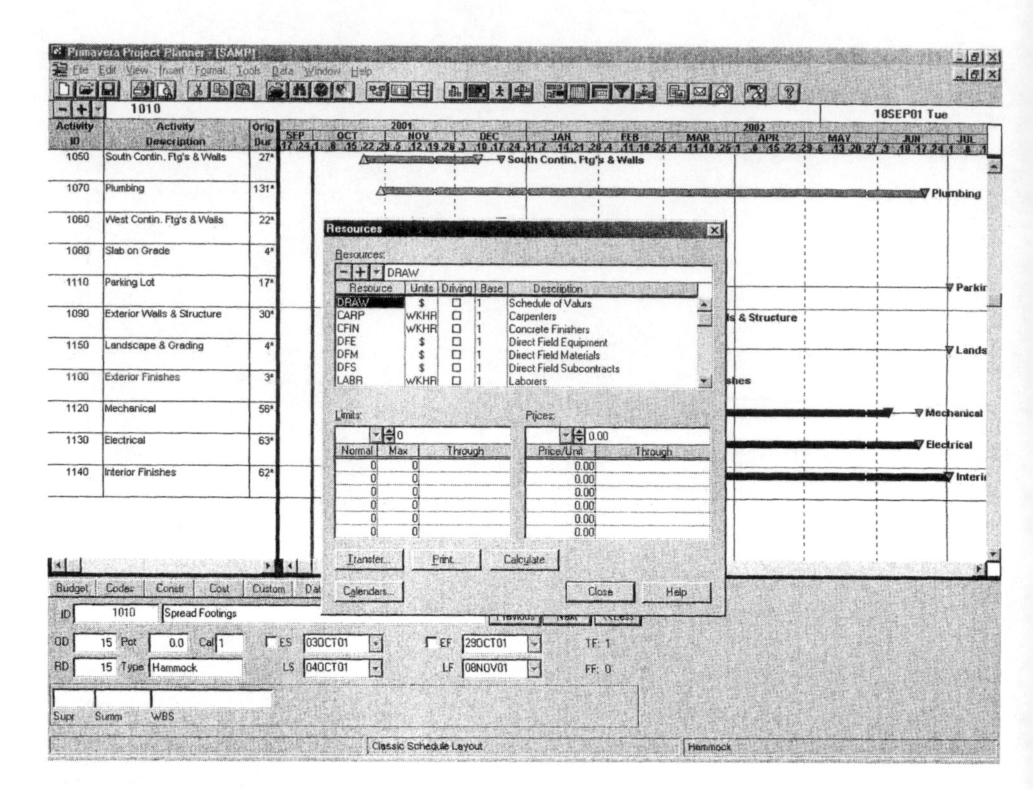

Screen Image 12-4
Defining DRAW as a Resource

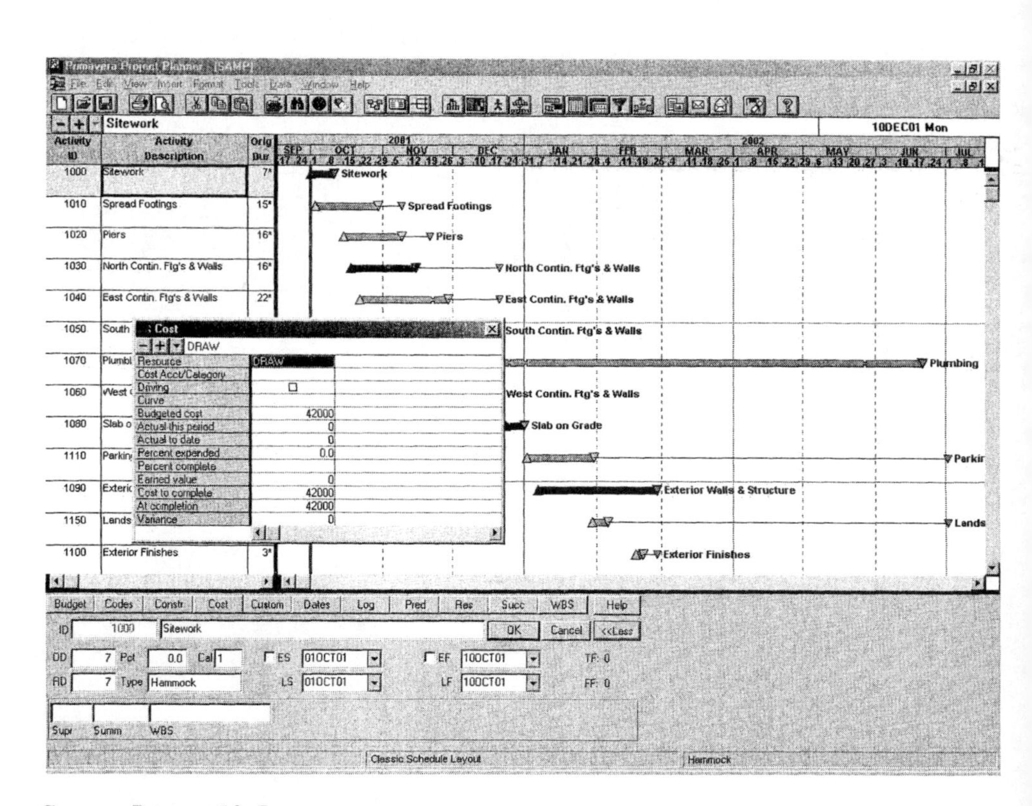

Screen Image 12-5
Adding Revenue to Schedule of Value Activities

and *At completion* were automatically generated using the rules specified on the *Autocost* window. All of the hammock activities need to be loaded with the corresponding amount from the schedule of values.

Cost Loading Report

Once the data concerning the schedule of values has been added, a tabular cost loading report can be produced. This report is basically identical to the resource loading reports that were presented in Chapter 11. Since the focus of this report is cost, the tabular cost options need to be selected. This is done by selecting the *Tools* command from the main menu, followed by selecting *Tabular Reports, Cost*, and *Loading* options as shown in Screen Image 12-6. The *Cost Loading Reports* window as shown in Screen Image 12-7 contains a list of all of the previously defined report specifications. To add a new set of report specifications click on the *Add* button and proceed with accepting the creation of these specifications. When the *Cost Loading Reports* window opens, there are four tabs that need to be processed, the first of which is the *Resource Selection.* This tab is where the resource to include in the report is specified. In Screen Image 12-8 the *DRAW* resource has been selected. This will cause only the budgeted cost items to be included for DRAW resources, which in this example correspond to the schedule of values.

Screen Image 12-9 details the *Format* tab, which is the same as with the resource loading report. However, for this report the resources are *Organized by ACT - Activity.* Since only one resource is selected and the report should list all of the activities down the page, the *Sort by* box is by early start, so the earliest activities will be listed first. For this report the *Column width* field has been increased to 7 characters to accommodate dollar amounts that are in the hundreds of thousands. Since there are no details for this report, the summary report option has also been selected. The *Selection* tab in actuality does not require an entry since the resource selection tab is controlling. However, if the other activities are not selected out, there will be an entry for every activity with no corresponding value. The bottom line on the report will be the same but it will require many additional pages. Screen Image 12-10 shows

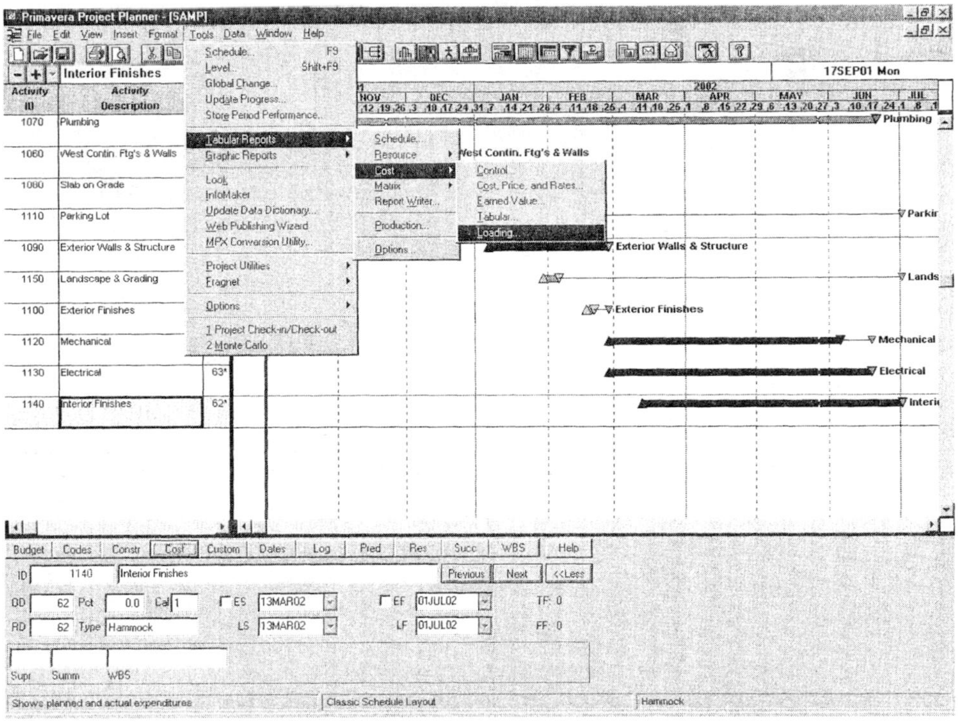

Screen Image 12-6
Opening the Cost Loading Reports Window

Primavera Project Planner - [SAMP]
File Edit View Insert Format Tools Data Window Help
Interior Finishes 08OCT01 Mon

Activity ID	Activity Description	Orig Dur
1070	Plumbing	131*
1060	West Contin. Ftg's & Walls	22*
1080	Slab on Grade	4*
1110	Parking Lot	17*
1090	Exterior Walls & Structure	30*
1150	Landscape & Grading	4*
1100	Exterior Finishes	3*
1120	Mechanical	56*
1130	Electrical	63*
1140	Interior Finishes	62*

2001 SEP OCT NOV DEC 2002 JAN FEB MAR APR MAY JUN JUL
Plumbing
West Contin. Ftg's & Walls
Parkir
Structure
Lands
Mechanical
Electrical
Interi

SAMP - Cost Loading Reports

ID	Title
CL-01	Cost Loading - Summary by Cost Account
CL-02	Cost Loading - Detailed by Day

Run Modify... Add... Delete Close Help

Budget Codes Constr Cost Custom Dates Log Pred Res Succ WBS Help
ID 1140 Interior Finishes Previous Next <<Less
OD 62 Pct 0.0 Cal 1 ES 13MAR02 EF 01JUL02 TF: 0
RD 62 Type Hammock LS 13MAR02 LF 01JUL02 FF: 0
Supr Summ WBS
Classic Schedule Layout Hammock

Screen Image 12-7
Cost Loading Reports Window

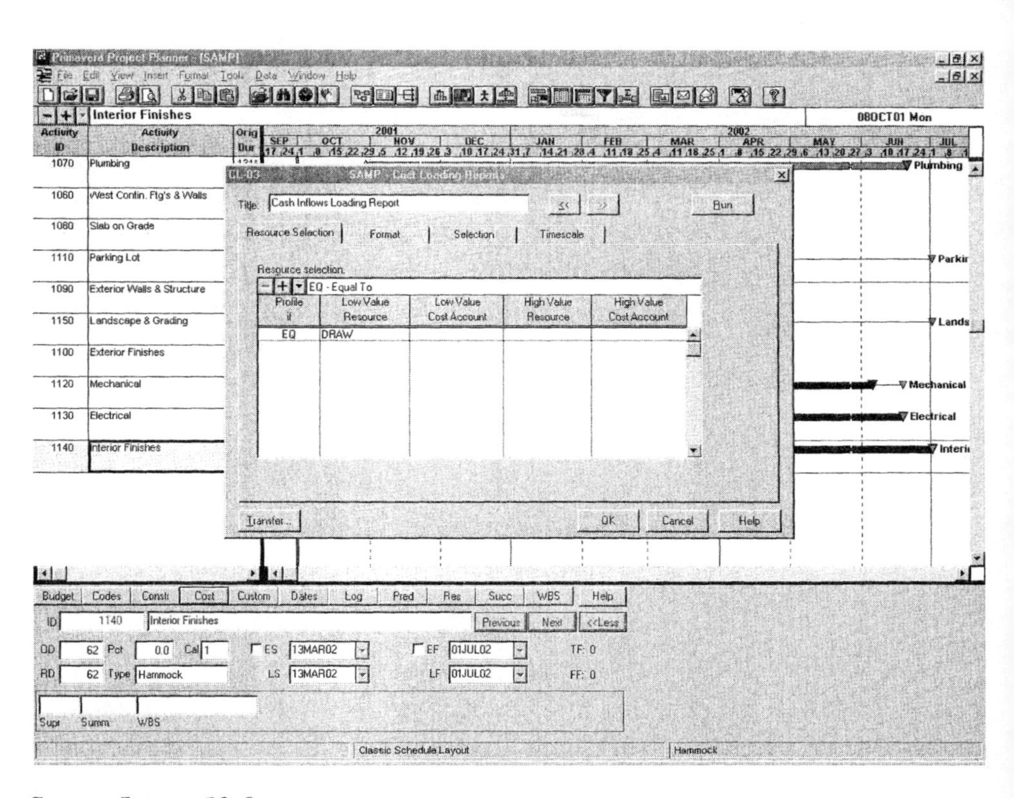

Screen Image 12-8
Resource Selection Tab

Primavera Project Planner - [SAMP]
File Edit View Insert Format Tools Data Window Help
07OCT01 Sun
Activity ID
Activity Description
Orig Dur
2001
SEP OCT NOV DEC
2002
JAN FEB MAR APR MAY JUN JUL
1070 Plumbing
1060 West Contin. Ftg's & Walls
1080 Slab on Grade
1110 Parking Lot
1090 Exterior Walls & Structure
1150 Landscape & Grading
1100 Exterior Finishes
1120 Mechanical
1130 Electrical
1140 Interior Finishes
Plumbing
Parkir
Lands
Mechanical
Electrical
Interi
CL-03 SAMP - Cost Loading Reports
Title: CASH INFLOWS LOADING REPORT
Run
Resource Selection
Format
Selection
Timescale
Generate report in P3
Font...
Arial, Regular, 8
Export report in CSV format to the file
C:\P3WIN\P3OUT\P3.csv
Browse
Organize
Organize by:
ACT - Activity
Group/subtotal by:
<none>
Summary report
Options
Truncate activity description
Repeat activity title on all horizontal sheets
Display row totals
Left
Right
Page width: 132
Column width: 7
Divide by: 1
Decimals: 0
Sort by:
Sort on
Order
Early start
Ascend
Transfer...
OK
Cancel
Help
Budget Codes Constr Cost Custom Dates Log Pred Res Succ WBS Help
ID
Previous
Next
<<Less
OD 0
Pct 0.0
Cal 1
ES
EF
TF:
RD 0
Type
LS
LF
FF:
Supr
Summ
WBS
Classic Schedule Layout
Hammock

Screen Image 12-9
Format Tab

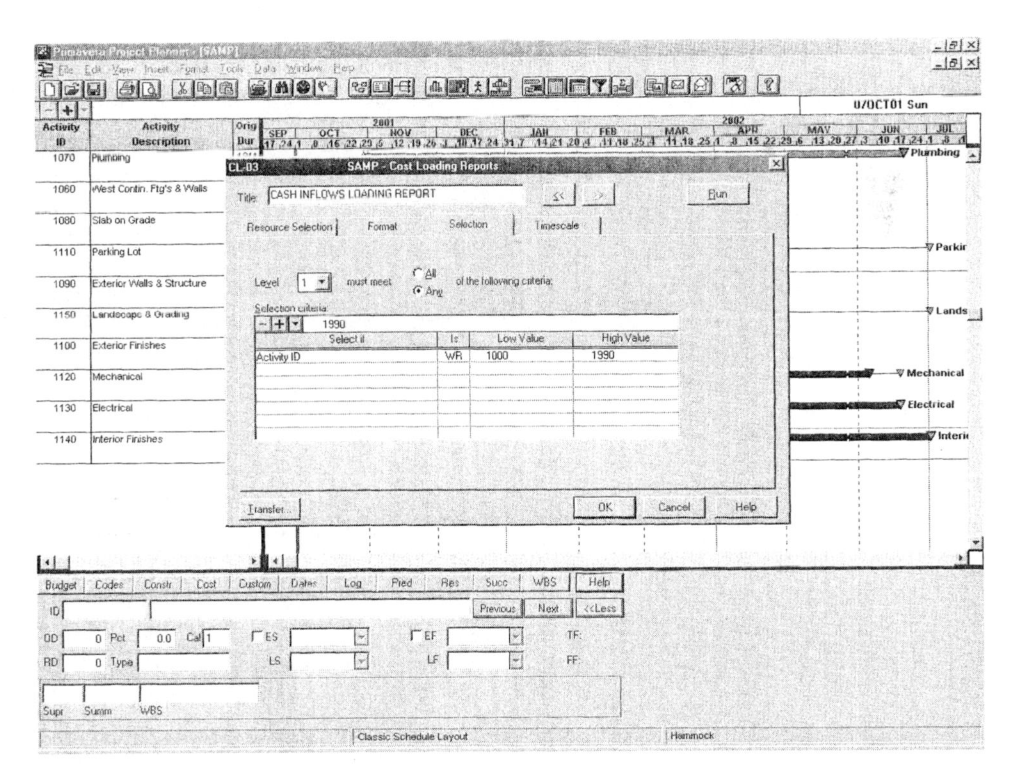

Screen Image 12-10
Selection Tab

All Star Developments | PRIMAVERA PROJECT PLANNER | Sample Construction Project

REPORT DATE 23DEC00 RUN NO. 83 23:21 | COST LOADING REPORT | START DATE 01OCT01 FIN DATE 01JUL02

CASH INFLOWS LOADING REPORT | TOTAL USAGE FOR WEEK | DATA DATE 01OCT01 PAGE NO. 1-1

ACT ID	DESC	TOTAL	1OCT 2001	8OCT 2001	15OCT 2001	22OCT 2001	29OCT 2001	5NOV 2001	12NOV 2001	19NOV 2001	26NOV 2001
1000	Sitework	42000	24000	18000							
1010	Spread Footings	18000	2400	4800	4800	4800	1200				
1020	Piers	32000			8000	8000	8000	8000			
1030	North Contin. Ftg's	30000			1875	7500	7500	7500	5625		
1040	East Contin. Ftg's &	31000				5636	5636	5636	5636	4227	4227
1050	South Contin. Ftg's	37000				4111	5481	5481	5481	4111	5481
1070	Plumbing	158000					3618	4824	4824	3618	4824
1060	West Contin. Ftg's &	31000							4227	4227	5636
1080	Slab on Grade	55000									
1110	Parking Lot	25000									
1090	Exterior Walls & Str	195000									
1150	Landscape & Grading	10000									
1100	Exterior Finishes	12000									
1120	Mechanical	70000									
1130	Electrical	55000									
1140	Interior Finishes	85000									
	REPORT TOTAL	886000	26400	22800	14675	30047	31436	31442	25795	16184	20170

All Star Developments | PRIMAVERA PROJECT PLANNER | Sample Construction Project

REPORT DATE 23DEC00 RUN NO. 83 23:21 | COST LOADING REPORT | START DATE 01OCT01 FIN DATE 01JUL02

CASH INFLOWS LOADING REPORT | TOTAL USAGE FOR WEEK | DATA DATE 01OCT01 PAGE NO. 1-2

ACT ID	3DEC 2001	10DEC 2001	17DEC 2001	24DEC 2001	31DEC 2001	7JAN 2002	14JAN 2002	21JAN 2002	28JAN 2002	4FEB 2002	11FEB 2002	18FEB 2002
1000												
1010												
1020												
1030												
1040												
1050	5481	1370										
1070	4824	4824	4824	3618	3618	4824	4824	4824	4824	4824	4824	4824
1060	5636	5636	5636									
1080				41250	13750							
1110					2941	5882	5882	5882	4412			
1090						26000	26000	26000	26000	26000	26000	26000
1150									5000	5000		
1100												12000
1120												
1130												
1140												
	15942	11831	10461	44868	20309	36707	36707	36707	40236	35824	30824	42824

Output 12-1
Cost Loading Report (pages 1 and 2)

the *Selection* window with only the hammock activities selected. The *Timescale* tab identical to the one that was discussed with the resource loading report. Once all of t specifications have been laid out, this report can be run. Output 12-1 is an example of t cost loading report.

The information from the cost loading report will form the basis of the revenue pl This loading report denotes the plan for when the contractor will earn money. Plotti this information as the cash inflow curve would ignore how the billing process works a retainage. The contractor performs work and then on some regular interval bills the ow for work in place. The owner then has a number of days to verify the accuracy of the b prior to paying it. Typically, the contractor bills once a month and the owner most lik would have between 10 and 30 days to pay. The amounts shown in the loading report gross earned amounts, not less retainage. For the purposes of this example the contrac will bill every 4 weeks and the owner will be allowed 2 weeks to pay. This arrangeme requires the information from the tabular cost loading report to be accumulated by mon as shown in Figure 12-7. That figure contains a small table for each of the billing perio

All Star Developments | PRIMAVERA PROJECT PLANNER | Sample Construction Project

REPORT DATE 08APR00 RUN NO. 76 15:12 | COST LOADING REPORT | START DATE 01OCT01 FIN DATE 01JUL02

DATA DATE 01OCT01 PAGE NO. 1-3

CASH INFLOWS LOADING REPORT | TOTAL USAGE FOR WEEK

ACT ID	25FEB 2002	4MAR 2002	11MAR 2002	18MAR 2002	25MAR 2002	1APR 2002	8APR 2002	15APR 2002	22APR 2002	29APR 2002	6MAY 2002	13MAY 2002
1000												
1010												
1020												
1030												
1040												
1050												
1070	4824	4824	4824	4824	4824	4824	4824	4824	4824	4824	4824	4824
1060												
1080												
1110												
1090	13000											
1150												
1100												
1120	2500	5000	5000	5000	5000	5000	5000	5000	5000	5000	5000	5000
1130	1746	3492	3492	3492	3492	3492	3492	3492	3492	3492	3492	3492
1140			2742	5484	5484	5484	5484	5484	5484	5484	5484	5484
	22070	13316	16058	18800	18800	18800	18800	18800	18800	18800	18800	18800

All Star Developments | PRIMAVERA PROJECT PLANNER | Sample Construction Project

REPORT DATE 08APR00 RUN NO. 76 15:12 | COST LOADING REPORT | START DATE 01OCT01 FIN DATE 01JUL02

DATA DATE 01OCT01 PAGE NO. 1-4

CASH INFLOWS LOADING REPORT | TOTAL USAGE FOR WEEK

ACT ID	20MAY 2002	27MAY 2002	3JUN 2002	10JUN 2002	17JUN 2002	24JUN 2002	1JUL 2002
1000							
1010							
1020							
1030							
1040							
1050							
1070	4824	3618	4824	4824	4824		
1060							
1080							
1110							
1090							
1150							
1100							
1120	5000	3750	3750				
1130	3492	2619	3492	3492	1746		
1140	5484	4113	5484	5484	5484	5484	1371
	18800	14100	17550	13800	12054	5484	1371

Output 12-1
Cost Loading Report (pages 3 and 4)

These tables can then be converted into a cash inflow table as shown in Figure 12-8. From that table a planned cash inflows curve can be plotted as shown in Figure 12-9. This graph was generated using a spreadsheet program and is stair-stepped to show how the contractor is actually paid. If the payments were connected with lines, there would be the implication that the cash inflows increased between payment dates. The contractor gets paid and the cash inflows curve is flat until the next payment is received. In that figure the inflows are shown as a dashed line, which is done so the actual cash inflows can be graphed as a solid line.

If both of these are plotted as cumulative curves, it will allow for quick comparisons between the inflows plan and what has actually transpired. If the actual line is above the planned line, more has been billed and paid for than planned. Conversely, if the actual line is below the planned line, cash is being received slower than planned. A deviation such as

Billing Period 1	
Week	
1	$26,400
2	$22,800
3	$14,675
4	$30,047
Gross Billing	$93,922
Retainage	$9,392
Net Billing	$84,530

Billing Period 2	
Week	
1	$31,436
2	$31,442
3	$25,795
4	$16,184
Gross Billing	$104,857
Retainage	$10,486
Net Billing	$94,371

Billing Period 3	
Week	
1	$20,170
2	$15,942
3	$11,831
4	$10,461
Gross Billing	$58,404
Retainage	$5,840
Net Billing	$52,564

Billing Period 4	
Week	
1	$44,868
2	$20,309
3	$36,707
4	$36,707
Gross Billing	$138,591
Retainage	$13,859
Net Billing	$124,732

Billing Period 5	
Week	
1	$36,707
2	$40,236
3	$35,824
4	$30,824
Gross Billing	$143,591
Retainage	$14,359
Net Billing	$129,232

Billing Period 6	
Week	
1	$42,824
2	$22,070
3	$13,316
4	$16,058
Gross Billing	$94,268
Retainage	$9,427
Net Billing	$84,841

Billing Period 7	
Week	
1	$18,800
2	$18,800
3	$18,800
4	$18,800
Gross Billing	$75,200
Retainage	$7,520
Net Billing	$67,680

Billing Period 8	
Week	
1	$18,800
2	$18,800
3	$18,800
4	$18,800
Gross Billing	$75,200
Retainage	$7,520
Net Billing	$67,680

Billing Period 9	
Week	
1	$18,800
2	$18,800
3	$14,100
4	$17,550
Gross Billing	$69,250
Retainage	$6,925
Net Billing	$62,325

Billing Period 10	
Week	
1	$13,800
2	$12,054
3	$5,484
4	$1,371
Gross Billing	$32,709
Retainage	$3,271
Net Billing	$29,438

Figure 12-7
Monthly Billings

CASH INFLOWS PLAN					
Billing Date	Payment Date	Gross Billings	Retainage	Net Billing	Cummulative
10/31/01	11/14/01	93,922	9,392	84,530	84,530
11/30/01	12/14/01	104,857	10,486	94,371	178,901
12/31/01	01/14/02	58,404	5,840	52,564	231,465
01/31/02	02/14/02	138,591	13,859	124,732	356,197
02/28/02	03/14/02	143,591	14,359	129,232	485,429
03/31/02	04/14/02	94,268	9,427	84,841	570,270
04/30/02	05/14/02	75,200	7,520	67,680	637,950
05/31/02	06/14/02	75,200	7,520	67,680	705,630
06/30/02	07/14/02	69,250	6,925	62,325	767,955
07/31/02	08/14/02	32,709	3,271	29,438	797,393
		885,992	88,599	797,393	

Figure 12-8
Cash Inflow Table

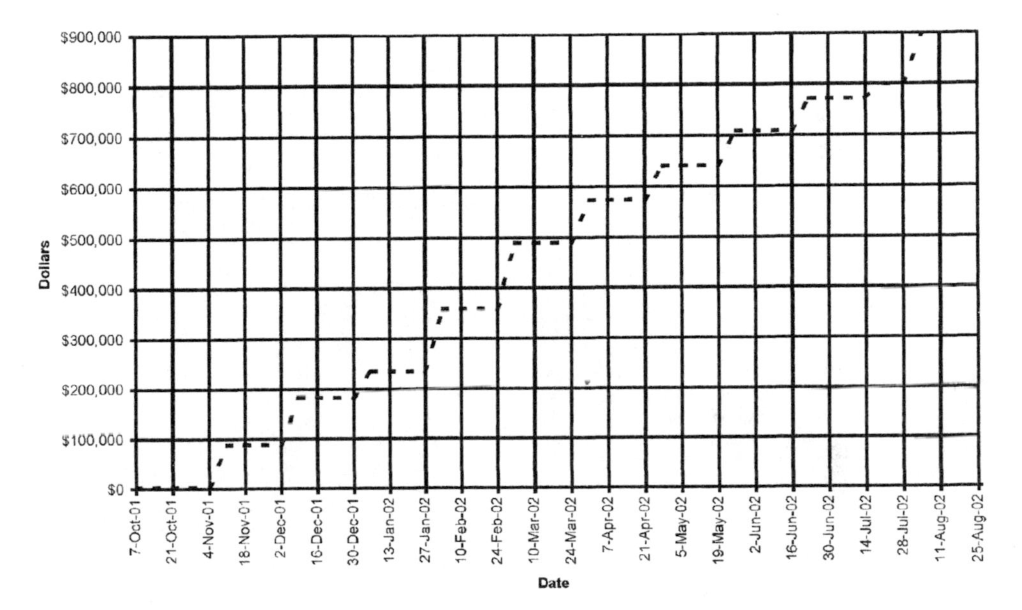

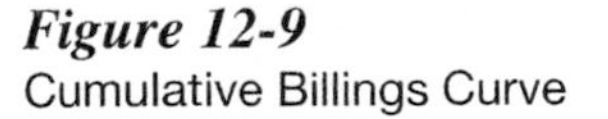
Figure 12-9
Cumulative Billings Curve

that could be the result of slower progress or slow payment by the owner—both of which are serious matters. A graph such as this will not tell what is the problem; it simply is an indicator of a problem. It is the responsibility of the individual people on the project team to dig deeper for the actual answer.

The cash inflow chart can also be produced by having Primavera Project Planner create a spreadsheet with the cost loaded information. This option will require manipulation of the data and revised formatting. By using the previous cost loading report and selecting the *Export record as CSV format to the file* (Screen Image 12-11), a comma delimited file will be created. This file can be opened using a variety of spreadsheet programs. Care needs to be taken when specifying the name and location of this file. In Figure 12-10 the *CSV* cost loaded file has been opened using a spreadsheet program. As one can see, the format and data type is quite rudimentary and in need of manipulation. In Figure 12-11, that information has been formatted and manipulated. In that figure 4 weeks of earned

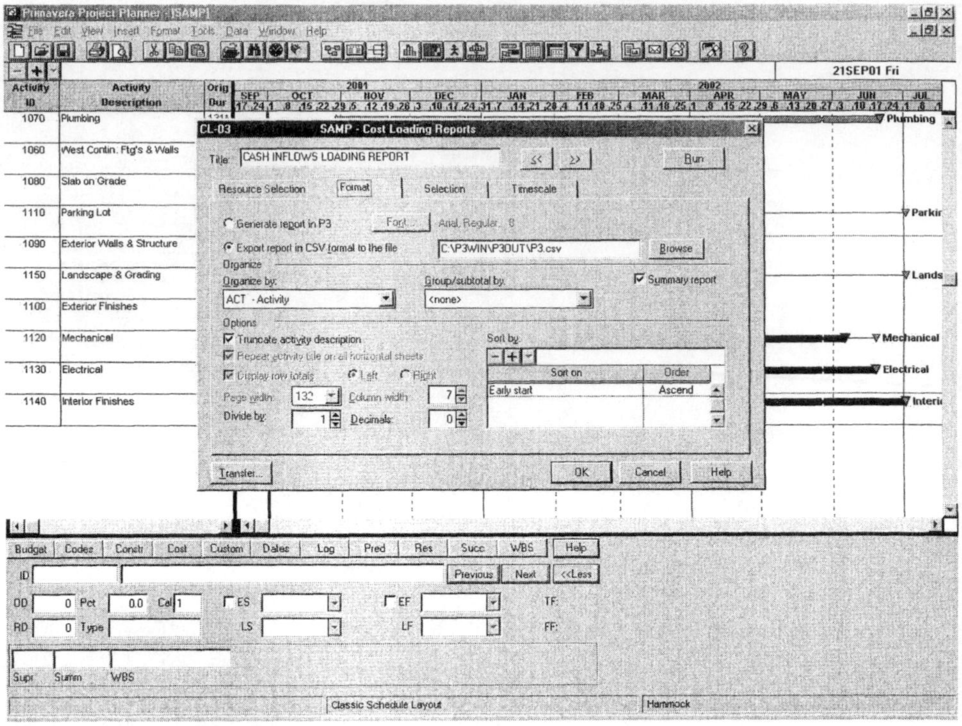

Screen Image 12-11
Exporting in CSV File Format

ACT	None	ACT ID	DESC	TOTAL	1OCT 2001	8OCT 2001	15OCT 2001	22OCT 2001	29OCT 2001	5NOV 2001	12NOV 2001
		1000	Sitework	42000	24000	18000					
		1010	Spread Footings	18000	2400	4800	4800	4800	1200		
		1020	Piers	32000			8000	8000	8000	8000	
		1030	North Contin. Ftg's	30000			1875	7500	7500	7500	5625
		1040	East Contin. Ftg's &	31000				5636	5636	5636	5636
		1050	South Contin. Ftg's	37000				4111	5481	5481	5481
		1070	Plumbing	158000					3618	4824	4824
		1060	West Contin. Ftg's &	31000							4227
		1080	Slab on Grade	55000							
		1110	Parking Lot	25000							
		1090	Exterior Walls & Str	195000							
		1150	Landscape & Grading	10000							
		1100	Exterior Finishes	12000							
		1120	Mechanical	70000							
		1130	Electrical	55000							
		1140	Interior Finishes	85000							

Figure 12-10
CSV File Opened Using Spreadsheet Program

Week Beginning		1-Oct-01	8-Oct-01	15-Oct-01	22-Oct-01	29-Oct-01	5-Nov-01	12-Nov-01	19-Nov-01	26-Nov-01
Week Ending		07-Oct-01	14-Oct-01	21-Oct-01	28-Oct-01	04-Nov-01	11-Nov-01	18-Nov-01	25-Nov-01	02-Dec-01
Gross Earned Work		26,400	22,800	14,675	30,047	31,435	31,441	25,793	16,183	20,168
Gross Monthly Billings					93,922				104,852	
Retainage					9,392				10,485	
Net Billings					84,530				94,367	
Cash Inflows								84,530		
Cummulative		0	0	0	0	0	0	84,530	84,530	84,530

Figure 12-11
Planned Cash Inflows Spreadsheet Table

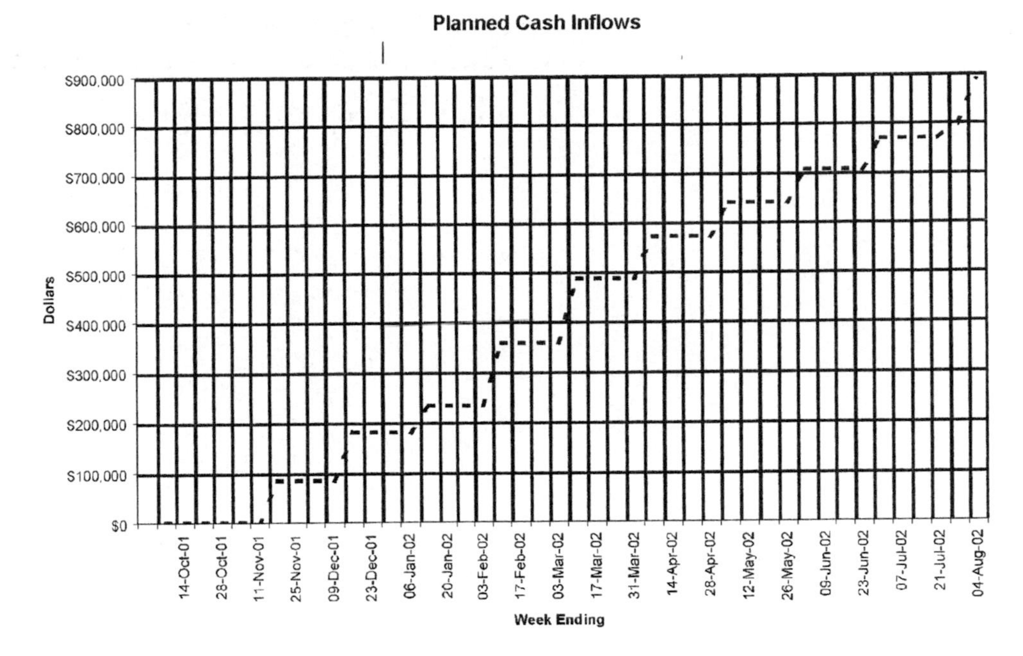

Figure 12-12
Planned Cash Inflows Using Data from CSV File

amounts have been accumulated into a row called Gross Monthly Billings; from that 10 percent retainage has been removed, and a Net Billings row created. The net billings are then accumulated. Note how the cumulative billing amounts have been entered every week until the next payment is received. This is done in order to get the stair-stepped curve. The information in Figure 12-11 is plotted in the graph found in Figure 12-12. That figure and the one found in Figure 12-9 are identical. The difference is the 12-9 figure had the information manually entered, whereas the latter had the information automatically transferred to the spreadsheet program.

Project Cost Plan

The project cost plan is developed in a fashion similar to the revenue plan. The major difference lies in how the resources are selected for the cost loading report and the lag between when the resources are consumed and when they are paid for. The four major elements of construction costs—direct labor, direct materials, direct subcontracts and direct equipment—must all be addressed in a different fashion. An expenditure plan needs to be developed for each of these items individually. This is done in Primavera Project Planner by creating a new cost loading report. This report requires all of the contractor's cost resources to be selected. In Screen Image 12-12 all of these resources for the sample project have been selected.

The *Format* tab as shown in Screen Image 12-13 has been organized by *CAT - Cost Categories*. This will group all of the resources by the cost categories that were defined when creating the chart of accounts. In this example it will group all of the labor costs together, all of the equipment costs together, all of the material costs together, and all of the subcontract costs together. Screen Image 12-14 shows the *Selection* tab with only the construction activities. This is optional since the *Resource Selection* tab took care of what to include. However, this command will reduce the size of the cost loading report.

Screen Image 12-12
Resource Selection Window

Screen Image 12-13
Format Window

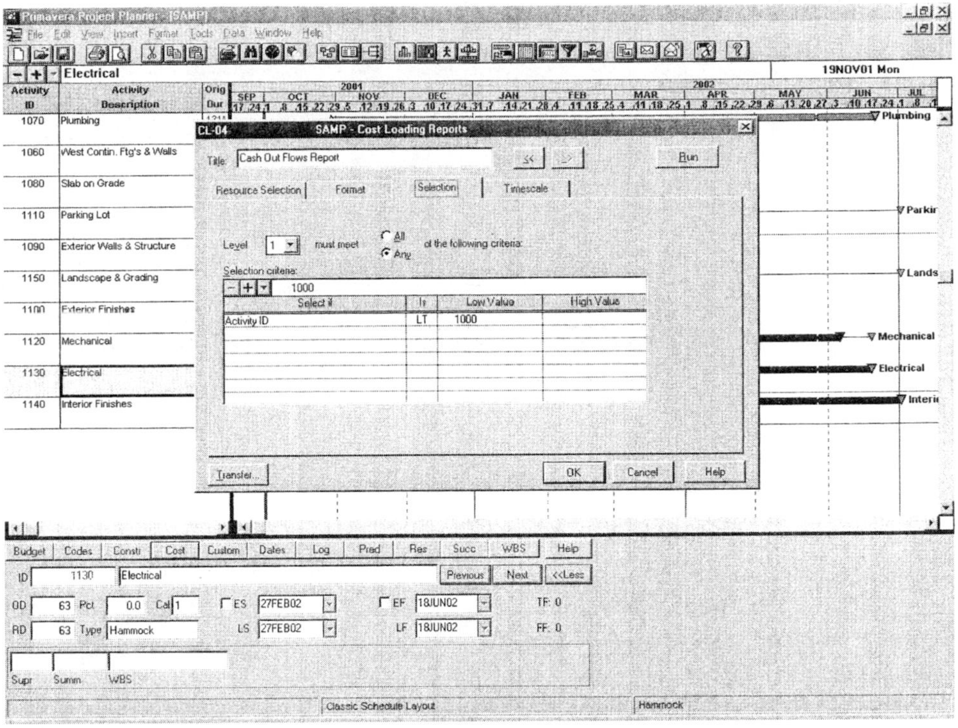

Screen Image 12-14
Selection Window

Output 12-2 is an excerpt of the report that was generated from the above report specifications. In that report all of the cost categories are shown, along with what activities contributed to those costs.

The length of the cash outflows loading report can be reduced by converting it into a summary report. This is done by selecting the *Summary report* option from the *Format* tab as shown in Screen Image 12-15. The corresponding output is found in Output 12-3. The bottom lines of these reports are identical, but the detail is lost with the summary report. If the project manager is trying to find an anomaly in the outflow plan, the detailed report will need to be run in order to investigate any anomalies. Just as with the owner's cost loading report, the contractor's cost loading report can be downloaded to a *CSV* file format as shown in Screen Image 12-16. Since this option is being used in conjunction with the summary option, the output will be small. Figure 12-13 is an excerpt of the *CSV* file opened using a spreadsheet program.

In Figure 12-14 the exported file has been manipulated to develop the needed information. The top labor, material, equipment and subcontract entries are straight from the exported file. The Labor Burdens row was generated by multiplying the labor costs by the planned burden rate. The Total Labor Costs row is now the labor costs plus the burdens. The second material and equipment rows have their costs shifted 3 weeks to show the lag between when these items are required and when they are paid for. The subcontract costs have been shifted 2 weeks to show the typical payment terms between contractors and subcontractors. The Total Direct Field Costs row is the timescaled projected direct field cost cash outflows. The remaining indirects are the total estimated indirects less the field labor burdens divided by the number of weeks in the project. Since these items are typically a function of time, this straight-line approach is adequate. The Home Office Costs allocation is simply distributed equally over the duration of the project. The Total Cost row is the timescaled planned cash outflows for the entire project. Finally these outflows are accumulated. The accumulated data can be graphed into a planned cash outflows diagram as shown in Figure 12-15. This diagram can be used as a basis of comparison between planned and

All Star Developments | PRIMAVERA PROJECT PLANNER | Sample Construction Project

REPORT DATE 08APR00 RUN NO. 79 23:56 | COST LOADING REPORT | START DATE 01OCT01 FIN DATE 01JUL02

Cash Out Flows Report | TOTAL USAGE FOR WEEK | DATA DATE 01OCT01 PAGE NO. 1-1

ACT ID	DESC	TOTAL	1OCT 2001	8OCT 2001	15OCT 2001	22OCT 2001	29OCT 2001	5NOV 2001	12NOV 2001	19NOV 2001	26NOV 2001
1 - Labor											
30	Fab. & Tie Steel Spr	424	141	283							
40	Excav. Spread Footin	728		182	546						
50	Form Spread Footings	164			164						
60	Reinforce Spread Foo	133			100	33					
70	Place Spread Footing	102			51	51					
80	Strip Forms - Spread	108				81	27				
90	Fab. & Tie Steel Pie	133			133						
100	Fab. & Tie Steel Con	1484			1484						
110	Fab & Tie Steel Fdn.	318				318					
120	Reinforce Piers	106				53	53				
130	Form Piers	1733				433	1300				
140	Place Piers	51					51				
150	Strip Forms - Piers	172					43	129			
160	Backfill Spread Foot	1730						1730			
170	North Side Excav. Co	840			840						
180	North Side Form Cont	215				215					
190	North Side Reinforce	265				265					
200	North Side Place Con	191				191					
210	North Side Strip Con	86					86				
220	North Side Reinforce	530					530				
230	North Side Form Fdn.	2188					729	1459			
240	North Side Place Fdn	26						26			
250	North Side Strip Fdn	430							430		
260	North Side Backfill	2465							2465		
270	East Side Excav. Con	604				604					
280	East Side Form Conti	132					132				
290	East Side Reinforce	265					265				
300	East Side Place Cont	128						128			
310	East Side Strip Cont	54						54			
320	East Side Reinforce	530							530		
330	East Side Form Fdn.	1313							875	438	
340	East Side Place Fdn.	26								26	
350	East Side Strip Fdn.	269									269
360	South Side Excav. Co	840				840					
370	South Side Form Cont	165							165		
380	South Side Reinforce	265							265		
390	West Side Excav. Con	550							550		
410	East Side Backfill &	1255									1255
420	South Side Place Con	191							191		
430	South Side Strip Con	97								97	
440	South Side Reinforce	530								530	
450	South Side Form Fdn.	1520									1520
460	South Side Place Fdn	26									26
470	West Side Form Conti	149								149	
480	West Side Reinforce	265									265
490	South Side Strip Fdn	430									
500	South Side Backfill	2465									
510	West Side Place Cont	115									115
520	West Side Strip Cont	65									
530	West Side Reinforce	530									
540	West Side Form Fdn.	1313									
550	West Side Place Fdn.	26									
560	West Side Strip Fdn.	269									
570	West Side Backfill &	1375									
580	Reinforce S.O.G.	2253									
590	Place & Finish S.O.G	6847									
600	CMU	15844									
690	Frame Interior Walls	429									
720	Form Parking Lot	858									
730	Reinforce Parking Lo	689									
740	Place & Finish Parki	5993									
750	Strip Forms - Parkin	215									

Output 12-2

Detailed Contractor's Cost Loading Report (page 1)

All Star Developments | PRIMAVERA PROJECT PLANNER | Sample Construction Project

REPORT DATE 08APR00 RUN NO. 79 23:56 | COST LOADING REPORT | START DATE 01OCT01 FIN DATE 01JUL02

DATA DATE 01OCT01 PAGE NO. 2- 1

Cash Out Flows Report | TOTAL USAGE FOR WEEK

ACT ID	DESC	TOTAL	1OCT 2001	8OCT 2001	15OCT 2001	22OCT 2001	29OCT 2001	5NOV 2001	12NOV 2001	19NOV 2001	26NOV 2001
1 - Labor											
760	Place Curbs	4820									
820	Hang & Trim Interior	8448									
840	Install Cabinets	2954									
TOTAL	1	79734	141	465	3318	3085	3216	3526	5471	1240	3450
2 - Material											
30	Fab. & Tie Steel Spr	377	126	251							
70	Place Spread Footing	781			391	391					
90	Fab. & Tie Steel Pie	2107			2107						
100	Fab. & Tie Steel Con	2036			2036						
110	Fab & Tie Steel Fdn.	424				424					
130	Form Piers	354				89	266				
140	Place Piers	258					258				
160	Backfill Spread Foot	995						995			
180	North Side Form Cont	345				345					
200	North Side Place Con	1561				1561					
230	North Side Form Fdn.	1271					424	847			
240	North Side Place Fdn	130						130			
260	North Side Backfill	1122							1122		
280	East Side Form Conti	218					218				
300	East Side Place Cont	976						976			
320	East Side Reinforce	795							795		
330	East Side Form Fdn.	795							530	265	
340	East Side Place Fdn.	81								81	
370	South Side Form Cont	221							221		
410	East Side Backfill &	712									712
420	South Side Place Con	1561							1561		
450	South Side Form Fdn.	1071									1071
460	South Side Place Fdn	130									130
470	West Side Form Conti	218								218	
500	South Side Backfill	1122									
510	West Side Place Cont	976									976
540	West Side Form Fdn.	795									
550	West Side Place Fdn.	81									
570	West Side Backfill &	712									
580	Reinforce S.O.G.	3054									
590	Place & Finish S.O.G	11148									
600	CMU	10593									
690	Frame Interior Walls	3584									
720	Form Parking Lot	214									
730	Reinforce Parking Lo	1149									
740	Place & Finish Parki	15943									
760	Place Curbs	1010									
820	Hang & Trim Interior	8726									
840	Install Cabinets	13645									
TOTAL	2	91291	126	251	4534	2809	1165	2948	4229	564	2889
3 - Equipmen											
40	Excav. Spread Footin	356		89	267						
160	Backfill Spread Foot	948						948			
170	North Side Excav. Co	394			394						
260	North Side Backfill	1069							1069		
270	East Side Excav. Con	250				250					
360	South Side Excav. Co	394				394					
390	West Side Excav. Con	250							250		
410	East Side Backfill &	678									678
500	South Side Backfill	1069									
570	West Side Backfill &	678									
TOTAL	3	6086		89	661	644		948	1319		678

Output 12-2

Detailed Contractor's Cost Loading Report (page 2)

All Star Developments | PRIMAVERA PROJECT PLANNER | Sample Construction Project

REPORT DATE 08APR00 RUN NO. 80 23:59 | COST LOADING REPORT | START DATE 01OCT01 FIN DATE 01JUL02

DATA DATE 01OCT01 PAGE NO. 1- 1

Cash Out Flows Report | TOTAL USAGE FOR WEEK

CATEGORY	CATEGORY DESCRIPTION	TOTAL	1OCT 2001	8OCT 2001	15OCT 2001	22OCT 2001	29OCT 2001	5NOV 2001	12NOV 2001	19NOV 2001
1	Labor	79734	141	465	3318	3085	3216	3526	5471	1240
2	Material	91291	126	251	4534	2809	1165	2948	4229	564
3	Equipmen	6086		89	661	644		948	1319	
4	Subs	434520	1833	574			21000	14000		
	REPORT TOTAL	**611631**	**2100**	**1379**	**8512**	**6538**	**25381**	**21422**	**11019**	**1804**

All Star Developments | PRIMAVERA PROJECT PLANNER | Sample Construction Project

REPORT DATE 08APR00 RUN NO. 80 23:59 | COST LOADING REPORT | START DATE 01OCT01 FIN DATE 01JUL02

DATA DATE 01OCT01 PAGE NO. 1- 2

Cash Out Flows Report | TOTAL USAGE FOR WEEK

CATEGORY	26NOV 2001	3DEC 2001	10DEC 2001	17DEC 2001	24DEC 2001	31DEC 2001	7JAN 2002	14JAN 2002	21JAN 2002	28JAN 2002	4FEB 2002	11FEB 2002
1	3450	1025	3804	1644	2253	6847	3167	9997	3234	7701	2881	1440
2	2889		1998	712	3054	11148	1997	18931	2156	2936	1926	963
3	678		1069	678								
4						1872	2808	14550	19400	24590	13991	
	7017	**1025**	**6871**	**3034**	**5307**	**19867**	**7972**	**43478**	**24789**	**35227**	**18798**	**2403**

All Star Developments | PRIMAVERA PROJECT PLANNER | Sample Construction Project

REPORT DATE 08APR00 RUN NO. 80 23:59 | COST LOADING REPORT | START DATE 01OCT01 FIN DATE 01JUL02

DATA DATE 01OCT01 PAGE NO. 1- 3

Cash Out Flows Report | TOTAL USAGE FOR WEEK

CATEGORY	18FEB 2002	25FEB 2002	4MAR 2002	11MAR 2002	18MAR 2002	25MAR 2002	1APR 2002	8APR 2002	15APR 2002	22APR 2002	29APR 2002	6MAY 2002
1					215	215						
2					1792	1792						
3												
4	46441	31660	29000	10589	1795	12333	29333	9333	10517	14069	10905	1415
	46441	**31660**	**29000**	**10589**	**3801**	**14340**	**29333**	**9333**	**10517**	**14069**	**10905**	**1415**

All Star Developments | PRIMAVERA PROJECT PLANNER | Sample Construction Project

REPORT DATE 08APR00 RUN NO. 80 23:59 | COST LOADING REPORT | START DATE 01OCT01 FIN DATE 01JUL02

DATA DATE 01OCT01 PAGE NO. 1- 4

Cash Out Flows Report | TOTAL USAGE FOR WEEK

CATEGORY	13MAY 2002	20MAY 2002	27MAY 2002	3JUN 2002	10JUN 2002	17JUN 2002	24JUN 2002	1JUL 2002
1				10417	985			
2				17823	4548			
3								
4	1415	1415	1061	28000	13500	43500	22419	1200
	1415	**1415**	**1061**	**56240**	**19033**	**43500**	**22419**	**1200**

Output 12-3
Summary Contractor's Cost Loading Report

Primavera Project Planner - [SAMP]

File Edit View Insert Format Tools Data Window Help

19NOV01 Mon

Electrical

Activity ID	Activity Description
1070	Plumbing
1060	West Contin. Ftg's & Walls
1080	Slab on Grade
1110	Parking Lot
1090	Exterior Walls & Structure
1150	Landscape & Grading
1100	Exterior Finishes
1120	Mechanical
1130	Electrical
1140	Interior Finishes

Orig Dur

2001 SEP OCT NOV DEC 2002 JAN FEB MAR APR MAY JUN JUL

Plumbing Parkir Lands Mechanical Electrical Interi

CL-04 SAMP - Cost Loading Reports

Title: Cash Out Flows Report

Run

Resource Selection Format Selection Timescale

Generate report in P3 Font... Arial, Regular, 8

Export report in CSV format to the file C:\P3WIN\P3OUT\p3.csv Browse

Organize

Organize by: CAT - Cost category

Group/subtotal by: <none>

Summary report

Options

Truncate activity description

Repeat activity title on all horizontal sheets

Display row totals: Left Right

Page width: 132 Column width: 7

Divide by: 1 Decimals: 0

Sort by:

Sort on Order

Transfer... OK Cancel Help

Budget Codes Constr Cost Custom Dates Log Pred Res Succ WBS Help

ID 1130 Electrical Previous Next <<Less

OD 63 Pct 0.0 Cal 1 ES 27FEB02 EF 18JUN02 TF: 0

RD 63 Type Hammock LS 27FEB02 LF 18JUN02 FF: 0

Supr Summ WBS

Classic Schedule Layout Hammock

Screen Image 12-15
Opting for Summary Report

Primavera Project Planner - [SAMP]

File Edit View Insert Format Tools Data Window Help

19NOV01 Mon

Electrical

Activity ID	Activity Description
1070	Plumbing
1060	West Contin. Ftg's & Walls
1080	Slab on Grade
1110	Parking Lot
1090	Exterior Walls & Structure
1150	Landscape & Grading
1100	Exterior Finishes
1120	Mechanical
1130	Electrical
1140	Interior Finishes

Orig Dur

2001 SEP OCT NOV DEC 2002 JAN FEB MAR APR MAY JUN JUL

Plumbing Parkir Lands Mechanical Electrical Interi

CL-04 SAMP - Cost Loading Reports

Title: Cash Out Flows Report

Run

Resource Selection Format Selection Timescale

Generate report in P3 Font... Arial, Regular, 8

Export report in CSV format to the file C:\P3WIN\P3OUT\p3.csv Browse

Organize

Organize by: CAT - Cost category

Group/subtotal by: <none>

Summary report

Options

Truncate activity description

Repeat activity title on all horizontal sheets

Display row totals: Left Right

Page width: 132 Column width: 7

Divide by: 1 Decimals: 0

Sort by:

Sort on Order

Transfer... OK Cancel Help

Budget Codes Constr Cost Custom Dates Log Pred Res Succ WBS Help

ID 1130 Electrical Previous Next <<Less

OD 63 Pct 0.0 Cal 1 ES 27FEB02 EF 18JUN02 TF: 0

RD 63 Type Hammock LS 27FEB02 LF 18JUN02 FF: 0

Supr Summ WBS

Classic Schedule Layout Hammock

Screen Image 12-16
Downloading the Contractor's Cost Loading Data

CAT None	CATEGORY	CATEGORY DESCRIPTION	TOTAL	1OCT 2001	8OCT 2001	15OCT 2001	22OCT 2001	29OCT 2001	5NOV 2001
1 - Labor		1 Labor	79734	141	465	3318	3085	3216	3526
2 - Material		2 Material	91291	126	251	4534	2809	1165	2948
3 - Equipmen		3 Equipmen	6086		89	661	644		948
4 - Subs		4 Subs	434520	1833	574			21000	14000

Figure 12-13
Summary CSV File Export of Contractor's Cost

		1	2	3	4	5	6	7	8	9
Week Beginning		01-Oct-01	08-Oct-01	15-Oct-01	22-Oct-01	29-Oct-01	05-Nov-01	12-Nov-01	19-Nov-01	26-Nov-01
Week Ending		07-Oct-01	14-Oct-01	21-Oct-01	28-Oct-01	04-Nov-01	11-Nov-01	18-Nov-01	25-Nov-01	02-Dec-01
Labor		141	465	3318	3085	3216	3526	5471	1240	3450
Material		126	251	4534	2809	1165	2948	4229	564	2889
Equipment		0	89	661	644	0	948	1319	0	678
Subcontracts		1833	574	0	0	21000	14000	0	0	0
Labor Burdens		28.2	93	663.6	617	643.2	705.2	1094.2	248	690
Total Labor Cost		169.2	558	3981.6	3702	3859.2	4231.2	6565.2	1488	4140
Material					126	251	4534	2809	1165	2948
Equipment					0	89	661	644	0	948
Subcontracts				1833	574	0	0	21000	14000	0
Total Direct Field Costs		169.2	558	5814.6	4402	4199.2	9426.2	31018.2	16653	8036
Remaining Field Indirects		3526	3526	3526	3526	3526	3526	3526	3526	3526
Home Office Costs		915	915	915	915	915	915	915	915	915
Total Cost		4610.2	4999	10255.6	8843	8640.2	13867.2	35459.2	21094	12477
Cummulative		4610.2	9609.2	19864.8	28707.8	37348	51215.2	86674.4	107768.4	120245.4

Figure 12-14
Manipulated CSV File Export of Contractors Cost

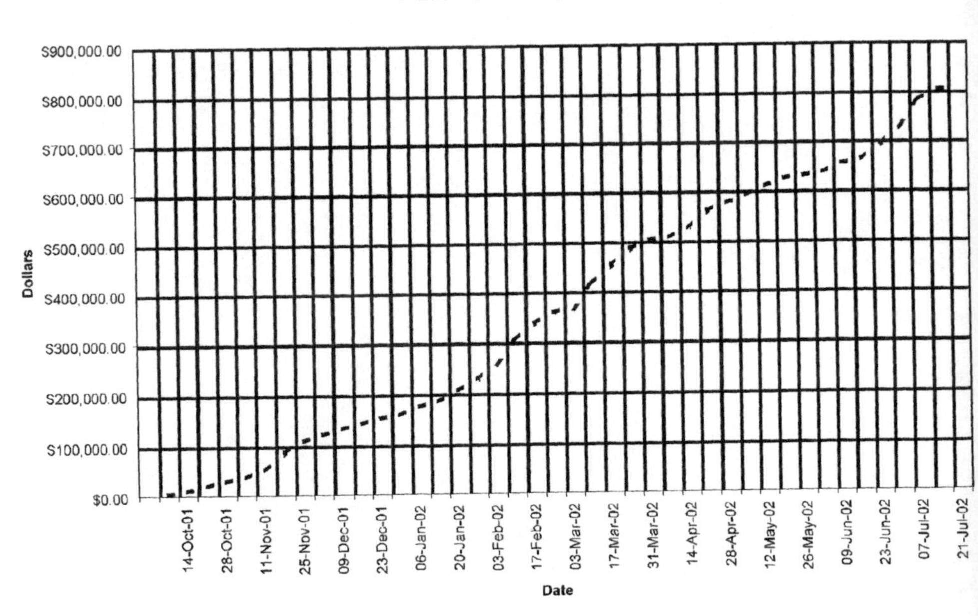

Figure 12-15
Planned Cash Outflows

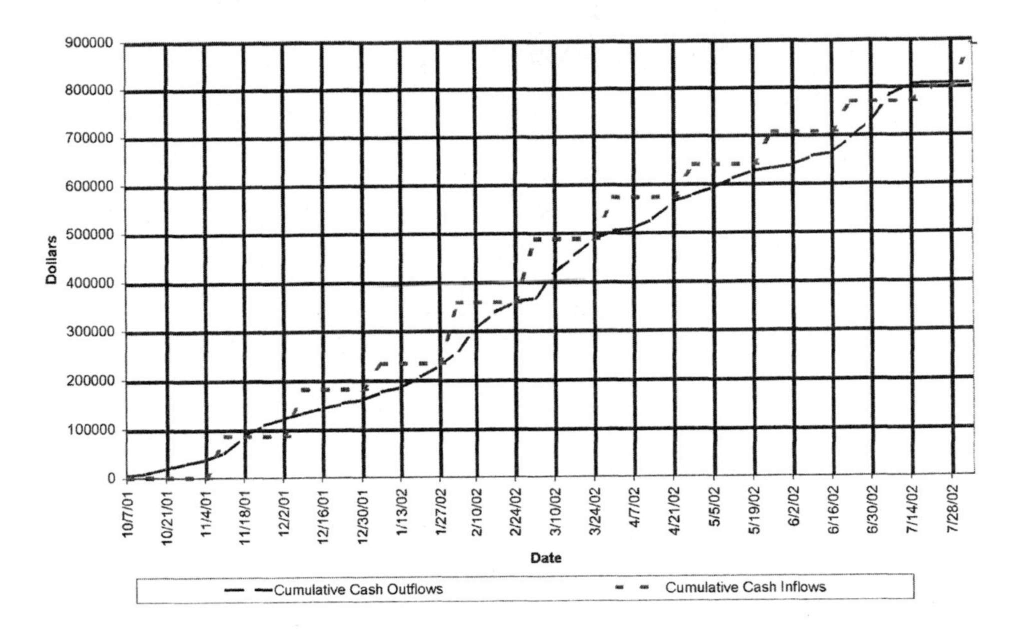

Figure 12-16
Planned Cash Inflows and Outflows

actual outflows. Furthermore, the graph can be superimposed on the inflow curve to generate a graph similar to the one found in Figure 12-16. In that figure, the planned positive and negative cash flow can easily be seen. If the inflows curve is above the outflows curve, there is a positive cash flow. Conversely if the inflows curve is below the outflow curve, there is a negative cash flow. By having an estimate of how much and when funds will be needed, the contractor can negotiate with various financial institutions to get the most advantageous arrangement.

It is important to remember that the cash flow plan is a dynamic document and that, as the project schedule changes, so will the cash flow plan. Also remember that a number of assumptions were made concerning when subcontractors and suppliers will be paid; with all of these assumptions, this plan is only an estimate.

Conclusion

By using hammock activities to describe the items on the schedule of values, an outline for cash inflows for the project can be developed. However, the raw data generated by Primavera Project Planner needs to be adjusted to reflect the payment terms between the contractor and the owner. This will enhance the accuracy of cash inflow projections. Just as the cash inflows can be projected over time, so can the costs to the contractor. However, the raw cost data needs to be adjusted to reflect the average payment terms between the contractor and the subcontractors and suppliers. Once all of the adjustments to both revenue and cost have been made, a plot can be developed to show when during the project there will be surpluses or shortages of cash. If shortages are anticipated, arrangements can be made to borrow funds to cover these short-term deficits. By having advance warning of the need to borrow funds, the contractor can have the flexibility to seek out a lender that offers the most competitive interest rate and terms. In addition, this curve can be used as the basis to modify the amount of the individual items on the schedule of values. If this is done in an iterative process, a schedule of values can be developed that will minimize the need to borrow funds. While the exact amount for the individual items on the schedule of values is often the subject of negotiations between the owner and the contractor, an advantageous schedule of values is a good place for the contractor to begin negotiations.

Suggested Exercises

1. Develop a schedule of values from the sample project developed in earlier chapters. The total amount of all of the items on the schedule of values must add up to the contract amount.
2. Establish hammock activities that match the schedule of values and load the schedule of values amount to each of these activities.
3. Produce a cost loading report using the information from the schedule of values activities.
4. Load all contractor direct costs to the construction activities.
5. Generate a cost loading report that includes only the contractor's direct field costs.
6. Assuming that the contractor can submit a billing every 2 weeks and the owner has 5 days to pay, develop a planned revenue curve (assume 10 percent retainage).
7. Assuming that labor is paid weekly and all other expenses are paid every 2 weeks, develop a planned expenditure curve.
8. Overlay the curves developed in Items 6 and 7 to determine when and how much additional capital will be needed to support the project.

Progress Planning and Control

Key Terms and Concepts
Developing the Progress Plan
Determining Actual Project Progress
Creating Planned Progress Curves with Primavera Project Planner
Superimposing Curves
Conclusion

KEY TERMS

Project Budget The amount of money that has been allocated by the contractor for the completion of the project.

Earned Workhours The quantity of workhours that should have been required to complete an amount of progress, based on the labor productivity from the estimate.

Earned Value The number of dollars that should have been expended based on a set amount of progress.

Actual Progress The completed percentage of the project based on actual field observations and quantification.

Contractor's Forces Those craft personnel who are employed by and under the control of the general contractor.

Developing the Progress Plan

Developing the overall project progress plan is one of the critical initial planning activities. This progress plan is typically represented as an S-shaped progress curve. Before developing the overall project progress curve, a number of different issues need to be explored. One is how will the project be executed. Is the majority of the work being performed by subcontractors or by the contractor's own forces? How and when will actual progress be measured? The answer to these questions will dictate the methodology used to develop the progress plan. When developing the progress plan and tracking actual progress, consistency is paramount. There are no perfect ways to plan or measure progress. Therefore, the consistency between how the progress plan is developed and how it is measured is critical. If a particular methodology or weighting process is used to develop the plan, that same process needs to be used for determining actual progress.

When planning or measuring progress, it is difficult to come up with a common unit of measure. Since some activities deal with cubic yards of concrete and others deal with tons of steel and others with linear feet of wall and so on it is difficult to come up with a common unit of measure. Progress on an individual activity does not have this problem since most of the units are consistent within a specific activity. The only apparent common unit of measure that will cross all activities is dollars. While there may be a wide variety of units of measurement, each activity has a cost. The problem with this measurement is the underlying assumption that every dollar generates an equal amount of progress. There are multitudes of examples that run contrary to that assumption. For example, if the schedule had an activity to set mechanical equipment, that activity requires a very short period of time and a very limited amount of effort. However, it generates a substantial cost. The impact of this assumption can be minimized by using the same weighting process for the actual and planned progress.

The development of a cost-weighted progress plan begins with spreading the cost of the activity over its duration, just as was done with a cost loading report. These costs can then be accumulated. Figure 13-1 shows a small sample schedule in which the costs have been spread and accumulated by week. Each of the weekly totals has been divided by the total to develop an incremental percent complete. These incremental percents complete have been accumulated to develop what percent complete the project should be every week. In

Activity	Labor $	Material $	Equipment $	Subcontract $	Total $	WEEK 1	2	3	4	5	6	7	8
A	$10,100	$7,400	$8,000		$25,500	$8,500	$8,500	$8,500					
B				$60,000	$60,000		$30,000	$30,000					
C				$90,000	$90,000			$30,000	$30,000	$30,000			
D	$32,000	$42,000	$13,900		$87,900				$29,300	$29,300	$29,300		
E	$5,200	$12,000			$17,200						$8,600	$8,600	
F	$2,000	$4,000			$6,000								$6,000
Weekly Total					$286,600	$8,500	$38,500	$68,500	$59,300	$59,300	$37,900	$8,600	$6,000
Incremental Planned Percent Complete						2.97%	13.43%	23.90%	20.69%	20.69%	13.22%	3.00%	2.09%
Cumulative Percent Complete						2.97%	16.40%	40.30%	60.99%	81.68%	94.91%	97.91%	100.00%

Figure 13-1
Planned Progress

Figure 13-2 this information has been graphed. The cumulative curve is S-shaped, showing that at beginning and end of the project progress is slow and the middle is where the majority of the work is being performed. When actual construction commences, the actual progress should be plotted. If the actual progress curve is above the planned curve, the project is ahead of schedule. Conversely, if the actual curve is below the planned curve, then the project is behind schedule. Just as with most of the other control graphics, this plot would be an indicator of the performance of the project. If the project is ahead of schedule, what has been the cost of getting ahead? Did labor productivity decline earning the extra progress? Was overtime worked, raising the cost of every workhour? These questions would need to be addressed prior to making any assessment about the future of the project. Conversely, if the project is behind schedule, what is the cause and what is the remedy? Has the reason been not enough craft persons, lower skill level of the craft persons, or perhaps material or

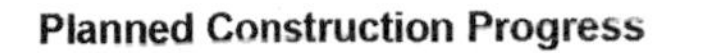

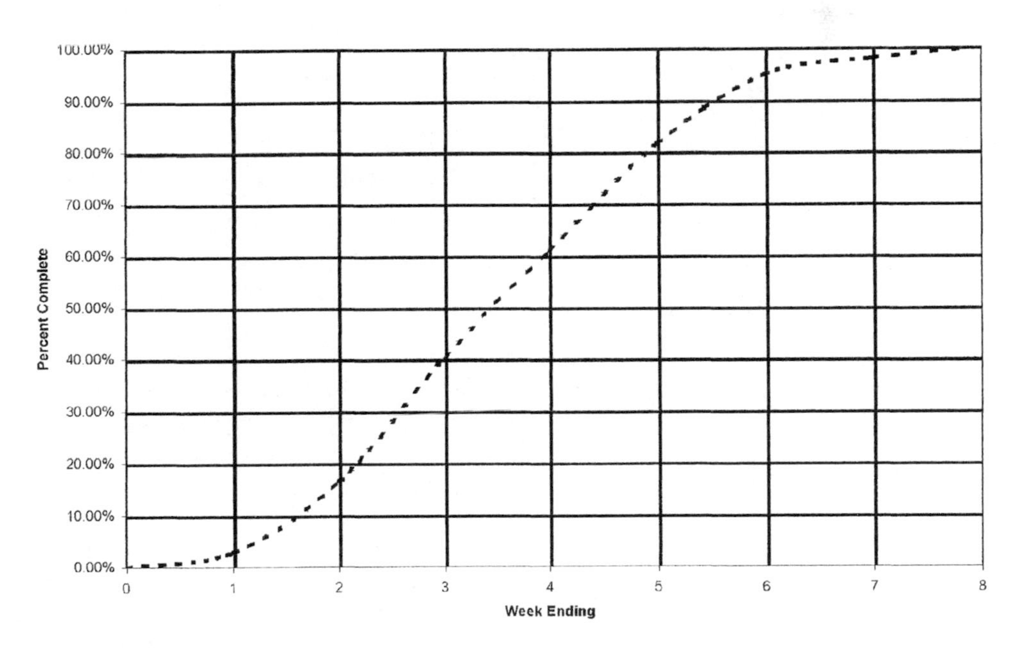

Figure 13-2
Planned Progress S-curve

equipment delays? The source of the problem will impact the strategy developed to remedy the situation.

One way that the planned progress graph could be improved would be to plot planned workhours or people on the same graph. By loading workhours on the activities, planned people per increment of time can be determined. In Figure 13-3, the planned workhours have been spread. In Figure 13-4, the equivalent persons have been plotted as a histogram. On that graph there is a common X-axis and two distinct Y-axes. The one on the left is for people and the one on the right is for percent complete. These graphs were produced using a spreadsheet program. In Figure 13-5, the project has been statused at the end of the second week. In that graph the project is ahead of schedule and overstaffed. The question that the project manager must ask is what is the cost for the added progress. If the extra people worked at the planned productivity rate, then it is possible that the project is ahead of schedule and on budget, a desirable situation.

Activity	Workhours	Labor $	Material $	Equipment $	Subcontract $	Total $	WEEK 1	2	3	4	5	6	7	8
A	654	$10,100	$7,400	$8,000		$25,500	218	218	218					
B					$60,000	$60,000								
C					$90,000	$90,000								
D	2601	$32,000	$42,000	$13,900		$87,900				867	867	867		
E	290	$5,200	$12,000			$17,200						145	145	
F	200	$2,000	$4,000			$6,000								200
Weekly Total						$286,600	218	218	218	867	867	1,012	145	200
Equivalent People							5.45	5.45	5.45	21.68	21.68	25.30	3.63	5.00

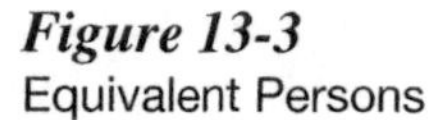

Figure 13-3
Equivalent Persons

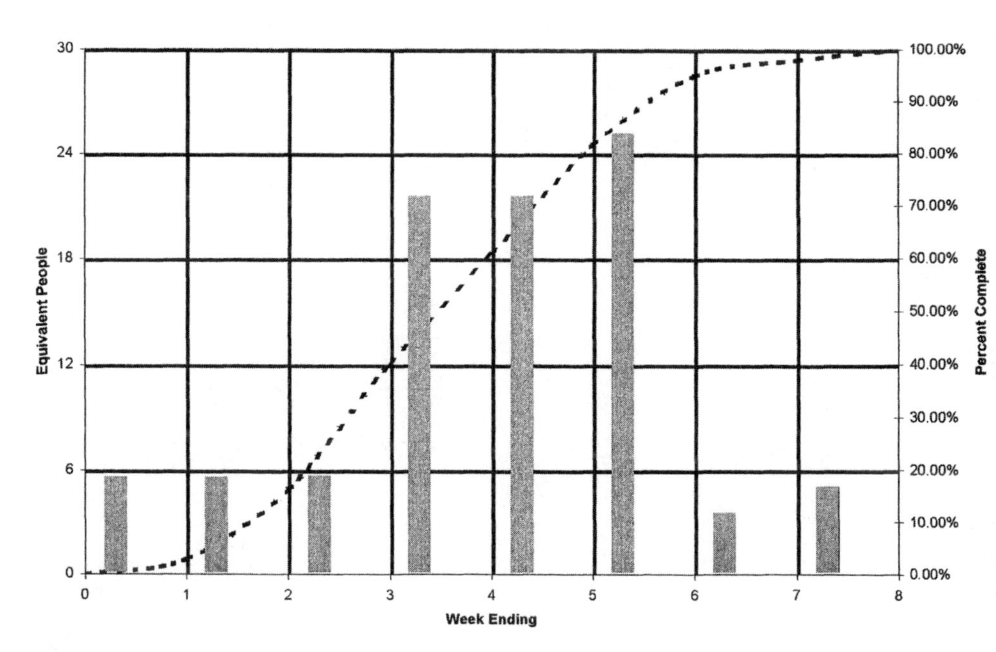

Figure 13-4
Planned Progress and People

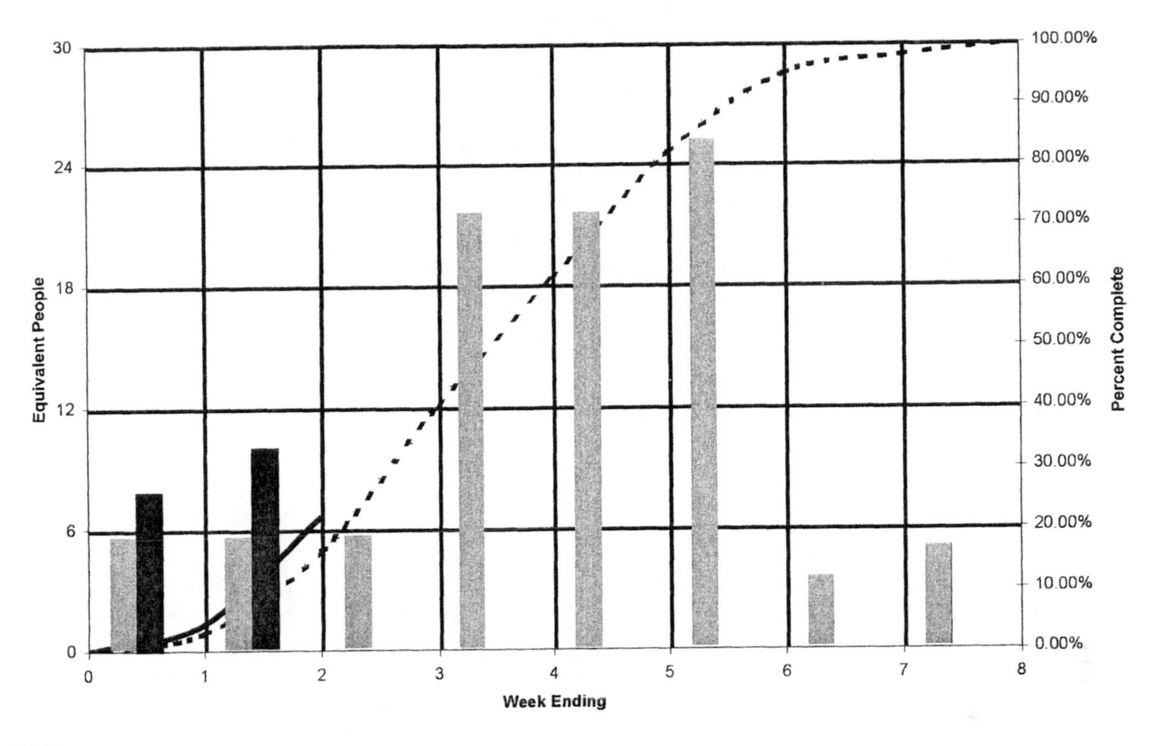

Figure 13-5
Updated Planned Progress and People

Determining Actual Project Progress

The determination of actual overall progress is directly related to how the planned progress was formulated. If dollars were used to weight each of the activities, then dollars need to be used to determine the actual overall percent complete. If that method was used, then the earned value of each of the activities needs to be determined. The earned value of an activity is the theoretical amount of value a completed activity adds to the project. The earned value is found by multiplying the percent complete by the estimated or budgeted cost as shown in Formula 13-1.

$$\text{Earned Value} = \text{Budgeted Amount} \times \text{Percent Complete}$$

Formula 13-1

The sum of all of the earned values can be divided by the budgeted amount to find the overall percent complete. In Figure 13-6 the first activity is 50 percent complete at the end of the first week. That yields an earned value of $12,750 for that activity, and since no other activities have started that is also the total earned value. If that amount is divided into the budget, the overall percent complete is of 4.45 percent. Figure 13-7 shows how the actual progress would be plotted. A quick glance at that graph clearly shows the project ahead of schedule. In Figures 13-8 and 13-9 the project has been statused at the end of the second week. The process and calculations are identical. Figure 13-9 shows the project behind schedule at the end of the second week.

One of the difficulties with using earned value is how to handle change orders. Since change orders add and delete items from the scope of the project, there is an actual change in the project budget. In the above example, suppose the scope of Activity D was increased and

Activity	Labor $	Material $	Equipment $	Subcontract $	Total $	WEEK 1	2	3	4	5	6	7	8	% Complete	Earned Value
A	$10,100	$7,400	$8,000		$25,500	$8,500	$8,500	$8,500						50.00%	$12,750
B				$60,000	$60,000		$30,000	$30,000						0.00%	$0
C				$90,000	$90,000			$30,000	$30,000	$30,000				0.00%	$0
D	$32,000	$42,000	$13,900		$87,900				$29,300	$29,300	$29,300			0.00%	$0
E	$5,200	$12,000			$17,200						$8,600	$8,600		0.00%	$0
F	$2,000	$4,000			$6,000								$6,000	0.00%	$0
		Weekly Total			$286,600	$8,500	$38,500	$68,500	$59,300	$59,300	$37,900	$8,600	$6,000	4.45%	$12,750
		Incremental Planned Percent Complete				2.97%	13.43%	23.90%	20.69%	20.69%	13.22%	3.00%	2.09%		
		Cumulative Percent Complete				2.97%	16.40%	40.30%	60.99%	81.68%	94.91%	97.91%	100.00%		
		Cumulative Actual Percent Complete				4.45%									

Figure 13-6
Updated Planned Progress at End of Week 1

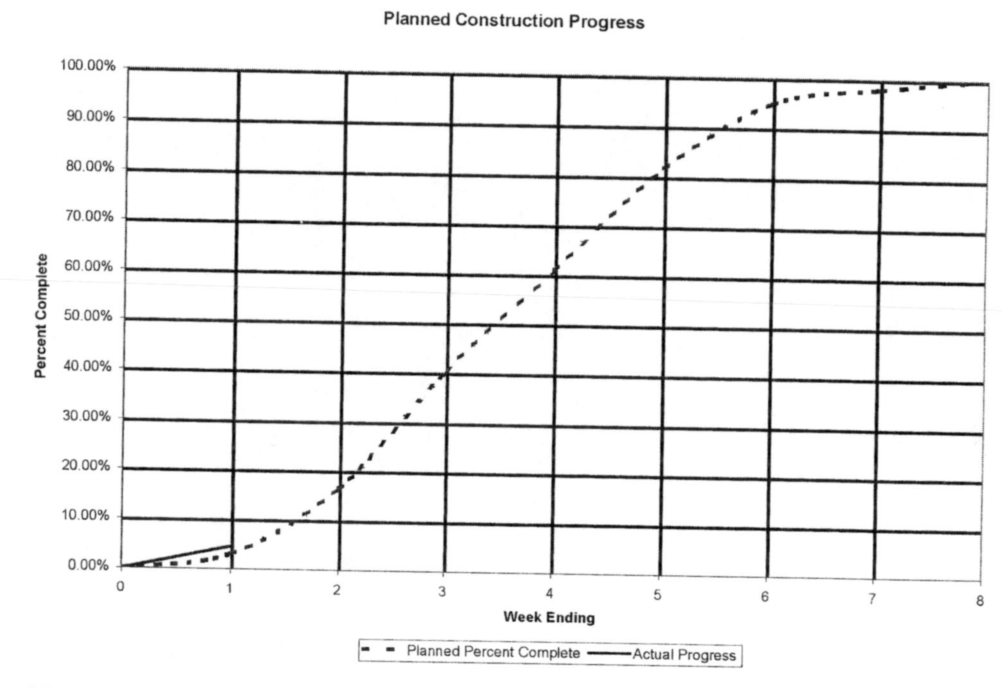

Figure 13-7
Updated Planned Progress at End of Week 1

Activity	Labor $	Material $	Equipment $	Subcontract $	Total $	WEEK 1	2	3	4	5	6	7	8	% Complete	Earned Value
A	$10,100	$7,400	$8,000		$25,500	$8,500	$8,500	$8,500						50.00%	$12,750
B				$60,000	$60,000		$30,000	$30,000						25.00%	$15,000
C				$90,000	$90,000			$30,000	$30,000	$30,000				10.00%	$9,000
D	$32,000	$42,000	$13,900		$87,900				$29,300	$29,300	$29,300			0.00%	$0
E	$5,200	$12,000			$17,200						$8,600	$8,600		0.00%	$0
F	$2,000	$4,000			$6,000								$6,000	0.00%	$0
		Weekly Total			$286,600	$8,500	$38,500	$68,500	$59,300	$59,300	$37,900	$8,600	$6,000	12.82%	$36,750
		Incremental Planned Percent Complete				2.97%	13.43%	23.90%	20.69%	20.69%	13.22%	3.00%	2.09%		
		Cumulative Percent Complete				2.97%	16.40%	40.30%	60.99%	81.68%	94.91%	97.91%	100.00%		
		Cumulative Actual Percent Complete				4.45%	12.82%								

Figure 13-8
Updated Planned Progress at End of Week 2

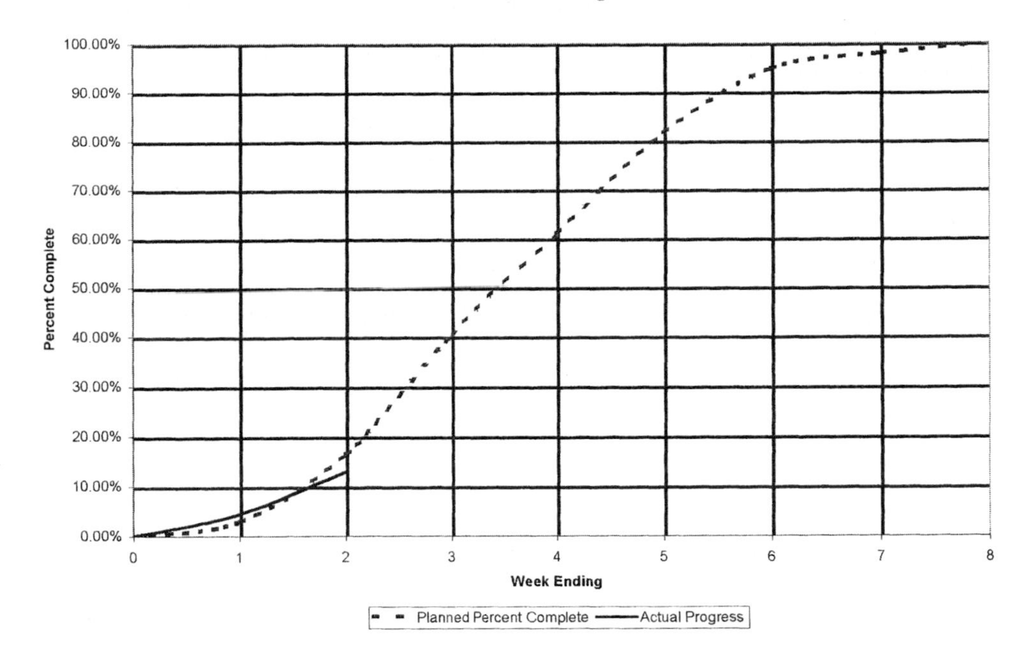

Figure 13-9
Updated Planned Progress at End of Week 2

the associated budget was increased by \$100,000. Now the total project budget is \$386,600 with the same earned value of \$46,750. With these new figures the project is 12.09 percent complete at the end of the second week. That would mean that the project percent complete would have dropped. It is a bad practice to change portions of the planned or actual progress curves that are in the past. Any changes should be shown in the future. In Figure 13-10 the percent complete table has been modified to show the original progress plan and the new revised plan. In Figure 13-11, this information is graphed.

Notice that none of the historical information was modified; instead, the revised planned progress curve is redrawn from the date of the change in scope and the actual percent complete is also recalculated from the date of the change in scope. The drop in the curve designates that a change in scope has happened, and the basis of progress has changed.

Activity	Labor $	Material $	Equipment $	Subcontract $	Total $	WEEK 1	2	3	4	5	6	7	8	% Complete	Earned Value
A	$10,100	$7,400	$8,000		$25,500	$8,500	$8,500	$8,500						100.00%	$25,500
B				$60,000	$60,000		$30,000	$30,000						75.00%	$45,000
C				$90,000	$90,000			$30,000	$30,000	$30,000				17.00%	$15,300
D	$32,000	$42,000	$13,900	$100,000	$187,900				$62,633	$62,633	$62,634			10.00%	$18,790
E	$5,200	$12,000			$17,200						$8,600	$8,600		0.00%	$0
F	$2,000	$4,000			$6,000								$6,000	0.00%	$0
		Weekly Total			$386,600	$8,500	$38,500	$68,500	$59,300	$59,300	$37,900	$8,600	$6,000	27.05%	$104,590
		Incremental Planned Percent Complete				2.97%	13.43%	23.90%	20.69%	20.69%	13.22%	3.00%	2.09%		
		Cumulative Percent Complete				2.97%	16.40%	40.30%	60.99%	81.68%	94.91%	97.91%	100.00%		
		Cumulative Actual Percent Complete				4.45%	12.82%								
		Revised Scope				8500	38500	68500	92633	92633	71234	8600	6000		
		Revised Incremental Percent Complete				2.20%	9.96%	17.72%	23.96%	23.96%	18.43%	2.22%	1.55%		
		Revised Cumulative Percent Complete				2.20%	12.16%	29.88%	53.84%	77.80%	96.22%	98.45%	100.00%		
		Revised Actual Cumulative Percent Complete				4.45%	9.51%	27.05%							

Figure 13-10
Updated Planned Progress at End of Week 3

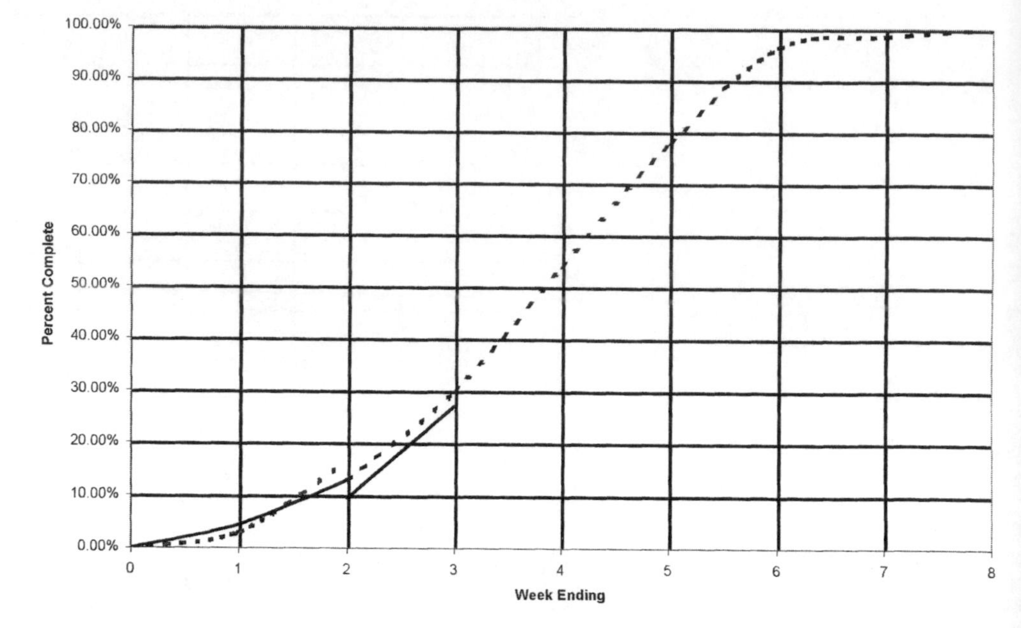

Figure 13-11
Updated Planned Progress at End of Week 3

If the majority of the work is being self-performed by the general contractor, workhours could become the basis of weighting the activities. In that scenario the earned value would be earned workhours rather than earned dollars. While workhours are a better measurement of effort required to complete an activity, there most likely still would be many major activities that would not have the available workhours. In that situation, those activities would have to be ignored for both the planned progress and actual progress. While this would be consistent, it would improperly skew the overall project progress curve. When there are multiple items of work within an activity, it is completely plausible to use workhours to determine the percent complete for the activity and then to use earned dollars to generate the overall percent complete. Figure 13-12 is the estimate for Activity A in the previous example. This information could then be used on a field progress report to determine what percent complete each of the items within the activity is. Figure 13-13 is an example of the field progress report at the end of the second week of the project. In that example the actual quantity is found from field observation and the activity percent complete is based

ACTIVITY A ESTIMATE								
Description	Quantity	Unit	Work Hours	Labor $	Material $	Equip $	Subcontract $	Total
Excavate Footings	950	CY	47	709		5,000		5,709
Form Footings	960	SFCA	48	620	720			1,340
Reinforce Footings	3	Tons	56	464	1,800			2,264
Place and Finish Footings	40	CY	21	250				250
Strip Forms	960	SFCA	24	188				188
Reinforce Piers	2	Tons	36	435	1,200			1,635
Form Piers	1200	SFCA	285	3,420	3,500			6,920
Place and Finish Piers	12	CY	6	63				63
Strip Forms	1200	SFCA	32	351				351
Backfill and Compact	900	CY	240	3,600	180	3,000		6,780
Total Activity A			795	10,100	7,400	8,000	0	25,500

Figure 13-12
Activity A Estimate

FIELD PROGRESS REPORT						
ACTIVITY A						
Description	Estimated Quantity	Unit	In-Place Quantity	% Installed	Estimate Work Hours	Earned Work Hours
Excavate Footings	950	CY	950	100.00%	47	47
Form Footings	960	SFCA	960	100.00%	48	48
Reinforce Footings	3	Tons	3	100.00%	56	56
Place and Finish Footings	40	CY	40	100.00%	21	21
Strip Forms	960	SFCA	960	100.00%	24	24
Reinforce Piers	2	Tons	2	100.00%	36	36
Form Piers	1200	SFCA	569	47.42%	285	135
Place and Finish Piers	12	CY	2	16.67%	6	1
Strip Forms	1200	SFCA	100	8.33%	32	3
Backfill and Compact	900	CY	100	11.11%	240	27
Total Activity A				50.00%	795	397

Figure 13-13
Activity A Field Progress Report

upon earned workhours. When the overall percent complete is figured, dollars are used to develop the overall percent complete.

Creating Planned Progress Curves with Primavera Project Planner

The planned and actual progress curves can be generated directly from Primavera Project Planner or with a combination of that program and a spreadsheet or graphics package. Primavera Project Planner graphics are quicker to generate but lack the presentation features of a spreadsheet program. If the overall planned progress curve is to be generated and updated using a spreadsheet program, the first step is to generate a cost loading report and either output the information to a tabular report or as a *CSV* file which can be opened using a spreadsheet program. The specifics of generating these reports can be found in the previous chapter. Output 13-1 is a portion of the cost loading report. From that report the planned

All Star Developments — PRIMAVERA PROJECT PLANNER — Sample Construction Project

REPORT DATE 20APR00 RUN NO. 83 12:06 — COST LOADING REPORT — START DATE 01OCT01 FIN DATE 01JUL02

DATA DATE 01OCT01 PAGE NO. 1- 1

Loading for Progress — TOTAL USAGE FOR WEEK

RESOURCE	RESOURCE DESCRIPTION	TOTAL	1OCT 2001	8OCT 2001	15OCT 2001	22OCT 2001	29OCT 2001	5NOV 2001	12NOV 2001	19NOV 2001
CARP	Carpenters	19472			99	648	1982	1100	825	479
CFIN	Concrete Finishers	18542			51	242	51	154	191	26
DFE	Direct Field Equipme	7208		89	661	644		948	2441	
DFM	Direct Field Materia	90169	126	251	4534	2809	1165	2948	3107	564
DFS	Direct Field Subcont	434520	1833	574			21000	14000		
LABR	Laborers	11216		137	1121	1210	335	2122	1290	205
MASN	Masons	10325								
MASNHLP	Mason Helper	5519								
OPENG	Operating Engineer	5940		45	330	315		150	2370	
ROD	Rodbuster	8720	141	283	1717	669	848		795	530
	REPORT TOTAL	611631	2100	1379	8512	6538	25381	21422	11019	1804

Output 13-1
Cost Loading Report

incremental percent complete for the first week would be 0.34 percent. That figure was found by dividing the following:

2,100 Incremental Dollars/611,631 Total Dollars

The incremental progress for the second week would be 0.23 percent, yielding a planned overall percent complete of 0.57 percent at the end of the second week. Figure 13-14 is the spreadsheet table with all of the incremental and cumulative percents complete. Figure 13-15 is a planned progress curve that was generated from the information in Figure 13-14.

A similar curve can be generated directly using Primavera Project Planner. This is done with the graphic output option. This process is started by selecting the *Tools* command followed by the *Graphic Reports/Resource and Cost* options as shown in Screen Image 13-1. These commands open the *Resources/Cost Graphics* window as shown in Screen Image 13-2. Clicking on the *Add* button will open the *Add a New Report* window as shown in Screen Image 13-3. Just as with all of the previous reports, this window generates a report number. By clicking the *Add* button again the *Resource/Cost Graphics* window opens.

This window as seen in Screen Image 13-4 has a number of tabs that need to be worked through in order to generate the planned progress curve. The *Resource Selection* tab within that window is slightly different from the one used when generating tabular reports. The main difference is the *Group #* column. This feature allows for multiple resources to be grouped together into a single curve. For example, if the Carpenters and Mason resource were put in group 1 and the Laborers and Mason Helper were in group 2, two separate graphs would be created. The first would include all of the resources selected in group 1, and the second would show those resources assigned to group 2. In Screen Image 13-4 all of the contractor cost resources have been selected and assigned to group 1. This will group all of the resources together into a single graph. The *Content* tab in Screen Image 13-5 allows the user to specify the schedule to be the basis of the graph and to perform calculations.

Week Beginning	01-Oct-01	08-Oct-01	15-Oct-01	22-Oct-01	29-Oct-01	05-Nov-01	12-Nov-01	19-Nov-01	26-Nov-01	03-Dec-01	10-Dec-01
Week Ending	08-Oct-01	15-Oct-01	22-Oct-01	29-Oct-01	05-Nov-01	12-Nov-01	19-Nov-01	26-Nov-01	03-Dec-01	10-Dec-01	17-Dec-01
Loaded Dollars	$2,100	$1,379	$8,512	$6,538	$25,381	$21,422	$11,019	$1,804	$7,017	$1,025	$6,871
Incremental Progress	0.34%	0.23%	1.39%	1.07%	4.15%	3.50%	1.80%	0.29%	1.15%	0.17%	1.12%
Cumulative Progress	0.34%	0.57%	1.96%	3.03%	7.18%	10.68%	12.48%	12.78%	13.93%	14.09%	15.22%

Week Beginning	17-Dec-01	24-Dec-01	31-Dec-01	07-Jan-02	14-Jan-02	21-Jan-02	28-Jan-02	04-Feb-02	11-Feb-02	18-Feb-02	25-Feb-02
Week Ending	24-Dec-01	31-Dec-01	07-Jan-02	14-Jan-02	21-Jan-02	28-Jan-02	04-Feb-02	11-Feb-02	18-Feb-02	25-Feb-02	04-Mar-02
Loaded Dollars	$3,034	$5,307	$19,867	$7,972	$43,478	$24,789	$35,227	$18,798	$2,403	$46,441	$31,660
Incremental Progress	0.50%	0.87%	3.25%	1.30%	7.11%	4.05%	5.76%	3.07%	0.39%	7.59%	5.18%
Cumulative Progress	15.71%	16.58%	19.83%	21.13%	28.24%	32.29%	38.05%	41.13%	41.52%	49.11%	54.29%

Week Beginning	04-Mar-02	11-Mar-02	18-Mar-02	25-Mar-02	01-Apr-02	08-Apr-02	15-Apr-02	22-Apr-02	29-Apr-02	06-May-02	13-May-02
Week Ending	11-Mar-02	18-Mar-02	25-Mar-02	01-Apr-02	08-Apr-02	15-Apr-02	22-Apr-02	29-Apr-02	06-May-02	13-May-02	20-May-02
Loaded Dollars	$29,000	$10,589	$3,801	$14,340	$29,333	$9,333	$10,517	$14,069	$10,905	$1,415	$1,415
Incremental Progress	4.74%	1.73%	0.62%	2.34%	4.80%	1.53%	1.72%	2.30%	1.78%	0.23%	0.23%
Cumulative Progress	59.03%	60.76%	61.38%	63.73%	68.52%	70.05%	71.77%	74.07%	75.85%	76.08%	76.31%

Week Beginning	20-May-02	27-May-02	03-Jun-02	10-Jun-02	17-Jun-02	24-Jun-02	01-Jul-02
Week Ending	27-May-02	03-Jun-02	10-Jun-02	17-Jun-02	24-Jun-02	01-Jul-02	08-Jul-02
Loaded Dollars	$1,415	$1,061	$56,240	$19,033	$43,500	$22,419	$1,200
Incremental Progress	0.23%	0.17%	9.20%	3.11%	7.11%	3.67%	0.20%
Cumulative Progress	76.55%	76.72%	85.91%	89.03%	96.14%	99.80%	100.00%

Figure 13-14
Cost Loading Table

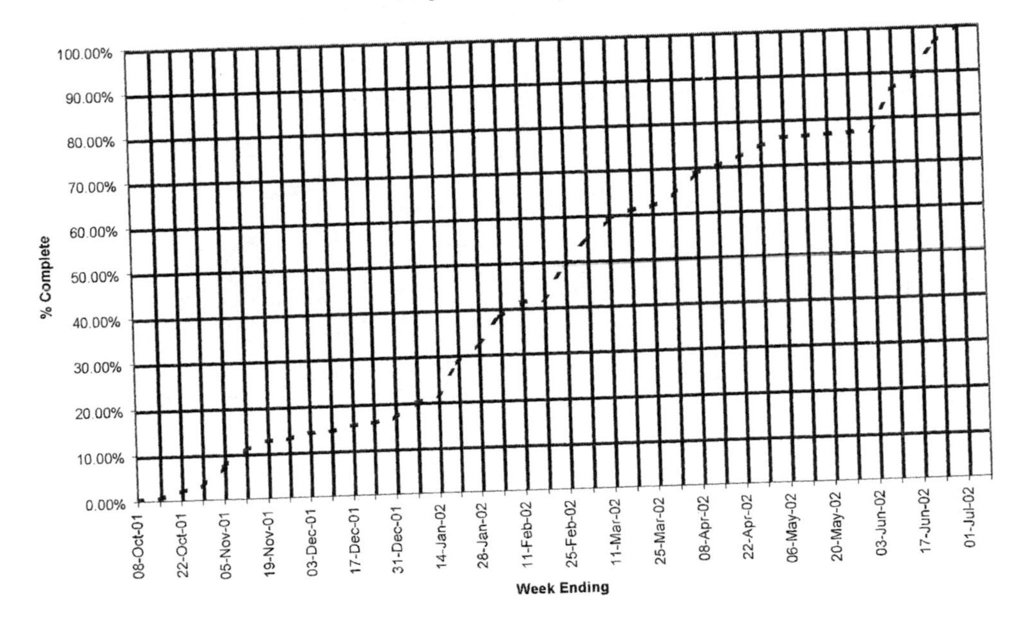

Figure 13-15
Planned Progress from Spreadsheet Table

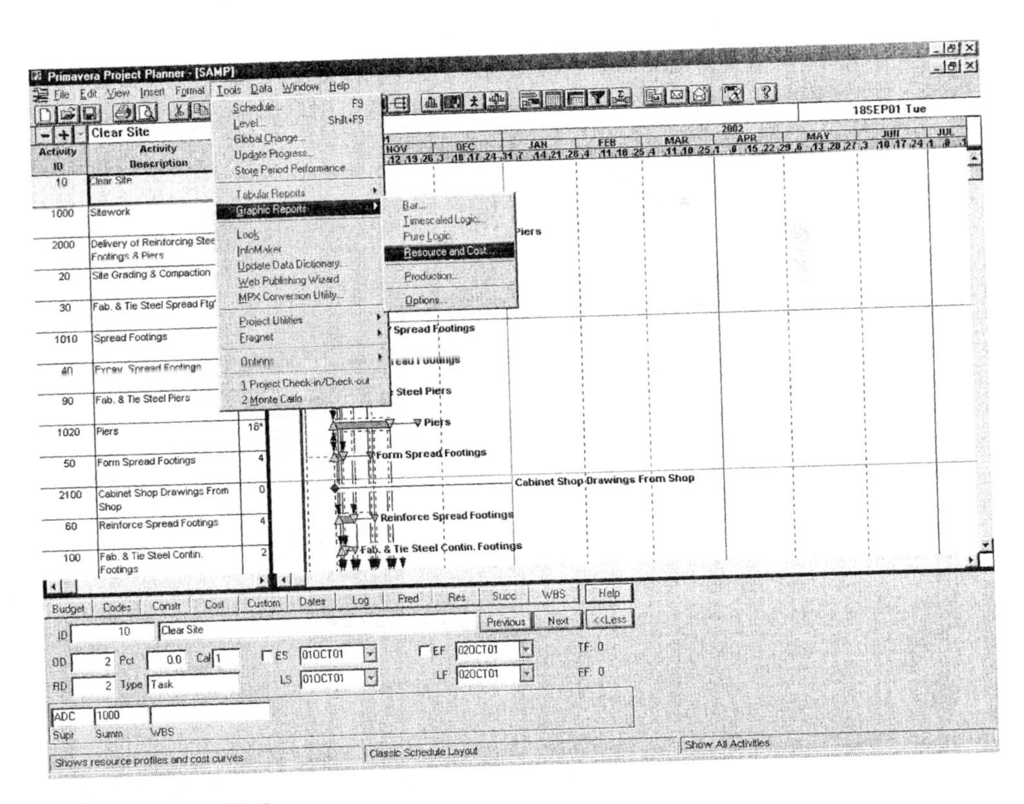

Screen Image 13-1
Opening Resource and Cost Window

Screen Image 13-2
Resource/Cost Graphics Window

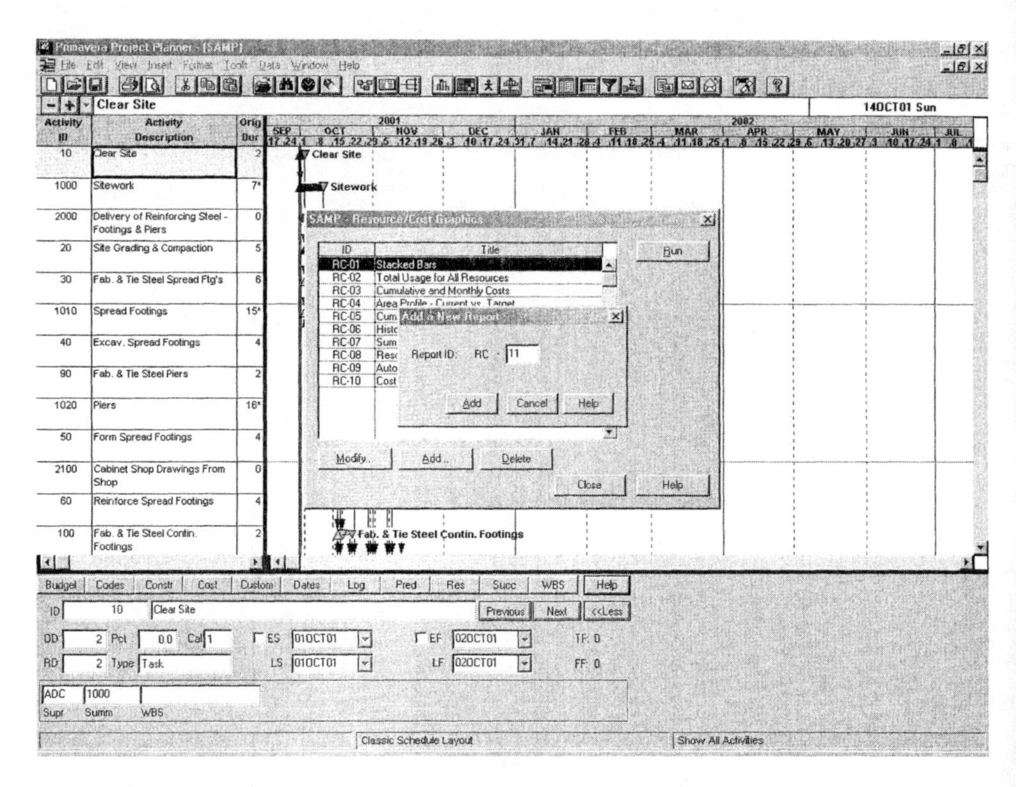

Screen Image 13-3
Add a New Report Window

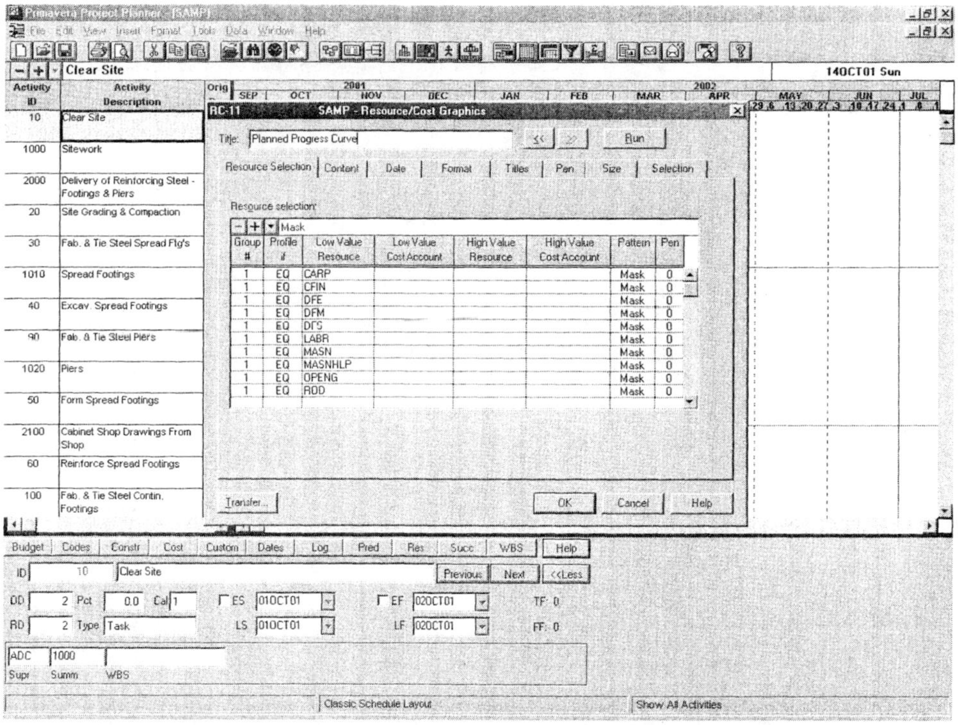

Primavera Project Planner - [SAMP]
File Edit View Insert Format Tools Data Window Help
Clear Site
14OCT01 Sun

Activity ID	Activity Description
10	Clear Site
1000	Sitework
2000	Delivery of Reinforcing Steel - Footings & Piers
20	Site Grading & Compaction
30	Fab. & Tie Steel Spread Ftg's
1010	Spread Footings
40	Excav. Spread Footings
90	Fab. & Tie Steel Piers
1020	Piers
50	Form Spread Footings
2100	Cabinet Shop Drawings From Shop
60	Reinforce Spread Footings
100	Fab. & Tie Steel Contin. Footings

Orig 2001 SEP OCT NOV DEC JAN FEB MAR 2002 APR MAY JUN JUL

RC-11 SAMP - Resource/Cost Graphics
Title: Planned Progress Curve << >> Run
Resource Selection | Content | Date | Format | Titles | Pen | Size | Selection
Resource selection:
Mask

Group #	Profile #	Low Value Resource	Low Value Cost Account	High Value Resource	High Value Cost Account	Pattern	Pen
1	EQ	CARP				Mask	0
1	EQ	CFIN				Mask	0
1	EQ	DFE				Mask	0
1	EQ	DFM				Mask	0
1	EQ	DFS				Mask	0
1	EQ	LABR				Mask	0
1	EQ	MASN				Mask	0
1	EQ	MASNHLP				Mask	0
1	EQ	OPENG				Mask	0
1	EQ	ROD				Mask	0

Transfer... OK Cancel Help

Budget | Codes | Constr | Cost | Custom | Dates | Log | Pred | Res | Succ | WBS | Help
ID 10 Clear Site Previous Next <<Less
OD 2 Pct 0.0 Cal 1 ES 01OCT01 EF 02OCT01 TF: 0
RD 2 Type Task LS 01OCT01 LF 02OCT01 FF: 0
ADC 1000
Supr Summ WBS
Classic Schedule Layout Show All Activities

Screen Image 13-4
Resource Selection Tab

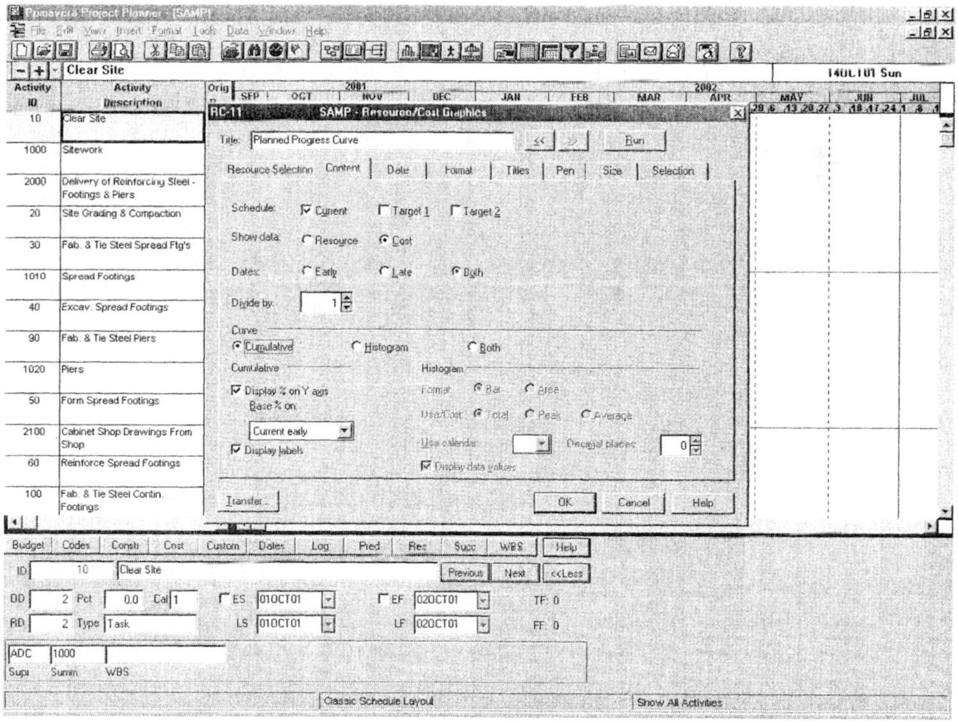

Primavera Project Planner - [SAMP]
File Edit View Insert Format Tools Data Window Help
Clear Site
14OCT01 Sun

Activity ID	Activity Description
10	Clear Site
1000	Sitework
2000	Delivery of Reinforcing Steel - Footings & Piers
20	Site Grading & Compaction
30	Fab. & Tie Steel Spread Ftg's
1010	Spread Footings
40	Excav. Spread Footings
90	Fab. & Tie Steel Piers
1020	Piers
50	Form Spread Footings
2100	Cabinet Shop Drawings From Shop
60	Reinforce Spread Footings
100	Fab. & Tie Steel Contin. Footings

Orig 2001 SEP OCT NOV DEC JAN FEB MAR 2002 APR MAY JUN JUL

RC-11 SAMP - Resource/Cost Graphics
Title: Planned Progress Curve << >> Run
Resource Selection | Content | Date | Format | Titles | Pen | Size | Selection
Schedule: ☑ Current ☐ Target 1 ☐ Target 2
Show data: ○ Resource ● Cost
Dates: ○ Early ○ Late ● Both
Divide by: 1
Curve
● Cumulative ○ Histogram ○ Both
Cumulative
☑ Display % on Y axis
Base % on:
Current early
☑ Display labels
Histogram
Format: ● Bar ○ Area
Use/Cost: ● Total ○ Peak ○ Average
Use calendar:
Decimal places: 0
☑ Display data values
Transfer... OK Cancel Help

Budget | Codes | Constr | Cost | Custom | Dates | Log | Pred | Res | Succ | WBS | Help
ID 10 Clear Site Previous Next <<Less
OD 2 Pct 0.0 Cal 1 ES 01OCT01 EF 02OCT01 TF: 0
RD 2 Type Task LS 01OCT01 LF 02OCT01 FF: 0
ADC 1000
Supr Summ WBS
Classic Schedule Layout Show All Activities

Screen Image 13-5
Resource/Cost Content Tab

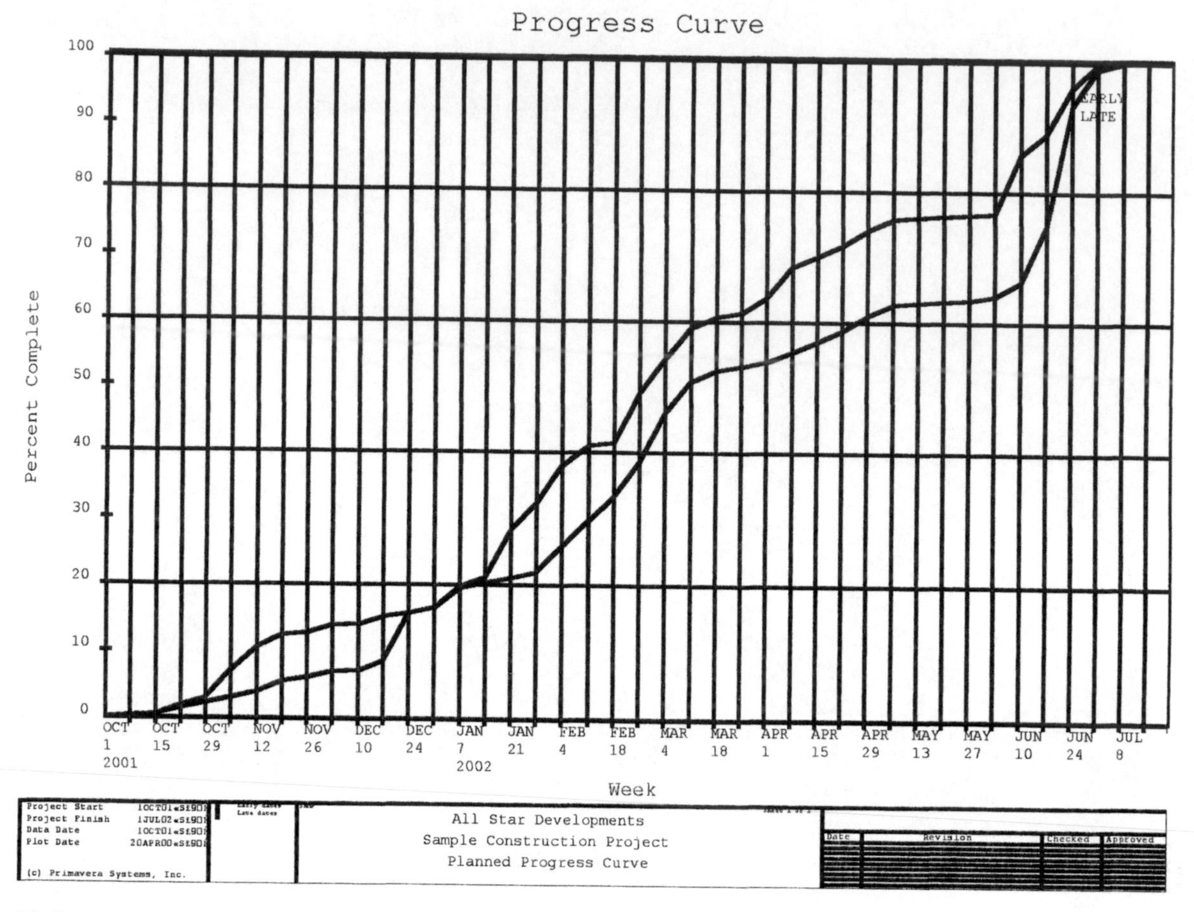

Output 13-2
Early and Late Progress Plan

The *Schedule* option at the top asks the user to designate which schedule to use as the basis of the graphic. While either of the target schedules can be selected, the current is the most common since it contains the most current information about the project. The *Show Data* options ask the user to use either *Resources* or *Cost* as the basis of the graphic. Since this is a progress curve and resources were only loaded for labor activities, the cost option is preferable: it will include information from every activity. The *Dates* option allows for the costs to be spread from early start to early finish or from late start to late finish across an activity. When the entire progress curve is plotted, either of these options will have the progress curve starting and stopping on the same date. The differences will occur between the end points. There is also the option to graph both of these curves. Output 13-2 (page 302) is the chart that was generated from this set of specifications; it shows both an early and a late progress curve. The *Divide by* option allows for the user to divide the values by a constant. This can be used to increase the readability of the graph by making the numbers smaller, say if they were divided by one thousand. The *Curve* option allows the user to specify the type of graphic to be plotted. The *Cumulative* is a curve that will add all of the increments together and plot the next point. The *Histogram* is used to show incremental amounts. The last two options are suboptions that are refinements to the type of curve selected. Screen Image 13-6 is the *Dates* tab. This tab is identical to the date tab that was introduced with the graphic bar chart schedules.

The *Format* tab (Screen Image 13-7) is where the user can enhance the presentation of the information. The top half of the screen allows the user to specify the units for both horizontal and vertical sight lines as well as their increment. The middle portion allows the

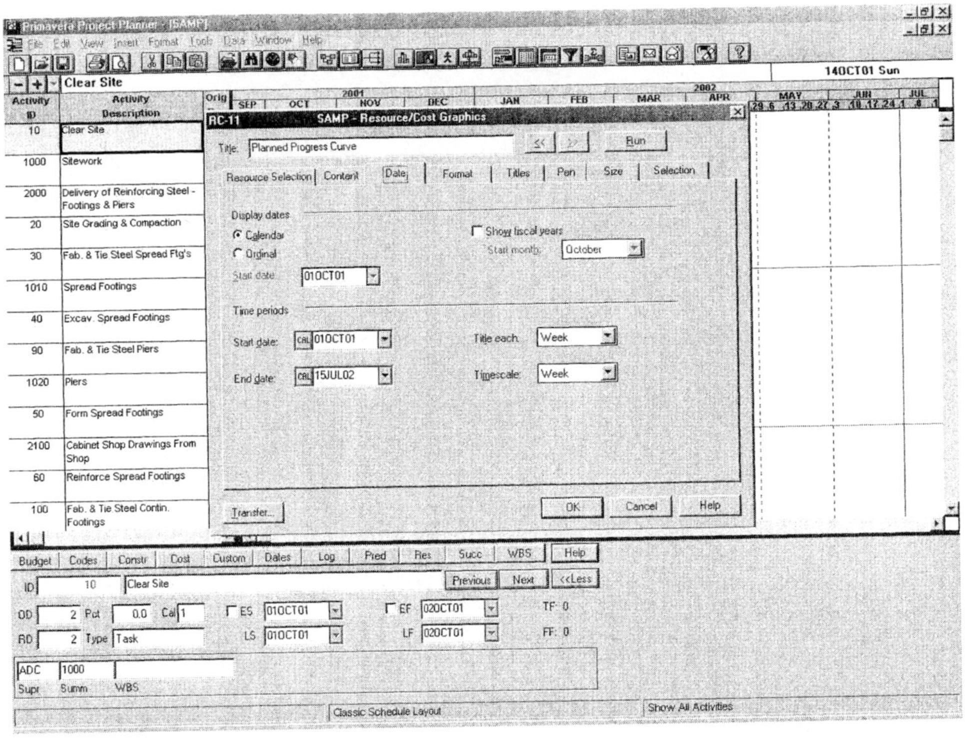

Primavera Project Planner - [SAMP]
File Edit View Insert Format Tools Data Window Help
14OCT01 Sun
Clear Site

Activity ID	Activity Description
10	Clear Site
1000	Sitework
2000	Delivery of Reinforcing Steel - Footings & Piers
20	Site Grading & Compaction
30	Fab. & Tie Steel Spread Ftg's
1010	Spread Footings
40	Excav. Spread Footings
90	Fab. & Tie Steel Piers
1020	Piers
50	Form Spread Footings
2100	Cabinet Shop Drawings From Shop
60	Reinforce Spread Footings
100	Fab. & Tie Steel Contin. Footings

Orig SEP OCT 2001 NOV DEC JAN FEB MAR 2002 APR MAY JUN JUL

RC-11 SAMP - Resource/Cost Graphics
Title: Planned Progress Curve << >> Run
Resource Selection | Content | Date | Format | Titles | Pen | Size | Selection
Display dates
Calendar
Ordinal
Start date: 01OCT01
Show fiscal years
Start month: October
Time periods
Start date: CAL 01OCT01
End date: CAL 15JUL02
Title each: Week
Timescale: Week
Transfer... OK Cancel Help

Budget | Codes | Constr | Cost | Custom | Dates | Log | Pred | Res | Succ | WBS | Help
ID 10 Clear Site Previous Next <<Less
OD 2 Pct 0.0 Cal 1 ES 01OCT01 EF 02OCT01 TF: 0
RD 2 Type Task LS 01OCT01 LF 02OCT01 FF: 0
ADC 1000
Supr Summ WBS
Classic Schedule Layout Show All Activities

Screen Image 13-6
Resource/Cost Date Window

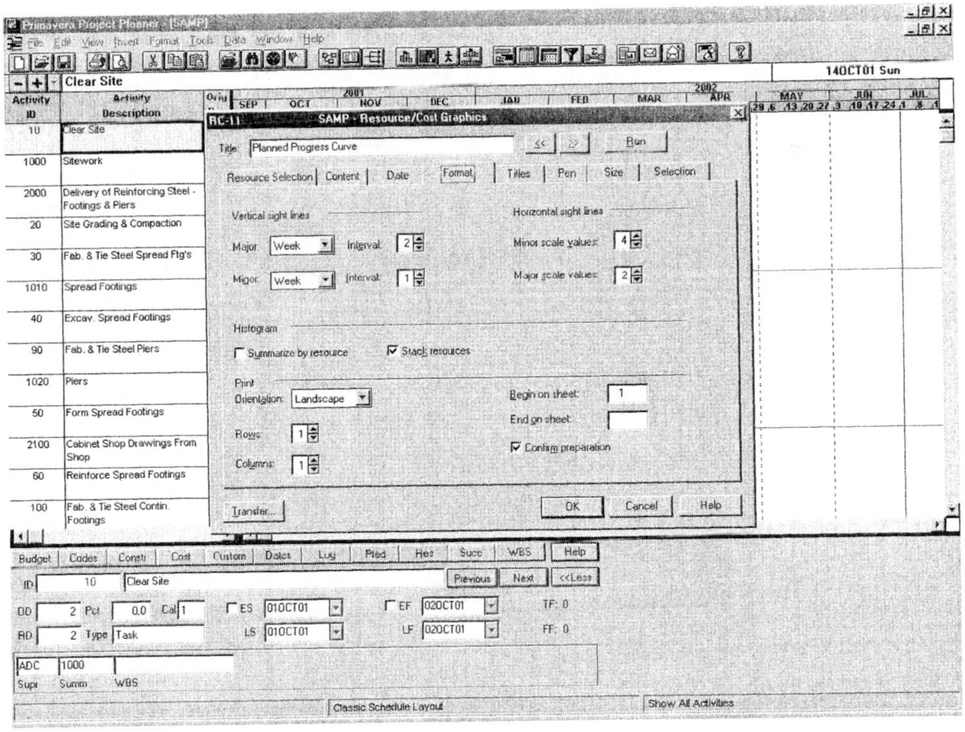

Primavera Project Planner - [SAMP]
File Edit View Insert Format Tools Data Window Help
14OCT01 Sun
Clear Site

Activity ID	Activity Description
10	Clear Site
1000	Sitework
2000	Delivery of Reinforcing Steel - Footings & Piers
20	Site Grading & Compaction
30	Fab. & Tie Steel Spread Ftg's
1010	Spread Footings
40	Excav. Spread Footings
90	Fab. & Tie Steel Piers
1020	Piers
50	Form Spread Footings
2100	Cabinet Shop Drawings From Shop
60	Reinforce Spread Footings
100	Fab. & Tie Steel Contin. Footings

Orig SEP OCT 2001 NOV DEC JAN FEB MAR 2002 APR MAY JUN JUL

RC-11 SAMP - Resource/Cost Graphics
Title: Planned Progress Curve << >> Run
Resource Selection | Content | Date | Format | Titles | Pen | Size | Selection
Vertical sight lines
Major: Week Interval: 2
Minor: Week Interval: 1
Horizontal sight lines
Minor scale values: 4
Major scale values: 2
Histogram
Summarize by resource
Stack resources
Print
Orientation: Landscape
Rows: 1
Columns: 1
Begin on sheet: 1
End on sheet:
Confirm preparation
Transfer... OK Cancel Help

Budget | Codes | Constr | Cost | Custom | Dates | Log | Pred | Res | Succ | WBS | Help
ID 10 Clear Site Previous Next <<Less
OD 2 Pct 0.0 Cal 1 ES 01OCT01 EF 02OCT01 TF: 0
RD 2 Type Task LS 01OCT01 LF 02OCT01 FF: 0
ADC 1000
Supr Summ WBS
Classic Schedule Layout Show All Activities

Screen Image 13-7
Resource/Cost Format Window

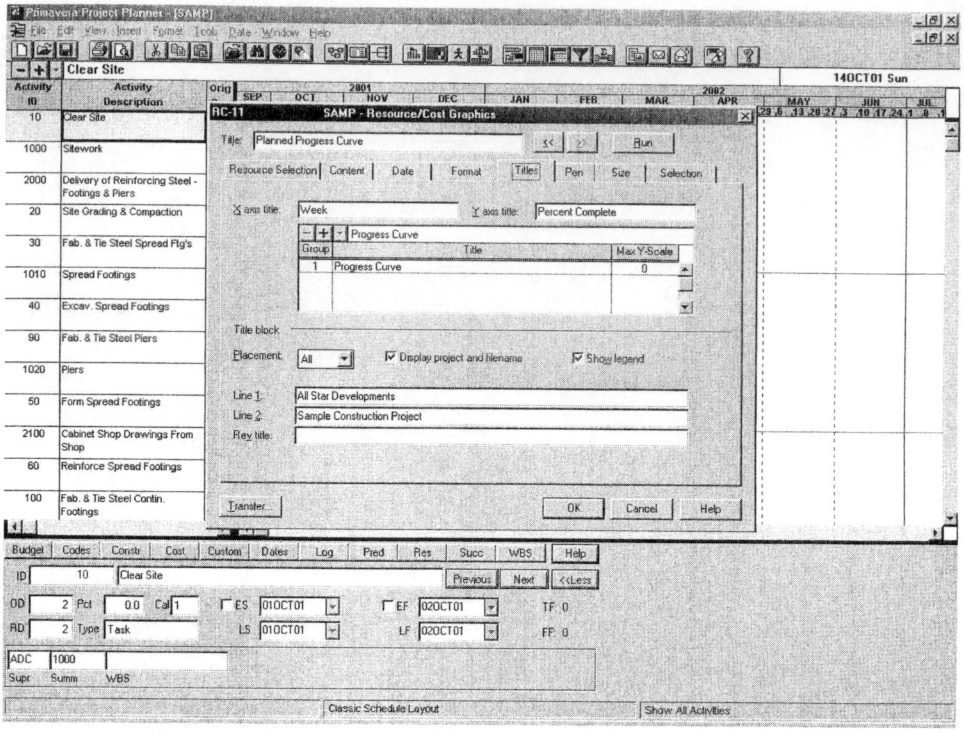

Screen Image 13-8
Resource/Cost Titles Window

user to specify a stacked histogram. With this option the individual resources are stacked one upon the other. The advantage of this option is that the relationship between the items being graphed can be seen. If this option is used, the histogram option needs to be selected on the *Content* tab. Furthermore, a different color or pattern needs to be specified in the *Pattern* and *Pen* columns of the *Resource Selection* tab. For the cumulative progress curve, which is not a histogram, any entry in that field will have no impact on the report. The bottom third of the tab allows the user to specify the placement of the chart on the page.

The *Title* tab allows the user to specify the chart title, axis names, and title block information. In Screen Image 13-8, the *X-axis* is labeled *Weeks* and the *Y-axis* has been labeled *Percent Complete*. The *Group* title box is where the group designation from the *Resource Selection* window is given a name. This name becomes the title on the individual graph. The bottom portion of the screen addresses information to be included in the chart title box. By looking at Output 13-2, these items can easily be seen.

The *Pen* tab as shown in Screen Image 13-9 is the means by which the color and thickness of the individual lines and items are defined. These features, while they work with any color printer, are described in terms of how they would perform with a pen plotter. The *Size* tab in Screen Image 13-10 is where the font size of the items is detailed. Finally, the *Selection* tab as shown in Screen Image 13-11 is identical to the selection criteria that has been used extensively throughout the program. When the run button is clicked, the user will be prompted about where to route the output. Once this is selected, a message appears telling the number of pages that are about to be generated. Screen Image 13-12 shows a sample of the window that informs the user of the number of pages and gives the user the opportunity to cancel the print job. This feature is included to prevent meaningless output that if plotted takes a substantial amount of time to print.

The Primavera Project Planner graphic curve is quicker and easier to generate than was the one generated with a spreadsheet program. However, this program has no way to handle changes in the budget. With Primavera Project Planner the curve is recalculated and there

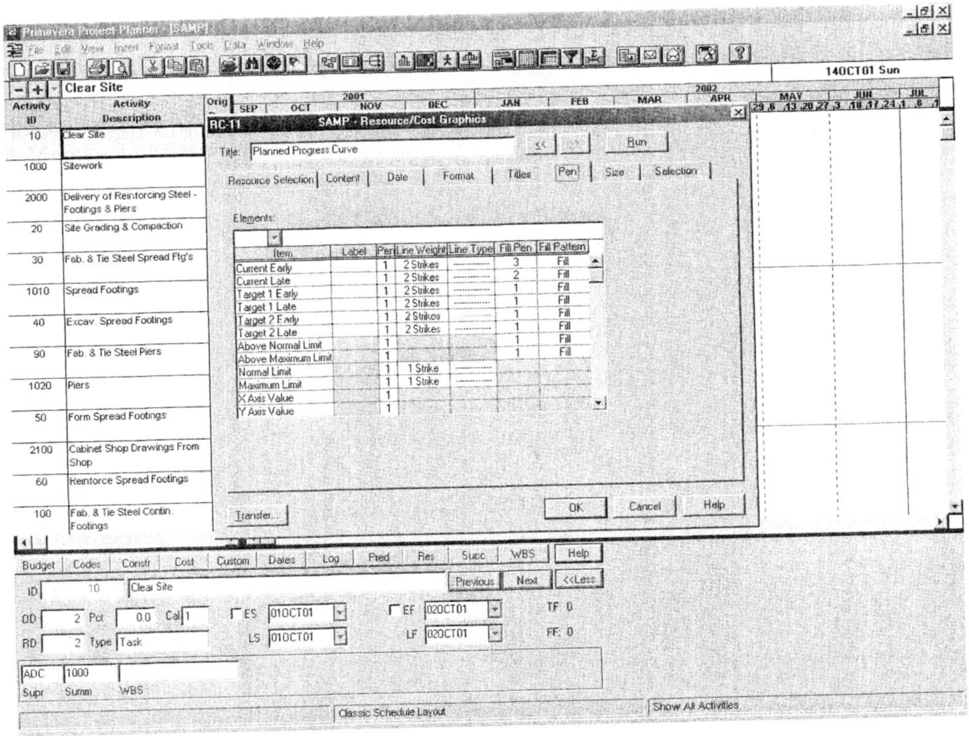

Primavera Project Planner - [SAMP]
File Edit View Insert Format Tools Data Window Help
14OCT01 Sun
Clear Site

Activity ID	Activity Description
10	Clear Site
1000	Sitework
2000	Delivery of Reinforcing Steel - Footings & Piers
20	Site Grading & Compaction
30	Fab. & Tie Steel Spread Ftg's
1010	Spread Footings
40	Excav. Spread Footings
90	Fab. & Tie Steel Piers
1020	Piers
50	Form Spread Footings
2100	Cabinet Shop Drawings From Shop
60	Reinforce Spread Footings
100	Fab. & Tie Steel Contin. Footings

Orig 2001 SEP OCT NOV DEC JAN FEB MAR 2002 APR MAY JUN JUL

RC-11 SAMP - Resource/Cost Graphics
Title: Planned Progress Curve << >> Run
Resource Selection | Content | Date | Format | Titles | Pen | Size | Selection
Elements:

Item	Label	Pen	Line Weight	Line Type	Fill Pen	Fill Pattern
Current Early		1	2 Strikes	....	3	Fill
Current Late		1	2 Strikes	....	2	Fill
Target 1 Early		1	2 Strikes	....	1	Fill
Target 1 Late		1	2 Strikes	....	1	Fill
Target 2 Early		1	2 Strikes	....	1	Fill
Target 2 Late		1	2 Strikes	....	1	Fill
Above Normal Limit		1			1	Fill
Above Maximum Limit		1			1	Fill
Normal Limit		1	1 Strike	....		
Maximum Limit		1	1 Strike	....		
X Axis Value		1				
Y Axis Value		1				

Transfer... OK Cancel Help

Budget Codes Constr Cost Custom Dates Log Pred Res Succ WBS Help
ID 10 Clear Site Previous Next <<Less
OD 2 Pct 0.0 Cal 1 ES 01OCT01 EF 02OCT01 TF: 0
RD 2 Type Task LS 01OCT01 LF 02OCT01 FF: 0
ADC 1000
Supr Summ WBS
Classic Schedule Layout Show All Activities

Screen Image 13-9
Resource/Cost Pens Window

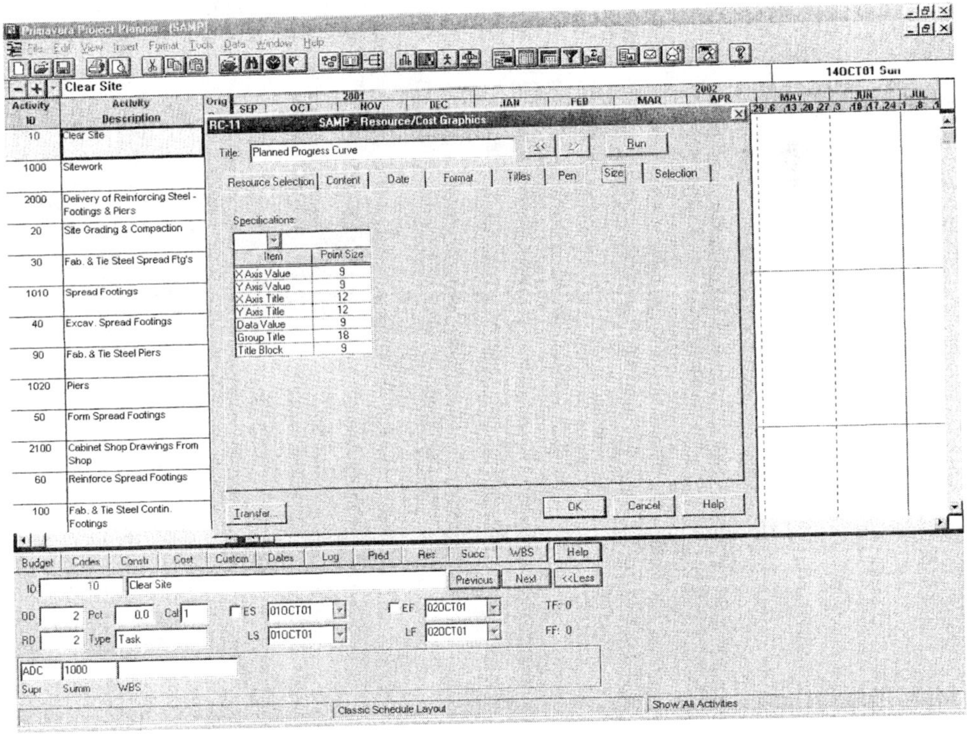

Primavera Project Planner - [SAMP]
File Edit View Insert Format Tools Data Window Help
14OCT01 Sun
Clear Site

Activity ID	Activity Description
10	Clear Site
1000	Sitework
2000	Delivery of Reinforcing Steel - Footings & Piers
20	Site Grading & Compaction
30	Fab. & Tie Steel Spread Ftg's
1010	Spread Footings
40	Excav. Spread Footings
90	Fab. & Tie Steel Piers
1020	Piers
50	Form Spread Footings
2100	Cabinet Shop Drawings From Shop
60	Reinforce Spread Footings
100	Fab. & Tie Steel Contin. Footings

Orig 2001 SEP OCT NOV DEC JAN FEB MAR 2002 APR MAY JUN JUL

RC-11 SAMP - Resource/Cost Graphics
Title: Planned Progress Curve << >> Run
Resource Selection | Content | Date | Format | Titles | Pen | Size | Selection
Specifications:

Item	Point Size
X Axis Value	9
Y Axis Value	9
X Axis Title	12
Y Axis Title	12
Data Value	9
Group Title	18
Title Block	9

Transfer... OK Cancel Help

Budget Codes Constr Cost Custom Dates Log Pred Res Succ WBS Help
ID 10 Clear Site Previous Next <<Less
OD 2 Pct 0.0 Cal 1 ES 01OCT01 EF 02OCT01 TF: 0
RD 2 Type Task LS 01OCT01 LF 02OCT01 FF: 0
ADC 1000
Supr Summ WBS
Classic Schedule Layout Show All Activities

Screen Image 13-10
Resource/Cost Size Window

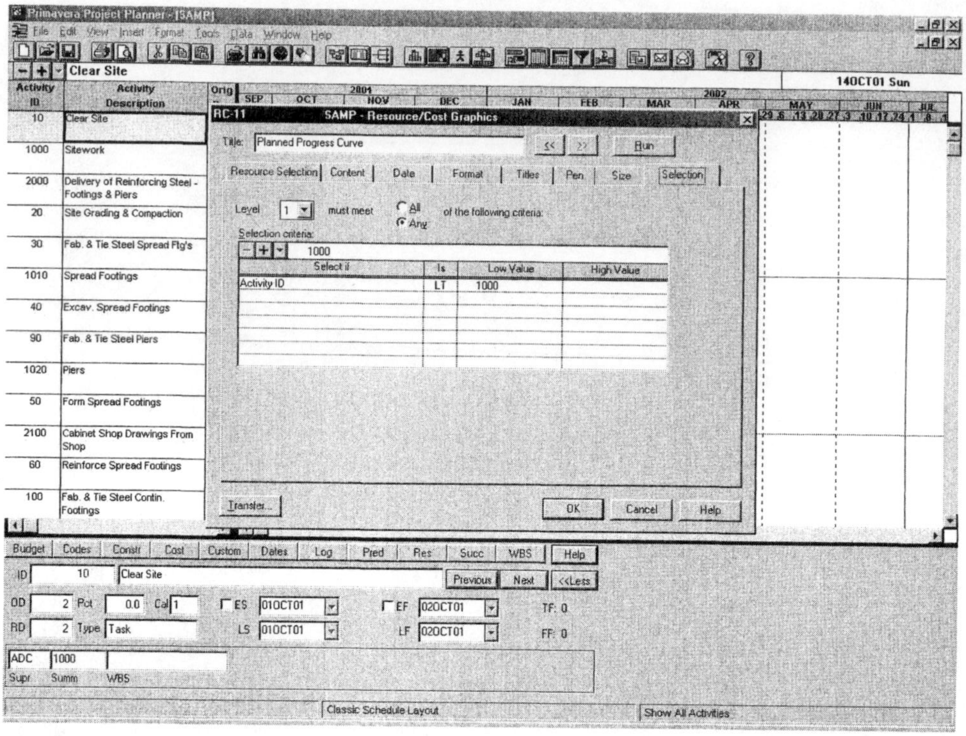

Primavera Project Planner - [SAMP]

File Edit View Insert Format Tools Data Window Help

Clear Site

14OCT01 Sun

Activity ID	Activity Description
10	Clear Site
1000	Sitework
2000	Delivery of Reinforcing Steel - Footings & Piers
20	Site Grading & Compaction
30	Fab. & Tie Steel Spread Ftg's
1010	Spread Footings
40	Excav. Spread Footings
90	Fab. & Tie Steel Piers
1020	Piers
50	Form Spread Footings
2100	Cabinet Shop Drawings From Shop
60	Reinforce Spread Footings
100	Fab. & Tie Steel Contin. Footings

Orig 2001 SEP OCT NOV DEC 2002 JAN FEB MAR APR MAY JUN JUL

RC-11 SAMP - Resource/Cost Graphics

Title: Planned Progress Curve << >> Run

Resource Selection | Content | Date | Format | Titles | Pen | Size | Selection

Level 1 must meet All / Any of the following criteria:

Selection criteria:

1000

Select if	Is	Low Value	High Value
Activity ID	LT	1000	

Transfer... OK Cancel Help

Budget Codes Constr Cost Custom Dates Log Pred Res Succ WBS Help

ID 10 Clear Site Previous Next <<Less

OD 2 Pct 0.0 Cal 1 ES 01OCT01 EF 02OCT01 TF: 0

RD 2 Type Task LS 01OCT01 LF 02OCT01 FF: 0

ADC 1000

Supr Summ WBS

Classic Schedule Layout Show All Activities

Screen Image 13-11
Resource/Cost Selection Window

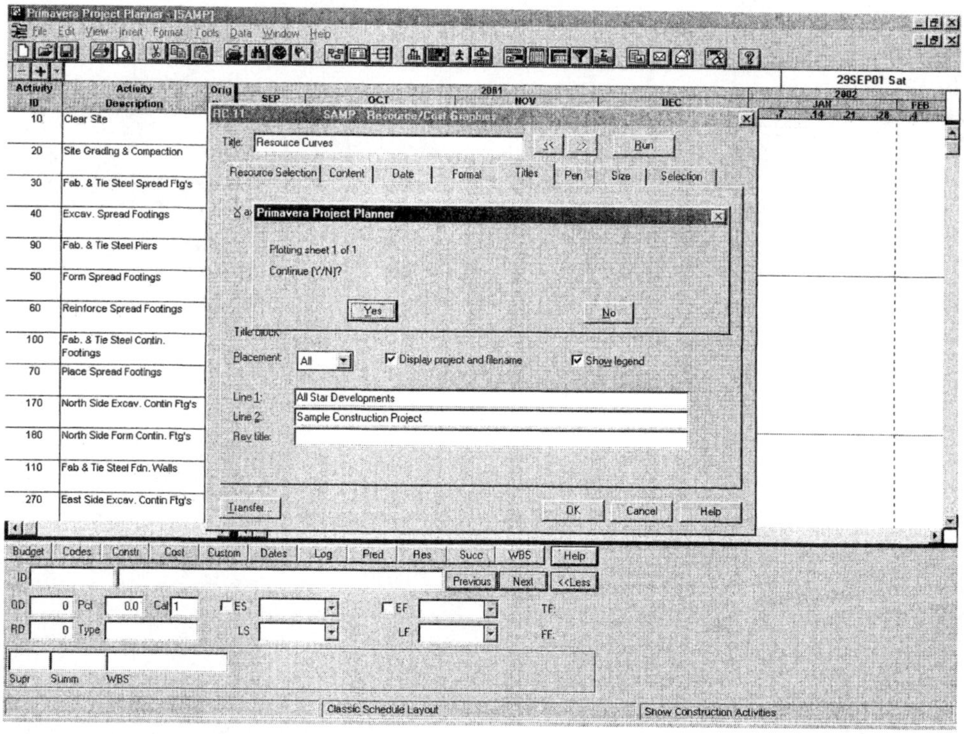

Primavera Project Planner - [SAMP]

File Edit View Insert Format Tools Data Window Help

29SEP01 Sat

Activity ID	Activity Description
10	Clear Site
20	Site Grading & Compaction
30	Fab. & Tie Steel Spread Ftg's
40	Excav. Spread Footings
90	Fab. & Tie Steel Piers
50	Form Spread Footings
60	Reinforce Spread Footings
100	Fab. & Tie Steel Contin. Footings
70	Place Spread Footings
170	North Side Excav. Contin Ftg's
180	North Side Form Contin. Ftg's
110	Fab & Tie Steel Fdn. Walls
270	East Side Excav. Contin Ftg's

Orig 2001 SEP OCT NOV DEC 2002 JAN FEB

RC-11 SAMP - Resource/Cost Graphics

Title: Resource Curves << >> Run

Resource Selection | Content | Date | Format | Titles | Pen | Size | Selection

Primavera Project Planner

Plotting sheet 1 of 1

Continue (Y/N)?

Yes No

Title block

Placement All Display project and filename Show legend

Line 1: All Star Developments

Line 2: Sample Construction Project

Rev title:

Transfer... OK Cancel Help

Budget Codes Constr Cost Custom Dates Log Pred Res Succ WBS Help

ID Previous Next <<Less

OD 0 Pct 0.0 Cal 1 ES EF TF:

RD 0 Type LS LF FF:

Supr Summ WBS

Classic Schedule Layout Show Construction Activities

Screen Image 13-12
Graphics Plot Verification Window

Screen Image 13-13
Graphic Bar Chart Content Window

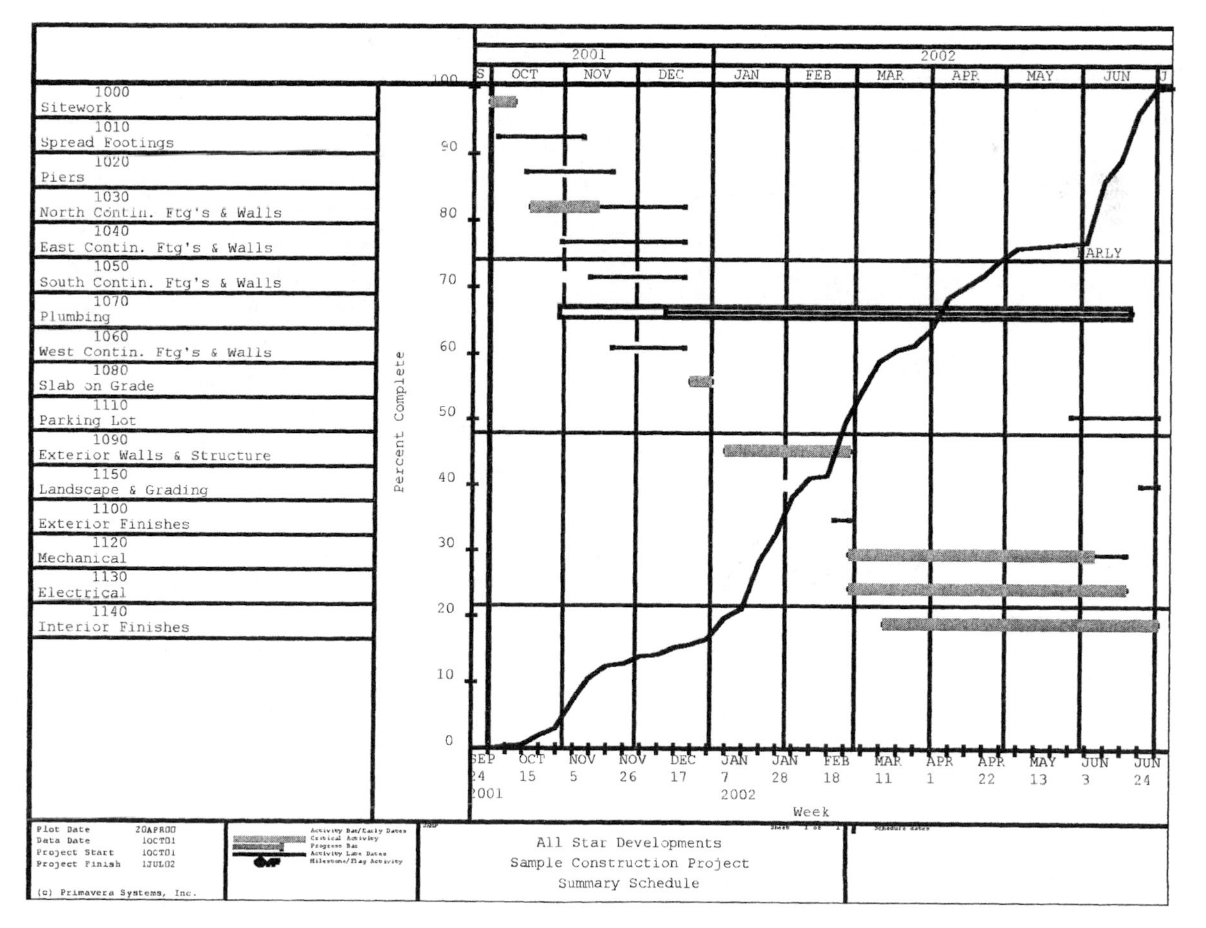

Output 13-3
Superimposed progress curve

is no drop in the curve to show when there is a change in the project scope. In addition, the plotting of actual does not work well and most likely would require the actual curve to be manually drawn.

Superimposing Curves

One of the features of Primavera Project Planner graphic output is the ability to superimpose curves on top of graphs. In the previous example a project progress curve was produced. When superimposing curves it is necessary to remember the *RC* number that is assigned to the desired overlay cost/graphic. In the previous example it was RC11. It is common to superimpose the progress curve over a summary bar chart schedule. The bar chart schedule shows which activities are ahead or behind schedule, but it does not show how they impact the overall project progress. Some activities may be ahead of schedule and some may be behind schedule, but there is no clear way of telling how they impact overall progress. If there is a deviation in the planned progress, the underlying bar chart schedule will add insight into where the problems exist.

The first step is to generate a graphic bar chart schedule that it is one page long. The curves are superimposed by going to the *Content* tab on the *Graphic Bar Chart* window (Screen Image 13-13). Entering the resource curve number in the *Overlay* option is then all that is required.

In Screen Image 13-13 Resource Curve 11 is being overlaid onto the summary schedule that was developed in Chapter 9. Output 13-3 is the new schedule graphic.

Conclusion

The development of planned overall project progress curves are essential for effective project management. As the number of activities increases, the ability to visually tell if the project is ahead of, behind, or on schedule is diminished. However, with the S-shaped overall progress curve it is easy to identify this information. Superimposed curves not only point out how the project is doing in relation to planned progress, but shed light on where deviations have occurred.

Suggested Exercises

1. Using your sample project, develop an overall project progress S-curve.
2. Take the first 20 activities and status them. Make some of them complete and some partially completed.
3. After the above activities have been statused, plot this information on the planned progress curve and determine if the project is ahead of, behind, or on schedule.

Project Analysis and Forecasting

KEY TERMS

Craft Labor Report A report that compares the actual cost of labor for a specific operation to its estimated cost.

Labor Performance A measurement of how fast craft personnel are placing work in comparison to the estimated rate.

Project Costs

The methods for gathering actual project costs vary from contractor to contractor. This variation is the result of differing accounting systems, levels of detail captured, percentage of work that is self-performed, and cost control philosophy. In spite of these differences, each contractor has the underlying objective of comparing the planned or estimated cost for a specific item of work to its actual cost. This comparison requires actual cost information that is timely and accurate. This is further complicated since financial accounting procedures often are not complementary to cost control. For example, in a financial accounting system an expense is occurred when the material is received. From a cost control perspective, when the purchase order is processed the material is as good as bought. If the actual cost information is processed from the financial accounting system, there could be many months between the ordering and the delivery of an item. During the intervening months these committed funds are not reflected within the financial accounting system. This would make it look as if the item was under budget when in reality, the information simply is not timely. The measurement of progress must coincide with actual cost information. For example, if progress was measured through the first 10 weeks of the project and the actual cost information only spans the first 9 weeks, it would appear that the actual cost per percent was lower than the planned. This would give the appearance that the project is under budget, when in reality there is a timing problem. This timeliness issue is complex because of the dynamics of a construction project. The project is never stopped for the purpose of measuring progress or cost. The progress of the project and its cost changes from moment to moment. A similar issue stems from multiple sources of information. The labor costs typically come from a payroll system, while the other costs come from an accounts payable system. With all of these variables, the challenge becomes to integrate timely information into the project schedule to develop a meaningful project status and forecast.

Most financial accounting systems that target the construction industry have some type of interface that will allow the actual cost data to be automatically transferred to the project cost accounting system. If this interface is automatic, the transfer of this information is rapid and accurate, allowing for more timely cost reporting information. If the project cost information can be downloaded to a database or spreadsheet format, it can automatically be loaded into Primavera Project Planner. The process for importing project cost information is identical to the importing process that was detailed in Chapter 7.

The cost control and forecasting operation relies on generating reports that present actual and planned cost information in a clear format. The estimated costs for an activity need to be compared to the actual costs. This is one of the areas where Primavera Project Planner surpasses many less expensive scheduling software packages.

Throughout the previous chapters in this text there have been multiple examples of how the report generation features are flexible enough to produce output that is effective for project management. However, Primavera Project Planner has a free-form report writer that allows the user to develop reports in virtually any format. This is extremely helpful

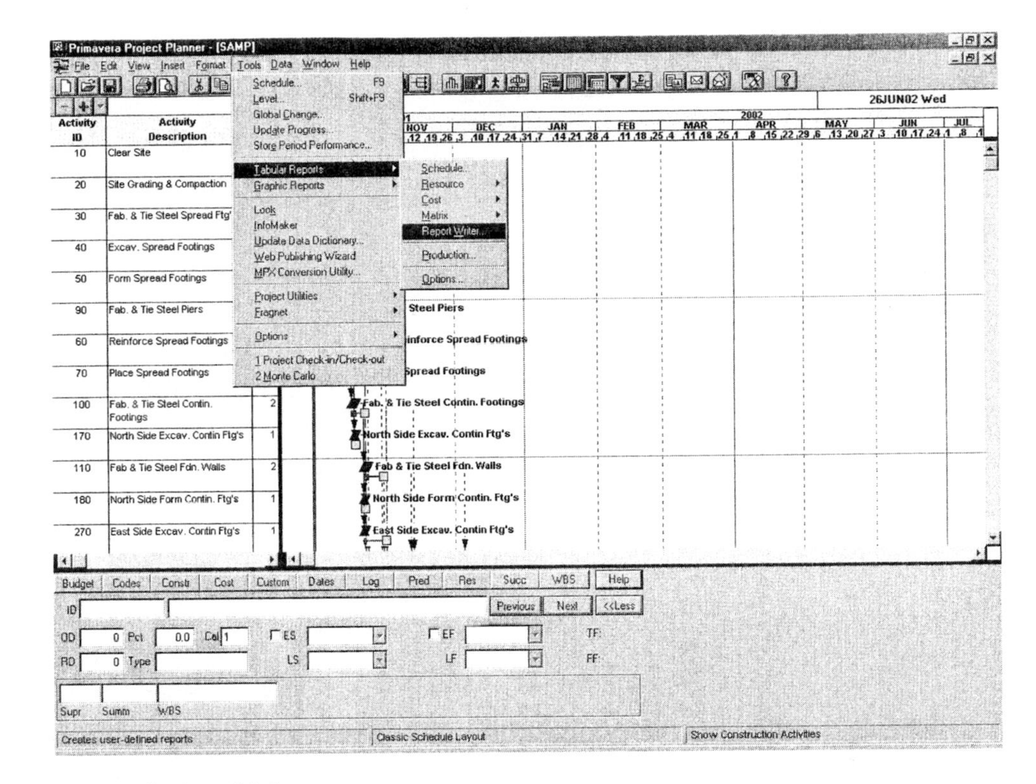

Screen Image 14-1

Opening the Report Writer Window

when generating reports that are not scheduling in nature. This report writer is accessed by selecting the *Tools* command from the main command line followed by selecting the *Tabular Reports* and *Report Writer* options as shown in Screen Image 14-1. Those commands will open the *Report Writer Reports* window as shown in Screen Image 14-2, which contains a list of report specifications for previously created reports.

To add a new report specification, click the *Add* button to open the *Add a New Report* window, as shown in Screen Image 14-3. If the *Add* button is clicked on this window, the report specification process can begin. In Screen Image 14-4 is the *Arithmetic* tab, the first of six such tabs. This tab allows the user to include if-then statements, along with mathematic computations. The allowed computations are addition, subtraction, multiplication, division, and concatenation. When fields are concatenated the content of one field is tagged onto the end of another. The *Content* tab as shown in Screen Image 14-5 is perhaps the most powerful.

From this window the actual columns that are to appear on the reports are specified. The columns must be accurately identified; therefore, it is best to use the pull-down menu option. To include an item in the report, go to the desired column and click in the *Data item* field and then use the pull-down menu. In Screen Image 14-6 the first column *Data item* was selected and the pull-down menu activated. The potential columns of information are categorized on the left, and the items within those categories are on the right. If the first column of the report is to contain any of those fields, simply click on that field. Screen Image 14-7 shows the content specifications with a number of different columns specified. Once the *Data items* have been specified, the column heading, field width, and format can be modified. Output 14-1 (page 315) is the report that was generated from this *Content* window. A review of that output will show the relationship between this tab and the columns of information on the report. Since the purpose of this report is for labor operations, the first column is logically for resource, followed by the specific item of work or activity, which is followed by a comparison of planned to actual work hours and labor dollars.

Screen Image 14-2
Report Writer Reports

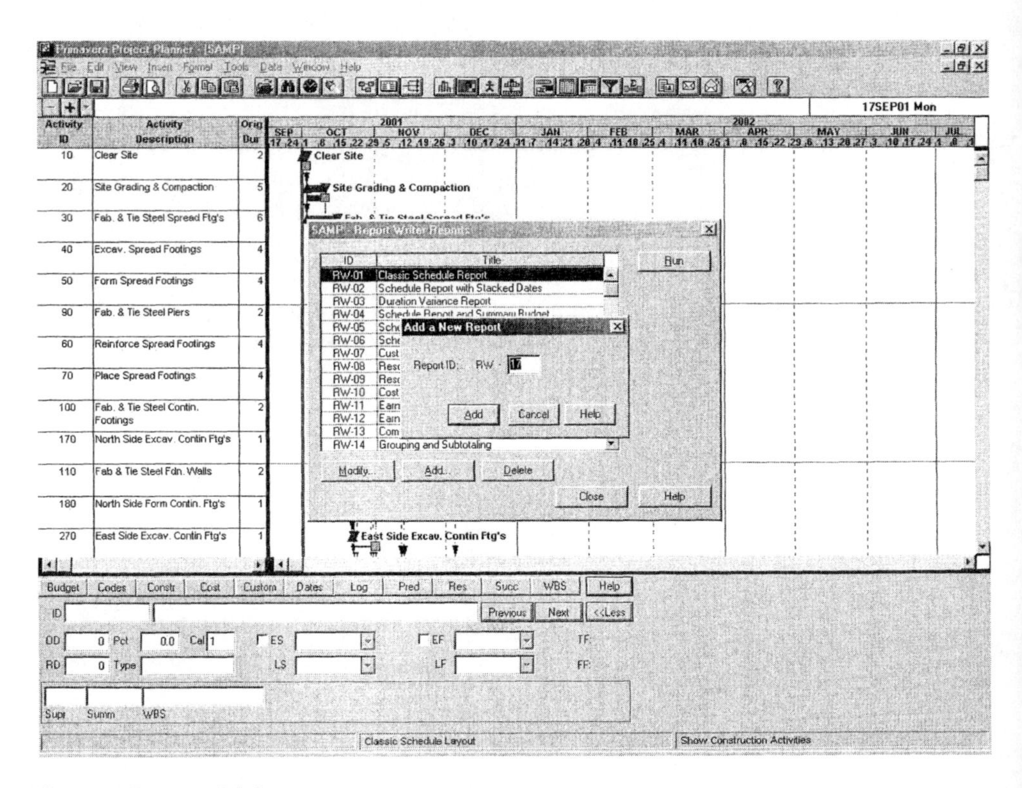

Screen Image 14-3
Add a New Report Window

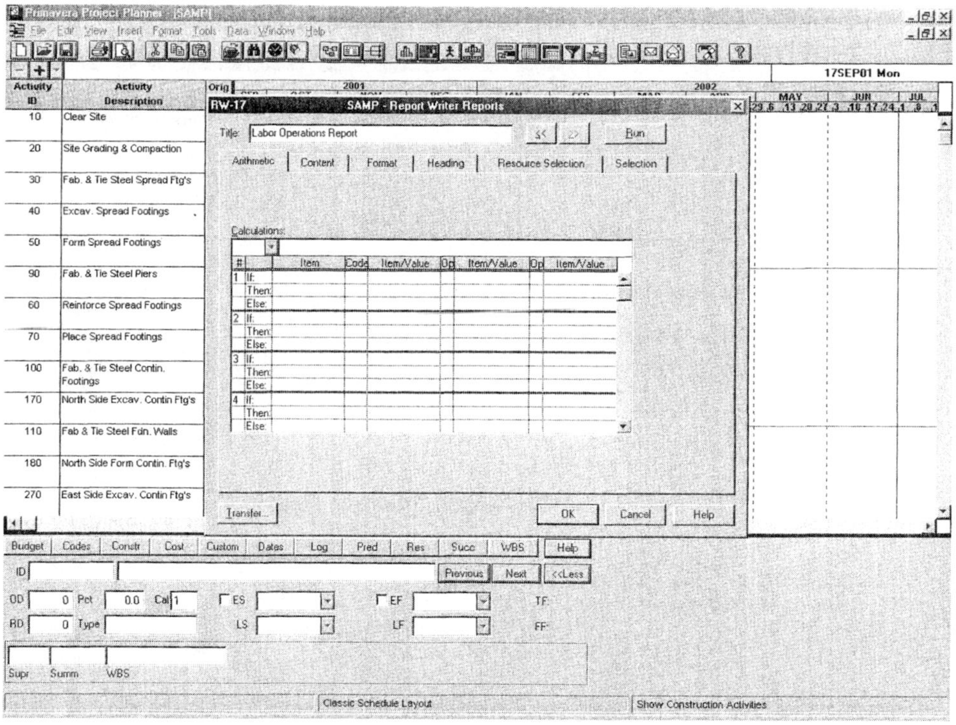

Primavera Project Planner - [SAMP]
File Edit View Insert Format Tools Data Window Help
17SEP01 Mon

Activity ID	Activity Description
10	Clear Site
20	Site Grading & Compaction
30	Fab. & Tie Steel Spread Ftg's
40	Excav. Spread Footings
50	Form Spread Footings
90	Fab. & Tie Steel Piers
60	Reinforce Spread Footings
70	Place Spread Footings
100	Fab. & Tie Steel Contin. Footings
170	North Side Excav. Contin Ftg's
110	Fab & Tie Steel Fdn. Walls
180	North Side Form Contin. Ftg's
270	East Side Excav. Contin Ftg's

Orig 2001 2002 MAY JUN JUL
RW-17 SAMP - Report Writer Reports
Title: Labor Operations Report << >> Run
Arithmetic | Content | Format | Heading | Resource Selection | Selection
Calculations:

#		Item	Code	Item/Value	Op	Item/Value	Op	Item/Value
1	If:							
	Then:							
	Else:							
2	If:							
	Then:							
	Else:							
3	If:							
	Then:							
	Else:							
4	If:							
	Then:							
	Else:							

Transfer... OK Cancel Help
Budget | Codes | Constr | Cost | Custom | Dates | Log | Pred | Res | Succ | WBS | Help
ID Previous Next <<Less
OD 0 Pct 0.0 Cal 1 ES EF TF
RD 0 Type LS LF FF
Supr Summ WBS
Classic Schedule Layout Show Construction Activities

Screen Image 14-4
Arithmetic Window

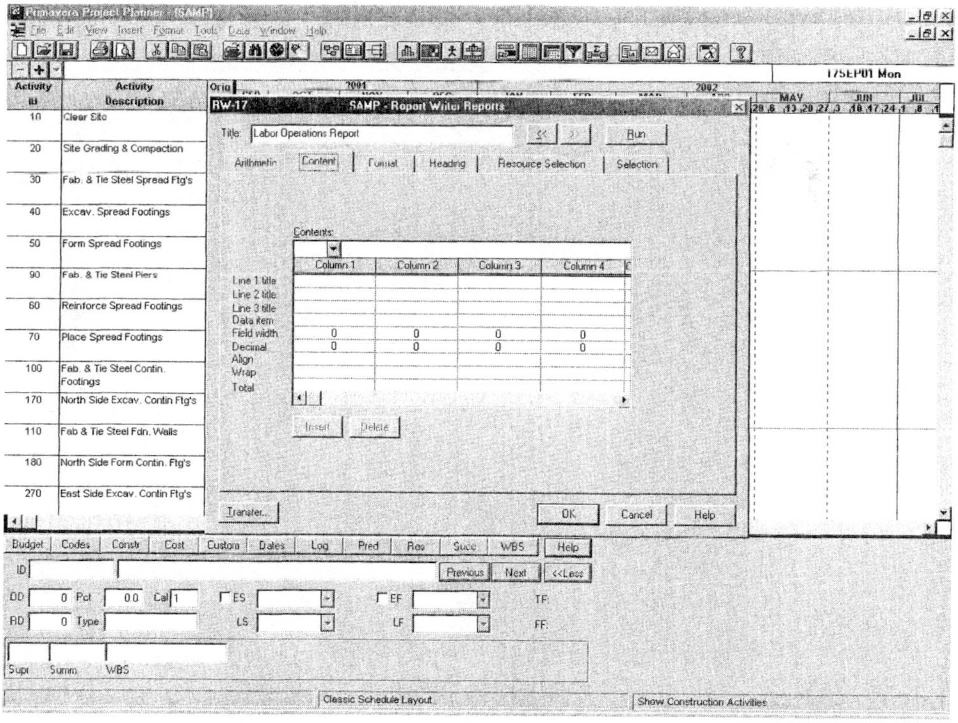

Primavera Project Planner - [SAMP]
File Edit View Insert Format Tools Data Window Help
17SEP01 Mon

Activity ID	Activity Description
10	Clear Site
20	Site Grading & Compaction
30	Fab. & Tie Steel Spread Ftg's
40	Excav. Spread Footings
50	Form Spread Footings
90	Fab. & Tie Steel Piers
60	Reinforce Spread Footings
70	Place Spread Footings
100	Fab. & Tie Steel Contin. Footings
170	North Side Excav. Contin Ftg's
110	Fab & Tie Steel Fdn. Walls
180	North Side Form Contin. Ftg's
270	East Side Excav. Contin Ftg's

Orig 2001 2002 MAY JUN JUL
RW-17 SAMP - Report Writer Reports
Title: Labor Operations Report << >> Run
Arithmetic | Content | Format | Heading | Resource Selection | Selection
Contents:

	Column 1	Column 2	Column 3	Column 4
Line 1 title				
Line 2 title				
Line 3 title				
Data item				
Field width	0	0	0	0
Decimal	0	0	0	0
Align				
Wrap				
Total				

Insert Delete
Transfer... OK Cancel Help
Budget | Codes | Constr | Cost | Custom | Dates | Log | Pred | Res | Succ | WBS | Help
ID Previous Next <<Less
OD 0 Pct 0.0 Cal 1 ES EF TF
RD 0 Type LS LF FF
Supr Summ WBS
Classic Schedule Layout Show Construction Activities

Screen Image 14-5
Initial Content Window

Primavera Project Planner - [SAMP]
File Edit View Insert Format Tools Data Window Help
17SEP01 Mon

Activity ID	Activity Description
10	Clear Site
20	Site Grading & Compaction
30	Fab. & Tie Steel Spread Ftg's
40	Excav. Spread Footings
50	Form Spread Footings
90	Fab. & Tie Steel Piers
60	Reinforce Spread Footings
70	Place Spread Footings
100	Fab. & Tie Steel Contin. Footings
170	North Side Excav. Contin Ftg's
110	Fab & Tie Steel Fdn. Walls
180	North Side Form Contin. Ftg's
270	East Side Excav. Contin Ftg's

RW-17 SAMP - Report Writer Reports
Title: Labor Operations Report << >> Run
Arithmetic | Content | Format | Heading | Resource Selection | Selection
Contents:

Data Item
Activity data
Constraints
Resource/cost data
Target data
Comparison data
Variances
Global data

NACT - # Activities
DES - Activity Description
ACT - Activity ID
BLNK - Blank column
STYP - Activity Type
ADUR - Actual Duration
AD - Actual Dates
AS - Actual Start
AF - Actual Finish

OK Cancel Help
Insert Delete
Transfer... OK Cancel Help

Budget | Codes | Constr | Cost | Custom | Dates | Log | Pred | Res | Succ | WBS | Help
ID Previous Next <<Less
OD 0 Pct 0.0 Cal 1 ES EF TF
RD 0 Type LS LF FF
Supr Summ WBS
Classic Schedule Layout | Show Construction Activities

Screen Image 14-6
Data Item Pull-down Menu

Primavera Project Planner - [SAMP]
File Edit View Insert Format Tools Data Window Help
17SEP01 Mon

Activity ID	Activity Description
10	Clear Site
20	Site Grading & Compaction
30	Fab. & Tie Steel Spread Ftg's
40	Excav. Spread Footings
50	Form Spread Footings
90	Fab. & Tie Steel Piers
60	Reinforce Spread Footings
70	Place Spread Footings
100	Fab. & Tie Steel Contin. Footings
170	North Side Excav. Contin Ftg's
110	Fab & Tie Steel Fdn. Walls
180	North Side Form Contin. Ftg's
270	East Side Excav. Contin Ftg's

RW-17 SAMP - Report Writer Reports
Title: Labor Operations Report << >> Run
Arithmetic | Content | Format | Heading | Resource Selection | Selection
Contents: RES

	Column 1	Column 2	Column 3	Column 4
Line 1 title				QTY.
Line 2 title		ACTIVITY		THIS
Line 3 title	RES	ID	DESCRIPTION	PRD.
Data item	RES	ACT	DES	QTP
Field width	8	10	48	10
Decimal	0	0	0	2
Align	Left	Left	Left	Right
Wrap	No	No	No	No
Total	No	No	No	Yes

Insert Delete
Transfer... OK Cancel Help

Budget | Codes | Constr | Cost | Custom | Dates | Log | Pred | Res | Succ | WBS | Help
ID Previous Next <<Less
OD 0 Pct 0.0 Cal 1 ES EF TF
RD 0 Type LS LF FF
Supr Summ WBS
Classic Schedule Layout | Show Construction Activities

Screen Image 14-7
Content Tab with Columns Specified

All Star Developments | PRIMAVERA PROJECT PLANNER | Sample Construction Project

REPORT DATE 09JUN00 RUN NO. 133 09:43 | START DATE 01OCT01 FIN DATE 08JUL02

Labor Operations Report | DATA DATE 05NOV01 PAGE NO. 1

RES	ACTIVITY ID	DESCRIPTION	QTY. THIS PRD.	QTY. TO DATE	BUDGET QTY.	COST THIS PERIOD	COST TO DATE	BUDGET COST
CARP - Carpenters								
CARP	50	Form Spread Footings	12.00	12.00	6.00	122.00	122.00	99.00
CARP	130	Form Piers	40.00	40.00	105.00	1000.00	1000.00	1733.00
CARP	180	North Side Form Contin. Ftg's	15.00	15.00	13.00	200.00	200.00	215.00
CARP	230	North Side Form Fdn. Walls	0.00	0.00	100.00	0.00	0.00	1650.00
CARP	280	East Side Form Contin. Ftg's	0.00	0.00	8.00	0.00	0.00	132.00
CARP	330	East Side Form Fdn. Walls	0.00	0.00	60.00	0.00	0.00	990.00
CARP	370	South Side Form Contin. Ftg's	0.00	0.00	10.00	0.00	0.00	165.00
CARP	450	South Side Form Fdn. Walls	0.00	0.00	40.00	0.00	0.00	660.00
CARP	470	West Side Form Contin. Ftg's	0.00	0.00	9.00	0.00	0.00	149.00
CARP	540	West Side Form Fdn. Walls	0.00	0.00	60.00	0.00	0.00	990.00
CARP	690	Frame Interior Walls	0.00	0.00	26.00	0.00	0.00	429.00
CARP	720	Form Parking Lot	0.00	0.00	52.00	0.00	0.00	858.00
CARP	820	Hang & Trim Interior Doors	0.00	0.00	512.00	0.00	0.00	8448.00
CARP	840	Install Cabinets	0.00	0.00	179.00	0.00	0.00	2954.00
			67.00	**67.00**	**1180.00**	**1322.00**	**1322.00**	**19472.00**
CFIN - Concrete Finishers								
CFIN	70	Place Spread Footings	12.00	12.00	8.00	120.00	120.00	102.00
CFIN	140	Place Piers	0.00	0.00	4.00	0.00	0.00	51.00
CFIN	200	North Side Place Contin. Ftg's	15.00	15.00	15.00	200.00	200.00	191.00
CFIN	240	North Side Place Fdn. Walls	0.00	0.00	2.00	0.00	0.00	26.00
CFIN	300	East Side Place Contin. Ftg's	0.00	0.00	10.00	0.00	0.00	128.00
CFIN	340	East Side Place Fdn. Walls	0.00	0.00	2.00	0.00	0.00	26.00
CFIN	420	South Side Place Contin. Ftg's	0.00	0.00	15.00	0.00	0.00	191.00
CFIN	460	South Side Place Fdn. Walls	0.00	0.00	2.00	0.00	0.00	26.00
CFIN	510	West Side Place Contin. Ftg's	0.00	0.00	9.00	0.00	0.00	115.00
CFIN	550	West Side Place Fdn. Walls	0.00	0.00	2.00	0.00	0.00	26.00
CFIN	590	Place & Finish S.O.G.	0.00	0.00	537.00	0.00	0.00	6847.00
CFIN	740	Place & Finish Parking Lot	0.00	0.00	470.00	0.00	0.00	5993.00
CFIN	760	Place Curbs	0.00	0.00	378.00	0.00	0.00	4820.00
			27.00	**27.00**	**1454.00**	**320.00**	**320.00**	**18542.00**
LABR - Laborers								
LABR	40	Excav. Spread Footings	62.00	62.00	51.00	672.00	672.00	548.00
LABR	50	Form Spread Footings	12.00	12.00	6.00	100.00	100.00	65.00
LABR	80	Strip Forms - Spread Footings	20.00	20.00	10.00	400.00	400.00	108.00
LABR	150	Strip Forms - Piers	0.00	0.00	16.00	0.00	0.00	172.00
LABR	160	Backfill Spread Footings	0.00	0.00	147.00	0.00	0.00	1580.00
LABR	170	North Side Excav. Contin Ftg's	65.00	65.00	60.00	672.00	672.00	645.00
LABR	210	North Side Strip Contin. Ftg's Forms	0.00	0.00	8.00	0.00	0.00	86.00
LABR	230	North Side Form Fdn. Walls	0.00	0.00	50.00	0.00	0.00	538.00
LABR	250	North Side Strip Fdn. Wall Forms	0.00	0.00	40.00	0.00	0.00	430.00
LABR	260	North Side Backfill & Compact	0.00	0.00	20.00	0.00	0.00	215.00
LABR	270	East Side Excav. Contin Ftg's	40.00	40.00	45.00	400.00	400.00	484.00
LABR	310	East Side Strip Contin. Ftg's Forms	0.00	0.00	5.00	0.00	0.00	54.00
LABR	330	East Side Form Fdn. Walls	0.00	0.00	30.00	0.00	0.00	323.00
LABR	350	East Side Strip Fdn. Wall Forms	0.00	0.00	25.00	0.00	0.00	269.00
LABR	360	South Side Excav. Contin Ftg's	70.00	70.00	60.00	700.00	700.00	645.00
LABR	390	West Side Excav. Contin Ftg's	0.00	0.00	40.00	0.00	0.00	430.00
LABR	410	East Side Backfill & Compact	0.00	0.00	100.00	0.00	0.00	1075.00
LABR	430	South Side Strip Contin. Ftg's Forms	0.00	0.00	9.00	0.00	0.00	97.00
LABR	450	South Side Form Fdn. Walls	0.00	0.00	80.00	0.00	0.00	860.00
LABR	490	South Side Strip Fdn Wall Forms	0.00	0.00	40.00	0.00	0.00	430.00
LABR	500	South Side Backfill & Compact	0.00	0.00	20.00	0.00	0.00	215.00
LABR	520	West Side Strip Contin. Ftg's Forms	0.00	0.00	6.00	0.00	0.00	65.00
LABR	540	West Side Form Fdn. Walls	0.00	0.00	30.00	0.00	0.00	323.00
LABR	560	West Side Strip Fdn. Wall Forms	0.00	0.00	25.00	0.00	0.00	269.00
LABR	570	West Side Backfill & Compact	0.00	0.00	100.00	0.00	0.00	1075.00
LABR	750	Strip Forms - Parking Lot	0.00	0.00	20.00	0.00	0.00	215.00
			269.00	**269.00**	**1043.00**	**2944.00**	**2944.00**	**11216.00**

Output 14-1
Labor Operations Report (page 1)

All Star Developments — PRIMAVERA PROJECT PLANNER — Sample Construction Project

REPORT DATE 09JUN00 09:43 RUN NO. 133 — START DATE 01OCT01 FIN DATE 08JUL02

Labor Operations Report — DATA DATE 05NOV01 PAGE NO. 2

RES	ACTIVITY ID	DESCRIPTION	QTY. THIS PRD.	QTY. TO DATE	BUDGET QTY.	COST THIS PERIOD	COST TO DATE	BUDGET COST
MASN - Masons								
MASN	600	CMU	0.00	0.00	631.00	0.00	0.00	10325.00
			0.00	**0.00**	**631.00**	**0.00**	**0.00**	**10325.00**
MASNHLP - Mason Helper								
MASNHLP	600	CMU	0.00	0.00	421.00	0.00	0.00	4105.00
MASNHLP			0.00	0.00	145.00	0.00	0.00	1414.00
			0.00	**0.00**	**566.00**	**0.00**	**0.00**	**5519.00**
OPENG - Operating Engineer								
OPENG	40	Excav. Spread Footings	16.00	16.00	12.00	200.00	200.00	180.00
OPENG	160	Backfill Spread Footings	0.00	0.00	10.00	0.00	0.00	150.00
OPENG	170	North Side Excav. Contin Ftg's	15.00	15.00	13.00	200.00	200.00	195.00
OPENG	260	North Side Backfill & Compact	0.00	0.00	150.00	0.00	0.00	2250.00
OPENG	270	East Side Excav. Contin Ftg's	0.00	0.00	8.00	0.00	0.00	120.00
OPENG	360	South Side Excav. Contin Ftg's	13.00	13.00	13.00	200.00	200.00	195.00
OPENG	390	West Side Excav. Contin Ftg's	0.00	0.00	8.00	0.00	0.00	120.00
OPENG	410	East Side Backfill & Compact	0.00	0.00	12.00	0.00	0.00	180.00
OPENG	500	South Side Backfill & Compact	0.00	0.00	150.00	0.00	0.00	2250.00
OPENG	570	West Side Backfill & Compact	0.00	0.00	20.00	0.00	0.00	300.00
			44.00	**44.00**	**396.00**	**600.00**	**600.00**	**5940.00**
ROD - Rodbuster								
ROD	30	Fab. & Tie Steel Spread Ftg's	35.00	35.00	32.00	475.00	475.00	424.00
ROD	60	Reinforce Spread Footings	20.00	20.00	10.00	220.00	220.00	133.00
ROD	90	Fab. & Tie Steel Piers	12.00	12.00	10.00	150.00	150.00	133.00
ROD	100	Fab. & Tie Steel Contin. Footings	200.00	200.00	112.00	1365.00	1365.00	1484.00
ROD	110	Fab & Tie Steel Fdn. Walls	30.00	30.00	24.00	600.00	600.00	318.00
ROD	120	Reinforce Piers	2.00	2.00	8.00	20.00	20.00	106.00
ROD	190	North Side Reinforce Contin. Ftg's	20.00	20.00	20.00	265.00	265.00	265.00
ROD	220	North Side Reinforce Fdn. Walls	0.00	0.00	40.00	0.00	0.00	530.00
ROD	290	East Side Reinforce Contin. Ftg's	0.00	0.00	20.00	0.00	0.00	265.00
ROD	320	East Side Reinforce Fdn. Walls	0.00	0.00	40.00	0.00	0.00	530.00
ROD	380	South Side Reinforce Contin. Ftg's	0.00	0.00	20.00	0.00	0.00	265.00
ROD	440	South Side Reinforce Fdn. Walls	0.00	0.00	40.00	0.00	0.00	530.00
ROD	480	West Side Reinforce Contin. Ftg's	0.00	0.00	20.00	0.00	0.00	265.00
ROD	530	West Side Reinforce Fdn. Walls	0.00	0.00	40.00	0.00	0.00	530.00
ROD	580	Reinforce S.O.G.	0.00	0.00	170.00	0.00	0.00	2253.00
ROD	730	Reinforce Parking Lot	0.00	0.00	52.00	0.00	0.00	689.00
			319.00	**319.00**	**658.00**	**3095.00**	**3095.00**	**8720.00**
			726.00	**726.00**	**5928.00**	**8281.00**	**8281.00**	**79734.00**

Output 14-1
Labor Operations Report (page 2)

Screen Image 14-8 is the *Format* tab. The arrangement of information on this window is different from that in previous *Format* tabs; however, its operation or process is the same. For the purpose of generating a labor operations report, *Subtotal on Res - Resource* is selected. This will give a total for each of the labor resources. Another feature of interest is the *Number of lines in column titles*. The entry in this field must be 1, 2, or 3; it details the number of title lines to be printed on the report. This is used in conjunction with the input on the *Content* tab; if three lines of column headings were specified on that window, then this field needs to be set to 3. The importance here is consistency between this entry and the columnar layout on the *Content* tab. As with the *Content* tab, a review of the report in Output 14-1 will show the implication of the subtotal selection and number of column heading lines. Screen Image 14-9, the *Heading* tab, is where report heading information, placement, and alignment is specified. This is the information that will be printed out on the top of every page of the report.

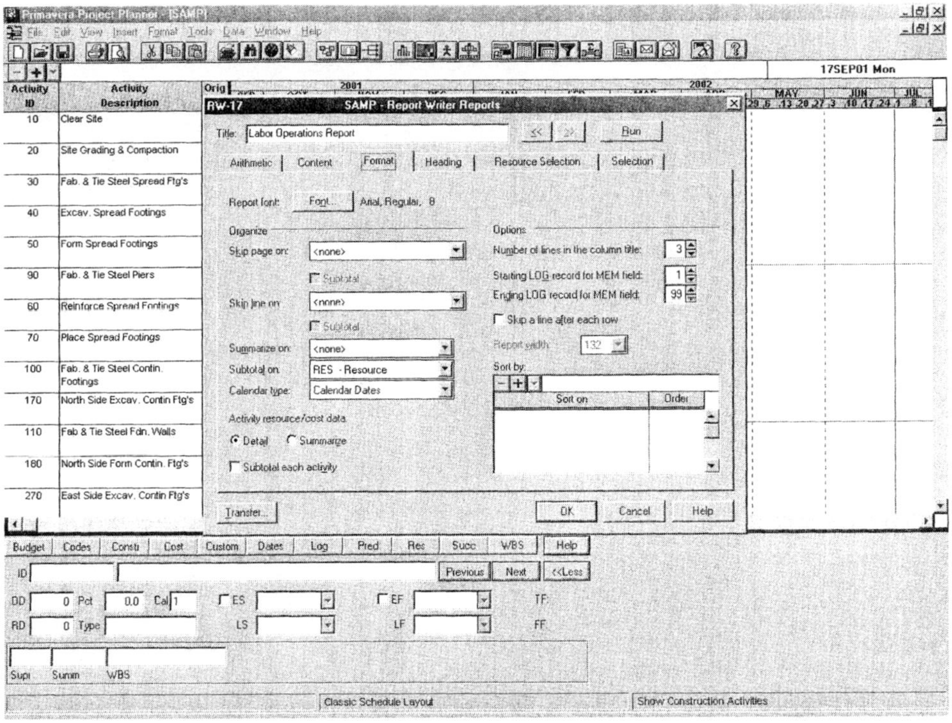

Primavera Project Planner - [SAMP]
File Edit View Insert Format Tools Data Window Help
17SEP01 Mon

Activity ID	Activity Description
10	Clear Site
20	Site Grading & Compaction
30	Fab. & Tie Steel Spread Ftg's
40	Excav. Spread Footings
50	Form Spread Footings
90	Fab. & Tie Steel Piers
60	Reinforce Spread Footings
70	Place Spread Footings
100	Fab. & Tie Steel Contin. Footings
170	North Side Excav. Contin Ftg's
110	Fab & Tie Steel Fdn. Walls
180	North Side Form Contin. Ftg's
270	East Side Excav. Contin Ftg's

RW-17 SAMP - Report Writer Reports
Title: Labor Operations Report << >> Run
Arithmetic | Content | Format | Heading | Resource Selection | Selection
Report font: Font... Arial, Regular, 8
Organize
Skip page on: <none>
Subtotal
Skip line on: <none>
Subtotal
Summarize on: <none>
Subtotal on: RES - Resource
Calendar type: Calendar Dates
Activity resource/cost data
Detail Summarize
Subtotal each activity
Options
Number of lines in the column title: 3
Starting LOG record for MEM field: 1
Ending LOG record for MEM field: 99
Skip a line after each row
Report width: 132
Sort by:
Sort on | Order
Transfer... OK Cancel Help

Budget | Codes | Constr | Cost | Custom | Dates | Log | Pred | Res | Succ | WBS | Help
ID Previous Next <<Less
OD 0 Pct 0.0 Cal 1 ES EF TF
RD 0 Type LS LF FF
Supr Summ WBS
Classic Schedule Layout | Show Construction Activities

Screen Image 14-8
Format Tab

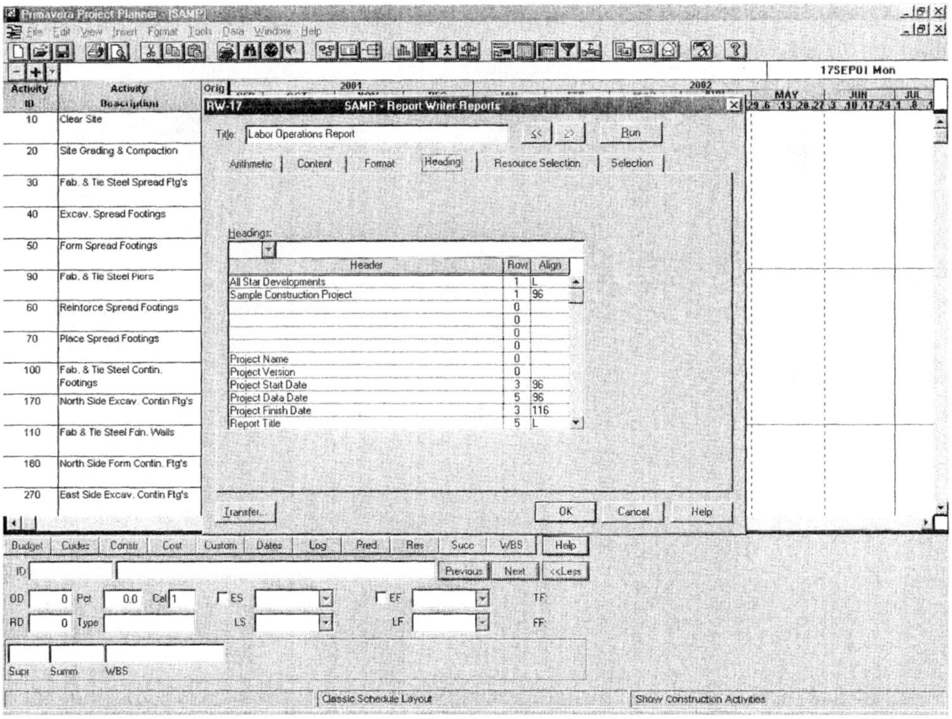

Primavera Project Planner - [SAMP]
File Edit View Insert Format Tools Data Window Help
17SEP01 Mon

Activity ID	Activity Description
10	Clear Site
20	Site Grading & Compaction
30	Fab. & Tie Steel Spread Ftg's
40	Excav. Spread Footings
50	Form Spread Footings
90	Fab. & Tie Steel Piers
60	Reinforce Spread Footings
70	Place Spread Footings
100	Fab. & Tie Steel Contin. Footings
170	North Side Excav. Contin Ftg's
110	Fab & Tie Steel Fdn. Walls
180	North Side Form Contin. Ftg's
270	East Side Excav. Contin Ftg's

RW-17 SAMP - Report Writer Reports
Title: Labor Operations Report << >> Run
Arithmetic | Content | Format | Heading | Resource Selection | Selection
Headings:

Header	Row	Align
All Star Developments	1	L
Sample Construction Project	1	96
	0	
	0	
	0	
	0	
Project Name	0	
Project Version	0	
Project Start Date	3	96
Project Data Date	5	96
Project Finish Date	3	116
Report Title	5	L

Transfer... OK Cancel Help

Budget | Codes | Constr | Cost | Custom | Dates | Log | Pred | Res | Succ | WBS | Help
ID Previous Next <<Less
OD 0 Pct 0.0 Cal 1 ES EF TF
RD 0 Type LS LF FF
Supr Summ WBS
Classic Schedule Layout | Show Construction Activities

Screen Image 14-9
Heading Window

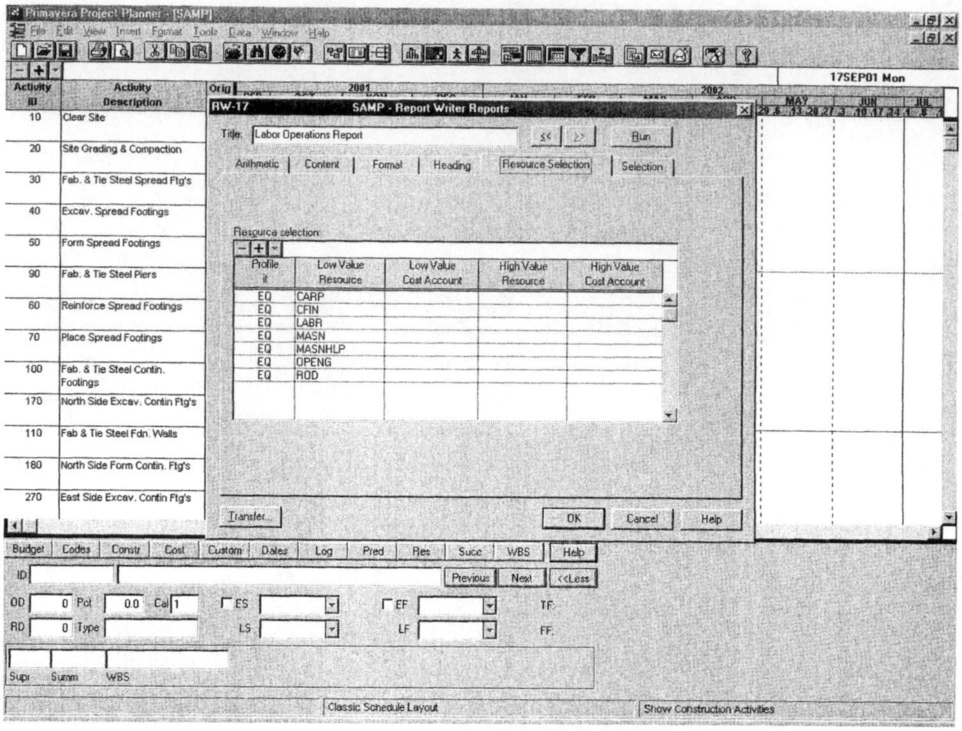

Primavera Project Planner - [SAMP]

File Edit View Insert Format Tools Data Window Help

17SEP01 Mon

Activity ID	Activity Description
10	Clear Site
20	Site Grading & Compaction
30	Fab. & Tie Steel Spread Ftg's
40	Excav. Spread Footings
50	Form Spread Footings
90	Fab. & Tie Steel Piers
60	Reinforce Spread Footings
70	Place Spread Footings
100	Fab. & Tie Steel Contin. Footings
170	North Side Excav. Contin Ftg's
110	Fab & Tie Steel Fdn. Walls
180	North Side Form Contin. Ftg's
270	East Side Excav. Contin Ftg's

RW-17 SAMP - Report Writer Reports

Title: Labor Operations Report << >> Run

Arithmetic | Content | Format | Heading | Resource Selection | Selection

Resource selection:

Profile #	Low Value Resource	Low Value Cost Account	High Value Resource	High Value Cost Account
EQ	CARP			
EQ	CFIN			
EQ	LABR			
EQ	MASN			
EQ	MASNHLP			
EQ	OPENG			
EQ	ROD			

Transfer... OK Cancel Help

Budget | Codes | Constr | Cost | Custom | Dates | Log | Pred | Res | Succ | WBS | Help

ID Previous Next <<Less

OD 0 Pct 0.0 Cal 1 ES EF TF

RD 0 Type LS LF FF

Supr Summ WBS

Classic Schedule Layout | Show Construction Activities

Screen Image 14-10
Resource Selection Window

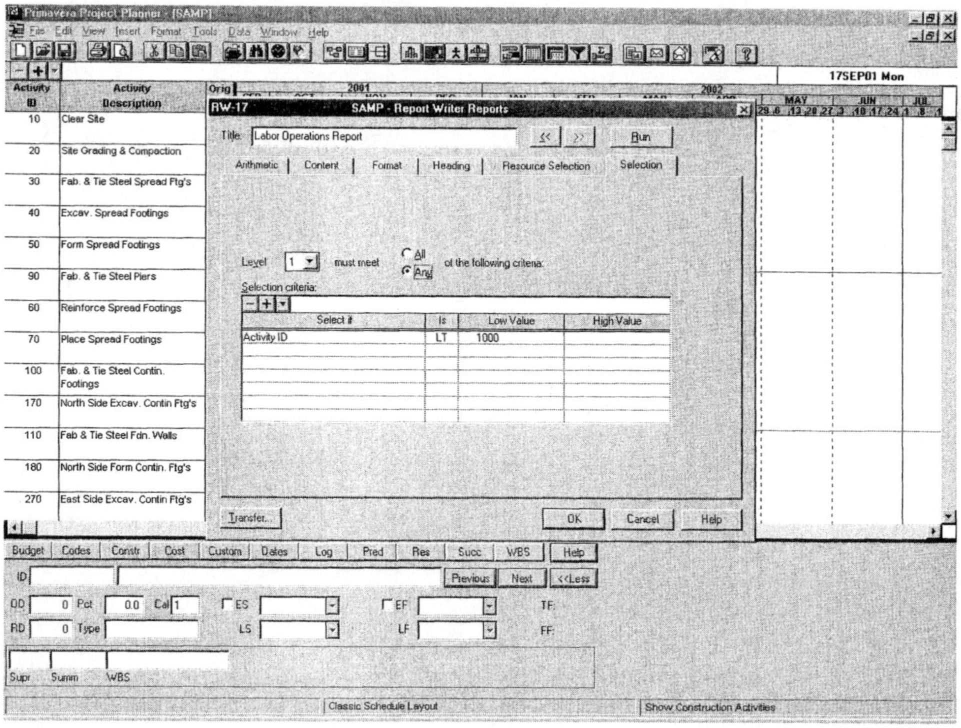

Primavera Project Planner - [SAMP]

File Edit View Insert Format Tools Data Window Help

17SEP01 Mon

Activity ID	Activity Description
10	Clear Site
20	Site Grading & Compaction
30	Fab. & Tie Steel Spread Ftg's
40	Excav. Spread Footings
50	Form Spread Footings
90	Fab. & Tie Steel Piers
60	Reinforce Spread Footings
70	Place Spread Footings
100	Fab. & Tie Steel Contin. Footings
170	North Side Excav. Contin Ftg's
110	Fab & Tie Steel Fdn. Walls
180	North Side Form Contin. Ftg's
270	East Side Excav. Contin Ftg's

RW-17 SAMP - Report Writer Reports

Title: Labor Operations Report << >> Run

Arithmetic | Content | Format | Heading | Resource Selection | Selection

Level 1 must meet All / Any of the following criteria:

Selection criteria:

Select if	Is	Low Value	High Value
Activity ID	LT	1000	

Transfer... OK Cancel Help

Budget | Codes | Constr | Cost | Custom | Dates | Log | Pred | Res | Succ | WBS | Help

ID Previous Next <<Less

OD 0 Pct 0.0 Cal 1 ES EF TF

RD 0 Type LS LF FF

Supr Summ WBS

Classic Schedule Layout | Show Construction Activities

Screen Image 14-11
Selection Window

Screen Image 14-10 is the *Resource Selection* window, which is identical to the *Resource Selection* window that has previously been used. In that screen image all of the labor resources have been selected. The nonlabor resources are of no use in a labor operations report. The final tab is *Selection*, which is also identical to the previously used *Selection* windows. In Screen Image 14-11 all activities less than 1000 have been selected, since they represent the actual construction activities for the selected project.

Analysis of Labor Operations Report

The labor operations report as presented is of minimal use if that is the only source of information. As a stand-alone document it can be used only to perform a simplistic wage rate analysis. This analysis will simply give the manager insight into how the planned wage rates compare to the actual wage rates. In Figure 14-1 the wage rates are compared and

SIMPLE WAGE RATE FORECAST											
	Planned		Actual		Work Hours To Go	Planned Wage Rate	Actual Wage Rate	Revised Wage Rate	Labor $ To Go	Indicated Labor Cost	Cost Variance
Craft	Work Hours	Dollars	Work Hours	Dollars							
Carpenters	1,180	$19,472	67	$1,322	1,113	$16.50	$19.73	$19.73	$21,960.99	$23,282.99	$3,811
Concrete Finishers	1,454	$18,542	27	$320	1,427	$12.75	$11.85	$11.85	$16,912.59	$17,232.59	-$1,309
Laborers	1,043	$11,216	269	$2,944	774	$10.75	$10.94	$10.94	$8,470.84	$11,414.84	$199
Masons	631	$10,325			631	$16.36		$16.36	$10,325.00	$10,325.00	$0
Mason Helpers	566	$5,519			566	$9.75		$9.75	$5,519.00	$5,519.00	$0
Operating Engineers	396	$5,940	44	$600	352	$15.00	$13.64	$13.64	$4,800.00	$5,400.00	-$540
Rod Busters	658	$8,720	319	$3,095	339	$13.25	$9.70	$9.70	$3,289.04	$6,384.04	-$2,336
Total	5,928	$79,734	726	$8,281		$13.45	$11.41		$71,277.46	$79,558.46	-$176

Figure 14-1
Wage Rate Analysis

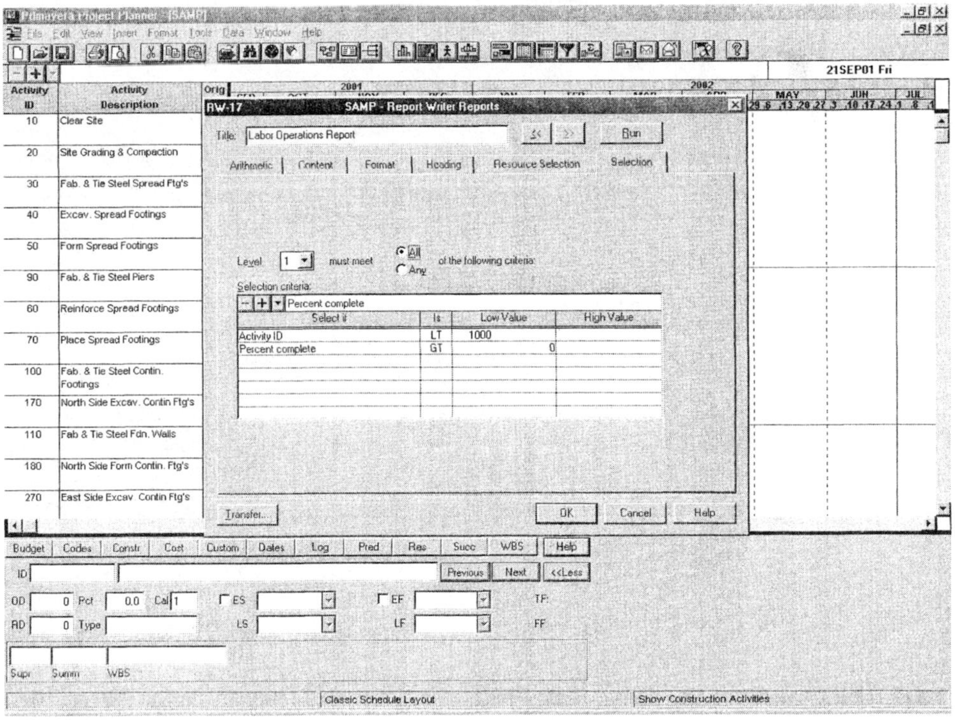

Screen Image 14-12
Selection Window

the impact of the variation is quantified. The cost variance is the wage rate variance multiplied by the expended-to-date workhours. This cost variance can be somewhat misleading since it compares the wage rate on the completed or partially completed activities to the overall average for that craft. This can be overcome by rerunning the report and including only activities that have begun. This is done by modifying the *Selection* window to include the construction activities that have some progress associated with them. Screen Image 14-12 is an example of the changes required to the *Selection* window. Output 14-2 is the labor operations report with only activities that have started.

If there is a sense that the wage rates experienced to date are what will be experienced in the future, a forecast of labor costs could be performed. However, other variables such as labor performance should be considered prior to making that forecast. Figure 14-2 shows the wage rate analysis based upon activities that have started. This information has more credibility since nonstarted activities are not included. If there is a high level of confidence that these wage rates will hold true throughout the project, then a labor cost forecast can

All Star Developments — **PRIMAVERA PROJECT PLANNER** — **Sample Construction Project**

REPORT DATE 20MAY00 RUN NO. 99 20:49 — **START DATE 01OCT01 FIN DATE 01JUL02**

Labor Operations Report — **DATA DATE 01OCT01 PAGE NO. 1**

RES	ACTIVITY ID	DESCRIPTION	QTY. THIS PRD.	QTY. TO DATE	BUDGET QTY.	COST THIS PERIOD	COST TO DATE	BUDGET COST
CARP - Carpenters								
CARP	50	Form Spread Footings	12.00	12.00	6.00	122.00	122.00	99.00
CARP	130	Form Piers	40.00	40.00	105.00	1000.00	1000.00	1733.00
CARP	180	North Side Form Contin. Ftg's	15.00	15.00	13.00	200.00	200.00	215.00
			67.00	67.00	124.00	1322.00	1322.00	2047.00
CFIN - Concrete Finishers								
CFIN	70	Place Spread Footings	12.00	12.00	8.00	120.00	120.00	102.00
CFIN	200	North Side Place Contin. Ftg's	15.00	15.00	15.00	200.00	200.00	191.00
			27.00	27.00	23.00	320.00	320.00	293.00
LABR - Laborers								
LABR	40	Excav. Spread Footings	62.00	62.00	51.00	672.00	672.00	548.00
LABR	50	Form Spread Footings	12.00	12.00	6.00	100.00	100.00	65.00
LABR	80	Strip Forms - Spread Footings	20.00	20.00	10.00	400.00	400.00	108.00
LABR	170	North Side Excav. Contin Ftg's	65.00	65.00	60.00	672.00	672.00	645.00
LABR	270	East Side Excav. Contin Ftg's	40.00	40.00	45.00	400.00	400.00	484.00
LABR	360	South Side Excav. Contin Ftg's	70.00	70.00	60.00	700.00	700.00	645.00
			269.00	269.00	232.00	2944.00	2944.00	2495.00
OPENG - Operating Engineer								
OPENG	40	Excav. Spread Footings	16.00	16.00	12.00	200.00	200.00	180.00
OPENG	170	North Side Excav. Contin Ftg's	15.00	15.00	13.00	200.00	200.00	195.00
OPENG	270	East Side Excav. Contin Ftg's	0.00	0.00	8.00	0.00	0.00	120.00
OPENG	360	South Side Excav. Contin Ftg's	13.00	13.00	13.00	200.00	200.00	195.00
ROD - Rodbuster								
ROD	30	Fab. & Tie Steel Spread Ftg's	35.00	35.00	32.00	475.00	475.00	424.00
ROD	60	Reinforce Spread Footings	20.00	20.00	10.00	220.00	220.00	133.00
ROD	90	Fab. & Tie Steel Piers	12.00	12.00	10.00	150.00	150.00	133.00
ROD	100	Fab. & Tie Steel Contin. Footings	200.00	200.00	112.00	1365.00	1365.00	1484.00
ROD	110	Fab & Tie Steel Fdn. Walls	30.00	30.00	24.00	600.00	600.00	318.00
ROD	120	Reinforce Piers	2.00	2.00	8.00	20.00	20.00	106.00
ROD	190	North Side Reinforce Contin. Ftg's	20.00	20.00	20.00	265.00	265.00	265.00
			319.00	319.00	216.00	3095.00	3095.00	2863.00
			726.00	726.00	641.00	8281.00	8281.00	8388.00

Output 14-2
Labor Operations Report—Started Activities

SIMPLE WAGE RATE FORECAST

Craft	Planned		Actual		Work Hours To Go	Planned Wage Rate	Actual Wage Rate	Revised Wage Rate	Labor $ To Go	Indicated Labor Cost	Cost Variance
	Work Hours	Dollars	Work Hours	Dollars							
Carpenters	1,180	$19,472	67	$1,322	1,113	$16.50	$19.73	$19.73	$21,960.99	$23,282.99	$3,811
Concrete Finishers	1,454	$18,542	27	$320	1,427	$12.75	$11.85	$11.85	$16,912.59	$17,232.59	-$1,309
Laborers	1,043	$11,216	269	$2,944	774	$10.75	$10.94	$10.94	$8,470.84	$11,414.84	$199
Masons	631	$10,325			631	$16.36		$16.36	$10,325.00	$10,325.00	$0
Mason Helpers	566	$5,519			566	$9.75		$9.75	$5,519.00	$5,519.00	$0
Operating Engineers	396	$5,940	44	$600	352	$15.00	$13.64	$13.64	$4,800.00	$5,400.00	-$540
Rod Busters	658	$8,720	319	$3,095	339	$13.25	$9.70	$9.70	$3,289.04	$6,384.04	-$2,336
Total	5,928	$79,734	726	$8,281		$13.45	$11.41		$71,277.46	$79,558.46	-$176

Figure 14-2
Wage Rate Analysis—Started Activities

SIMPLE WAGE RATE FORECAST

Craft	Planned		Actual		Work Hours To Go	Planned Wage Rate	Actual Wage Rate	Revised Wage Rate	Labor $ To Go	Indicated Labor Cost	Cost Variance
	Work Hours	Dollars	Work Hours	Dollars							
Carpenters	1,180	$19,472	67	$1,322	1,113	$16.50	$19.73	$19.73	$21,960.99	$23,282.99	$3,811
Concrete Finishers	1,454	$18,542	27	$320	1,427	$12.75	$11.85	$11.85	$16,912.59	$17,232.59	-$1,309
Laborers	1,043	$11,216	269	$2,944	774	$10.75	$10.94	$10.94	$8,470.84	$11,414.84	$199
Masons	631	$10,325			631	$16.36		$16.36	$10,325.00	$10,325.00	$0
Mason Helpers	566	$5,519			566	$9.75		$9.75	$5,519.00	$5,519.00	$0
Operating Engineers	396	$5,940	44	$600	352	$15.00	$13.64	$13.64	$4,800.00	$5,400.00	-$540
Rod Busters	658	$8,720	319	$3,095	339	$13.25	$9.70	$9.70	$3,289.04	$6,384.04	-$2,336
	5,928	$79,734	726	$8,281		$13.45	$11.41		$71,277.46	$79,558.46	-$176

Figure 14-3
Wage Rate Analysis and Forecast

be produced as shown in Figure 14-3. If this is done and the wage rates are applied from the resource window, the variance will be identical to the methodology shown in Figure 14-1. If the assumption that the currently experienced wage rates will prevail throughout the project is accepted, then there is an anticipated cost savings of $176, with a planned labor cost of $79,734.

Labor Performance

Labor performance is a measurement of how fast the craft personnel are working in comparison to the estimate. In the estimate, the labor productivity rate was used to convert the quantity takeoff into required workhours. During the construction process, it is necessary to compare that planned rate with an actual rate. If these rates are the same, labor performance is quantified as 1. The formula for labor performance is shown in Formula 14-1.

$$\text{Labor Performance} = \frac{\text{Actual Workhours}}{\text{Earned Workhours}}$$

Formula 14-1

In Formula 14-1 a labor performance that is greater than 1 means that the actual workhours are greater than the earned workhours, which translates into the statement that it is taking more workhours than planned. Some construction firms invert the labor performance. If the inverse is used, then a labor performance greater than 1 means it is taking fewer hours than planned. Since the formula for labor performance may vary from contractor to contractor, it is essential to know how labor performance is being calculated in order to know whether it is good and bad. The development of the raw data to calculate labor performance on an activity by activity basis can be found by using the Primavera Project Planner report writer. Output 14-3 is an example of a report that can be used to determine labor performance. A couple of unique operations were required to generate that report. The first was with

All Star Developments — PRIMAVERA PROJECT PLANNER — Sample Construction Project

REPORT DATE 24MAY00 23:21 RUN NO. 129 — START DATE 01OCT01 FIN DATE 20JUN02

Earned Workhour Report — DATA DATE 01OCT01 PAGE NO. 1

ACTIVITY ID	DESCRIPTION	PCT COMPL	BUDGET QTY.	EARNED WORK HOURS	QUANTITY TO DATE
10	Clear Site	100			
20	Site Grading & Compaction	100			
30	Fab. & Tie Steel Spread Ftg's	100	32.00	32	35.00
40	Excav. Spread Footings	100	63.00	63	78.00
50	Form Spread Footings	100	12.00	12	24.00
60	Reinforce Spread Footings	100	10.00	10	20.00
70	Place Spread Footings	100	8.00	8	12.00
80	Strip Forms - Spread Footings	100	10.00	10	20.00
90	Fab. & Tie Steel Piers	100	10.00	10	12.00
100	Fab. & Tie Steel Contin. Footings	100	112.00	112	200.00
110	Fab & Tie Steel Fdn. Walls	100	24.00	24	30.00
120	Reinforce Piers	30	8.00	2	2.00
130	Form Piers	50	105.00	53	40.00
170	North Side Excav. Contin Ftg's	100	73.00	73	80.00
180	North Side Form Contin. Ftg's	100	13.00	13	15.00
190	North Side Reinforce Contin. Ftg's	100	20.00	20	20.00
200	North Side Place Contin. Ftg's	100	15.00	15	15.00
270	East Side Excav. Contin Ftg's	100	53.00	53	40.00
360	South Side Excav. Contin Ftg's	100	73.00	73	83.00
		91	**641.00**	**583**	**726.00**

Output 14-3
Earned Workhour Report—Started Activities

the *Arithmetic* tab as shown in Screen Image 14-13. Two formulas are specified. The first is *&PC = PCT/100*. This formula creates the variable *&PC* and determines its value by dividing the percent complete by 100. Since the percent complete is not entered as a decimal, it needs to be converted into a decimal. The second formula takes the new, decimal percent complete and multiplies it by the budgeted quantity. The result of that multiplication is expressed as variable *&EWH*. Within Screen Image 14-14, the columns are organized across the page and column *&EWH* is specified and header descriptions are supplied. In order to have only started activities within the report the *Selection* tab needs to be configured to include only activities that have some physical progress as shown in Screen Image 14-15. In Figure 14-4 the information from Primavera Project Planner is used to determine the labor performance for each of the activities. From the labor performance in Figure 14-4 it is taking a total of 13 percent more workhours than estimated. If the project is sufficiently along, a forecast similar to the one found in Figure 14-5 can be performed.

The other item of interest in this analysis is the progress curve. Progress curves have been discussed in several of the previous chapters. For the above example the schedule status date was November 5, 2001. From the previous chapter in Figure 13-14, the planned progress for that date is 7.18 percent complete. Output 14-4 is a schedule update report that has had the *Budget* cost line added to its *Content* tab. From the last page of that report, the earned value is $19,056.30 of a total value of $611,631, yielding an overall percent complete of 3.16 percent complete. From a review of all of the above information the following conclusions can be drawn:

1. The project's labor costs are greater than planned.
2. The craft persons are producing work at a rate slower than planned.
3. The project is behind schedule.
4. There is the potential for a wage rate savings, but it is too early in the project to make that adjustment.

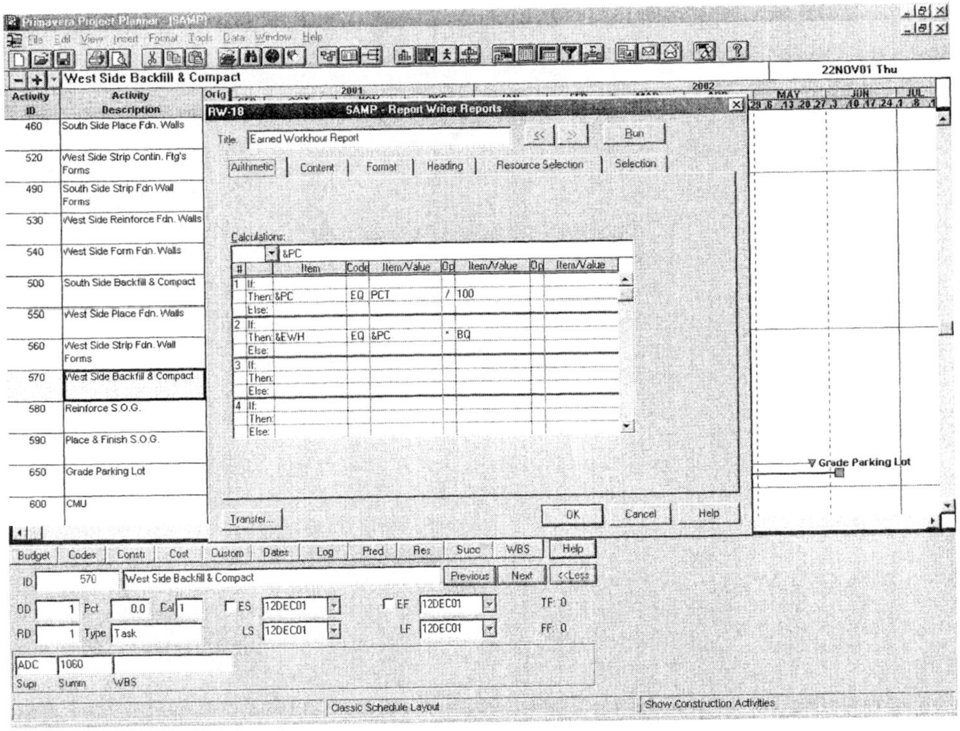

Primavera Project Planner - [SAMP]
File Edit View Insert Format Tools Data Window Help
West Side Backfill & Compact
22NOV01 Thu

Activity ID	Activity Description
460	South Side Place Fdn. Walls
520	West Side Strip Contin. Ftg's Forms
490	South Side Strip Fdn Wall Forms
530	West Side Reinforce Fdn. Walls
540	West Side Form Fdn. Walls
500	South Side Backfill & Compact
550	West Side Place Fdn. Walls
560	West Side Strip Fdn. Wall Forms
570	West Side Backfill & Compact
580	Reinforce S.O.G.
590	Place & Finish S.O.G.
650	Grade Parking Lot
600	CMU

RW-18 SAMP - Report Writer Reports

Title: Earned Workhour Report << >> Run

Arithmetic | Content | Format | Heading | Resource Selection | Selection

Calculations:
&PC

#		Item	Code	Item/Value	Op	Item/Value	Op	Item/Value
1	If:							
	Then:	&PC	EQ	PCT	/	100		
	Else:							
2	If:							
	Then:	&EWH	EQ	&PC	*	BQ		
	Else:							
3	If:							
	Then:							
	Else:							
4	If:							
	Then:							
	Else:							

Transfer... OK Cancel Help

Grade Parking Lot

Budget | Codes | Constr | Cost | Custom | Dates | Log | Pred | Res | Succ | WBS | Help

ID 570 West Side Backfill & Compact Previous Next <<Less

OD 1 Pct 0.0 Cal 1 ES 12DEC01 EF 12DEC01 TF: 0
RD 1 Type Task LS 12DEC01 LF 12DEC01 FF: 0

ADC 1060
Supr Summ WBS

Classic Schedule Layout | Show Construction Activities

Screen Image 14-13
Arithmetic Window

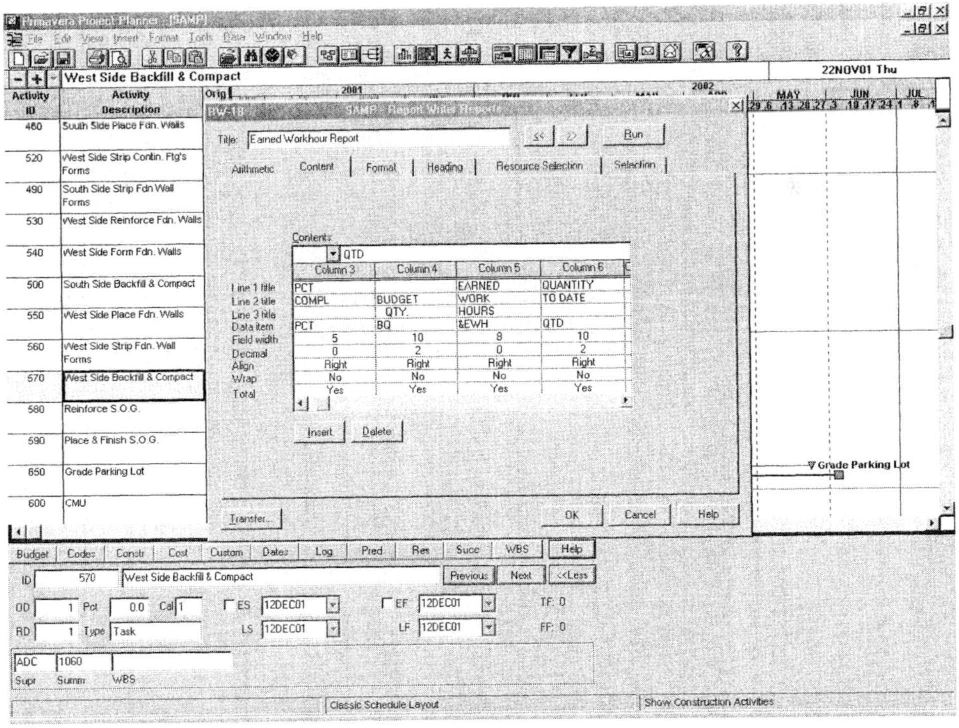

Primavera Project Planner - [SAMP]
File Edit View Insert Format Tools Data Window Help
West Side Backfill & Compact
22NOV01 Thu

Activity ID	Activity Description
460	South Side Place Fdn. Walls
520	West Side Strip Contin. Ftg's Forms
490	South Side Strip Fdn Wall Forms
530	West Side Reinforce Fdn. Walls
540	West Side Form Fdn. Walls
500	South Side Backfill & Compact
550	West Side Place Fdn. Walls
560	West Side Strip Fdn. Wall Forms
570	West Side Backfill & Compact
580	Reinforce S.O.G.
590	Place & Finish S.O.G.
650	Grade Parking Lot
600	CMU

RW-18 SAMP - Report Writer Reports

Title: Earned Workhour Report << >> Run

Arithmetic | Content | Format | Heading | Resource Selection | Selection

Contents:
QTD

	Column 3	Column 4	Column 5	Column 6
Line 1 title	PCT		EARNED	QUANTITY
Line 2 title	COMPL	BUDGET	WORK	TO DATE
Line 3 title		QTY.	HOURS	
Data item	PCT	BQ	&EWH	QTD
Field width	5	10	8	10
Decimal	0	2	0	2
Align	Right	Right	Right	Right
Wrap	No	No	No	No
Total	Yes	Yes	Yes	Yes

Insert Delete

Transfer... OK Cancel Help

Grade Parking Lot

Budget | Codes | Constr | Cost | Custom | Dates | Log | Pred | Res | Succ | WBS | Help

ID 570 West Side Backfill & Compact Previous Next <<Less

OD 1 Pct 0.0 Cal 1 ES 12DEC01 EF 12DEC01 TF: 0
RD 1 Type Task LS 12DEC01 LF 12DEC01 FF: 0

ADC 1060
Supr Summ WBS

Classic Schedule Layout | Show Construction Activities

Screen Image 14-14
Content Window

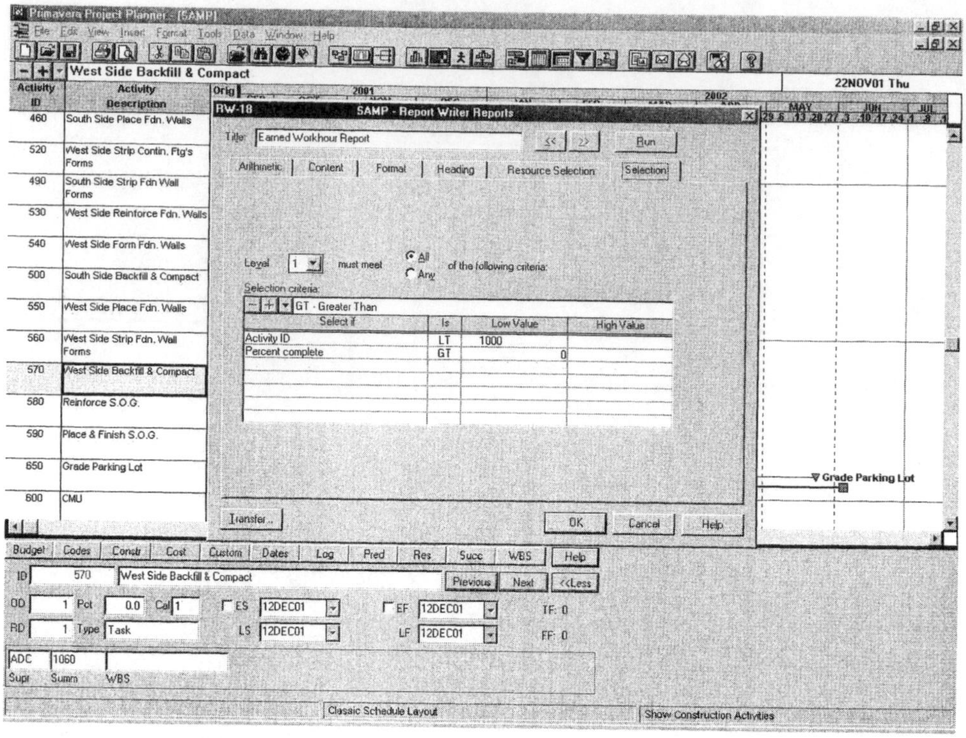

Screen Image 14-15
Selection Window

Labor Performance Analysis					
Activity	% Complete	Actual Workhours	Planned Workhours	Earned Workhours	Labor Performance
10	100.00%			0	
20	100.00%			0	
30	100.00%	35	32	32	1.09
40	100.00%	78	63	63	1.24
50	100.00%	24	12	12	2.00
60	100.00%	20	10	10	2.00
70	100.00%	12	8	8	1.50
80	100.00%	20	10	10	2.00
90	100.00%	12	10	10	1.20
100	100.00%	200	112	112	1.79
110	100.00%	30	24	24	1.25
120	30.00%	2	8	2.4	0.25
130	50.00%	40	105	52.5	0.38
170	100.00%	80	73	73	1.10
180	100.00%	15	13	13	1.15
190	100.00%	20	20	20	1.00
200	100.00%	15	15	15	1.00
270	100.00%	40	53	53	0.75
360	100.00%	83	73	73	1.14
		726	641	582.9	1.13

Figure 14-4
Labor Performance Analysis

Labor Cost Forecast	
Estimate Work Hours	5,928
Labor Performance	1.13
Revised Estimated Work Hours	6,714
Estimated Wage Rate	$13.45
Revised Estimated Labor Cost	$90,303.00
Estimated Labor Cost	$79,734.00
Forecasted Labor over / under run	$10,569.00

Figure 14-5
Labor Forecast

All Star Developments | PRIMAVERA PROJECT PLANNER | Sample Construction Project

REPORT DATE 25MAY00 RUN NO. 132 8:21 | START DATE 1OCT01 FIN DATE 8JUL02

Schedule Update Report | DATA DATE 5NOV01 PAGE NO. 1

Albert Concrete

ACTIVITY ID	ORIG DUR	REM DUR	CAL	%	CODE	ACTIVITY DESCRIPTION	BUDGET	EARNED	EARLY START	EARLY FINISH	LATE START	LATE FINISH	TOTAL FLOAT
10	2	0	1	100	ADC	Clear Site	1450.00	1450.00	1OCT01A	2OCT01A			
20	5	0	1	100	ADC	Site Grading & Compaction	957.00	957.00	3OCT01A	10OCT01A			
30	6	0	1	100	ADC	Fab. & Tie Steel Spread Ftg's	801.00	801.00	3OCT01A	15OCT01A			
40	4	0	1	100	ADC	Excav. Spread Footings	1084.00	1084.00	11OCT01A	17OCT01A			
50	4	0	1	100	ADC	Form Spread Footings	164.00	164.00	15OCT01A	25OCT01A			
90	2	0	1	100	ADC	Fab. & Tie Steel Piers	2240.00	2240.00	15OCT01A	16OCT01A			
60	4	0	1	100	ADC	Reinforce Spread Footings	133.00	133.00	16OCT01A	1NOV01A			
70	4	0	1	100	ADC	Place Spread Footings	883.00	883.00	17OCT01A	23OCT01A			
100	2	0	1	100	ADC	Fab. & Tie Steel Contin. Footings	3520.00	3520.00	17OCT01A	18OCT01A			
170	1	0	1	100	ADC	North Side Excav. Contin Ftg's	1234.00	1234.00	18OCT01A	18OCT01A			
110	2	0	1	100	ADC	Fab & Tie Steel Fdn. Walls	742.00	742.00	22OCT01A	23OCT01A			
180	1	0	1	100	ADC	North Side Form Contin. Ftg's	560.00	560.00	22OCT01A	22OCT01A			
270	1	0	1	100	ADC	East Side Excav. Contin Ftg's	854.00	854.00	22OCT01A	22OCT01A			
80	4	0	1	100	ADC	Strip Forms - Spread Footings	108.00	108.00	23OCT01A	29OCT01A			
190	1	0	1	100	ADC	North Side Reinforce Contin. Ftg's	265.00	265.00	23OCT01A	23OCT01A			
360	1	0	1	100	ADC	South Side Excav. Contin Ftg's	1234.00	1234.00	23OCT01A	23OCT01A			
200	1	0	1	100	ADC	North Side Place Contin. Ftg's	1752.00	1752.00	24OCT01A	24OCT01A			
120	4	3	1	30	ADC	Reinforce Piers	106.00	31.80	24OCT01A	7NOV01		15NOV01	5
130	4	2	1	50	ADC	Form Piers	2087.00	1043.50	25OCT01A	8NOV01		19NOV01	5
210	1	1	1	0	ADC	North Side Strip Contin. Ftg's Forms	86.00	.00	5NOV01	5NOV01	5NOV01	5NOV01	0
140	4	4	1	0	ADC	Place Piers	309.00	.00	5NOV01	12NOV01	13NOV01	20NOV01	5
220	1	1	1	0	ADC	North Side Reinforce Fdn. Walls	530.00	.00	6NOV01	6NOV01	6NOV01	8NOV01	0
280	1	1	1	0	ADC	East Side Form Contin. Ftg's			6NOV01	6NOV01	7NOV01	7NOV01	1

Output 14-4
Update Report with Budget Cost Line (page 1)

All Star Developments | PRIMAVERA PROJECT PLANNER | Sample Construction Project

REPORT DATE 25MAY00 RUN NO. 132 8:21 | START DATE 1OCT01 FIN DATE 8JUL02

Schedule Update Report | DATA DATE 5NOV01 PAGE NO. 2

Albert Concrete

ACTIVITY ID	ORIG DUR	REM DUR	CAL	%	CODE	ACTIVITY DESCRIPTION / BUDGET	EARNED	EARLY START	EARLY FINISH	LATE START	LATE FINISH	TOTAL FLOAT
						350.00	.00					
230	3	3	1	0	ADC	North Side Form Fdn. Walls		7NOV01	12NOV01	7NOV01	12NOV01	0
						3459.00	.00					
290	1	1	1	0	ADC	East Side Reinforce Contin. Ftg's		7NOV01	7NOV01	8NOV01	8NOV01	1
						265.00	.00					
300	1	1	1	0	ADC	East Side Place Contin. Ftg's		8NOV01	8NOV01	12NOV01	12NOV01	1
						1104.00	.00					
150	4	4	1	0	ADC	Strip Forms - Piers		8NOV01	14NOV01	19NOV01	26NOV01	5
						172.00	.00					
160	4	4	1	0	ADC	Backfill Spread Footings		12NOV01	15NOV01	20NOV01	27NOV01	5
						3673.00	.00					
240	1	1	1	0	ADC	North Side Place Fdn. Walls		13NOV01	13NOV01	13NOV01	13NOV01	0
						156.00	.00					
310	1	1	1	0	ADC	East Side Strip Contin. Ftg's Forms		14NOV01	14NOV01	15NOV01	15NOV01	1
						54.00	.00					
320	1	1	1	0	ADC	East Side Reinforce Fdn. Walls		15NOV01	15NOV01	19NOV01	21NOV01	1
						1325.00	.00					
370	1	1	1	0	ADC	South Side Form Contin. Ftg's		15NOV01	15NOV01	19NOV01	19NOV01	1
						386.00	.00					
250	1	1	1	0	ADC	North Side Strip Fdn. Wall Forms		19NOV01	19NOV01	19NOV01	19NOV01	0
						430.00	.00					
380	1	1	1	0	ADC	South Side Reinforce Contin. Ftg's		19NOV01	19NOV01	20NOV01	20NOV01	1
						265.00	.00					
390	1	1	1	0	ADC	West Side Excav. Contin Ftg's		19NOV01	19NOV01	28NOV01	28NOV01	5
						800.00	.00					
330	3	3	1	0	ADC	East Side Form Fdn. Walls		20NOV01	26NOV01	20NOV01	26NOV01	0
						2108.00	.00					
420	1	1	1	0	ADC	South Side Place Contin. Ftg's		20NOV01	20NOV01	21NOV01	21NOV01	1
						1752.00	.00					
260	1	1	1	0	ADC	North Side Backfill & Compact		20NOV01	20NOV01	27DEC01	27DEC01	20
						4656.00	.00					
340	1	1	1	0	ADC	East Side Place Fdn. Walls		27NOV01	27NOV01	27NOV01	27NOV01	0
						107.00	.00					
430	1	1	1	0	ADC	South Side Strip Contin. Ftg's Forms		27NOV01	27NOV01	28NOV01	28NOV01	1
						97.00	.00					
470	1	1	1	0	ADC	West Side Form Contin. Ftg's		28NOV01	28NOV01	29NOV01	29NOV01	1
						367.00	.00					
440	1	1	1	0	ADC	South Side Reinforce Fdn. Walls		28NOV01	28NOV01	3DEC01	3DEC01	2
						530.00	.00					
480	1	1	1	0	ADC	West Side Reinforce Contin. Ftg's		29NOV01	29NOV01	3DEC01	3DEC01	1
						265.00	.00					
350	1	1	1	0	ADC	East Side Strip Fdn. Wall Forms		3DEC01	3DEC01	3DEC01	3DEC01	0
						269.00	.00					
510	2	2	1	0	ADC	West Side Place Contin. Ftg's		3DEC01	4DEC01	4DEC01	5DEC01	1
						1091.00	.00					

Output 14-4
Update Report with Budget Cost Line (page 2)

All Star Developments — PRIMAVERA PROJECT PLANNER — Sample Construction Project

REPORT DATE 25MAY00 RUN NO. 132 START DATE 1OCT01 FIN DATE 8JUL02
8:21
Schedule Update Report DATA DATE 5NOV01 PAGE NO. 3

Albert Concrete

ACTIVITY ID	ORIG DUR	REM DUR	CAL	%	CODE	ACTIVITY DESCRIPTION	BUDGET	EARNED	EARLY START	EARLY FINISH	LATE START	LATE FINISH	TOTAL FLOAT
450	1	1	1	0	ADC	South Side Form Fdn. Walls	2591.00	.00	4DEC01	4DEC01	4DEC01	4DEC01	0
410	1	1	1	0	ADC	East Side Backfill & Compact	2645.00	.00	4DEC01	4DEC01	27DEC01	27DEC01	13
460	1	1	1	0	ADC	South Side Place Fdn. Walls	156.00	.00	5DEC01	5DEC01	5DEC01	5DEC01	0
520	1	1	1	0	ADC	West Side Strip Contin. Ftg's Forms	65.00	.00	10DEC01	10DEC01	11DEC01	11DEC01	1
490	2	2	1	0	ADC	South Side Strip Fdn Wall Forms	430.00	.00	11DEC01	12DEC01	11DEC01	12DEC01	0
530	1	1	1	0	ADC	West Side Reinforce Fdn. Walls	530.00	.00	11DEC01	11DEC01	12DEC01	17DEC01	1
540	3	3	1	0	ADC	West Side Form Fdn. Walls	2108.00	.00	13DEC01	18DEC01	13DEC01	18DEC01	0
500	1	1	1	0	ADC	South Side Backfill & Compact	4656.00	.00	13DEC01	13DEC01	27DEC01	27DEC01	7
550	1	1	1	0	ADC	West Side Place Fdn. Walls	107.00	.00	19DEC01	19DEC01	19DEC01	19DEC01	0
560	1	1	1	0	ADC	West Side Strip Fdn. Wall Forms	269.00	.00	26DEC01	26DEC01	26DEC01	26DEC01	0
570	1	1	1	0	ADC	West Side Backfill & Compact	2765.00	.00	27DEC01	27DEC01	27DEC01	27DEC01	0
580	3	3	1	0	ADC	Reinforce S.O.G.	5307.00	.00	31DEC01	3JAN02	31DEC01	3JAN02	0
590	1	1	1	0	ADC	Place & Finish S.O.G.	17995.00	.00	7JAN02	7JAN02	7JAN02	7JAN02	0
650	5	5	1	0	ADC	Grade Parking Lot	4680.00	.00	8JAN02	15JAN02	3JUN02	10JUN02	82
720	3	3	1	0	ADC	Form Parking Lot	1072.00	.00	16JAN02	21JAN02	11JUN02	18JUN02	82
730	5	5	1	0	ADC	Reinforce Parking Lot	1838.00	.00	17JAN02	24JAN02	12JUN02	19JUN02	82
740	1	1	1	0	ADC	Place & Finish Parking Lot	21936.00	.00	21JAN02	28JAN02	19JUN02	20JUN02	85
750	1	1	1	0	ADC	Strip Forms - Parking Lot	215.00	.00	24JAN02	30JAN02	25JUN02	25JUN02	85
760	2	2	1	0	ADC	Place Curbs	5830.00	.00	31JAN02	4FEB02	26JUN02	27JUN02	82
SUBTOTAL					ADC		119975.00	19056.30					

Output 14-4
Update Report with Budget Cost Line (page 3)

All Star Developments — PRIMAVERA PROJECT PLANNER — Sample Construction Project

REPORT DATE 25MAY00 RUN NO. 132 8:21 — START DATE 1OCT01 FIN DATE 8JUL02

Schedule Update Report — DATA DATE 5NOV01 PAGE NO. 4

Joe Bob Steel

ACTIVITY ID	ORIG DUR	REM DUR	CAL	%	CODE	ACTIVITY DESCRIPTION	BUDGET	EARNED	EARLY START	EARLY FINISH	LATE START	LATE FINISH	TOTAL FLOAT
610	12	12	1	0	JBS	Erect Structural Steel	58200.00	.00	15JAN02	20FEB02	31JAN02	20FEB02	10
680	3	3	1	0	JBS	Ceiling Grid	5384.00	.00	19MAR02	21MAR02	19MAR02	21MAR02	0
690	6	6	1	0	JBS	Frame Interior Walls	4013.00	.00	25MAR02	2APR02	25MAR02	2APR02	0
800	8	8	1	0	JBS	Hang Drywall	28137.00	.00	24APR02	7MAY02	24APR02	7MAY02	0
810	16	16	1	0	JBS	Float, Tape & Paint	5661.00	.00	8MAY02	5JUN02	8MAY02	5JUN02	0
840	6	6	1	0	JBS	Install Cabinets	16599.00	.00	6JUN02	17JUN02	6JUN02	17JUN02	0
820	2	2	1	0	JBS	Hang & Trim Interior Doors	17174.00	.00	6JUN02	10JUN02	25JUN02	26JUN02	10
870	3	3	1	0	JBS	Ceiling Tiles	5667.00	.00	27JUN02	2JUL02	27JUN02	2JUL02	0
880	1	1	1	0	JBS	Carpet	16752.00	.00	3JUL02	3JUL02	3JUL02	3JUL02	0
890	1	1	1	0	JBS	Final Cleaning	1200.00	.00	8JUL02	8JUL02	8JUL02	8JUL02	0
SUBTOTAL					JBS		158787.00	.00					

Output 14-4
Update Report with Budget Cost Line (page 4)

All Star Developments — PRIMAVERA PROJECT PLANNER — Sample Construction Project

REPORT DATE 25MAY00 RUN NO. 132 8:21 — START DATE 1OCT01 FIN DATE 8JUL02

Schedule Update Report — DATA DATE 5NOV01 PAGE NO. 5

John Doe

ACTIVITY ID	ORIG DUR	REM DUR	CAL	%	CODE	ACTIVITY DESCRIPTION	BUDGET	EARNED	EARLY START	EARLY FINISH	LATE START	LATE FINISH	TOTAL FLOAT
400	5	5	1	0	JED	Underground Plumbing	35000.00	.00	5NOV01	12NOV01	19DEC01	27DEC01	25
780	2	2	1	0	JED	Final Grading	4800.00	.00	5FEB02	6FEB02	1JUL02	2JUL02	82
770	1	1	1	0	JED	Stripe Parking Lot	390.00	.00	5FEB02	5FEB02	8JUL02	8JUL02	85
790	2	2	1	0	JED	Landscape	9141.00	.00	7FEB02	11FEB02	3JUL02	8JUL02	82
660	5	5	1	0	JED	Mechanical Rough-in	25000.00	.00	5MAR02	12MAR02	5MAR02	18MAR02	0
670	8	8	1	0	JED	Overhead Electrical Rough-in	28000.00	.00	5MAR02	18MAR02	5MAR02	18MAR02	0
710	12	12	1	0	JED	Electrical Rough-in	28000.00	.00	3APR02	23APR02	3APR02	23APR02	0
700	3	3	1	0	JED	Plumbing Top-out	30000.00	.00	3APR02	8APR02	24JUN02	26JUN02	45
830	3	3	1	0	JED	Mechanical Trim	28000.00	.00	6JUN02	11JUN02	19JUN02	24JUN02	7
850	4	4	1	0	JED	Electrical Trim	27000.00	.00	18JUN02	24JUN02	18JUN02	24JUN02	0
860	2	2	1	0	JED	Plumbing Trim	30000.00	.00	25JUN02	26JUN02	25JUN02	26JUN02	0
SUBTOTAL					JED		245331.00	.00					

Output 14-4
Update Report with Budget Cost Line (page 5)

All Star Developments — PRIMAVERA PROJECT PLANNER — Sample Construction Project

REPORT DATE 25MAY00 RUN NO. 132 8:21 — START DATE 1OCT01 FIN DATE 8JUL02

Schedule Update Report — DATA DATE 5NOV01 PAGE NO. 6

WIlliam Roe

ACTIVITY ID	ORIG DUR	REM DUR	CAL	%	CODE	ACTIVITY DESCRIPTION	BUDGET	EARNED	EARLY START	EARLY FINISH	LATE START	LATE FINISH	TOTAL FLOAT
600	22	22	1	0	WER	CMU			10JAN02	18FEB02	10JAN02	18FEB02	0
							26437.00	.00					
620	6	6	1	0	WER	Roofing / Flashing			21FEB02	4MAR02	21FEB02	4MAR02	0
							43981.00	.00					
640	3	3	1	0	WER	Overhead Doors			21FEB02	26FEB02	27FEB02	4MAR02	3
							8000.00	.00					
630	1	1	1	0	WER	Exterior Glass & Glazing			21FEB02	21FEB02	4MAR02	4MAR02	5
							9120.00	.00					
SUBTOTAL					WER		87538.00	.00					
REPORT TOTAL							611631.00	19056.30					

Output 14-4

Update Report with Budget Cost Line (page 6)

These are disturbing trends that would require further analysis into the cause and would require that an immediate recovery plan be developed. Some of the potential scenarios could be that the cheaper labor is less trained and working slower because they lack the skill required for the task; or that the field progress is understated; or that there is a discrepancy between when the actual cost data was gathered and when progress was measured. It should be clear that there are any number of potential explanations for these deviations. The project manager must use this information as an indicator and promptly begin to investigate the cause of the deviation. The solution to the problem must address the problem. Simply reacting to the information may result in developing a remediation plan that does more damage than good. Meeting with the field staff will often add insight into the root cause of the problem. Once the problem is identified, it must be fixed and a plan developed to recover from the lost time and money.

Material and Subcontract Costs

Material and subcontract cost reports are of questionable value if the basis of comparison is using expended-to-date figures to compare to the estimate. Since there is a lag of weeks between the time that the materials are delivered and when they are paid for, using this information can often lead to faulty forecasting of the final material costs. The best time to monitor these costs is when the purchase orders are issued. The purchase-order quantity should match the quantity in the estimate. If there is a deviation, it needs to be addressed immediately, not when the cost data hits a project cost report. For example, after a concrete pour has been completed, the actual number of yards should be compared to the amount in the estimate. If fewer cubic yards were required, that savings should be recognized and recorded. On the other hand, if more concrete was required, the reason needs to be determined. Two likely reasons for such a deviation would be that the quantity was wrong in the estimate or the thickness of the concrete was greater than what was called for on the drawings. Regardless of the reason, this deviation should be noted and reflected in the anticipated final project cost.

To effectively control the cost of materials and subcontracts, these costs have to be recognized when the purchase order is issued or when the contract is signed. The difficulty with this scheme is that most financial accounting systems are not designed to deal with this matter. Typically, financial systems will not recognize a cost until title to the material passes to the contractor. From a planning perspective, this is too late to make any changes in the project.

Conclusion

The process of effectively controlling a construction project must begin with thoughtful planning from the inception of the project and continue until the project is turned over to the owner. By having a plan, the contractor has a baseline of comparison between what was planned and what has actually happened. This allows for early detection of deviations and the opportunity to develop corrective plans that minimize the impact of deviations.

Suggested Exercise

1. Enter some actual cost data for the activities that have been completed on your sample project. Then determine the overall status of the project and forecast the completed project cost.

Appendix A

Start-up Screen Menu

File	Tools	Help
File New Open Send Mail Receive Mail	Tools Look Web Publishing Wizard MPX Conversion Utility Project Utilities Copy Delete Merge Summarize Back Up Restore Options Default Autocost Rules Summarization Earned Value Activity Inserting Tool Bar Set Language Change Password Default Activity Codes File Extension	Help Contents Search Tutorial Technical Support World Wide Web Support Network Setup About P3

Main Input Screen

- File
  - New
  - Open
  - Close
  - Project Overview
  - Page Setup
  - Print Preview
  - Print
  - Print Setup
  - Preview Picture
  - Send Mail
  - Receive Mail
  - Save as Web Page
  - Exit

- Edit
  - Cut
  - Copy
  - Paste
  - Paste Special
  - Copy Picture
  - Edit Activity
  - Fill Cell
  - Delete
  - Dissolve Activity
  - Extract Activity
  - Find Activity
  - Find Objects
  - Select All
  - Unselect
  - Relationships
  - Link Activities
  - Unlink Activities
  - Object
  - Paste Link
  - Links

- View
  - PERT
  - Layout
    - New
    - Open
    - Save
    - Save As
    - Transfer
    - Delete
    - Options
    - Make Default
  - Activity Form
  - Activity Detail
    - Budget
    - Codes
    - Constraints
    - Cost
    - Custom Data
    - Dates
    - Log
    - Predecessors
    - Resources
    - Successors
    - WBS
    - Close All
  - Resource Profile
  - Resource Table
  - Relationships
  - Progress Line
  - Progress Spotlight
  - Datometer
  - Toolbar
  - Page Breaks
  - Thousands Separator
  - Attachment Tools
  - Objects
  - Current Users

- Insert
  - Activity
  - Resource Assignment
  - Autolink
  - Page Break
  - Clear All Page Breaks
  - Object

- Format
  - Columns
  - Bars
  - Selected Bars
    - Modify Bar Forma
    - Copy Bar Format
    - Paste Bar Format
    - Use Default Bar
  - Summary Bars
  - Organize
  - Reorganize Now
  - Filter
  - Run Filter Now
  - Summarize
  - Summarize All
  - Timescale
  - Sight Line
  - Row Height
  - Screen Colors
  - Fonts
  - Dates

Tools	Data	Windows	Help
Schedule	Calendars	Cascade	Contents
Level	Activity Codes	Tile	Search
Global Change	Project Codes	Arrange Icons	Tutorial
Update Progress	WBS	Activity Columns Split	Technical Support
Store Period Performance	Resources	Vertical Split	World Wide Web Support
Tabular Reports	Resource Curves	Bar Area Split	Network Setup
Schedule	Cost Accounts	Horizontal Split	About P3
Resource	Custom Data Items	Activity Form Split	
Control			
Productivity			
Earned Value			
Tabular			
Loading			
Cost			
Control			
Cost, Price, and Rates			
Earned Value			
Tabular			
Loading			
Matrix			
Activity			
Resource / Cost			
Report Writer			
Production			
Option			
Graphic Reports			
Bar			
Timescaled Logic			
Pure Logic			
Resource and Cost			
Production			
Options			
Look			
Infomaker			
Update Data Dictionary			
Web Publishing Wizard			
MPX Conversion Utility			
Project Utilities			
Copy			
Delete			
Merge			
Summarize			
Backup			
Restore			
Targets			
Import			
Export			
Fragnet			
Store Fragnet			
Retrieve Fragnet			
Delete Fragnet			
Options			
Autocost Rules			
Summarization			
Earned Value			
Activity Inserting			
Critical Activities			
Toolbar			
Set Language			
Change Password			
Default Activity Codes			

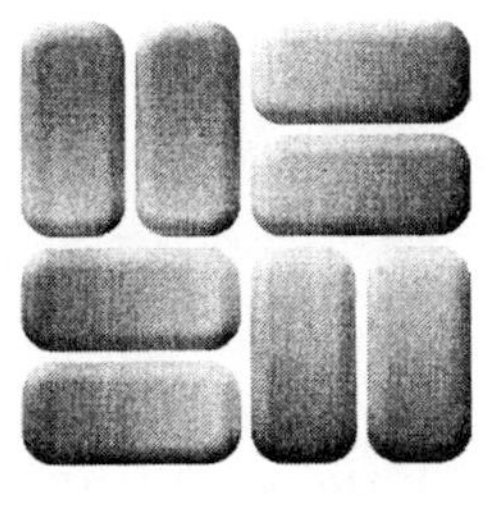

Subject Index